DRAINAGE AND WATER TABLE CONTROL

Proceedings of the
Sixth International Drainage
Symposium

13-15 December 1992
Nashville, Tennessee

Published by
American Society of Agricultural Engineers
2950 Niles Rd., St. Joseph, Michigan 49085-9659 USA

Copyright © 1992 by
American Society of Agricultural Engineers
All rights reserved

Library of Congress Card Number (LCCN) 92-74577
International Standard Book Number (ISBN) 0-929355-34-2
ASAE Publication 13-92

The American Society of Agricultural Engineers is not responsible for statements and opinions advanced in its meetings or printed in its publications. They represent the views of the individual to whom they are credited and are not binding on the Society as a whole.

Table of Contents

	Page
FOREWORD	xi
TO THE AUTHORS	xiv

GENERAL SESSION

Keynote - Drainage and Water Management Modeling Technology
R. W. Skaggs 1

Management Support System for Conjunctive Irrigation and Drainage
L. A. Garcia, T. K. Gates, M. E. Jensen, T. H. Podmore 12

Evaluation of Stress Day Index Models to Predict Corn and Soybean Yield Response to Soil-Water Stresses
R. O. Evans, R. W. Skaggs 20

Validation of COSSMO: A Combined Crop Growth and Water Management Model
C. D. Perry, D. L. Thomas, R. O. Evans 30

WATER MANAGEMENT MODELING

Modeling Water Table Control Systems with High Head Losses Near the Drain
G. M. Chescheir, C. Murugaboopathi, R. W. Skaggs, R. H. Susanto, R. O. Evans 38

Drain Outlet Water Level Control: A Simulation Model
J. L. Fouss, J. S. Rogers 46

Design of Tube-Well Drainage and Water Management with
the Help of a Large-Scale Groundwater Model
W. Bogacki, Th. Höddinghaus 62

Economic Modelling of Soil Workability Criteria
K. E. Oskoui 70

Drainage Flow Patterns and Modeling in a Seasonally
Waterlogged Heavy Clay Soil
D. Zimmer 80

SUSTAINABILITY OF DRAINAGE SYSTEMS

Drainage Sustainability in the Imperial Valley of California
R. H. Brooks 88

Corn Yields, Soil Properties, and Drainage Effects on Crosby-Kokomo Soil
R. Lal, N. R. Fausey 96

Effects of Cultural Practices on Recovery of Permeability in
Compacted Subsoil
N. R. Fausey 105

Durability of Agro-Ameliorative Measures on Mineral
Lowland Soils with Heterogeneous Substrate and Hydrologic
Conditions
L. Müller, U. Schindler 112

ENVIRONMENTAL IMPACTS

Environmental Sustainability of Drainage Projects
C. A. Madramootoo 119

Subirrigation and Controlled Drainage: Management Tools for
Reducing Environmental Impacts of Nonpoint Source
Pollution
P. K. Kalita, R. S. Kanwar, S. W. Melvin 129

Assessment of Soil Water Regime Requirements of Wild Plant
Species
G. Spoor, J. M. Chapman, P. B. Leeds-Harrison 137

Environmental Consequences of Major Irrigation
Development - A Case Study from Sri Lanka
N. Amarasekara, K. Mohtadullah 145

ENVIRONMENT IMPACTS AND WETLANDS ISSUES

Hydraulic Effects of Upstream Drainage Reconstruction on
Existing Downstream Floodplains for Small Watersheds in
Delaware (Abstract)
R. T. Smith, R. F. Gronwald, J. O. Kelley 153

Bacterial Density Changes Upstream and Downstream of a
Sludge Disposal Facility
L. Cheng, B. Buras, D. M. Griffin, Jr., J. D. Nelson 154

Pesticides in Storm Water Runoff from Agricultural Chemical
Facilities
E. O. Ackerman, A. G. Taylor 161

Drainage Effects in Marsh Soils - I. Effects on Water Table
and Drainflow
S. M. Ibrahim 169

CONTRACTOR CONCERNS

Contractor Panel Member Remarks
C. Schwieterman 177

Subsurface Drainage Design
O. R. Row 178

Contractors Comments
F. Galehouse 179

The Future of Drainage Contractors
R. D. Cornwell 182

ONE-ON-ONE PRESENTATIONS

Long Term Performance of Geotextiles on Horizontal
Subsurface Drainage Systems
R. B. Bonnell, R. S. Broughton, J. Mlynarek 183

Drainage Needs of the Small Inland Valley Swamps of
West Africa
Y. M. Mohamoud 193

Eliminating Downstream Sediment Transport by Pumping
Base Flows Around Construction Areas (Abstract)
R. T. Smith 200

Palustrine Wetland Retention, Restoration and Creation as
Part of Delaware's Tax Ditch Program
R. T. Smith, T. G. Barthelmeh, S. L. Griffith-Kepfer 201

Low Cost Channel Side Slope Scarification and Seeding
(Abstract)
R. T. Smith, R. F. Gronwald, J. O. Kelley 209

Status of the Water Table Management-Water Quality
Research Project in the Lower Mississippi Valley
G. H. Willis, J. L. Fouss, J. S. Rogers, C. E. Carter,
L. M. Southwick 210

In-Situ Hydraulic Head Indicator System for Monitoring Field
Water Table Elevation
J. L. Fouss, M. Mahler, N. A. Ellis 219

A Water-Table - Water Quality Management Research Facility
for the Southeastern Coastal Plain: Progress Report
C. R. Camp, K. C. Stone, P. G. Hunt 229

Field Estimation of Drainable Pore Space and Application in
Drainage Design
S. K. Gupta 236

Subsoil Workability of Heavy Soils with Shallow Water Tables
L. Müller, U. Schindler 244

Trafficability and Crop Yields on Groundwater Influenced
Heavy Soils
L. Müller 251

Effect of Drain Envelopes on the Water Acceptance of
Wrapped Subsurface Drains
L. C. P. M. Stuyt 257

Nitrate Leaching in Drained Seasonally Waterlogged Shallow
Soils
M. P. Arlot, D. Zimmer 264

Effect of Deep Seepage on Drainage Functioning and Design
in Shallow Soils
D. Zimmer 272

Water Table Management Rainshelter Project Research
H. W. Belcher, T. L. Loudon, G. E. Merva 280

Water Conveyance Capacity of Corrugated Plastic Drain Pipe
with Internal Diameters Between 38 and 75 mm
R. S. Broughton, B. W. Fuller, R. B. Bonnell 288

Weathering Tests on U-PVC Agricultural Drains: An Artificial Versus a Natural Test
J. C. Benoist, P. Beccaria 297

Boundary Modeling of Steady Saturated Flow to Drain Tubes
S. C. Yu, K. D. Konyha 305

GENERAL SESSION

Keynote - Research Needs for Water Quality / Salinity / Wetlands Management
D. A. Bucks 314

DRAINMOD-N: A Nitrogen Model for Artificially Drained Soils
M. A. Breve, R. W. Skaggs, H. Kandil, J. E. Parsons, J. W. Gilliam 327

An Evaluation of the ADAPT Water Table Management Model
S. O. Chung, A. D. Ward, N. R. Fausey, W. G. Knisel, T. J. Logan 337

Analytical Method of Determining Wetland Hydrology
S. Palalay, F. Geter 345

WATER QUALITY / SALINITY CONTROL

Concentration of Agricultural Chemicals in Drainage Water
S. Abdel-Dayem, M. Abdel-Ghani 353

A Field Monitoring System to Evaluate the Impact of Agricultural Production Practices on Surface and Ground Water Quality
R. S. Kanwar, D. G. Baker, G. F. Czapar, K. W. Ross, D. Shannon, M. Honeyman 361

Drainage Effects on Water Quality in Irrigated Lands
J. E. Ayars, R. A. Schoneman 371

Monitoring System for Water Table Management
Water Quality Research
H. W. Belcher, G. E. Merva, A. C. Fogiel 379

Efficacy of Acid Reclaimants in Combination with
Nonconventional Fertilizers for Salinity Control
M. H. K. Niazi, N. Hussain, S. M. Mehdi, M. Rashid,
G. D. Khan 387

Solute Concentration Prediction in Agricultural Drainage
Lines Under a Structured Soil
G. Shalit, T. S. Steenhuis, J. Boll, L. D. Geohring,
H. A. M. Hakvoort, H. van Es 395

MEASUREMENT TECHNIQUES AND DRAINAGE TECHNOLOGY

Evaluation of an Ultrasonic Ranging System for Monitoring
Water Table Levels
J. D. Eigel, S. A. Marquie, B. A. Vorst 403

Comparison of the Performance of Thick and Thin Envelope
Materials
H. El-Sadany Salem, L. S. Willardson 411

Hydraulic Head Losses Near Agricultural Drains During
Drainage and Subirrigation
R. H. Susanto, R. W. Skaggs 419

Crop Yield Increases Required to Justify Subsurface Drainage
Installation Costs in the Lower Mississippi Valley
C. E. Carter, R. L. Bengtson, C. R. Camp, J. L. Fouss,
J. S. Rogers 428

An Infiltration Method to Assess the Function of Old
Subdrains
U. Schindler, R. Dannowski, L. Müller 440

A Decade of Subsurface Drainage Environmental Research in
Southern Louisiana
R. L. Bengtson, C. E. Carter, J. L. Fouss 448

INTEGRATED SYSTEMS

Water Table Effects on Soybean Yield and Moisture and
Nitrate Distribution in the Soil Profile
C. A. Madramootoo, S. Broughton, A. Papadopoulos 458

Crop Management to Maximize the Yield Response of
Soybeans to a Subirrigation / Drainage System
R. L. Cooper, N. R. Fausey, J. G. Streeter 466

Some Results of a 12 Year Experiment with Different
Subirrigation Levels in a Young Marine Clay Soil in the
Ijsselmeerpolders in The Netherlands
J. Visser 474

Agroforestry Systems for On Farm Drain Water Management
G. S. Jorgensen, K. H. Solomon, V. Cervinka 484

Hydrological Impact of Farm Water Management Alternatives
in the Canadian Prairies
N. D. MacAlpine, D. W. Cooper, R. D. Neilson 491

Sixth International Drainage Symposium Summary
J. van Schilfgaarde 499

Foreword

The Sixth Drainage Symposium, sponsored by the ASAE and cooperating organizations, continued the 25-year tradition of promoting detailed technical reporting and communications between science, industry, and practice in agricultural drainage and water management. This popular series began in 1965, and over the years has attracted good attendance and participation from North America and other parts of the world. By the Fifth Symposium in 1987 (following the Third International Workshop on Land Drainage held at The Ohio State University), the growth in international participation was noteworthy.

The first in the series to be announced and conducted as an International Drainage Symposium, the two-full-day program included presenters from the United States, Canada, England, The Netherlands, Germany, France, Egypt, Belgium, India, Pakistan, Nigeria, and Sri Lanka. Also for the first time, a combination of oral and one-on-one presentations permitted the program committee to accommodate a total of 46 oral presentations and 20 one-on-one presentations/exhibits. This made the Sixth Symposium the largest in the series in terms of number of papers in the published proceedings. Papers for the one-on-one presentations are also printed in the proceedings, and participants were able to view the one-on-one displays during the entire symposium.

The enhanced symposium format included an opening night session to exhibit/demonstrate up-to-date technology on the drainage and water management simulation model, DRAINMOD, which is widely distributed and used throughout the world. The topic of "Environmental Impacts of Land Drainage: A World Overview" was addressed following the luncheon on the first day. Because of its past popularity, the evening session, featuring new developments from industry, was included once again and was expanded to the one-on-one format. Two other first-time exhibits at the Sixth International Drainage Symposium were: (1) the 1898 Model No. 0 Buckeye Tile Trencher exhibited by drainage contractors from Ohio; and (2) the Drainage Hall of Fame display from The Ohio State University.

Session topics reflected both continuing and emerging issues in drainage and water management. Environment impacts, wetlands, and sustainability of drainage systems have emerged as major issues and concerns since the Fifth Drainage Symposium. These and other topics are being emphasized in much of the newer research in drainage and water management technology. High priority research involving drainage as a part of total water management systems, particularly as it relates to water quality issues, continues to receive major emphasis in many regions of North America and other parts of the world. The development and validation of computer simulation models to predict the performance of integrated water management and crop production systems, and the transport of soil, plant nutrients, and agrochemicals in drainage waters, continues to be of high priority. Several papers are included on these subjects as well as a session that reported on new measurement techniques and drainage research instrumentation. Concerns and questions from drainage contractors were covered in a special "Contractors' Concerns" panel discussion included in the formal program.

The planning for this Symposium began almost immediately after the Fifth National Drainage Symposium. The organization and technical committees worked efficiently and cooperatively to produce another outstanding program and symposium. The committees included:

Program:

    Carl R. Camp, Chairman
    James S. Rogers, Co-Chairman and One-On-One Coordinator
    Richard D. Wenberg, Chandra Madramootoo,
    George E. Merva, Adel Shimmoahmmadi, John E. Parsons,
    Daniel L. Thomas, James E. Ayars, Harold R. Duke,
    Cy Schwieterman, and John F. Rice.

Proceedings:

    Richard L. Bengtson, Chairman
    Cade E. Carter, Co-Chairman

Publicity:

    Andrew D. Ward, Chairman

Local Arrangements:

    Fredrick H. Galehouse, Chairman
    Cy Schwieterman

Opening Night:

    Robert O. Evans, Chairman
    Kevin D. Robbins, John E. Parsons

Luncheon:

    Norman R. Fausey, Chairman

Industry Night:

    William E. Altermatt, Chairman

Publicity, finances, and symposium management were handled by ASAE Headquarters under the direction of Mr. Jon Hiler and Ms. Linda Fritsch, and publication of the pre-printed proceedings by Ms. Pam DeVore-Hansen and Ms. Julia Costello. The Sixth International Drainage Symposium committee owes a special note of thanks and appreciation to these ASAE Headquarters staff members for their extra effort to make this drainage symposium a success.

The Symposium Committee also wishes to give a special note of thanks and appreciation to our Keynote Speakers, Dr. R. Wayne Skaggs and Dr. Dale A. Bucks; to our Luncheon Speaker, Dr. Robert S. Broughton; and to Dr. Jan van Schilfgaarde for a fine Symposium Summary. Thanks is also expressed to all Session Moderators and Coordinators for their assistance, and to all Industry Exhibitors for their participation in the Drainage Symposium.

Many members of the Society have contributed and given assistance. The fact that they are not listed by name does not diminish the importance of their ideas and help, and the Symposium Committee expresses thanks and appreciation for their efforts. Other societies and organizations who agreed to co-sponsor the Symposium are:

> American Society of Agronomy
> American Society of Civil Engineers
> American Water Resources Association
> Canadian Society of Agricultural Engineering
> Corrugated Plastic Pipe Association
> Crop Science Society of America
> Land Improvement Contractors of America
> Soil and Water Conservation Society
> Soil Science Society of America
> U. S. Committee on Irrigation and Drainage

I am very thankful for the assistance and support of all who have contributed and shared ideas in making the Sixth International Drainage Symposium a success. Their efforts and cooperation have made it an honor for me to serve as the general chairman, and the recognition belongs to those who gave so willingly and freely of their time and effort.

James L. Fouss
General Chairman

To The Authors

A goal of the Steering Committee for the Sixth International Drainage Symposium was to produce a Symposium Proceedings which would be available during the symposium. In order to achieve this goal, rigid guidelines and dates were established in regard to the drafting, review, and completion of the papers. I wish to thank all of the authors with papers in this Proceedings for meeting the deadlines.

To ensure a high quality publication, most of the papers were reviewed. I wish to thank the authors and those whose evaluations helped to ensure that the papers met high technical standards.

Richard L. Bengtson
Proceedings Chairman

DRAINAGE AND WATER MANAGEMENT MODELING TECHNOLOGY

R. Wayne Skaggs
Fellow ASAE

It is my pleasure to open this Sixth National Drainage Symposium. The emphasis of these meetings and the image of drainage has changed dramatically since the first symposia in the 1960's and 70's. At that time drainage was considered a good and honorable profession. Drainage technology was in a stage of rapid development. The advent of corrugated plastic tubing, laser grade control, the drainage plow and associated developments resulted in impressive increases in both the rate and quality of drainage installations. As a profession we were proud of these advances and reported statistics on the high rates of installation in the symposium proceedings.

Now the mere mention of the word drainage is greeted with an angry response in many quarters. Drainage is held responsible for the large loss of wetlands in this country with consequent detrimental effects on water quality and the reduction of habitat for birds and wildlife. This has come as a great surprise to drainage contractors, technicians, and engineers who have spent their lives promoting drainage as a soil and water conservation practice. These professionals see drainage as an effective method of reducing crop losses and improving the efficiency and profitability of farming. This difference in perception of drainage is due to a number of factors. A large part of the problem is the definition of the term "drainage". To many people, "drainage" is inseparably connected to the act of "draining the swamp" to enable changes in land use from wetlands to agricultural, forestry, recreational or other uses. To these people "drainage" means land use conversion. While drainage was necessary to convert wetlands to farmlands, it was only one of the essential elements. Furthermore, such land use conversion is nearly nonexistent today. Economic conditions and government regulations to protect wetlands have combined to essentially stop the development of new agricultural lands in this country.

While the image of drainage has been largely influenced by its role in land use conversion, this role has little to do with our interests or activity in these symposia. The large majority of agricultural drainage is conducted on lands that have been farmed for many years. In this context, drainage is an agricultural rather than a land conversion practice. It is aimed at increasing efficiency of production and profitability on poorly drained agricultural lands. These lands comprise approximately 25% of the cropland in the USA (Pavilis, 1987) and 23% in Canada (Shady, 1989). They are generally capable of high levels of production with low losses of sediment and other pollutants and are an important part of our food production base, on both national and global scales. This is not to imply that agricultural drainage does not impact the environment. It does, sometimes severely.

R. Wayne Skaggs, William Neal Reynolds Professor and Distinguished University Professor, Biological and Agricultural Engineering Department, North Carolina State University, Raleigh, NC.

Impacts of drainage on hydrology and water quality have been the focus of sessions at these symposia since 1976. A cursory review of the program for this symposium shows that the emphasis in drainage research today is on improving water quality and reducing environmental impacts. Increasingly, farmers will be asked to satisfy environmental as well as agricultural goals. In general this task will be easier and more successful on drained soils, which are relatively flat, than on naturally well drained uplands. However, losses of sediment and other pollutants may still be unacceptable. We have made progress in this area. Methods have been developed to dramatically reduce pollutant loads from drained lands in some locations (Gilliam, 1987; Evans et al., 1990). However, application of methods to improve water quality is not always straightforward. For example, increasing the intensity of subsurface drainage will reduce losses to receiving waters of some constituents while increasing others (Baker and Johnson, 1975; Bottcher et al., 1981; Gilliam, 1987). These losses depend on fertility, pest control, and other cultural practices, as well as the design and operation of the drainage system. Furthermore, novel designs for reducing pollutant losses may be counterproductive if they are not properly managed (Skaggs and Gilliam, 1981).

The design and operation of drainage systems to satisfy both agricultural and environmental objectives will require a different approach than the traditional "install it and forget it" routine. The goal of a design should be to remove only the least amount of water necessary to satisfy the drainage objectives: trafficability, protection of the crop from excessive soil water conditions, salinity control, etc. To drain away more than the least amount necessary may increase drought stresses and almost certainly will increase pollutant loads to the receiver. In some cases it will increase the volume of water that must be treated before release to receiving streams. Since drainage intensities required to satisfy the multiple objectives vary during the year and from year-to-year due to both function and the vagaries of the weather, adjustable controls on the outlets may be required to remove only the least amount of water necessary. Proper design and operation of systems and controls will require a better understanding of drainage and water table management by all concerned, from the design engineer to the farmer-operator. This understanding involves not only the mechanisms controlling the operation of the system with the ability to carefully tailor the water management system to soil, site, crop and climatological conditions, but also the interactions of fertility, cultural and water management practices and their impact on water quality and productivity. This finally brings me to my subject of modeling.

Models are not important because they enable us to understand fundamental mechanisms governing movement and fate of water and solutes in soils. They don't. It is generally not possible to model phenomena that we don't understand. Models are important and useful in quantifying these phenomena because they allow us to handle large numbers of calculations to predict actions and interactions of multiple processes over long periods of time. I believe the use of models will be essential in designing and developing drainage and related water management systems to satisfy both environmental and agricultural goals. There are so many system components and operational factors that affect both crop production and drainage water quality that it will be nearly impossible to find optimum, or even acceptable, solutions without the use of simulation models. In the past, models have mostly been used in research, but they are rapidly finding their way into the practice of design and analysis on a practical basis. This has already happened in areas where the stakes are higher. The use of sophisticated models on a production scale is common in the analysis and treatment of contamination of groundwater aquifers, for example. A comprehensive review of drainage models is clearly beyond the scope of this paper. Rather, I will present an overview of approaches for modeling drainage and water table control systems. The paper will conclude with a discussion of barriers to the application of models in the field of drainage and water table management.

APPROACHES FOR MODELING POORLY DRAINED SOILS

Models have been used to relate drainage system performance to design parameters such as drain spacing and depth since shortly after Darcy developed his fundamental law in 1856. Youngs (1989) reviewed the early history of the development of methods for drainage design. He noted that a formula developed by Colding in 1872 is still widely used to calculate maximum water table heights in lands drained by rows of parallel sinks. Drainage models span the wide spectrum from very approximate guidelines that relate drainage requirements to soil type, to drain spacing equations such as the Hooghoudt equation, to computer models that simulate the performance of multicomponent systems over long periods of weather record. In this broad context the term "model" is defined as a set of guidelines, equations or computer programs that can be used to quantify the performance of a system in terms of an objective function or functions.

In recent years, computer models have been developed to simulate the day-by-day performance of drainage and water table control systems (see reviews by Feddes et al., 1988; Skaggs, 1987). They range from very complex numerical solutions of differential equations to approximate methods for conducting a water balance in the soil profile. Many of these models have been field tested and provide means of describing the hydrology of shallow water table soils, including effects of drainage and related water management practices on yields. Models have also been developed to predict the losses of salts, fertilizer nutrients, sediment and other pollutants from drained lands. Prediction of the movement and fate of solutes and other pollutants depends on, and is more complex than, quantifying the hydrology of the system. These models are generally not as advanced, nor uniformly accepted in practice, as the drainage simulation models. However, development and testing of such models is the focus of much of the current research in the field of drainage, as indicated by several papers on the program of this symposium.

The most theoretically rigorous method for modeling soil water movement and storage is the so-called exact approach. This method involves the solution of the Richards equation for combined saturated-unsaturated flow in two or three dimensions. It is based on the Darcy-Buckingham equation for soil water flux and the principle of conservation of mass. By solving the Richards equation subject to the appropriate boundary and initial conditions the system can be described for most conditions of interest. The solution provides soil water contents and pressure heads as a function of time and space, the water table position, drainage or subirrigation rates, infiltration and evapotranspiration rates and excess water available for surface runoff. Flow paths can be determined and the time that water resides in various sections of the soil profile can be predicted. This capability is important in modeling effects of water management practices on solute transport and drainage water quality.

Early applications of the exact approach to drainage were confined to events of short duration. As faster computers have reduced computational requirements, simulations based on solutions to the 2-D Richards equation for periods of months in duration were obtained (e.g. Zaradny, 1986). General programs for solving the 2-D Richards equation subject to changing surface conditions are now available. Harmsen et al. (1991) and Munster (1992) modified the USGS program VS2DT (Lappala et al., 1987) to simulate drainage and subirrigation. Solutions to the flow equations were coupled to numerical solutions to the advective-dispersive-reactive equations by Healy (1990) to simulate the transport and fate of nitrogen (Harmsen et al., 1991) and the pesticide aldicarb (Munster, 1992). Results in good agreement with field measurements were obtained in both cases. Kamra et al. (1991) developed a semidiscrete model for water and solute movement in tile drained soils. They assumed steady state conditions for 2-dimensional water flow and solved the governing equations for solute movement. The approach allowed long-term prediction of concentrations in drainage effluents and salt distributions in soil and groundwater. By employing analytical solutions for steady water flow, numerical solutions for solute transport yielded explicit expressions for concentrations at any future time without having to compute concentrations at intermediate times.

The biggest problem with the exact approach is that the equations are nonlinear and numerical methods must be used to obtain solutions. Such solutions are both difficult and expensive. Formerly, computer time required for these solutions was prohibitive. This problem has been nearly eliminated with the wide availability of fast, modern computers. Computational requirements still remain a problem for some users and applications, however. A more basic limitation is the requirement of detailed descriptions of the unsaturated soil properties for each profile layer. These functions are not generally known and must be measured or estimated for each site. While the application of this approach to water table management problems is useful and will probably continue to increase, its use will likely be confined to research scientists.

An approach less complex than the one discussed above, but, which still predicts water contents in the unsaturated zone with good resolution, is based on numerical solutions to the one-dimensional Richards equation for vertical flow. Lateral water movement due to drainage is evaluated using approximate equations that are imposed as boundary conditions on the solutions to the Richards equation. The SWATRE model developed by Feddes and colleagues (Feddes et al., 1978) is an example of this approach. The model can consider up to five soil layers having different physical properties. Climatological data including daily rainfall and potential evaporation and transpiration are used to specify the boundary condition at the top of the profile. Flux in the saturated zone at the bottom of the profile is calculated by a steady state equation developed by Ernst. Six other bottom boundary conditions can be considered. SWATRE has been linked with other models to predict trafficability, crop emergence, growth and production (Feddes and Van Wijk, 1990). Karvonen (1988) introduced very efficient numerical techniques and added the capability of simulating the effects of soil temperature differences and frozen conditions. This approach was also used by Bronswijk (1988) to model the effects of soil swelling and shrinkage and by Workman and Skaggs (1990) to include the effects of macropores.

Numerical solutions to the Richards equation for vertical flow have formed the basis of the LEACHM model (Wagenet and Hutson, 1987) for predicting the movement and fate of solutes. LEACHM was used by Bigger et al. (1990) to predict soil profile salinity in drained lands. Clemente and Prasher (1992) combined numerical solutions for equations governing convection, adsorption, diffusion and degradation of pesticides with SWATRE for soil water movement, and other components for surface runoff and heat flow to form the model PESTFADE, which may also be applied to drained lands.

The advantage of the 1-D Richards equation approach is that it is based on sound theory for vertical water movement in the unsaturated zone. Since most of the unsaturated water movement tends to be in the vertical direction in drained soils, this approach should provide reliable predictions of the soil water conditions above the water table. Another advantage is that it can also be applied for soils without water tables. A disadvantage of this approach is the requirement of the unsaturated soil water properties as discussed above for the two-dimensional model. Because execution of the model involves numerical solutions to the 1-D Richards equation, problems with convergence and stability may occur for some cases and extensive user training is necessary for its application.

A third approach to modeling drainage systems is based on numerical solutions to the Boussinesq equation. This approach is normally applied for watershed scale systems or where the horizontal water table variation is critical. Examples are the models developed by de Laat et al. (1981) and Parsons et al. (1991). Approximate methods are used to conduct a water balance in the unsaturated zone above the water table. These models may be used in large nonuniform areas where lateral differences in soils, surface elevations and water table depths are important. This approach can also be used to describe soil water conditions during the transition from drainage to subirrigation or vice versa. This is especially important in the use of models for simulating effects of various control strategies for water table management systems. Fouss and Rogers (1992) will use this approach in a paper on this program.

The first computer simulation models were based on a water balance in the soil profile (van Schilfgaarde, 1965; Young and Ligon, 1972; Skaggs, 1975; Chiang et al., 1978). An example is DRAINMOD which uses functional algorithms to approximate the hydrologic components of shallow water table soils. DRAINMOD was developed for design and evaluation of mutlicomponent water management systems. Approximate methods are used to simulate infiltration, drainage, surface runoff, evapotranspiration, and seepage processes on an hour-by-hour, day-by-day basis. Input data include soil properties, crop parameters, drainage system parameters, climatological and irrigation data. Water table position and factors such as the ET deficit are calculated to quantify stresses due to excessive and deficient soil water conditions. Stress-day-index methods (Hiler, 1969) are used to predict relative yields as affected by excessive soil water, deficit or drought conditions and planting date delay.

Parsons et al. (1989) combined DRAINMOD with CREAMS to predict the effect of water table control practices on surface runoff and the losses of sediment and associated contaminants. Recent modifications in DRAINMOD provide daily average fluxes as a function of depth. This has allowed the development of methods to predict soil salinity as affected by drainage system design and management (Kandil et al., 1992). Algorithms have also been incorporated to predict nitrogen movement and fate in drained soils as will be reported on this program (Breve et al. 1992).

The water balance approach has also been used to develop other models. The ADAPT model, which will be discussed at this symposium (Chung et al., 1992), uses functional relationships for drainage processes in GLEAMS to evaluate pesticide losses from drained soils. Another functional model is SIDRA (LeSaffre and Zimmer, 1988) which was developed in France to predict peak flow rates from drained lands. Recent developments in the modeling research of this group will also be discussed at these meetings by Zimmer and Lorre (1992).

BARRIERS TO PROGRESS

The major challenge in the management of poorly drained soils is the development of methods to satisfy both agricultural and environmental objectives. In many cases it will not be possible to find a solution that will completely satisfy both objectives. In those cases it will be necessary to quantify the benefits or costs to each objective, in order to provide a rational basis for determining trade offs and making difficult decisions. This will require the use of simulation models. Models have been developed and successfully applied, especially to design systems to satisfy agricultural objectives. However, there are a number of barriers to the application of models toward satisfying objectives at all levels for poorly drained soils.

Probably the most fundamental barrier to our progress is the lack of recognition that drainage is only a part of the production system. It will be necessary to address the whole to solve some of the more difficult problems. An example is in irrigated lands where drainage is but a part of the water management system. The drainage intensity necessary to control water logging and salinity may be greatly reduced by increasing the irrigation efficiency or by modifying the frequency and rates of irrigation. That is, it may be possible to partially solve the drainage problem by modifying the irrigation component. Likewise it must be recognized that pollutant loads from drained soils depend on fertility, cultural and pest control practices, as well as the design and management of the drainage system.

The lack of reliable models that can be used <u>on an applied scale</u> to predict the interactive effects of the many factors that affect drainage water quality is currently a barrier to progress in this important area. As discussed above, a number of models have been developed and applied in research, and good progress has been made on others that promise to be easier to apply on a field scale. Some of these will be discussed at this symposium. However, at this point there is no single model, nor even a suite of models, that could be recommended for the job. Many of us are working toward filling this need, but at present we must recognize it as a barrier to progress. This deficiency has partly been due to the lack of experimental data for developing and testing such models. Hopefully, this gap will be filled as a result of field experiments currently being conducted in several states and provinces; some of these efforts will be discussed at this symposium.

While models have been developed and applied to predict the effects of drainage and irrigation system design and operation on yields, there are barriers to this application for some cases. Crop inputs and response functions are only available for a limited number of crops. Furthermore, models for predicting the effects of drainage design on soil salinity and the consequent effects on yields have not been completely tested and accepted in practice. Again, good progress has been made in both areas, as will be discussed at these meetings, but there are gaps which limits progress for many applications.

One of the first barriers encountered in using a model, whether it be a drain spacing equation or a complex simulation model, is the need for soil property inputs. These inputs include saturated hydraulic conductivity (K), drainable porosity and depth to the impermeable layer for the simpler models. Soil water characteristics, $h(\theta)$, unsaturated hydraulic conductivity relationships, $K(\theta)$ (where θ is soil water content and h is pressure head), and anisotropic ratios, are required for each profile horizon by the more complex models. For models that include movement and fate of solutes, additional inputs for hydrodynamic dispersion and coefficients for various reactions such as nitrification and denitrification are required. In a research setting it may be possible to measure the necessary inputs. On an applied scale, the lack of soil property inputs, or the cost of obtaining them, may be an insurmountable barrier preventing the application of the most appropriate model.

This problem can be addressed by first conducting sensitivity analyses to determine the consequences of errors in the individual inputs for various applications of the model. Such analyses should be conducted on a general basis by the model developer, and included in the user's manual and other references. Additional sensitivity analyses should be conducted by the user for his particular application. This will identify those input properties and parameters that should be measured for each site and those that can be estimated from soil series information, textural data, etc. Sensitivity analyses of models applied to drainage and water table management systems typically show that results are most sensitive to K and profile depth with only moderate or low sensitivity to the unsaturated $h(\theta)$ and $K(\theta)$. This is fortuitous because the unsaturated properties are more difficult and expensive to measure. However, they do vary widely among soils and are required inputs for many simulation models. Two approaches have been most commonly used to estimate the unsaturated soil properties for simulation models. One is to simply use data that has been measured and tabulated by soil series. The best example is in the State of Florida where soil water characteristic data for many of the state's soil series have been measured and are available in both published (e.g. Carlisle et al., 1989) and computer data base forms. The other approach is to predict the unsaturated properties in term of texture, density and other properties that are more easily determined. The method by Van Genuchten (1980) is frequently used for this purpose. Baumer and Rice (1988) developed methods specifically for predicting the unsaturated inputs for DRAINMOD. Development of fast, easily applied methods for measuring the soil water characteristic in the field is a research need that seems achievable in view of recent advances in methods to measure soil water content.

Determination of K for use in drainage models has been a longstanding problem in drainage design. The measurement of K with standard methods is complicated by the fact it may vary by several fold within a small area. This is illustrated by results given in Fig. 1 for a hypothetical field made up of uniform blocks of soil, 4m x 4 m in surface area, having K values of 5.0 m/d (shaded areas) and 0.5 m/d (remainder of the area) (Fig. 1a). The water table response predicted by a numerical solution to the 2-D Boussinesq equation was a uniform surface (Fig 1b). That is, this very heterogeneous field responds to drainage as a homogeneous block with a field effective K nearly equal to the arithmetic average of the conductivities of the individual units (Fig. 1c). The field effective value could be obtained directly by back-calculation from drawdown measurements or from simultaneous measurements of water table and drain flow as suggested by several of us (e.g. Dielemann and Trafford, 1976).

Determination of field effective values of soil properties as discussed above lumps the effects of small scale variabilities and effects of heterogeneities such as macropores into an average value that can be used to describe the drainage process on a field scale. Macropores and the effects of preferential flow are clearly important, and in some cases may be the dominant processes controlling pollutant movement into agricultural drains. The failure to properly account for preferential flow represents another barrier to the use of models. The importance of preferential flow processes and the research emphasis being placed in this area are indicated by the fact that ASAE sponsored a symposium on the subject in 1990. Casual observations indicate that preferential flow has replaced hysteresis as the most often cited (blamed) cause of the failure of model predictions to agree with experimental observations. While our understanding of preferential flow processes has improved in

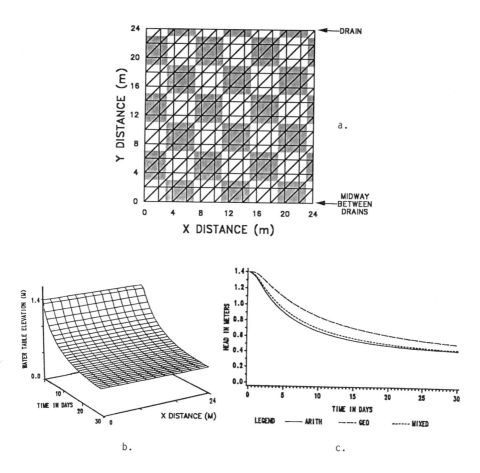

Figure 1. Effect of bimodal variation in hydraulic conductivity on water table drawdown midway between drains. a. Finite element grid where K=5.0 m/d in shaded area and K=0.5 m/d in remainder of grid. b. 3-D plot of water table drawdown midway between drains. c. Drawdown midway between drains comparing solutions for K-arithmetic (ARITH) and geometric (GEO) means to solutions for grid in Fig. 1A. (From Bentley et al., 1989).

the last few years and models have been developed to simulate them for drainage (Bronswijk, 1988; Leeds-Harrison and Jarvis, 1986; Workman and Skaggs, 1990), direct application requires more detailed soil input data than are normally available. An alternative approach may be to determine field effective values of the properties affecting both solute and water movement from field scale measurements. The effect of preferential flow paths on solute flow to drains will be discussed in several papers at this symposium.

Another barrier to progress in drainage modeling is the problem of quantifying head losses near the drain, and the effects of various envelopes on these losses. Stuyt (1992) has conducted impressive studies on the mechanisms controlling water entry into drains and I look forward to hearing his results at this meeting. Several other papers on the program will also make valuable contributions to our understanding in this area.

SUMMARY

Drainage and related water management systems should be designed and operated to satisfy both agricultural and environmental goals. This will require improved understanding of the mechanisms controlling the operation of water table control systems, as weell as their interactions with fertility and cultural practices. Simulation models have been and are being developed to quantify these processes and interactions in terms of soil properties, crop parameters, site conditions and climatological variables. Approaches currently being used to develop and apply simulation models for drained lands were discussed in this paper. New simulation models, together with results of field experiments that can be used to test and further develop such models will be presented at this symposium. In addition, new information on drainage processes and the mechanisms controlling outflow water quality will be presented and debated during these meetings. Thus, the proceedings of this symposium will represent a step forward in our ability to model drainage systems for the purpose of improving their design and operation. I congratulate the organizers of this symposium and look forward to the program.

REFERENCES

Baker, J.L. and H.P. Johnson. 1976. Impact of subsurface drainage on water quaity. Proc. of 3rd National Drainage Symposium, ASAE Publ. 77-1, ASAE, St. Joseph, MI.

Baumer, O. and J. Rice. 1988. Methods to predict soil input data for DRAINMOD. ASAE Paper No. 88-2564, ASAE, St. Joseph, MI.

Bentley, W.J., R.W. Skaggs, and J.E. Parsons. 1989. Effect of variation in hydraulic conductivity on water table drawdown. N.C. Agri. Res. Serv. Tech. Bull. 288, N.C. State University, Raleigh, 23 p.

Bigger, J.W., R.J. Wagenet, J.L. Hutson, and D.E. Rolston. 1990. Predicting soil profile salinity using LEACHMS for drainage design. Symp. on Land Drain. for Salinity Control in Cairo, Egypt, p. 157-164.

Bottcher, A.B., E.J. Monke, and L.F. Huggins. 1981. Nutrient and sediment loadings from a subsurface drainage system. Trans. ASAE, 24(5):1221-1226.

Breve, M.A., R.W. Skaggs, H. Kandil, J.E. Parsons, and J.W. Gilliam. 1992. DRAINMOD-N: A nitrogen model for artificially drained soils. This issue.

Bronswijk, J.J.B. 1988. Effect of swelling and shrinkage on the calculation of water balance and water transport in clay soils. Agricultural Water Management 14:185-193.

Carlisle, V.W., F. Sodek, III, M.E. Collins, L.C. Hammond, and W.G. Harris. 1989. Characterization data for selected Florida soils. Soil Sci. Res. Report No. 89-1, Soil Sci. Dept., Univ. of FL, Gainesville.

Chieng, S.T., R.S. Broughton, and N. Foroud. 1978. Drainage rates and water table depths. J. Irr. & Drain. Div., ASCE 104(IR4):413-433.

Chung, S.O., A. Ward, N.R. Fausey, W.G. Knisel, and T.J. Logan. 1992. An evaluation of the ADAPT Water Table Management Model. This issue.

Clemente, R.S. and S.O. Prasher. 1992. PESTFADE, a model for simulating pesticide fate and transport in soils. Symp. XVII on Ag. Water Quality Priorities, Beltsville, MD.

de Laat, P.J.M., R.H.C.M. Atwater, and P.J.T. van Bakel. 1981. GELGAM - A model for regional water management. Proc. of Tech. Meeting 37. Versl. Meded. Comm. Hydrol. Onderz. TN027. The Hague, Netherlands: 25-53.

Dieleman, P.J. and B.D. Trafford. 1976. Drainage testing. Irrigation and Drainage Paper No. 28, FAO, Rome, pp.. 63-84.

Evans, R.O., J.W. Gilliam, and R.W. Skaggs. 1990. Controlled drainage management guidelines to improve drainage water quality. N. C. Agri. Ext. Serv. Bull., AG-443, Raleigh, NC, 16 p.

Feddes, R.A. and A.L.M. van Wijk. 1990. Dynamic land capability model: a case history. Phil. Trans. Royal Society of London, Vol. 329:411-419.

Feddes, R.A., P. Kabat, P.J.T. van Bavel, J.J.B. Bronswijk, and J. Halbertsma. 1988. Modeling soil water dynamics in the unsaturated zone-state of the art. Journal of Hydrology, Vol. 100:69-111.

Feddes, R.A., P.J. Kowalik, and H. Zaradny. 1978. Simulation of water use and crop yield. PUDOC, Wageningen, Simulation monograph. 189 p.

Fouss, J.L. and J.S. Rogers. 1992. Drain outlet water level control: a simulation model. This issue.

Gilliam, J.W. 1987. Drainage water quality and the environment. Keynote address. proc. Fifth Nat. Drain. Symp., ASAE Publ. 7-87, pp. 19-28.

Harmsen, E.W., J.W. Gilliam, R.W. Skaggs, and C.L. Munster. 1991. Variably saturated 2-dimensional nitrogen transport. 1991. Winter meeting of the ASAE Paper No. 91-2630.

Healy, R.W. 1990. Simulation of solute transport in variably saturated porous media with supplemental information on modifications to the US Geological Survey's computer program VS2DT. Water-Resources Investigations Report 90-4052. Denver, CO.

Hiler, E.A. 1969. Quantitative evaluation of crop drainage requirements. Transactions of the ASAE, 12(4):499-505.

Kandil, H., C.T. Miller, and R.W. Skaggs. 1992. Modeling long-term solute transport in drained unsaturated zones. Water Resources Research, In Press.

Karma, S.K., S.R. Singh, K.V.G. Rao, and M. TH. van Genuchten. 1991. A semidiscrete model for solute movement in tile-drained soils. 1. Governing equations and solution; and 2. Field validation and applications. Water Resources Research, Vol. 27(9):2439-2456.

Karvonen, T. 1988. A model for predicting the effect of drainage on soil moisture, soil temperature and crop yield. PhD thesis, Helsinki Univ. of Tech., Helsinki, Finland, 215 p.

Lappala, R.W., R.W. Healy, and E.P. Weeks. 1987. Documentation of computer program VS2D to solve the equations of fluid flow in variably saturated porous media. USGS Water-Resources Investigations Report 83-4099. Denver, CO.

Leeds-Harrison, P.B. and N.J. Jarvis. 1986. Drainage modeling of heavy clay soils. Proc. of Intl. Seminar on Land Drainage, Helsinki Univ. of Tech., 198-220 pp.

LeSaffre, B. and D. Zimmer. 1988. Subsurface drainage peak flows in shallow soils. J. of Irrigation and Drainage Engineering, Vol. 114(3):387-406.

Munster, C.L. 1992. Effects of water table management on transport of the pesticide aldicarb. PhD Dissertation, North Carolina State University, Raleigh, NC.

Pavelis, G.A. 1987. Chapter 11: Economic survey of farm drainage. In: G.A. Pavelis (ed.). Farm Drainage in the U. S.: History, Status, and Prospects. USDA-ERS, Misc. Publ. 1455, pp. 110-136.

Parsons, J.E., R.W. Skaggs, and C.W. Doty. 1991. Development and testing of a water management simulation model (WATRCOM). Trans. ASAE, Vol. 34(1):120-128.

Parsons, J.E., R.W. Skaggs, and J.W. Gilliam. 1989. Pesticide fate with DRAINMOD/CREAMS. Proc. CREAMS/GLEAMS Symp., Publ. 4, Agric. Engr. Dept., Univ. of GA, Tifton, GA pp. 123-135.

Shady, A.M. 1989. Irrigation drainage and flood control in Canada. Canadian International Development Agency, Ottawa, Canada, 309 p.

Skaggs, R.W. 1975. A water management model for shallow water table soils. Univ. of North Carolina Water Resour. Res. Inst. Tech. Rep. 134.

Skaggs, R.W. 1987. Model development, selection and use. Proc. of Third Intl. Drainage Workshop on Land Drainage. The Ohio State University, Columbus:29-39.

Skaggs, R.W. and J.W. Gilliam. 1981. Effect of drainage system design and operation on nitrate transport. Trans. ASAE, 24(4):929-934.

van Genuchten, M.T. 1980. A closed-form equation for predicting the hydraulic conductivity of unsaturated soils. Soil Sci. Soc. Am. J. 44:892-898.

van Schilfgaarde, J. 1965. Transient design of drainage systems. J. of the Irrig. and Drain. Div. ASCE, 91(IR3)9-22.

Wagenet, R.J. and J.L. Hutson. 1987. LEACHM, Leaching estimation and chemistry models. Continuum Vol. 2, Water Resources Research Institute, Cornell Univ., Ithaca, NY.

Workman, S.R. and R.W. Skaggs. 1990. PREFLO: A water management model capable of simulating preferential flow. Trans. ASAE, Vol. 33(6):1939-1948.

Young, T.C. and J.T. Ligon. 1972. Water table and soil moisture probabilities with tile drainage. Trans. ASAE 15(3):448-451.

Youngs, E.G. 1989. The interaction between groundwater and unsaturated soil-water flow regions. Internat. Workshop on Meth. for Develop. and Management of Groundwater Resources IGW-89, Hyderabad, India, Vol. III:139-154.

Zaradny, H. 1986. A method for dimensioning drainage in heavy soils considering reduction in PET. Proc. Intl. Seminar Land Drainage, Helsinki Univ. Tech.:258-265.

Zimmer, D. and E. Lorre. 1992. Use of SIDRA model to assess the impact of deep seepage on drainage design. This issue.

Management Support System for Conjunctive Irrigation and Drainage

Luis A. Garcia[1], Timothy K. Gates[2], Marvin E. Jensen[3] Terrence H. Podmore[4]

Abstract: This paper describes a computer-based management support system (MSS) for the design and management of conjunctive irrigation and drainage that is under development at Colorado State University. The system will provide advanced technology to assist professionals in analyzing field-scale irrigation and drainage processes in semiarid and arid areas. Modules in the system include: irrigation and drainage simulation, a geographic information system (GIS), and a graphical user interface (GUI). The irrigation and drainage simulation modules form the heart of the MSS. They enable considering the major process of irrigation scheduling and application; precipitation; unsaturated and saturated flow and salinity transport; flow to subsurface drains; drainage effluent in collector drains, crop growth, consumptive use and relative yield as affected by waterlogging, salinity and water deficits.

The system framework is a two-dimensional vertical plane analytical model for preliminary analyses and a quasi three-dimensional numerical model for more detailed consideration. The GUI and the GIS facilitate running the models at different levels of detail. The GIS enables spatially-based input and output. It takes spatial data and performs computations needed for input to the irrigation and drainage models. The spatially-based output is displayed on a screen to enable the user to interactively query the system and modify parameters and boundary conditions as needed for sensitivity analyses and for evaluating alternative design and management scenarios.

INTRODUCTION

Irrigated agriculture has been essential in this century to provide food and fiber for an expanding population. Production per unit area of irrigated land will become more important in the future because it has been estimated that world population has a doubling time of only 32 years (Pop. Ref. Bur., 1989) and the rate of expansion of irrigated land has decreased to less than 1% per year (FAO, 1979, 1989a, 1989b). At the same time, the productivity of many irrigation projects has been declining due to waterlogging, salinity and poor irrigation management practices.

This paper is to describes a computer-based management support system (MSS) for the design and management of conjunctive irrigation and drainage that is under development at Colorado State University (CSU). The MSS can be used to improve the design and management of new irrigation projects and the rehabilitation of existing projects.

NEED FOR IMPROVED IRRIGATION DESIGN AND MANAGEMENT

Irrigation Development

In 1990, FAO (1990) indicated that the total irrigated land area in the world was 235 million hectares. About two-thirds of this area was in Asia (FAO, 1979, 1989a). World Bank/UNDP (1990) estimated the "gross irrigated area" to be 253 million hectares. During the past four decades, development of irrigated agriculture provided a major part of the increase in production necessary to meet population food demands. About 1/3 of the total crop production came from irrigated land

---

1. Assistant Professor, Dept. of Agricultural and Chemical Engineering, Colorado State University, Fort Collins, CO.
2. Assistant Professor, Dept. of Civil Engineering, Colorado State University, Fort Collins, CO.
3. Director, Colorado Institute for Irrigation Management, Colorado State University, Fort Collins, CO.
4. Professor, Dept. of Agricultural and Chemical Engineering, Colorado State University, Fort Collins, CO.

which makes up about 15 percent of the total arable land. The rate of expansion of irrigated land reached a peak of 2.3 percent per year from 1972 to 1975. It has declined since and is now less than 1 percent per year (FAO, 1990). The decline in rate of expansion is due to higher costs and lower performance than was expected.

Waterlogging, Soil Salinity and Drainage

The greatest cause of decreased production on many irrigated projects is waterlogging followed by soil salinization. Waterlogging and salinity can be prevented by better water control by assuring that all irrigation projects have adequate drainage. In the mid-1980s, the United Nations predicted that by 1990, 52 Mha of irrigated land would require drainage to control soil salinity (Oosterbaan, 1988).

Integrated Irrigation and Drainage

Historically, most irrigation and drainage systems were designed separately and the responsibility for the management of irrigation and drainage systems has generally been assigned to different agencies. Optimal use of irrigated agricultural lands require irrigation and drainage systems to be designed, constructed and managed as an integrated unit. A combined system can be very complex, requiring modeling the system to fully understand and predict long-term performance. Irrigation practices have direct effects on the water table, drain spacing is dependent on excess water applied and rainfall, and the costs and benefits of irrigation and drainage need to be mutually considered.

Modeling Irrigation and Drainage Systems

Many of the variables affecting the irrigation and drainage system are stochastic in nature. Irrigation scheduling is based on ET, crop growth stage and available soil water. The stochastic nature of meteorological variables can be simulated using a weather generator. Generated meteorological time series can be used as input into the scheduling model. The resulting stochastic schedules generated creates uncertainty in the system behavior. Similarly, system boundary conditions, soil flow and formation properties (hydraulic conductivity, pressure-saturation characteristics, etc.), irrigation application efficiency and other parameters can be modeled as cross-correlated spatial-temporal random fields.

GENERAL DESCRIPTION OF THE MODEL DEVELOPMENT

The Colorado State University Irrigation and Drainage Model (CSUID) is being developed with three main modules. The modules include irrigation and drainage simulation, a geographic information system (GIS), and a graphical user interface (GUI). The irrigation and drainage simulation modules form the heart of the MSS. They enable considering the major processes of irrigation scheduling and application; precipitation; unsaturated and saturated flow and transport; flow to subsurface drains; drainage effluent in collector drains; and crop growth, consumptive use and relative yield as affected by waterlogging, salinity and water deficits.

The system requires historical time series of meteorological data for the irrigation scheduling module. These data are used to create a set of daily reference crop evapotranspiration values using one of the following equations: Penman-Montheith, FAO Penman, or Jensen-Haise (Jensen, et. al. 1990). Reference crop evapotranspiration values are multiplied by experimentally derived crop coefficients to provide actual evapotranspiration (ET) estimates for each crop.

When modeling irrigation and drainage as a conjunctive system, the inputs are stochastic in nature. Irrigation scheduling is based on estimated ET and rainfall which is a function of meteorological variables, crop growth stage, and soil water. The stochastic nature of meteorological time series can be simulated using a weather generation model or input as measured data. In this work a model called WGEN (Richardson, 1984), is used to generate the stochastic time series of meteorological data. Generated meteorological time series are then used as input to an irrigation scheduling component of the MSS. The irrigation schedule can be input directly into the system if the user is interested in evaluating a particular scenario.

CSUID allows the user to evaluate drainage requirements initially using a two-dimensional vertical plane analytical model (Dumm, 1960). The analytical two-dimensional model allows the user to perform an initial screening of alternative irrigation schedules and drain spacings. This portion of the MSS contains an ET, crop growth, and upward flow components, and estimated relative yield as affected by waterlogging, salinity and soil water deficits. The user can generate a number of synthetical time series and model different irrigation practices on a two-dimensional plane. After evaluating the results of these runs, the user can access a numerical model to evaluate in more detail the performance of a particular system.

The numerical model allows the user to study the effects of spatial variations in the inputs. The quasi-three dimension model (El-Hessy, 1991) used as the basis for the numerical part of the system computes spatial and temporal distributions of soil water and salinity as affected by field-scale practices of irrigation and drainage in the presence of a saline shallow water table. The model solves the depth-averaged Boussinessq equation for areal flow in the saturated zone below the water table and the Richard's equation for one-dimensional vertical flow in the unsaturated zone above the water table. The mixing cell concept is used to predict advection-dominated salinity transport. Solutions are obtained via finite difference approximations of the equations at discrete grid points in the domain. In addition to calculating salinity and water distributions, the model predicts depth to the water table, upward flux from the water table, leaching efficiency, volume and salinity of drainage effluent collected, and relative crop yield. The control variables describing irrigation and drainage alternatives are the amount, timing, and salinity of irrigation water applied and the depth, spacing and slope of lateral subsurface drains. The model explicitly considers variability due to the diverse soil and crop properties and irrigation practices on multiple fields in an area. Future enhancements will allow a fully stochastic option where selected parameters (representing system boundary conditions and properties) may be modeled as cross-correlated spatial-temporal random fields.

Evaluation of Design Module

The stochastic nature of the inputs to the irrigation system creates uncertainty in the conjunctive system behavior. This uncertainty makes it impossible for the designer to forecast conjunctive system behavior with accuracy over its design life. Because of the stochastic inputs, a set of simulations are conducted to generate a distribution of net benefits for conjunctive design.

The MSS allows the user to simulate the plant-water-soil system on a daily basis over the life of a conjunctive system. In order to simulate the behavior of the system over long periods of time the user is required to specify the expected lifetime of the system. After CSUID computes the daily status of the plant-water-soil system, fixed and/or variable costs, benefits and net benefit for each growing season are computed. The MSS computes an annual net benefit based on the net benefits generated by each crop and the number of crops grown in a year.

The mean value of each net benefit distribution is computed. In order to generate a distribution of mean lifetime net benefit, CSUID simulates the behavior of the conjunctive system for a user specified number of lifetimes. At the end of the simulation CSUID has generated the user specified number of distributions, each representing a lifetime of the conjunctive system. CSUID then creates a distribution which contains the annual net benefit from all the lifetimes simulated, creating one distribution with all the values of annual net benefit (number of lifetime simulations multiplied by the number of years in each lifetime e.g. 40 simulation x 30 years = 1200 values). Currently, the simulation description is given for a single drainage spacing and depth and one set of soil properties. When an existing drainage system is being analyzed, drain spacing is known, and the performance of that spacing can be simulated. However, when a new irrigation and drainage system is being designed, the optimal drain spacing is not known. For this reason, a range of drain spacings needs to be simulated in order to determine the optimal design. CSUID requires that the user specify a minimum spacing, a maximum spacing, and a spacing increment. For each spacing CSUID simulates conjunctive system performance over the specified number of lifetimes.

The criteria used in analyzing conjunctive system performance for each design are the following: the minimum net benefit for all simulations, the two year minimum net benefit based on a moving average, the 90% reliable net benefit, the mean net benefit for all simulations, and the maximum net benefit.

CSUID Graphical User Interface

CSUID was written to run on a Sun UNIX workstation. The graphical user interface (GUI) was developed using the "C" programming language and "Motif" and "X Intrinsic Libraries" for the graphics. The GUI is based on a mouse driven approach that allows the user to select the options from the program by pressing the different mouse buttons (3 mouse buttons are present on a Sun Workstation mouse). The user friendly interface frees users from the normal tedium associated with analysis of numerical output in the form of large output files, file input/output, and computer program execution. The GIS component provides tools for analysis and display of spatial data. The GIS contains the ability to create and/or manipulate the spatial data associated with the three dimensional grid for the finite difference model. The user has the ability to select what information is required by pressing on menu items. If the user wants to run the quasi-three dimensional irrigation and drainage model, the user makes that selection from the initial screen. The screen that the system creates to interact with the finite different numerical model is displayed on Figure 1. This screen allows the user to create and/or manipulate the information for the finite difference

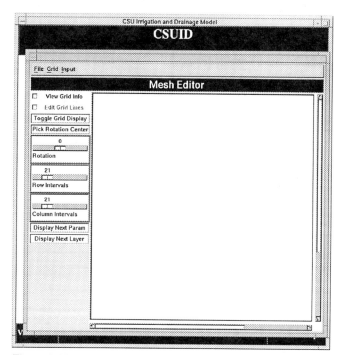

Figure 1. First screen for the quasi-three dimensional module

model. The screen is divided into two major active areas with the menu bar at the top of the screen and the buttons displayed on the left hand side of the screen. The top menu contains three pull-down menus. The pull-down menus allow the user to select the input file (if one exists), to edit the different types of input data by selecting the type of input data the user would like to edit/enter. When the user selects the "Input" button on the top menu a pull-down menu is displayed containing all the different types of data that are available for the map (unsaturated physical, saturated physical, irrigation, crops, salts, global parameters). The user can select any of these menu items and a form with the data (as shown in Figure 3), or can edit the color scheme used to display the results (by selecting color edit from the "Grid" button). On the left hand side of the screen the buttons allow the user to create and/or view the grid values. If the user selects an existing data file, a grid that is associated with that data file is displayed automatically as shown in Figure 2. This grid can also be created by the user by selecting the upper left hand corner and dragging the mouse across the screen. The box that is created by this operation is divided into the number of rows and columns that the user has set in the "Row Interval" and "Column Interval" slider bars. After the user has created a grid the grid can be rotated by selecting the "Pick Rotation Center" button. On the left hand side of the interface the user can select a rotation point for the grid. Then the user can select the "Rotation" slide bar and rotate the mesh to any particular angle. This feature will be useful when a picture/map of a particular area is displayed on the background and the user is trying to match the grid to particular features on the picture/map. The user can modify the grid by selecting the "Edit Grid Lines" button which allows the user to toggle on/off any line on the grid.

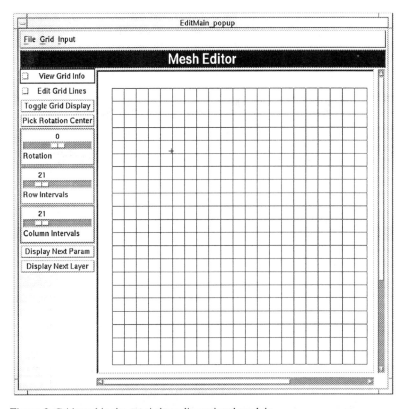

Figure 2. Grid used in the quasi-three dimensional module.

After the user has created the grid the input for the model can be entered by pressing the "Input" button at the top of the screen and a pull down menu is displayed with all the different types of data (unsaturated physical, saturated physical, irrigation, crops, salts, global parameters) required to run the model. The user then selects the information that is required and the user is prompted for the information. The information can be entered in several ways. If the data is a single value the program will provide an entry field for that data as shown in Figure 3

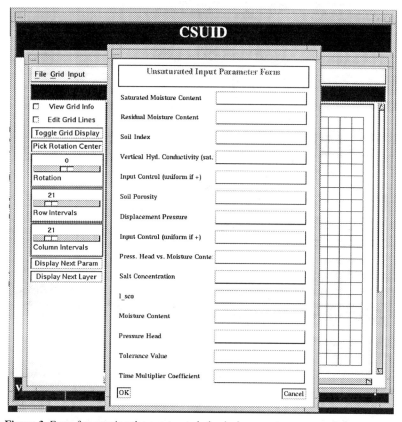

Figure 3. Form for entering the unsaturated physical parameters.

The user then has the ability to compute the results of the simulation and display the results in graphical form. The present implementation of the model allows the user to display the results by pressing the "Display Next Param" button on the left hand side of the screen. The first parameter the program solved for is displayed on the grid as shown in Figure 4. The pop-up window on the right-hand side displays the name of the parameter being displayed along with the horizontal layer those results apply to. The user can then select the "View Grid Info" button and select any grid cell on the display and a second pop-up window is display providing the values for all the parameter that are available for that cell. Any cell on the screen is selected the parameters are re-displayed.

17

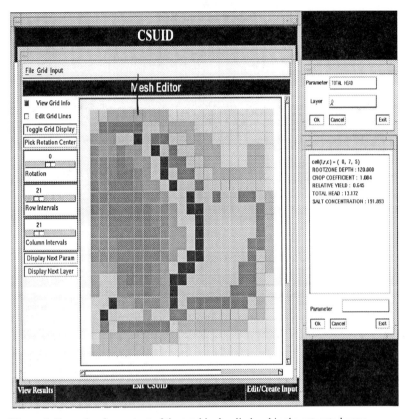

Figure 4. Example of an output of the total hydraulic head in the saturated zone.

Summary:

The MSS that is being developed at Colorado State University allows analysts to manipulate a large amount of spatial information required to run a irrigation and drainage problem when the analyst is interested in studying the spatial variability of data and the impacts of design and management decision on an irrigation and drainage system. The system allows the user to visualize the spatial distribution of the input and output from the model, significantly reduces the amount of effort involved in the creation and/or debugging of a input data set, and facilitates improved understanding of the output.

Future development of the system will include the ability to compare the results from different runs of the model, extension of some of the GIS capabilities such as interactive assignment of values to the grid graphically, and the ability to extract information into the grid from existing GIS maps or generate surfaces from the results of the system.

The system will be used to perform research on many of the unanswered question regarding the characterization, design and management of irrigation and drainage in regions affected by saline shallow ground water tables.

References

Dumm, L. D. (1960), "Validity and use of the transient-flow concept in subsurface drainage.", ASAE Winter Meeting, Memphis.

El-Ashry, M. T., Van Schilfgaarde, J., and Schiffman, S. (1985), "Salinity pollution from irrigated agriculture.", *Journal of Soil and Water Conservation* 40(1):48-52.

El-Hessy, F. A. M., (1991), "Irrigation and drainage management in the presence of a saline shallow water table", Ph.D. Dissertation, Department of Civil Engineering, Colorado State University.

FAO. 1979. Production 1978. Food and Agriculture Organization of the United Nations, Yearbook, Vol. 32, 294 pp.

FAO. 1989a. Production 1988. Food and Agriculture Organization of the United Nations, Yearbook, Vol. 42, 356 pp.

FAO. 1989b. The State of Food and Agriculture 1989: World and regional reviews; Sustainable development and natural resources management. Food and Agriculture Organization of the United Nations, Rome, 191 pp.

FAO. 1990. An International Action Programme on Water and Sustainable Agricultural Development. Food and Agriculture Organization of the United Nations, Rome, 42 pp.

Jensen, M.E., Burman, R.D. and Allen, R.G. (1990), "Evapotranspiration and Irrigation Water Requirements.", *A manual prepared by the Committee on Irrigation and Drainage Division of the American Society of Civil Engineers.*, ASCE, New York, NY.

Oosterbaan, R.J. (1988), "Effectiveness and environmental impacts of irrigation projects: A review", *Paper presented at the 3rd National Drainage Symposium, 20 to 23 September 1988, Izmir, Turkey.*

Rangeley, W.R. (1986), "Scientific advances most needed for progress in irrigation", *Phil. Trans. R. Soc. Lond.*, 316:355-368.

Richardson, C.W., and Wright, D.A. (1984), "WGEN: A Model for Generating Daily Weather Variables", *U.S. Department of Agriculture*, Agricultural Research Service, ARS-8.

Tanji, K.K. (1990), "The nature and extent of agricultural salinity", Chapter 1. In: Agricultural Salinity Assessment and Management, ASCE Manuals and Reports on Engineering Practice, No. 71, ASCE, NY, 10017.

Walker, W.R., and Skogerboe, G.V. (1987), *Surface Irrigation Theory and Practice.* Prentice Hall INC, Englewood Cliffs, New Jersey. 386 pp.

World Bank/UNDP. 1990. Irrigation and drainage research: A proposal for an internationally-supported program to enhance research on irrigation and drainage technology in developing countries. Vol. 1, April, 32 pp.

EVALUATION OF STRESS DAY INDEX MODELS TO PREDICT CORN AND SOYBEAN YIELD RESPONSE TO SOIL-WATER STRESSES

R. O. Evans R. W. Skaggs
Member ASAE Member ASAE

ABSTRACT

Stress day index (SDI) models are presented which can be used to predict corn and soybean yield response to excessive and deficient soil-water conditions. The relative yield - SDI models developed along with other SDI models reported in the literature were tested using a comprehensive data based developed from field experiments providing an equivalent of 94 and 128 site-years of measured corn and soybean yield, respectively. The SDI models tested provided a relatively good fit of predicted to observed corn yield with an average error less than 3 percent. The correlation between predicted and observed overall yield was 79.5 percent when compared to the 1:1 line. The soybean yield - SDI model did a relatively good job predicting long-term average yield with an average error of -0.4 percent. But, the correlation between predicted and observed year to year soybean yield was only 47.9 percent when compared to the 1:1 line.

INTRODUCTION

The primary purposes of agricultural water management systems are to increase production efficiency and yield reliability by improving the soil- water environment. Crop yield is the ultimate measure of crop response to water stresses for the purpose of optimizing the water management system design. The stress-day-index approach (Hiler, 1969) was developed to quantify the cumulative effect of stresses imposed on a crop throughout the growing season. Evans et al. (1991) developed yield - SDI relationships to estimate corn and soybean yield response to excessive soil water conditions.

The purpose of the study presented herein was to test the yield - SDI relationships reported by Evans et al. (1991) against corn and soybean yields observed in field experiments conducted over a wide range of weather conditions. A comprehensive data base was developed from corn and soybean yield studies conducted in eastern North Carolina over the past 35 years. The data base provided the equivalent of 94 site-years of corn yield data and 128 site-years of soybean yield data. The soil-water conditions were sufficiently variable during the 35 year period to provide a thorough test and evaluation of the yield - SDI relationship.

STRESS DAY INDEX MODELS

The general form of the SDI concept described by Hiler (1969) may be expressed

$$SDI_x = \sum_{i=1}^{n} SD_i \times CS_i \quad \quad (1)$$

where n is the number of growth periods and SD and CS are the stress day and crop susceptibility factors for period i, respectively. The subscript x has been added herein and when replaced by w, d, or p is used to denote the specific yield reducing condition, either wet, dry or plant delay, respectively.

The water management simulation model, DRAINMOD, (Skaggs, 1978) simulates soil water conditions in high water table soils. The model considers rainfall, infiltration, surface runoff, drainage, storage and deep seepage to perform a water balance for the soil profile. Hardjoamidjojo and Skaggs (1982) incorporated approximate methods based on the stress-day-index concept to predict corn yield response to stresses caused by excessive and

---

The authors are R. O. Evans, Assistant Professor and Extension Specialist and R. W. Skaggs, William Neal Reynolds and Graduate Alumni Distinguished Professor, Biological and Agricultural Engineering, North Carolina State University, Raleigh, N.C.

deficient soil water conditions. Skaggs et al. (1982) incorporated an approximate method to predict yield reductions caused by planting delays. The general crop response model represented by these modifications was described by Skaggs et al. (1982) as

$$RY = RY_w * RY_d * RY_p \quad \quad \quad \quad \quad \quad \quad \quad \quad \quad \quad \quad \quad (2)$$

where RY is the overall relative yield for a given year, RY_w is the relative yield that would be obtained if only wet stresses occurred, RY_d is the relative yield that would be obtained if only dry stresses occurred, and RY_p is the relative yield resulting from plant delays only. To compare predicted yields to field measured yields, relative yield may also be expressed as

$$RY = Y / Y_o \quad \quad \quad \quad \quad \quad \quad \quad \quad \quad \quad \quad \quad \quad \quad \quad \quad (3)$$

where Y is the measured or observed yield for a given year and Y_o is the yield that would have occurred in the absence of any soil-water related stresses. Y_o refers to the base maximum yield that would occur for a consistent combination of agronomic inputs that were not limited by soil-water.

The relative yield components, RY_w, RY_d and RY_p, are assumed to be independent. Individual submodels are used to calculate each component. Evans et al. (1990, 1991) presented reviews of recent developments and applications of the yield - stress day index relationships for relating yields to excessive soil-water stresses for several crops. The relationships and submodels evaluated herein are summarized in Table 1.

EXPERIMENTAL PROCEDURES AND ANALYSIS

Identification and Description of Test Data (Observed Yields)

Field experiments have been conducted on the Tidewater Research Station near Plymouth, N. C. for over 50 years. The soils at the Tidewater Station are poorly drained. Drainage of most fields has been improved by the installation of parallel ditches or drain tile/tubing so that the site is conducive for evaluation by DRAINMOD. The drainage intensity varies from field to field as discussed by Evans (1991). Even with improved drainage, soil wetness is a problem in some years resulting in planting delays and high water table conditions during the growing season. Droughty conditions develop during periods with below normal rainfall. Corn and soybean are the predominant crops grown on the station.

Five independent studies were identified that provided data suitable to evaluate the SDI relations presented in Table 1. The source of the yield data evaluated in this Stress Day Index analysis is summarized in Table 2. Complete references and soil, site, and drainage system parameters for each study were reported by Evans (1991).

Predicted Yields

Overall relative yield (equation 2) was predicted using DRAINMOD with equations 4, 5, and 7 to predict the individual corn yield components and equations 8, 9, and 10 to predict the individual soybean yield components. The corn yield relationship given by equation 5 was evaluated both without (referred to herein as Method 1) and with (Method 2) the severe-stress dry weight factor described by Shaw (1974, 1978, 1983)(See footnote at bottom of Table 1).

Measured or observed inputs were used for the DRAINMOD simulations wherever possible. These included most of the drainage system parameters, including periods of controlled drainage and subirrigation, hydraulic conductivity, maximum root depth, and soil properties. Daily maximum and minimum temperatures and daily rainfall were available from station records. Daily rainfall values were converted to hourly values by the disaggregation methods described by Robbins (1988). Prior to predicting yield, the inputs were calibrated by comparing 12 site-years of measured water table data to predicted values. The calibration procedure involved starting with all known or estimated inputs, running simulations for those fields and periods with water table data, then comparing predicted to observed water tables. This procedure was continued while varying the estimated inputs, primarily surface storage, upflux and root depth vs time relationship,

Table 1. Relative yield - stress day index relationships used to predict corn and soybean yield response to excessive and deficient soil water conditions and to planting delays.

Stress	Submodel		Equation No.	Reference	Stress Day Factor	Reference	Crop Susceptibility	Reference
CORN								
Wet	$RY_w = 100 - 0.71 * SDI_w$ $RY_w = 0$	$SDI \leq 141$ $SDI > 141$	(4a) (4b)	Evans et al. (1991)	SEW_{30}	Seiben (1964)	6 - stages	Evans et al. (1990)
Dry	$RY_d = 100 - 1.22 * SDI_d$ $RY_d = 0$	$SDI < 82$ $SDI \geq 82$	(5a)* (5b)	Shaw (1974) normalized by Hardjoamidjojo and Skaggs (1982)	1 - (AET/PET) (eq. 6)	Shaw (1974)	17 5-day stages	Shaw (1974)
Plant - Delay	$RY_p = 100 - 0.88 * PD$ $RY_p = 130 - 1.60 * PD$ $RY_p = 0$	$PD < 40$ $40 \leq PD \leq 80$ $PD > 80$	(7a) (7b) (7c)	Seymour (1986) after Krenzer and Fike (1977)	Plant Delay Past Optimum (April 15)	Seymour (1986) after Krenzer and Fike (1977)	2 - stages	Seymour (1986) Krenzer and Fike (1977)
SOYBEAN								
Wet	$RY_w = 100 - 0.65 * SDI_w$ $RY_w = 0$	$SDI \leq 154$ $SDI > 154$	(8a) (8b)	Evans et al. (1991)	SEW_{30}	Seiben (1964)	6 - stages	Evans et al. (1990)
Dry	$RY_d = 100 - 7.2 * SDI_d$ $RY_d = 0$	$SDI < 13.9$ $SDI \geq 13.9$	(9a) (9b)	Evans (1991) after Sudar et al. (1979)	1 - (AET/PET)	Sudar et al. (1979)	10 - stages	Evans (1991) from Literature values
Plant - Delay	$RY_p = 100 - 0.5 * PD$ $RY_p = 140 - 1.8 * PD$ $RY_p = 0$	$PD \leq 30$ $30 < PD \leq 78$ $PD > 78$	(10a) (10b) (10c)	Evans (1991) after Fike (1974)	Plant Delay Past Optimum (May 15)	Evans (1991) after Fike (1974)	2 - stages	Evans (1991) after Fike (1974)

*Under severe stress conditions, the relationship given by equation 5 over predicted yield for the Iowa data evaluated by Shaw (1974). Shaw investigated several methods of weighting the stress day factor (eq. 6). The best fit for the Iowa conditions was obtained when an additional weighting factor of 1.5 was applied to the SD factor whenever two or more consecutive 5-day SD factors were 4.5 or greater. The corn data presented herein are evaluated both with and without this severe stress dry weight factor.

Table 2. Source of corn and soybean yield data used to test yield-SDI relationships. All studies conducted at Tidewater Research Station, near Plymouth, N.C.

Code Name	Main Treatment(s)	Reference
E4	Phosphorus	McCollum (1991)
H3	Potassium, Rotation	Dunphy (1990)
OVT	Variety	Lewis, Rice, Bowman*, (1966-1990)
N	Nitrogen, Row Widths, Variety Population	Nunez (1967), Nunez and Kamprath (1969), Raja (1971), Tonapa (1971),Kamprath et al. (1989), Kamprath et al. (1973), Chancy (1981), Kamprath (1986),
WTM	Water Table Management	Evans et al. (1991b)

*Results are reported annually in "Measured Crop Performance", Research Report No. (), Department of Crop Science, N. C. State University.

until the combination of inputs providing the best water table fit were identified. The RMSE between the observed and simulated water table depths ranged from 12.1 to 21.2 cm/day. The RMSE and AABE of prediction for the pooled data was 15.8 and 11.3 cm/day, respectively. These results indicate that predicted values were in relatively good agreement with observed values. Detailed input values used in the simulations were reported by Evans (1991).

Statistical Procedures

The adequacy of the SDI models was tested by computing average error (AE), average absolute error (AABE), root mean square error (RMSE), and correlation between predicted and observed RY using standard statistical procedures (Evans, 1991). The fit of the predicted yields to the 1:1 line (perfect model) was compared by first determining the best fit linear regression line between predicted and observed yield using the method of least squares (SAS, 1985; Sendecor and Cochran, 1967). The intercept and slope of the best fit regression line was compared to those of the 1:1 line (intercept = 0, slope = 1) using the methods described by Ostle (1963). Finally, the fit of the data (predicted vs observed relative yield) was compared to the 1:1 line. The coefficient of determination, r^2, of this comparison was determined by dividing the best fit regression model sum of squares by the corrected total sum of squares. The corrected total sum of squares for this comparison was the best fit regression model sum of squares plus the error sum of squares $(RY_i - RY'_i)^2$ where RY_i is the relative yield predicted by the simulation model (not regression model) and RY'_i is the observed relative yield.

RESULTS
Predicted Corn Yield

Predicted relative yield components (wet, dry, plant delay), overall predicted relative yield and observed yields were reported in detail by Evans (1991). Space constraints prohibit their presentation here. Observed and predicted relative yield covered the full range of values from 0 to 100 percent, although a majority of values (about 80 percent) occurred in the upper half of the range (RY values between 50 and 100 percent). Over the total period, wet and dry stresses reduced average predicted RY about equally (about 15 percent each). In some years, predicted RY reduction was due entirely to wet or dry stress, but in most years, both wet and dry stresses contributed to the predicted RY reduction.

The AE, AABE, and RMSE for the overall predicted RY is summarized in Table 3. The average error helps identify systematic errors in the prediction method. When the AE is greater than zero indicates that the model may be systematically overestimating observed values or underpredicting if the AE is less than zero. As seen in Table 3, there is variation from study to study, but relative yield was only slightly underpredicted for the pooled data.

The AABE and RMSE provide an indication of the overall performance of the model in terms of the variation between predicted and observed values. The AABE indicates the average magnitude (sign ignored) of the error of each predicted value, with all errors weighted the same. If all errors are about the same magnitude, the AABE and RMSE will be about the same. The RMSE increases above the AABE as the number and magnitude of the poorer predictions increase. Thus, the RMSE provides a better indication of the range of errors.

Combining the results of the AABE and RMSE indicates that both methods had prediction errors of similar magnitude on average (AABE of 7.95 vs 8.89), but, prediction errors involving the severe-stress WF_d (Method 2) had a larger variation within individual observations (RMSE 9.89 vs 13.05). This is demonstrated by comparing the individual year predicted and observed yields shown in Figs. 1 and 2. If both methods predict the same RY, the severe-stress WF_d was not activated. Yield reduction occurring in those years when both methods predict the same yield result primarily from wet stress or plant delay. Years where the predicted RY values diverge represent years with increasing amounts of dry stress. In years where Method 1 overpredicted RY, Method 2 usually predicted RY closer to the observed value. In years where Method 1 underpredicted RY, Method 2 provided an even lower prediction.

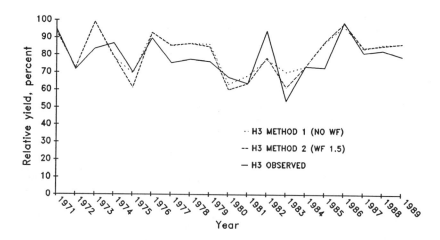

Figure 1 Observed and predicted corn relative yield for study H3.

The sensitivity of the prediction methods to year to year variation in soil-water stresses is best evaluated by comparing correlation between observed and predicted RY. These results are summarized in Table 4. The intercept of the best fit regression lines for Methods 1 and 2 did not test significantly different from 0 at the 5 percent level of significance. The regression lines shown in Table 4 were forced through the 0 intercept and the slope re-computed and tested against the slope of the 1:1 line (perfect model). The slopes of Method 1 was not significantly different from 1 while the slope involving a severe-stress dry weight factor was significantly less than 1 indicating that RY was underpredicted when the severe-stress weight factor was used. Relative yield predicted by Method 1 is plotted against observed RY in Figure 3. As seen in Table 4, Method 1 accounted for nearly 80 percent of the year to year variation in observed corn yield.

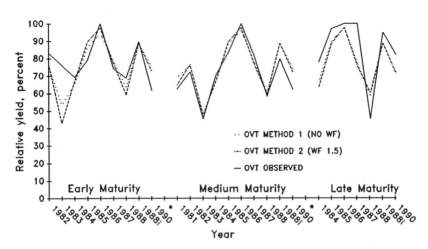

Figure 2 Observed and predicted corn relative yield for study OVT.

Table 3. Goodness of fit evaluation of predicted to observed overall RY

Study	Method 1	Method 2	Study	Method 1	Method 2
Average Error			Average Absolute Error		
OVT	-1.39	-2.22	OVT	7.62	7.94
E4	-8.95	-12.75	E4	11.86	15.71
H3	3.33	2.19	H3	6.84	6.56
N	0.74	-1.38	N	7.54	8.15
WTM	0.33	0.15	WTM	5.41	5.60
Pooled	-1.35	-2.84	Pooled	7.95	8.89
Root Mean Square Error			Correlation Coefficient r^2		
OVT	9.39	10.79	OVT	0.649	0.576
E4	13.68	22.52	E4	0.534	0.292
H3	8.78	8.26	H3	0.500	0.585
N	8.46	9.00	N	0.503	0.507
WTM	7.29	7.4	WTM	0.955	0.952
Pooled	9.89	13.05	Pooled	0.808	0.710

Table 4. Regression analysis of overall predicted and observed corn RY

Method	Slope	Intercept	r²	Intercept Significantly Different from 0		
				0.01	0.05	NS
1	0.889	6.71	0.808			x
2	0.891	4.87	0.710			x

Regression line forced through intercept = 0				1:1 line	Slope significantly different from 1
1	0.974	-	0.795		x
2	0.954	-	0.691	x	

Predicted Soybean Yield

Soybean RY was predicted by equation 2 using equations 8, 9, and 10 to compute the individual yield components (wet, dry, and plant delay). The individual relative yield components, overall relative yield, and observed relative yield were summarized by Evans (1991).

Predicted and observed RY are plotted in Figure 4. Over 90 percent of the RY values (both predicted and observed) lie between 60 and 100 percent. Relative yields less than 50 percent were observed for only 6 cases. Average error, AABE and RMSE are summarized in Table 5. The AE was -0.44 percent, a slight underprediction. The year to year variation in observed and predicted RY is plotted for one set of data in Figure 5. Note that predicted RY matched the year to year variation in observed RY except for years 1964 and 1965 where the trends were reversed. Similar results were observed for the other data. Predictions and trends were good for some years and completely reversed for others.

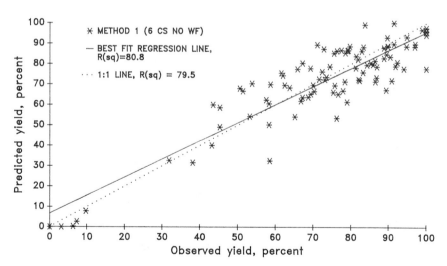

Figure 3 Observed and predicted corn relative yield, best fit regression line for pooled data from all studies (94 observations) and 1:1 line

Table 5. Goodness of fit evaluation of predicted to observed soybean overall RY

	AE	AABE	RMSE	r^2
Soybean model (128 observations)	-0.44	7.46	10.24	0.559

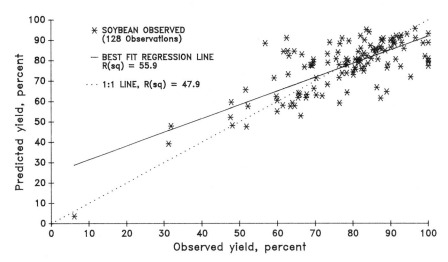

Figure 4   Observed and predicted soybean relative yield, best fit regression line for pooled data from all studies (128 observations) and 1:1 line

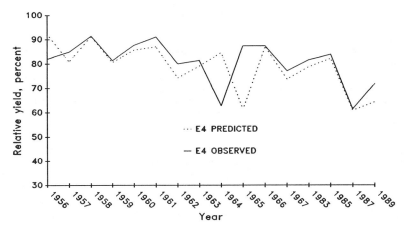

Figure 5   Year to year variation in observed and predicted soybean relative yield, study E4

The intercept of the best fit regression line between predicted and observed RY yield tested significantly greater than zero, Table 6. This could be due to the lack of data points at the lower end of the range. It could also be due to inadequacy of the prediction method. Regardless, the adequacy of the model is best described in terms of correlation to the 1:1 line. The slope of the regression line forced through the intercept did not test significantly differently from 1, but the model explained only 47.9 percent of observed RY variation when compared to the 1:1 line.

The poor correlation between predicted and observed soybean RY may be due to several factors. Soybean can tolerate short term stress with little effect on yield. For example, if dry conditions exist during the pod set period, fewer pods are set. If conditions become more favorable during the subsequent pod fill stage, larger beans can be produced in each pod resulting in about the same yield as would have occurred if more pods had been set but filled with smaller beans. During stressful periods, some physiological process can even be temporarily halted until more favorable conditions develop (Dunphy, not dated). The simple prediction methods evaluated herein are not capable of predicting these complex physiological recovery processes.

Table 6. Regression analysis of overall predicted and observed soybean relative yield

Method	Slope	Intercept	r^2	Intercept Significantly Different from 0		
				0.01	0.05	NS
Soybean Model	0.646	26.90	0.559	x		
Regression line forced through intercept = 0			1:1 line	Slope significantly different from 1		
Soybean Model	0.981	-	0.479			x

SUMMARY

Corn and soybean relative yield - stress day index models were tested with yields observed in field experiments conducted in eastern North Carolina. The field experiments provided an equivalent of 94 and 128 site-years of measured corn and soybean yield, respectively. The yields were observed over a wide range of weather and soil-water conditions. Daily soil-water stresses resulting from the variable weather conditions were predicted using DRAINMOD. Yield reduction resulting from excessive and deficient soil-water stresses and planting delays were then predicted. All predicted yields were compared to observed yields with the goodness of fit evaluated using several statistical indicators.

The results presented showed that long-term average corn yields can be predicted with DRAINMOD using SDI models to predict the individual yield components (wet, dry and plant delay). The data tested suggest that severe- stress dry weight factors are not necessary to predict corn yield response to deficient soil-water stresses under traditionally high water table conditions such as exist in eastern North Carolina. Minor modifications to DRAINMOD would facilitate omission of the severe-stress criteria. This would reduce the required yield inputs because several 5-day CS values with the same value could be combined as one input. The methodology used to describe the deficient yield-SDI would then parallel the methods currently used to describe the wet yield-SDI relationship.

The soybean yield-SDI model did a relatively good job predicting long- term average yield with an average error of -0.4 percent. The prediction methods were not as good for predicting year to year soybean yield variation. Correlation between predicted and observed yield was only 47.9 percent when compared to the 1:1 line.

REFERENCES

Dunphy, E. J. (Undated). Relating soybean growth and development to irrigation. In: Chapter 8, Irrigation Handbook. North Carolina Agricultural Extension Service, North Carolina State University, Raleigh. 3p.

Evans, R. O., R. W. Skaggs and R. E. Sneed. 1990. Normalized crop susceptibility factors for corn and soybean to excess water stress. TRANSACTIONS of the ASAE 33(4):1153-1161.

Evans, R. O. 1991. Development and evaluation of stress day index models to predict corn and soybean yield under high water table conditions. PhD Dissertation. N. C. State Univeristy.

Evans, R. O., R. W. Skaggs and R. E. Sneed. 1991. Stress day index models to predict corn and soybean relative yield under high water table conditions. TRANSACTIONS of the ASAE. 34(5):1997-2005.

Fike, W. T. 1974 (approximate). Soybean yields as affected by dates of planting. Unpublished Data. Department of Crop Science, North Carolina State University, Raleigh.

Hardjoamidjojo, S. and R. W. Skaggs. 1982. Predicting the effects of drainage systems on corn yields. Agricultural Water Management 5:127- 144. Amsterdam, The Netherlands.

Hardjoamidjojo, S., R. W. Skaggs and G. O. Schwab. 1982. Corn yield response to excessive soil water conditions. TRANSACTIONS of the ASAE 25(4):922-927,934.

Hiler, E. A., 1969. Quantitative evaluation of crop-drainage requirements. TRANSACTION of the ASAE 12(4):499-505.

Krenzer, E. G.,Jr. and W. T. Fike. 1977. Corn production guide-planting and plant population. North Carolina Agricultural Extension Service, North Carolina State University. Raleigh. 2p.

Ostle, B. 1963. Statistics in Research. Iowa State University Press, Ames, IA. pp 201-205.

Robbins, K. D. 1988. Simulated climate data inputs for DRAINMOD. Ph.D. Dissertation. Department of Biological and Agricultural Engineering, North Carolina State University, Raleigh.

SAS, Institute, Inc. 1985. Statistical Analysis Systems User's Guide: Statistic, Version 5 Edition. Cary, NC.

Seymour, R. M. 1986. Corn yield response to plant date. MS Thesis, N. C. State University.

Shaw, R. H. 1974. A weighted moisture-stress index for corn in Iowa. Iowa State Journal of Research 49:101-114.

Shaw, R. H. 1978. Calculation of soil moisture and stress conditions in 1976 and 1977. Iowa State Journal of Research 53(2):120-127.

Shaw, R. H. 1983. Soil moisture and moisture stress prediction for corn in western corn belt state. Korean Journal of Crop Science 28(1):1- 11.

Skaggs, R. W. 1978. A water management model for shallow water table soils. Report No. 134. Water Resources Research Institute of the University of North Carolina, N. C. State University, Raleigh.

Skaggs, R. W., S. Hardjoamidjojo, E. H. Wiser, and E. A. Hiler. 1982. Simulation of crop response to surface and subsurface drainage systems. TRANSACTIONS of the ASAE 25(6):1673-1678.

Sieben, W. H. 1964. Het verban tussen ontwatering en opbrengst bij de jonge zavelgronden in de Noordoostpolder. Van Zee tot Land. 40, Tjeenk Willink V. Zwolle, The Netherlands. (as cited by Wesseling, J., 1974. Crop growth and wet soils. Chapter 2 in: Drainage for Agriculture. J. van Schilfgaarde, ed. ASA Monograph No. 17. Madison, WI.

Snedecor, G. W. and W. G. Cochran. 1967. Statistical Methods. Sixth Edition. Iowa State University Press.

Sudar, R. A., K. E. Saxton, and R. G. Spomer. 1979. A predictive model of water stress in corn and soybeans. ASAE Paper No. 79-2004. ASAE. St. Joseph, MI.

VALIDATION OF COSSMO: A COMBINED CROP GROWTH AND WATER MANAGEMENT MODEL

Calvin D. Perry Daniel L. Thomas Robert O. Evans[*]
Member ASAE   Member ASAE   Member ASAE

ABSTRACT

The combined crop growth (SOYGRO 5.4) and water management (DRAINMOD 3.4) modeling system COSSMO (COmbined Subirrigated Soybean MOdel) was validated using soil, weather, and crop data from North Carolina. The two models, each modified to accept data from the other, are managed by an expert system. The expert system runs the models in an iterative manner after obtaining necessary input from the user.

Sixty-five sets of field data from Plymouth, North Carolina were used to compare observed yields to predicted yields from COSSMO and SOYGRO (stand-alone). Fifty-three of the data sets were soybean Official Variety Test (OVT) results from soybean grown on conventionally drained soils and 12 were results from soybean grown on soils with water table management (WTM) in place. A predetermined "allowable" range of error of 670 kg/ha (10 bu/ac) overprediction to 340 kg/ha (5 bu/ac) underprediction was set to gauge resulting predicted yields. Forty-four (68%) of the 65 COSSMO predicted yields were within the "allowable" range while only 16 (25%) SOYGRO predicted yields fell within this range. The mean difference between COSSMO predicted and observed yields for all data sets was +340 kg/ha (+5 bu/ac) with a sample standard deviation of 590 kg/ha (9 bu/ac). The mean difference between SOYGRO predicted and observed yields for all data sets was 690 kg/ha (-10 bu/ac) with a sample standard deviation of 1170 kg/ha (17.5 bu/ac). This preliminary validation indicates the modeling system, with some enhancements, has the potential to become a tool for evaluating water management effects on crop growth. Further validation is needed under a variety of soil conditions before a general release of COSSMO is feasible. **KEYWORDS.** Water Table Management, Soybean, Yields, Models, Expert Systems.

---

[*] **Calvin D. Perry**, Research Engineer II, and **Daniel L. Thomas**, Associate Professor and Research Leader, Biological and Agricultural Engineering Dept., University of Georgia, Coastal Plain Experiment Station, Tifton, GA 31793; and **Robert O. Evans**, Extension Specialist, Biological and Agricultural Engineering Dept., North Carolina State University, Raleigh.
  The authors wish to acknowledge the support of Drs. E.J. Dunphy, R.E. McCollum, Daryl Bowman, and technician Mark Langdon, N.C. State University, who were instrumental in supplying validation data.

INTRODUCTION

The water management model DRAINMOD (Skaggs, 1978) was developed as a design and management tool for use with drainage and subirrigation systems in humid regions. The model has been shown to perform reliably in California, Florida, Virginia, Georgia, North Carolina, and Ohio (Chang et al., 1983; Rogers, 1985; McMahon et al., 1987; Shirmohammadi et al., 1991; Skaggs, 1978; Skaggs et al., 1981). A limiting aspect of DRAINMOD is the inability to simulate crop growth directly. Since crop growth is affected by water management strategies, a method is needed to predict the crop response to these various strategies. Improper management can lead to plant injury by allowing either too much or too little water to reach the plant's roots, thereby reducing yields. Stanley et al. (1980) found that soybean top growth and yields were decreased more by lack of water than by root damage caused by flooded conditions. They did note that some root damage occurred in soybean when water tables rose after the pre-flowering stage of growth.

Many efforts have been made to enable DRAINMOD to simulate crop growth. One approach has been to combine DRAINMOD with existing crop simulation models such as EPIC (Sabbagh et al., 1991), CERES-MAIZE (Brink, 1986), and SOYGRO (Perry et al., 1990). Each of these physiologically-based models predict growth and development based on weather data and soil type. Another recent approach is the use of the stress-day-index concept to predict relative crop yield (Evans et al., 1991). The stress-day-index approach attempts to utilize the stresses imposed upon a crop during the growing season and the susceptibility of the crop to these stresses to determine a relative yield that is a percentage of potential yield.

Perry et al. (1990) attempted to add crop growth capabilities to DRAINMOD by developing an "expert simulation system" - COSSMO (COmbined Subirrigated Soybean MOdel). This system utilizes both DRAINMOD (version 3.4) and SOYGRO (version 5.4), executing the two in an iterative fashion. DRAINMOD was modified to accept evapotranspiration and effective rooting depth from SOYGRO, which in turn was modified to accept water movement from a water table from DRAINMOD and also to account for effects of flooded conditions.

This paper indicates the results of validating the COSSMO system with soil, weather data, and observed yields from studies in North Carolina. Simulation results from COSSMO are also compared to simulation results from SOYGRO stand-alone.

METHODS

Sixty-five sets of field data from the Tidewater Research Station near Plymouth, North Carolina were obtained. Fifty-three of the data sets were from soybean Official Variety Test (OVT) results. The OVT trials, held since 1966, are conducted to provide variety information to growers. The tests are designed in a randomized block with variety being the main effect. Each treatment is replicated 3 or 4 times. Cultural practices have changed over the years, but are the same for all varieties at a given test site each year. Cultural practices each test year were selected for producing high yields. The OVT procedures and results are reported annually in: "Measured Crop Yields", Research Report, Department of Crop Science, N.C. State University. Twelve of the 65 data sets were from water table management studies initiated in 1989 at the Tidewater Research Station by the Biological and Agricultural Engineering Department, N.C. State University. The research study is designed to evaluate the effect water table management systems have on crop yields. Evans et al. (1991) fully described the water management system.

A Portsmouth fine sandy loam soil (fine sandy over sandy or sandy - skeletal, mixed, thermic Typic Umbraquult) was used in all simulations. Complete soil properties measured at the Tidewater station were reported by Gilliam et al. (1978) and Skaggs (1978). A soil file for both SOYGRO and DRAINMOD (each has its own soil data format) was developed. For each year, variety, and study combination, actual field drainage system parameters were used in the COSSMO simulations. The reader is referred to Evans (1991) for detailed information on the field and system characteristics.

Rainfall and temperature data were obtained from weather records collected by a weather station at the site. Solar radiation was not available for the Tidewater station, therefore monthly mean daily insolation values were obtained (de Jong, 1973) and converted to solar radiation values.

Four varieties grown in the OVT studies were in the SOYGRO variety database: Bragg, Davis, Forrest, and Ransom. The WTM study used only Ransom variety. Thus, these four varieties were used in the simulations.

Once the necessary data files were generated, COSSMO simulations were run using a row spacing of 91.4 cm (36 inches) and a plant spacing of 7.6 cm (3 inches). The SOYGRO model was executed stand-alone using the same soil, weather, and variety information.

A predetermined "allowable" range of error of 670 kg/ha (10 bu/ac) overprediction to 340 kg/ha (5 bu/ac) underprediction was assumed to gauge resulting predicted yields. This range is based on a ±340 kg/ha (5 bu/ac) allowable range (17 % of maximum yield). The additional +340 kg/ha allowing for overprediction was based on the assumption that the model would overpredict in most cases, because factors (such as weed and insect pest effects) which would tend to reduce observed yields are not accounted for in the simulations.

RESULTS AND DISCUSSION

Figures 1 through 4 represent the COSSMO and SOYGRO results from simulating the OVT study with Bragg, Davis, Forrest, and Ransom varieties. Figure 5 shows the results of the WTM study simulation which involved only the Ransom variety.

Forty-four (68%) of the 65 COSSMO predicted yields were within the "allowable" range while only 16 (25%) SOYGRO predicted yields fell within this range. The mean difference between COSSMO predicted and observed yields for all data sets was +340 kg/ha (+5 bu/ac) with a sample standard deviation of 590 kg/ha (9 bu/ac). The mean difference for the WTM study only was +9 kg/ha (<0.15 bu/ac) with a sample standard deviation of 370 kg/ha (5.5 bu/ac). The mean difference between SOYGRO predicted and observed yields for all data sets was -689 kg/ha (-10.2 bu/ac) with a sample standard deviation of 1170 kg/ha (17.5 bu/ac). The mean difference for the WTM study only was 550 kg/ha (-8 bu/ac) with a sample standard deviation of 400 kg/ha (6 bu/ac).

Figure 6 presents an overall comparison of the results of the two models by plotting both COSSMO and SOYGRO predicted yield versus observed yield for all the simulations. The data exhibit relatively large variability; however, the figure suggests that COSSMO yields tended to be at or above the 1:1 line. This result is in agreement with the earlier discussion concerning the absence from the model of factors which could potentially reduce yields. The SOYGRO yield results vary more widely and tend to be primarily below the 1:1 line.

For these simulations, COSSMO appears to predict yields more reliably than does SOYGRO alone. Thus, the combination of DRAINMOD and SOYGRO apparently reflects the soil-water interactions during water table management (conventional drainage, controlled

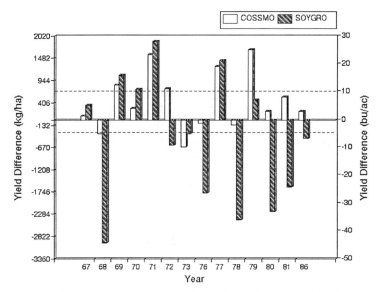

Figure 1. Difference between predicted and observed soybean yields for Bragg variety - OVT study.

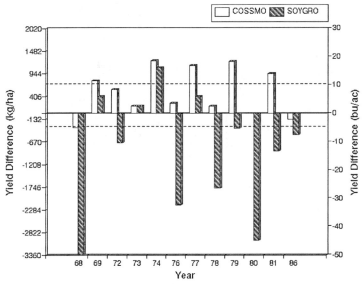

Figure 2. Difference between predicted and observed soybean yields for Davis variety - OVT study.

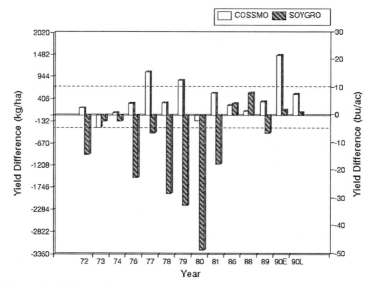

Figure 3. Difference between predicted and observed soybean yields for Forrest variety - OVT study.

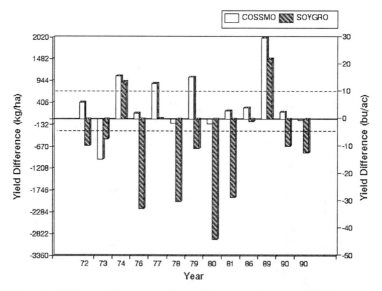

Figure 4. Difference between predicted and observed soybean yield for Ransom variety - OVT study.

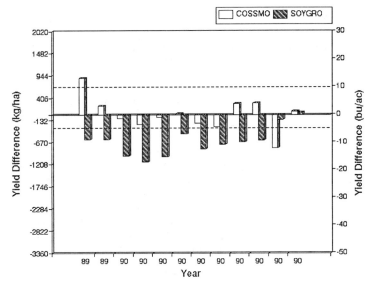

Figure 5. Difference between predicted and observed soybean yields for Ransom variety - WTM study.

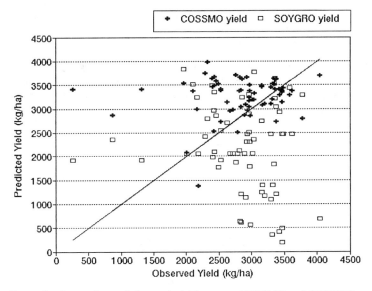

Figure 6. Comparison of observed yields versus COSSMO and SOYGRO predicted yields for all simulations.

drainage, or subirrigation) more accurately than SOYGRO alone does. Perry et al. (1990) notes that the SOYGRO water balance model is limited to rainfall and sprinkler irrigation and does not account for upward flux from a water table. By adding a rudimentary upward flux function to SOYGRO and using water table and upward flux values from DRAINMOD, SOYGRO could account for these water table effects.

In analyzing particular varieties, the Forrest variety was apparently simulated better (by both COSSMO and SOYGRO) while Bragg was not represented as well. SOYGRO excessively underpredicted the yields in several cases by not accounting for the water addition from the water table. The COSSMO system responded as would be expected by overpredicting slightly in most cases.

The eight lowest SOYGRO data points (between 2750 and 4250 kg/ha observed yield), representing the greatest yield errors, were further examined to try to determine an explanation for the extremely low simulated yields. Total rainfall, monthly rainfall distribution, planting date, variety, and drainage system were compared for the eight data points. No explanation for the error could be observed. The large overprediction by COSSMO at observed yields of 256 and 1312 kg/ha apparently resulted from the observed yields being dramatically reduced due to surface flooding during the growing season.

SUMMARY

The combined crop growth (SOYGRO 5.4) and water management (DRAINMOD 3.4) modeling system COSSMO was validated using soil, weather, and crop data from Plymouth, North Carolina. Sixty-five sets of field data were used to compare observed yields to predicted yields from COSSMO and SOYGRO (stand-alone). Sixty-eight percent of COSSMO predicted yields were within the "allowable" range (670 kg/ha (10 bu/ac) overprediction to 340 kg/ha (5 bu/ac) underprediction) while only 25% of SOYGRO predicted yields fell within the range. The difference between COSSMO predicted yields and observed yields (mean of +340 kg/ha (+5 bu/ac), sample standard deviation of 590 kg/ha (9 bu/ac)) as compared to SOYGRO differences (mean of -690 kg/ha (-10 bu/ac), sample standard deviation of 1170 kg/ha (17.5 bu/ac)) indicated that the combination of DRAINMOD and SOYGRO represented the soil-water interactions associated with a drainage system better than SOYGRO alone did.

This preliminary validation, while limited in soil types, indicated COSSMO has the potential to become a tool for evaluating water management effects on crop growth. Further validation is needed under a variety of soil conditions before a general release of COSSMO is feasible. Efforts are underway to obtain additional data for further validation.

REFERENCES

Brink, P. 1986. A water management model for predicting the effects of a drainage and subirrigation system on corn yields. MS Thesis, Michigan State University, East Lansing, MI.

Chang, A.C., R.W. Skaggs, L.F. Hermsmeier and W.R. Johnston. 1983. Evaluation of a water management model for irrigated agriculture. *Transactions of the ASAE* 26(2):412-418, 422.

de Jong, B. 1973. Net radiation received by a horizontal surface at the earth. Deft University Press. Rotterdam, Netherlands.

Evans, R.O. 1991. Development and evaluation of stress day index models to predict corn and soybean yield under high water table conditions. PhD Thesis, North Carolina State University, Raleigh, NC.

Evans, R.O., R.W. Skaggs and J.W. Gilliam. 1991. A field experiment to evaluate the water quality impacts of agricultural drainage and production practices. In: *Proceedings of the National Conference on Irrigation and Drainage Engineering*, Irrig. and Drain. Div. of ASCE, July 22-26, 1991, Honolulu, Hawaii.

Gilliam, J.W., R.W. Skaggs, and S.B. Weed. 1978. An evaluation of the potential for using drainage control to reduce nitrate loss from agricultural fields to surface waters. Report No. 128. Water Resources Research Institute of the University of North Carolina, Raleigh.

McMahon, P.C., S. Mostaghimi and F.S. Wright. 1987. Simulation of corn yield by a water management model for a coastal plain soil in Virginia. ASAE Paper No. 87-2047. St. Joseph, MI: ASAE.

Perry, C.D., D.L. Thomas, M.C. Smith, and R.W. McClendon. 1990. Expert system-based coupling of SOYGRO and DRAINMOD. *Transactions of the ASAE* 33(3):991-997.

Rogers, J.S. 1985. Water management model evaluation for shallow sandy soils. *Transactions of the ASAE* 26(2):412-418, 422.

Sabbagh, G.J., R.L. Bengtson and J.L. Fouss. 1991. Modification of EPIC to incorporate drainage systems. *Transactions of the ASAE* 34(2):467-472.

Shirmohammadi, A., D.L. Thomas, and M.C. Smith. 1991. Drainage-subirrigation design for Pelham loamy sand. *Transactions of the ASAE* 34(1):73-80.

Skaggs, R.W. 1978. A water management model for shallow water table soils. Report No. 134. Water Resources Research Institute of the University of North Carolina, Raleigh.

Skaggs, R.W., N.R. Fausey, and B.H. Nolte. 1981. Water management model evaluation for north central Ohio. *Transactions of the ASAE* 24(4):922-928.

Stanley, C.D., T.C. Kaspar, and H.M. Taylor. 1980. Soybean top and root response to temporary water tables imposed at three different stages of growth. *Agronomy Journal* 72:341-346.

MODELING WATER TABLE CONTROL SYSTEMS WITH HIGH HEAD LOSSES NEAR THE DRAIN

G. M. Chescheir, C. Murugaboopathi, R. W. Skaggs, R. H. Susanto, and R. O. Evans

ABSTRACT

Approximate methods for calculating flow to drains were modified to accommodate both the drainage and subirrigation cases. The methods were incorporated into the water management model, DRAINMOD, and tested against the numerical solutions to the Boussinesq equation. Water table predictions by the modified version of DRAINMOD were in good agreement with the Boussinesq equation. The modified version of DRAINMOD was used to evaluate the sensitivity of predicted relative corn yields to head loss at the drains for a site in eastern North Carolina. Radial head loss near the drain had a larger effect on predicted relative corn yield for subirrigation conditions than for drainage conditions. However, reductions in relative yield due to head losses were small compared to yield reductions caused by decreasing soil hydraulic conductivity or increasing drain spacing. **Keywords.** Drainage, Subirrigation, Water table, Modeling.

INTRODUCTION

High head losses near the drain will affect the performance of water table control systems (Susanto and Skaggs, this volume). Since many models for evaluating drainage and subirrigation systems are based on the Dupuit-Forchheimer assumptions, a correction for head loss at the drain is required when drain tubes are used in the system. Normally, the effect of head losses at the drain is considered in equations for calculating drainage rates by using methods developed by the Dutch scientist Hooghoudt (van Schilfgaarde, 1974) to reduce the depth between the drain and the impermeable layer. Moody (1966) reviewed the original calculations of Hooghoudt and presented equations for calculating this equivalent depth (d_e).

The equivalent depth concept only considers the effect of head losses on flow occurring in the bottom half of the drain, which is satisfactory for most drainage situations. However, Skaggs (1991) showed that the equivalent depth concept may lead to errors during subirrigation when the effect of head losses on flow through the top half of the drain may be more significant. Skaggs (1991) developed numerical solutions to the Boussinesq equation that consider head losses due to radial flow both above and below the drain tube. Fipps and Skaggs (1991) developed simple analytic methods for predicting flow to drains based on the Hooghoudt and radial flow equations.

The purpose of this paper is to adapt the methods of Fipps and Skaggs (1991) and install them in DRAINMOD for application to both drainage and subirrigation. The modified model will be tested against the numerical solutions to the Boussinesq equation. Then the model will be used to evaluate the sensitivity of predicted corn yield to head loss at the drains for a drainage and subirrigation system in eastern North Carolina.

G. M. Chescheir, Research Associate; C. Murugaboopathi, Research Associate; R. W. Skaggs, William Neal Reynolds Professor and Distnguished University Professor; R. H. Susanto, Ph.D. Candidate: and R. O. Evans, Assistant Professor; Department of Biological and Agricultural Engineering, North Carolina State University, Raleigh, NC.

MODEL DESCRIPTION

The Hooghoudt equation is used in DRAINMOD, and other drainage design models, to relate drainage rates to water table elevation at the midpoint between parallel drains. Referring to Fig. 1 the equation is written

$$q = \frac{4Km^2 + 8Kdm}{L^2} \quad (1)$$

Where q = drainage flux, K = saturated horizontal hydraulic conductivity, m = midpoint water table elevation above the drains, d = depth from the drain to the impermeable layer, and L = distance between the drains. The Hooghoudt equivalent depth, d_e, is normally substituted for d in equation 1 to correct for convergence head losses near the drain.

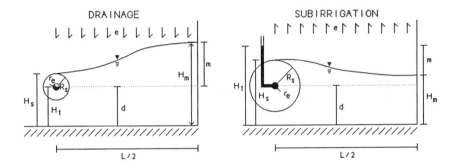

Figure. 1 Schematic of radial flow near drains during drainage and subirrigation.

Rather than using the equivalent depth concept, Fipps and Skaggs (1991) expressed drainage flux (q) in terms of the predicted elevation of the water table above the drain, R_s. This method required simultaneous solution of the Houghoudt equation and the radial flow equation. For this solution the Houghoudt equation is written in terms of R_s as,

$$q = \frac{((8Kd + 4K(m + R_s))(m - R_s)}{L^2} \quad (2)$$

Radial flow to the drains may be expressed as,

$$q = (2\pi K (R_s - r_e))/(\ln(R_s/r_e))/L \quad (3)$$

where r_e is effective radius of drain and R_s is assumed to be small compared to L.

Adapting equation 2 for both drainage and subirrigation conditions yields:

$$q = \frac{((8KH_t + 4K(m + R_s))(m - R_s)}{L^2} \quad (4)$$

Where m = H_m - H_t, H_t is the hydraulic head in the drain tube referenced to the impermeable layer, and H_m is the hydraulic head or water table elevation midway between the drains. For drainage with the tube half full, H_t = d; while for subirrigation H_t is as shown in Fig. 1.

The procedure for solving equations 3 and 4 is listed below.

1. Given a known midpoint water level, m, the drainage flux (q) is estimated using Eq. 1.
2. Using q, the elevation of the water table above the drain R_s is estimated from Eq. 3. This equation is solved numerically using the Newton-Raphson method.
3. The R_s value is used in Eq. 4 to estimated a new value for q.
4. Steps 2 and 3 are repeated until the value of q converges to a constant value.

The above solution yields a steady state flow rate to the drain for a given midpoint water table elevation. To achieve transient response of the water table and drain flow rates, the solution was incorporated into DRAINMOD replacing the standard Hooghoudt equation. Transient water table response is determined by using a quasi-steady state assumption, employing the steady state solutions at successive times as programmed in the original DRAINMOD.

MODEL VALIDATION

Predictions by the modified DRAINMOD were compared to solutions to the Boussinesq equation (Eq. 5) for both steady state and transient conditions. The solutions presented by Skaggs (1991) were used for these comparisons.

$$f \frac{\partial h}{\partial t} = K \frac{\partial}{\partial x}\left[h \frac{\partial h}{\partial x} \right] + e \tag{5}$$

Where, referring to Fig. 1, h is the elevation of the water table above the impermeable layer, t is time, x is horizontal position, e is the rate of recharge to the water table (e is positive for infiltration and negative for ET), K is the lateral hydraulic conductivity, and f is drainable porosity.

The boundary conditions for Equation 5 may be written as (see Fig. 1),

$$h = h_1(x) \qquad 0 \le x \le L/2 \qquad t = 0 \tag{6a}$$

$$\frac{\partial h}{\partial x} = \pi(h-H_t)/\{\ln[(h-d)/r_e]h\}, \qquad x = 0 \qquad t > 0 \tag{6b}$$

$$\frac{\partial h}{\partial x} = 0, \qquad x = L/2 \qquad t > 0 \tag{6c}$$

Where $h_1(x)$ is the initial water table elevation. The hydraulic head is set at H_t for t > 0.

<u>Steady State Conditions</u>

Predictions by the approximate methods (equations 3 and 4) were compared to the Boussinesq solutions for steady state drainage and subirrigation conditions. Twelve

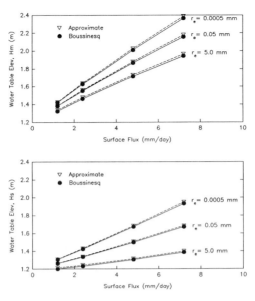

Figure 2. Comparison of Boussinesq and approximate methods for steady state drainage conditions. Comparisons are for water table elevations at the midpoint (top) and at the drain (bottom).

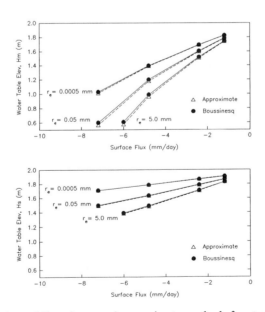

Figure 3. Comparison of Boussinesq and approximate methods for steady state subirrigation conditions. Comparisons are for water table elevations at the midpoint (top) and at the drain (bottom).

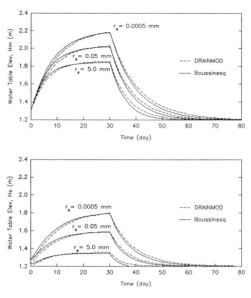

Figure 4. Comparison of Boussinesq equation and DRAINMOD for transient drainage conditions. Comparisons are for water table elevations at the midpoint (top) and at the drain (bottom).

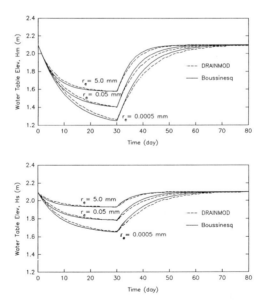

Figure 5. Comparison of Boussinesq equation and DRAINMOD for transient subirrigation conditions. Comparisons are for water table elevations at the midpoint (top) and at the drain (bottom).

combinations of different values for e (1.2, 2.4, 4.8, and 7.2 mm/day) and for r_e (0.0005, 0.05, and 5.0 mm) were compared for drainage. Other variables for these comparisons were L = 23 m, d = 1.2 m, H_t = 1.2 m, and K = 0.5 m/day. Conditions for the twelve subirrigation comparisons were the same as for drainage except H_t = 1.95 m and the values for e were negative. For r_e = 0.0005 mm, the e values of -7.2 mm/day were changed to -6.0 mm/day to keep the midpoint water table elevations above zero. Steady state solutions to Boussinesq equation were achieved after 100 days of simulation at which time drain flow rate equaled surface flux.

The approximate methods calculated water table elevations at the midpoint (H_m) and at the drain (H_s) that were very similar to those calculated by the Boussinesq solutions. Compared to the Boussinesq solutions H_m and H_s values calculated by the simple methods were higher for drainage conditions (Fig. 2) and lower for subirrigation conditions (Fig. 3). The largest difference in H_m calculations between the two methods was 0.035 m. The largest difference in H_s calculations between the two methods was 0.014 m.

<u>Transient Conditions</u>

The transient response of the water table predicted by the modified version of DRAINMOD was compared to solutions to the Boussinesq equation for both drainage and subirrigation conditions. The initial water table elevation (h_1) was 1.3 m and H_t was 1.2 m for the drainage comparison. Steady rainfall (e) was 6.1 mm/day until day 30 of simulation when e was set to 0.0 mm/day until the end of the 80 day simulation. For the subirrigation comparisons h_1 and H_t were 2.1 m and e was -4.8 mm/day until day 30 when e was set to 0.0 mm/day until the end of the 80 day simulation. The other variables for these comparisons were the same as for the steady state comparisons. The drainable porosity was constant at f = 0.06 for the comparisons. Upward flux and infiltration variables used in DRAINMOD were set high so as not to limit surface flux.

The modified version of DRAINMOD predicted water table responses very similar to those calculated by the Boussinesq solutions for both drainage and subirrigation conditions (Fig. 4 and 5). For both water table drawdown and recharge the modified DRAINMOD predicted a slower response than calculated by the Boussinesq solutions; however, differences in water table elevations calculated by the two methods were always less than 5 cm. The relative effects of r_e on the response of the water table were nearly the same for the two solution methods. Based on these comparisons it was concluded that the modified version of DRAINMOD could be reliably used to determine the effects of radial head loss on the performance of both drainage and subirrigation systems.

MODEL APPLICATION

The modified version of DRAINMOD was used to evaluate how radial head loss at the drain affects corn yield under drainage and subirrigation conditions. The soil, weather and system design parameters for the simulations were similar to those of the water table management study at Plymouth NC described by Munster et al. (1990), Harmsen et al. (1991), and Susanto and Skaggs (this volume). The simulations used the drain depth (d = 1.2 m) and spacing (L = 23 m) of the new drain lines installed at the research site in the fall of 1991. In order to simplify the analysis, a field effective soil hydraulic conductivity of 0.48 m/day was used for the simulations rather than a depth dependent conductivity. This value was determined in the field by measuring flow between drain lines for different constant hydraulic heads (unpublished data). Susanto and Skaggs (this

Table 1. The effects of effective radius (r_e) on predicted relative corn yield for drainage and subirrigation conditions. The yields are predicted by the modified version of DRAINMOD for 30 yr (1951 - 1980) simulations for soil, weather, and drainage design conditions similar to those at Plymouth NC.

K m/day	L m	d m	No Head Loss	r_e (mm)				Percent Change
				50.0	5.0	0.05	0.0005	
			Conventional Drainage					
0.12	23.0	1.20	76.5	75.2	73.7	70.6	68.3	10.7
0.24	23.0	1.20	81.5	80.9	80.1	77.9	76.0	6.7
0.48	23.0	1.20	83.0	83.0	82.7	82.1	81.3	2.0
0.96	23.0	1.20	83.2	83.2	83.1	83.0	82.9	0.4
0.48	11.5	1.20	83.1	83.1	83.0	82.8	82.5	0.7
0.48	23.0	1.20	83.0	83.0	82.7	82.1	81.3	2.0
0.48	46.0	1.20	76.5	75.8	75.3	73.4	71.8	6.1
0.48	92.0	1.20	61.1	60.8	60.6	60.1	59.0	3.4
0.48	23.0	0.75	80.8	79.9	78.9	76.6	74.5	7.8
0.48	23.0	0.90	82.5	81.8	81.1	78.9	77.3	6.3
0.48	23.0	1.20	83.0	83.0	82.7	82.1	81.3	2.0
0.48	23.0	1.50	82.5	82.5	82.4	82.2	82.0	0.6
			Subirrigation					
0.12	23.0	1.20	67.5	64.1	63.2	60.5	58.5	13.3
0.24	23.0	1.20	75.8	72.1	71.3	68.1	65.3	13.9
0.48	23.0	1.20	82.5	79.3	78.5	75.7	73.4	11.0
0.96	23.0	1.20	87.9	85.0	84.7	82.4	80.4	8.5
0.48	11.5	1.20	91.5	88.7	87.7	85.6	83.2	9.1
0.48	23.0	1.20	82.5	79.3	78.5	75.7	73.4	11.0
0.48	46.0	1.20	67.5	65.3	65.1	63.3	61.5	8.9
0.48	92.0	1.20	54.6	53.9	54.0	53.8	53.1	2.7
0.48	23.0	0.75	82.0	79.2	78.2	74.7	71.6	12.7
0.48	23.0	0.90	82.2	79.3	78.6	75.1	72.2	12.2
0.48	23.0	1.20	82.5	79.3	78.5	75.7	73.4	11.0
0.48	23.0	1.50	82.5	79.1	78.4	75.7	73.4	11.0

volume) reported much lower K values in the vicinity of the drain, resulting in increased head losses. These effects are not considered in this analysis.

The effects of head loss on corn yield were evaluated over a range of conditions using different values for soil hydraulic conductivity (K = 0.12, 0.24, 0.48, and 0.96 m/day), drain spacing (L = 11.5, 23.0, 46.0 and 92.0 m), and drain depth (d = 0.75, 0.90, 1.20, and 1.50 m). The corn yield factors for 130 day corn as presented by Evans et al. (1990 and 1991) were assumed for the simulation. For the subirrigation simulations, the hydraulic head in the drain was maintained 0.6 m below soil surface from planting until maturity.

Radial head loss at the drain had only a small effect on predicted relative corn yield for conventional drainage conditions (Table 1). Decreasing r_e to 0.0005 mm from no head loss conditions reduced relative corn yield by less than 11 percent. Greater reductions in yield due to increased head loss occurred for marginal drainage designs. The effect of head loss on relative yield was less for well drained conditions (i.e. L = 11.5 m) and for poorly drained conditions (i.e. L = 92.0 m).

The effects of radial head loss on corn yield were greater for subirrigation conditions. Predicted relative yields fell by as much as 13.9 percent in response to decreasing r_e from no head loss conditions to 0.0005 mm. The larger yield reductions occurred for the lower soil hydraulic conductivity (k ≤ 0.24 m/day) and for the smaller drain depth (d ≤ 0.9 m). These reductions in relative yield, however, were small compared to yield reductions in response to decreasing K or increasing L.

REFERENCES

1. Evans, R.O., R.W. Skaggs, and R.E. Sneed. 1990. Normalized crop susceptibility factors for corn and soybean to excess water stress. Transactions of the ASAE, 33(4): 1153-1161.

2. Evans, R.O., R.W. Skaggs, and R.E. Sneed. 1991. Stress day index models to predict corn and soybean relative yield under high water table conditions. Transactions of the ASAE, 34(5): 1997-2005.

3. Fipps, G. and R.W. Skaggs. 1991. Simple methods for predicting flow to drains. J. Irrig. and Drain. Div., ASCE, 117(6): 881-896.

4. Harmsen, E.W., J.W. Gilliam, R.W. Skaggs, and C.L. Munster. 1991. Variably saturated 2-dimensional nitrogen transport. ASAE, St Joseph, MI, Paper No. 91-2630.

5. Moody, W.T. 1966. Nonlinear differential equation of drain spacing. J. Irrig. and Drain. Div., ASCE, 92(2): 1-9.

6. Munster, C.L., J.E. Parsons, R.W. Skaggs, R.O. Evans, and J.W. Gilliam. 1990. Using the personal computer for water table management research. ASAE, St Joseph, MI, Paper No. 90-2527.

7. Skaggs, R.W. 1991. Modeling water table response to subirrigation and drainage. Transactions of the ASAE, 34(1): 169-175.

8. Susanto, R.H. and R.W. Skaggs. 1992. Hydraulic head losses near agricultural drains during drainage and subirrigation. Proceedings of the Sixth International Drainage Symposium. ASAE, St Joseph, MI.

9. van Schilfgaarde, J. 1974. Nonsteady flow to drains. In: Drainage for Agriculture, ed. J. van Schilfgaarde. Madison, WI: American Society of Agronomy.

DRAIN OUTLET WATER LEVEL CONTROL: A SIMULATION MODEL [1]

James L. Fouss   James S. Rogers [2]
Member ASAE   Assoc. Member ASAE

ABSTRACT

Technology is still developing on various methods of operating dual purpose controlled-drainage and subirrigation systems for water table management. Automatic control of such systems is merited in many cases, however, the operational system must be designed to match the dynamic response characteristics of the water table. The design method in this paper is based upon a computer model which, by iteratively solving the Boussinesq subsurface drainage equation, simulates the response of the field water table between subsurface drains to control of the water level at the drainage outlet. The model allows the designer to select various criteria, such as timing and magnitude of adjustments of the drainage outlet water level, to maintain the field water table depth within desired limits. This simulation model was developed and validated with field research data. The field data were acquired with a microprocessor-based system that automatically monitored and controlled field water table depth by feedback adjustments of the water level in sump-type drainage outlet structures. The validated models can be applied to simulate the operation of various types of automated control systems, for proposed or existing subsurface conduit water table management systems. Thus the designer can determine the most critical operational criteria, and select the "best" method of system control. The design process is illustrated with simulation results for some example system design proposals and scenarios.

INTRODUCTION

The management of soil-water by water table control is becoming widely adopted in the humid regions of the United States and Canada. One of the more troublesome aspects of designing and operating such water management systems is insuring that when the water table is being held high for subirrigation purposes, periods of wet weather do not cause crop injury because of excessive soil-water conditions. Another difficulty is determining the proper duration of drainage following significant rainfall events. Drainage should be "controlled" so that the

---

[1] Contribution from the Soil and Water Research Unit, USDA-ARS, Baton Rouge, Louisiana in cooperation with the Louisiana Agricultural Experiment Station, Louisiana State University Agricultural Center, Baton Rouge, LA.

[2] The authors are Agricultural Engineers, USDA-ARS, Soil and Water Research Unit, P. O. Box 25071 - University Station, Baton Rouge, LA 70894-5071.

soil profile is not overdrained, requiring subsequent subirrigation to maintain the desired water table depth. Subirrigation should be "suppressed" if appreciable rainfall is forecast, thus reducing the potential risk of an excess soil-water event.

The objectives of this research were to: (a) Develop or adapt a computer model to simulate the operation and performance of an automated water table control system; (b) validate the model for soil and climatic conditions in the southern humid region; and (c) develop approaches to use the model in water table management system design, and to select the "best" method of system operation or control.

This paper presents the computer modeling-simulation approach developed to predict and evaluate the dynamic performance of "controlled" water table management systems. The modeling-simulation method can be used for overall water management system design, or can be applied to develop or select operational parameters for a given system design. The resulting design can optimize performance in terms of soil-water control and water-use efficiency for the range of weather events in the climatological record for the geographical area of interest. Example simulation results are given for an automated water table control system in a Commerce silt loam alluvial soil typical of the Mississippi Delta area. The approach can be applied to predict and evaluate the performance of controlled water table management systems, and to develop rational recommendations for farmers to follow in operating similar systems in other geographic areas.

AUTOMATED WATER TABLE CONTROL

Automation of dual-purpose controlled-drainage and subirrigation system operation for water table control provides an alternative to labor-intensive manual system management methods. Control of the outlet water level may be fully automatic or semi-automatic. For purposes of discussion here, it is assumed that the drainlines are connected into a sump-structure for control of the outlet water level. The water level in the sump during drainage cycles is controlled by pumping from the sump into surface drainage channels. During subirrigation, the water level in the sump is maintained by pumping from an external source. The starting or "standard" water levels maintained in the sump for controlled-drainage and subirrigation are based upon operational experience, field calibration of the water table control system, or computer simulations for the specific system design. Typically for most water table management systems, the "standard" water levels maintained in the sump for subirrigation are somewhat higher than those used for controlled-drainage. Automatic control of the sump water level (SWL) by controlling subdrainage and subirrigation flows may include the options "with" or "without" feedback of the monitored field water table depth (WTD) between drainlines in the field. In this paper only control <u>with</u> feedback is discussed.[3] A cross-sectional schematic of the outlet sump structure and WTD sensor in the subirrigation mode of operation is shown in Fig. 1. The sump operation is automatically switched from the

---

[3] For a discussion of automated water table control "without" feedback of the monitored field water table depth, refer to Fouss, et al., 1989.

controlled-drainage mode to subirrigation and vice versa, as needed to maintain the SWL between the MIN and MAX water levels (elevations) specified, and to control the field WTD within the desired range, WTDmin $\leq$ WTD $\leq$ WTDmax. The design of the automated control system was based upon the assumption that the occurrences of short periods of excess soil-water conditions (i.e., the water table is too shallow) may be more detrimental to crop roots and plant growth than the occurrences of the same or longer periods of deficit soil-water conditions (when the water table is too deep).

Automated Controlled-Drainage (CD): In the CD mode of operation, the SWL is maintained between the predetermined high-drainage (HD) and low-drainage (LD) water levels (Fig. 1) by pumping water from the sump to control the field WTD within the desired range, WTDmin $\leq$ WTD $\leq$ WTDmax. These HD and LD sump water levels ('on' and 'off' switch levels of the drainage pump) may be "stored" as software values in an electronic microprocessor controller system (Fouss et al., 1987c; 1990). In the feedback CD mode, if the monitored field WTD remains less than the WTDmin during a 24-h period (midnight-to-midnight), then the system operation is automatically switched to regular subsurface drainage (SD) so that the WTD will be increased more quickly. In the SD mode of operation, the SWL is maintained at approximately drain depth for a predetermined interval of time (e.g., 12 to 24 h), or optionally until the field WTD recedes into the desired range, and then the system operation is again switched back to the CD mode. The system operation is also switched to the SD mode anytime that rainfall causes the field WT to rise and remain very near the soil surface for a specified period of time (e.g., within 20 cm of the soil surface for a period of 2 h or more). Following such a high water table event the system is typically kept in the SD mode of operation for 24-hours or more before it is returned to the CD mode (this will vary with soil type and system design). When subsurface drainage flow ceases and the SWL recedes to the MIN water level (the maximum depth below the soil surface), the system operation is automatically switched to the subirrigation (SI) mode.

Automated Subirrigation (SI): In the SI mode of operation, the SWL is maintained between the predetermined low-irrigation (LI) and high-irrigation (HI) water levels by pumping water into the sump from a well or other source (Fig. 1). As gravity flow subirrigation from the sump lowers the SWL to the LI level, the irrigation pump is automatically operated to raise the SWL to the HI level. An electronic microprocessor controller system, as described above for the CD mode, can be used to store these "on" and "off" SWL values and operate the irrigation pump. In the feedback SI mode, if the monitored field WTD is deeper than the desired maximum, WTDmax, during a midnight-to-midnight 24-h period, the SWL control threshold water levels (HI and LI) are adjusted upward by a step amount "Y" (i.e., HI + Y, and LI + Y), and the system operated at the new SWL thresholds for the next 24-h period. For control reference purposes, the sump MAX control threshold water level (elevation) is adjusted upward by the same step "Y" (MAX + Y). If the monitored field WTD remains deeper than the WTDmax value for the next 24-h period, another feedback upward step of "Y" is implemented at midnight. Successive step-wise feedback adjustments are made, if needed, until the adjusted

MAX level in the sump is less than or equal to an absolute minimum depth below the soil surface. When the monitored field WT rises to the approximate center of the desired range, the SI control threshold water levels in the sump are returned to the "standard" elevations; the irrigation pump does not operate again until the SWL recedes to the LI level. In the SI mode, if the monitored field WTD is shallower than the desired minimum, WTDmin, for a specified short period of time (e.g., two hours), the irrigation is turned off by the controller until the field WTD recedes to the WTDmax depth. However, if the field WTD remains shallower than WTDmin for a 24-h period, the SWL control thresholds are returned to the "standard" elevations for MAX, HI, and LI, and the irrigation pump will not operate again until the WTD $\geq$ WTDmax.

A tipping-bucket type raingage on the site can be used to provide an input to the system controller microprocessor to make other control mode changes in response to rainfall events, as follows. If rainfall occurs and the cumulative amount exceeds a threshold (R_{max}) in a given time (e.g., 50 mm in 2 h), the system operation can be automatically switched to the CD mode. If the rainfall does not exceed this amount/time threshold (which is based on experience), but infiltration is sufficient to cause subdrainage into the sump which raises the SWL to the MAX level, the system operation will be switched to the CD mode. The system operation is automatically switched back to the SI mode when SWL is deeper than the MIN level in the sump.

If at any time during SI operation the field WT rises, due to rainfall, to a level very near the soil surface for a specified period of time (e.g., within 20 cm of the soil surface for two hours or more), the system operation is automatically switched to the SD mode. When the system operation has been switched from the SI mode to the SD mode, it is left in the SD mode until the controller is manually reset to the CD or SI modes. This manual reset procedure is used during the growing season to insure that current or predicted weather conditions will not cause additional excessive soil-water conditions soon after changing back to the CD or SI modes.[4]

SYSTEM MODELING AND SIMULATION

Two types of computer simulation models were modified by the authors to predict the performance of controlled water table management systems: (1) DRAINMOD, developed and validated by Skaggs (1980, 1982) to predict the performance of dual purpose CD-SI water management systems over a long period of climatological record, was modified (Fouss, 1985; Fouss et al., 1989) to simulate automated or feedback-operation of CD-SI systems; and (2) a computer program to solve the Boussinesq equation to predict subsurface water movement and the water table profile between adjacent drainlines (Smith, 1983; Smith et al., 1985), was significantly modified and enhanced by the authors (Fouss and Rogers, 1988) to

---

[4] Fouss, et al., 1990, discuss a means via telecommunications to remotely reset electronic data-logger/controller systems, and Fouss and Cooper, 1988, present a method to utilize the daily weather forecast to aid in management/operational decisions for subirrigation systems.

more accurately simulate various modes of automated operation for a sump-controlled water table management system. In the design/evaluation procedure described in this paper, the feedback-control version of DRAINMOD was first used to simulate the performance of a given or trial water table management system operated with one or more simple modes of operation [e.g., various constant or variable outlet water levels, or a two-stage weir controlled outlet water level, maintained during the growing season (Fouss et al., 1987b)]. These simulations were conducted for the total period of climatological record at a specific site. From these DRAINMOD simulation results, periods of two or more years (e.g., years with a "wet" and a "dry" growing season) were selected for study (simulate) in greater detail with the Boussinesq-based model.

Typically, the computer time required to run the revised Boussinesq-based model was up to 750 times greater than that to run DRAINMOD (ver. 3.0 with the Feedback Subroutine; Fouss, 1985). The computer time required to run the revised Boussinesq model depends upon the time-step used in the finite difference solution of the Boussinesq equation. The revised model included an option to use a very small time-step (i.e., less than one minute) for those outlet boundary conditions imposed by a small sump structure in which the water level can change quickly. The water table depth fluctuations were predicted for only the midpoint between drainlines with DRAINMOD, but the revised Boussinesq-based model provided predicted soil-water status and water table depths at selected positions [nodes] between drainlines. Fifteen (15) nodes at increasing spacings from the drainline to the midpoint (Fig. 2) were used in these simulations. The 15 nodes where the WTD and other hydrologic and water balance parameters are calculated in the model were located at 0, 2, 4, 6, 10, 15, 20, 25, 30, 40, 50, 60, 70, 80, and 100% of the distance between the drainline and the midpoint between drains. Node number 1 was over the drain and node number 15 was at the midpoint between drains. The change in water table in the soil directly over the drain (node 1) was assumed to be the same as the change in water level in the sump at the drain outlet, with no significant time delay.[5] The distance from node 1 to node 2 is critical. If the distance is too small, excessive errors occur in the calculation of drain flow (Rogers and Fouss, 1989). For the simulations presented here, a minimum distance of 15 cm gave satisfactory results for a 15 m drain spacing.

The Green-Ampt procedure was incorporated into the revised Boussinesq model to estimate infiltration at each node, and not at just the midpoint between drains as in DRAINMOD. Thus, the surface runoff could also be estimated at each node, and the cumulative or total runoff from the area (cross-section) between the drain and the midpoint was a weighted average. Surface runoff was assumed to occur in a direction parallel to the drainlines. With the multi-node solution of water table depth, the predicted subsurface drainage and subirrigation volumes, and changes in soil-water storage, represent the entire soil profile cross-section between drainlines, and not just the midpoint between drains as with DRAINMOD.

---

[5] Skaggs (1991) proposed a numerical solution of the Boussinesq equation for radial flow conditions near the drain, which will be incorporated into the Boussinesq model presented here.

The prediction of the water table depth at multi-nodes between drainlines with the revised Boussinesq model provided data to compute excess and deficit soil-water parameters during the growing season as a function of distance from the drainline. For example, SEW_{30} (excess soil-water within 30 cm of the soil surface) was computed separately at each node. Revisions of the Boussinesq-based model included a subroutine to simulate the creation of a "dry-zone" in the surface soil layer whenever the predicted water table exceeded a specified depth and the potential evapotranspiration (PET) could not be met by upward flux from the water table. This was similar to the procedure used in DRAINMOD, with two exceptions: (1) The dry zone created could vary in depth at different distances from the drainline; and (2) the calculated water table depth was not adjusted for the soil-water volume removed from the upper layer (dry zone) by ET; instead a dry zone deficit value was computed which represented the soil-water storage volume to be refilled by the next infiltrated rainfall. DRAINMOD outputs a sum of the dry days (i.e., days when upward flux, in combination with soil-water removed from the root zone, does not meet ET demand), plus a computed stress-day-index to quantify excess and deficit soil water conditions as they relate to crop growth and yield. These output features were not incorporated into the revised Boussinesq model. However, the revised Boussinesq model was provided with the DRAINMOD type subroutine to predict "relative" crop yield based upon soil-water status during the preplant and growing season periods. The relative yield could be predicted with the Boussinesq model as a function of distance from the drainline, or a weighted average yield for the spacing between drains, rather than only at the midpoint between drains as with DRAINMOD.

Simulations were conducted for an alluvial Commerce silt loam soil, typical of large areas in the Lower Mississippi Valley. Soil parameters needed for model inputs, namely lateral saturated conductivity, unsaturated conductivity, soil-water desorption characteristics, drained volume vs. water table depth, and infiltration characteristics, have been previously reported by Fouss, et al. (1987a, 1987b, 1987c) for the Commerce silt loam. Climatological data for the southern Louisiana area near the town of Paincourtville, covering the 5-year period, 1986 to 1990, were used for the simulations of water table management. Both DRAINMOD and the Boussinesq-based models require hourly rainfall as input; DRAINMOD requires daily maximum/minimum temperatures to estimate (by the Thornthwaite method) daily potential evapotranspiration (PET), or optionally daily pan evaporation, and the Boussinesq model requires daily pan evaporation.[6] These data were collected at a field experimental site on the Westfield Sugarcane Plantation, Paincourtville, LA, where an automated water table management system was installed in 1983 (Fouss et al., 1987c, 1989). Data acquired from this field site were used to validate predicted performance of the sump-controlled automated water table management system with both DRAINMOD (Fouss et al., 1989) and the Boussinesq models.

---

[6] The monthly adjustment factors for pan evaporation in southern Louisiana proposed by Bengtson, et al. (1985) were used to estimate PET for the Boussinesq-based model; the monthly PET adjustment factors for the Thornthwaite estimates were selected to closely approximate the adjusted pan values of PET.

MODEL EVALUATION

The performance of the revised Boussinesq model in simulating the dynamic response of a sump-controlled water table management system was evaluated by: (a) comparing predicted results with field observed performance; and (b) by comparing simulation results with those obtained from the Feedback version of DRAINMOD. The model was further evaluated for potential applications in design of water table management systems by simulation of system response for various design options and scenarios of operation.

The performance of the automated water table control system was quantified and evaluated in terms of the following parameters measured or computed from field experiments or predicted by the simulation models: (1) daily fluctuation in water table depth, (2) subsurface drainage and subirrigation flow volumes, (3) excess soil-water in the root zone (SEW_{30}), and (4) relative crop yields. When comparing simulation results generated by the models, the following additional parameters were considered: (5) infiltration and surface runoff, and (6) estimated ET. For a water table management system to meet accepted performance requirements for a silt loam soil, drainage should control excess soil-water such that SEW-30 is maintained less than the 100-150 cm-days range on a 5-year recurrence interval (R.I.), and subirrigation should provide sufficient soil-water so that no more than 8 to 10 dry days occur during the growing season on a 5-year R.I. (Skaggs, 1977).

<u>Comparison with Field Observed Performance:</u> A simulation was conducted with the revised Boussinesq model for an outlet sump and the corresponding field plot at the Westfield Sugarcane Plantation water table management system for the 1988 growing season. The predicted and observed fluctuations of the outlet sump water level and the water table depth at the midpoint between drains are shown in Fig. 3. The actual field water table management system was switched from regular subsurface drainage to automated subirrigation on about day 122 (1 May), but the feedback control option was not implemented. On about day 140 (19 May), the outlet sump water level was raised (manually) by resetting the system controller. The feedback control option was implemented about day 165 (15 Jun), and the controller was set for a 15 cm feedback adjustment step. When feedback was implemented, a 90 $\pm$ 15 cm desired midpoint water table depth was set into the controller. The feedback option was deactivated on day 193 (11 Jul), and the nominal sump water level control limits were reset to a range of 5 to 15 cm above the drain depth of 100 cm, for the rest of sugarcane season. System operation was reset to regular subsurface drainage on day 274 (30 Sep) for the rest of the year.

The observed field data plotted in Fig. 3 represent the daily average for both the sump water level (plotted as <u>depth</u> below the soil surface) and the midpoint water table depth. The data plotted from the Boussinesq model simulations are hourly average values for both the sump water level and the predicted midpoint water table. The simulated changes in the sump water level in the SI mode, after feedback control was implemented, did not match every observed change made by the system controller at the field site. This was due to a more sluggish response in the simulated WTD during the recession period after the SWL had been abruptly lowered.[7] The range of water table fluctuation simulated during the SI period (see days 165-193), however, was about the same as that observed. Following day 193, the simulated fluctuations in the midpoint WTD compared favorably with those observed; all significant rises in the water table due to infiltration from rainfall were predicted by the model.

This same observed water table depth data was used by the authors for comparison with the DRAINMOD simulation of daily water table fluctuation (see Fouss et al., 1989, Fig. 5, for sump 3); the observed daily average sump water level was used as an input to DRAINMOD for the drain outlet water level boundary condition. The DRAINMOD simulation of water table fluctuation during the period of feedback SI control was very satisfactory, however, after day 194 the model did not predict every significant rise in the water table caused by infiltration of rainfall.

In general, the revised Boussinesq model provided a very satisfactory prediction of the total performance of the existing sump-controlled water table management system. It is considered significant that the model was able to predict both the feedback control of the water level in the outlet sump as well as the fluctuations in field water table depth.

Comparison with DRAINMOD simulations: Simulations were made with both the revised Boussinesq-based and DRAINMOD models for the automatically controlled water table management system design at the Westfield Sugarcane Plantation. The period simulated was 1986-1990 for which climatological data had been measured at the site. Sugarcane was the assumed crop, with a maximum rooting depth of 45 cm, and the desired midpoint water table depth was selected as 85 $\pm$ 10 cm for the portion of the sugarcane growing season from day 100 to day 230. Based on experience at the site, two nominal or "standard" SWL control thresholds were chosen at 65 and 75 cm below the soil surface. The assumed SWL control range for the 65 cm nominal depth was 60-70 cm, and 70-80 cm for the 75 cm nominal depth. These standard SWL's were used for both the CD and SI modes of operation. The SI mode of operation was initiated on day 100, and a feedback adjustment step of + 15 cm was selected. The feedback option was activated in the simulation from days 100 to 220, which is the critical growth stage of sugarcane.

A preliminary analysis of the simulation results for the Boussinesq model indicated that the predicted values for infiltration (INFIL), surface runoff (RO),

---

[7] In earlier research (Fouss and Rogers, 1988) with a preliminary version of the Boussinesq model, observed SWL data collected during automated SI operation was used as a model input to define average hourly SWL boundary conditions. Predicted fluctuations in midpoint WTD compared closely to observed field WTD fluctuations, however, the predicted magnitudes of fluctuations were notably less. Note: The authors previously observed that the field WTD data (determined with an electrical sensor in a WT "well") often appeared to show a quick response to rapid changes in sump water level; the high lateral conductivity in the soil layer at drain depth (Fouss et al., 1989) may have caused the WT "well" to function more like a piezometer pipe.

evapotranspiration (ET), excess soil-water (SEW_{30}), and relative crop yield at node numbers 11 and 12 (i.e., at 50 and 60 percent of the distance from the drainline to the midpoint between drains) were approximately equal to the previously defined weighted average values for these parameters (detailed data are not presented here). A summary of the simulation results for the 5-year period (1986-1990) are presented in Table 1 for both models. The Boussinesq simulation results tabulated for the above-noted parameters are for node number 11, but the predicted subirrigation (IRR) and pumped subsurface drainage (DRN) volumes are for the total soil profile (cross-section) between drainlines. These same outputs for DRAINMOD are based on the water balance at the midpoint between drains only.

The predicted crop (sugarcane) yield for all runs listed in Table 1 was 100%, thus satisfactory water table control was achieved for each year in the 5-year period as simulated with both models.[8] The weighted average infiltration predicted with the Boussinesq model was about 10% less than that predicted with DRAINMOD. The weighted average ET predicted with the Boussinesq model was also about 12% less than that predicted with DRAINMOD for the midpoint between drains. For the 5-year simulated period, the predicted subsurface drainage flow was about 15% higher for the Boussinesq model than for DRAINMOD at the 65 cm nominal SWL, and about 6% higher at the 75 cm nominal SWL. Predicted subirrigation volume required was about double for the Boussinesq model in comparison to DRAINMOD. The large difference in the predicted subirrigation volumes was probably caused by the significant amounts of water required to develop the water table profile when switching mode of operation from drainage to subirrigation. These significant differences in volume of subirrigation, plus the noteworthy differences in subsurface drainage volumes, are important for a sump-controlled system where all water into and out of the sump must be pumped. The Boussinesq model provides a more realistic estimate of these volumes for water table management system design purposes, especially in sizing outlet sumps and pumps.

A graphical comparison of the DRAINMOD and Boussinesq-model predictions of SWL feedback control and field WTD fluctuations for an automated water table management system are shown in Fig. 4 for the 1989 season. The plotted data are daily averages for both SWL and WTD (the WTD is for the midpoint between drains only). For the Feedback version of DRAINMOD, the simulated SWL (assumed to be the same as the water table at the drain) cannot be deeper than the drain depth (100 cm), and SWL adjustments were not simulated to occur more frequently than about every three days. DRAINMOD was not intended to predict transient changes in WTD that may occur with large and/or frequent changes in the drainage outlet water level (SWL) (Fouss, 1985). For the Boussinesq model, the WT at the drain (WT1) can be simulated even when it recedes deeper than the drain in the soil profile, as in the 1989 simulation (Fig. 4); the SWL is assumed to

---

[8] Without subirrigation, that is with regular subsurface drainage only, the yields predicted with DRAINMOD were also 100% for all 5 years, primarily because of the 45 cm maximum sugarcane rooting depth assumed and the significant upward flux potential of the alluvial soil.

be the same as WT1. The Boussinesq-model-simulated SWL could be adjusted upward daily (at midnight), thus at the start of subirrigation (day 100 in Fig. 4) the SWL was raised, by feedback control, to a level above the "standard" or nominal SWL of 65 cm in 15 cm daily steps, to about the 30 cm depth. This forced the Boussinesq-simulated field WT to rise to the desired depth range (85 $\pm$ 10 cm) more quickly. The simulated adjustments of the SWL with DRAINMOD as subirrigation began was raised only to the "standard" level because the WT response was rapid enough in the first three days to avoid a second step. In DRAINMOD simulations for other years in the 1986-1990 period (not shown here), the simulated SWL included one additional upward adjustment step of 15 cm on day 103, and three days later the SWL was returned to the "standard" level.

After the initiation of subirrigation in 1989, the simulated WTD's with both models were very similar, except for some rainfall events. A rainfall event on about day 140 (Fig. 4) caused DRAINMOD to simulate a rise in WT that was not predicted with the Boussinesq model. This same event caused DRAINMOD to simulate a feedback corrective reaction by lowering the SWL for three days. For the rest of the season the simulated feedback adjustments of the SWL by the two models were nearly the same (Fig. 4), however, the simulated response in the WTD was somewhat more sluggish with the Boussinesq model than that predicted by DRAINMOD. The predicted volumes of subirrigation (IRR) and drainage (DRN) flows were also different, as shown in Table 1 (Nominal SWL = 65 cm) for 1989.

Design Applications: Some example simulations were conducted with the revised Boussinesq model to illustrate its use in the design of automated sump-controlled water table management systems. Particular attention is given to selecting the "best" operational parameters or options rather than determining drain depth and spacing design factors. All simulations were conducted for the Commerce silt loam soil with drains at 1.0 m depth and 15.0 m spacing.

Results of the revised Boussinesq model simulations of automated water table management, for the 1990 season (at the Westfield Sugarcane Plantation), are shown graphically in Figs. 5, 6, and 7, for two nominal or "standard" SWL control ranges, and two ranges for the control of field WTD. The simulated WTD's at four of the fifteen nodes are plotted: 1, 8, 11 and 15 (i.e., for 0, 25, 50, and 100% of the distance from the drainline to the midpoint between drains). In Fig. 5, the nominal SWL of 55-65 cm in SI was too shallow and thus the feedback controller repeatedly lowered the SWL to prevent the WTD from becoming too shallow. In contrast, the SWL of 70-80 cm in SI shown in Fig. 6 indicates a much more stable controlled WTD.[9] The desired range for field WTD was 85 $\pm$ 10 cm for both the simulations illustrated in Figs. 5 and 6. The differences in IRR and DRN volumes for the 1990 simulated year for these two simulations were: (a) for the 55-65 cm SWL range, IRR = 24 cm and DRN = 40 cm; and (b) for the 70-80 cm SWL

---

[9] In other research the authors have proposed a method of "Adaptive Feedback" to assist the system designer or operator determine a more suitable "standard" SWL range for an automated water table management system (Fouss and Rogers, 1992).

range, IRR = 14 cm and DRN = 35 cm. The simulation conducted and presented in Fig. 7 is the same as that in Fig. 6, except the 85 ± 5 cm desired range for control of field WTD. The simulation results show that the smaller desired control range (Fig. 7), in effect, made the water table control less stable than the wider desired WTD range (Fig. 6); this is typical with many feedback control systems. The IRR volume for the Fig. 7 run was 17 cm and the DRN volume was 37 cm. Thus, the smaller desired WTD control range required more pumping of both drainage and irrigation water, and the simulation provided a means of quantifying the increased pumping demand.

DISCUSSION and COMMENTS

The design and method of operation for water table management systems should be kept as simple as possible, and still meet performance objectives. Fully automated systems cannot be justified in economic terms for soils and crops in many humid region areas of the United States and Canada. However, the semi-automatic operation of controlled-drainage can be justified in many cases to prevent the occurrences of crop damaging excess soil-water events. The manual operation of controlled-drainage systems is often not satisfactory, because needed control changes in the drainage outlet water level are typically delayed longer than desired to prevent crop damage from excess soil-water.

Both DRAINMOD with the Feedback subroutine, and the Boussinesq-equation based model discussed in this paper, can aid the engineer or system designer in evaluating the potential use of automated control methods for water table management systems being considered. For the simpler system designs, such as a fixed or step-wise adjustable weir to control drainage flow at the outlet, DRAINMOD is often very adequate. For more complex designs, where sizing of outlet sumps and pumps is involved, the Boussinesq-based model can provide the system designer with more realistic predictions and evaluations of expected performance for different design options and operational parameters. Design of subirrigation systems with DRAINMOD is satisfactory for those cases where the outlet water level is held relatively fixed. If the subirrigation system is to be controlled by frequent (daily) feedback adjustments of the outlet water level, the Boussinesq-based model is recommended.

In future research, an alternative feedback signal from measured soil matric potential in the root zone [e.g., using instrumentation of the type described by Phene, et al. (1981)] should be investigated as a means to further enhance the subirrigation control mode. Feedback signals from such a sensor could be used to override controller commands to lower the SWL during either CD or SI operations, when the monitored WTD is shallower than the predetermined limits, but the soil-water status is still in the deficit range. The soil matric potential should not be a primary feedback signal for adjustment of the outlet water level in water table management, because changes in soil matric potential would lag behind fluctuations in water table depth.

Table 1. 5-Year Simulation Summary of Automated WT Control w/ Feedback CD-SI in the Lower Mississippi Valley; Commerce silt loam soil *_/

Year	RAIN (cm)	INFIL (cm)	RO (cm)	ET (cm)	SEW$_{30}$ (cm-da)	IRR/DRN (cm)	INFIL (cm)	RO (cm)	ET (cm)	SEW$_{30}$ (cm-da)	IRR/DRN (cm)
		\multicolumn{5}{c}{Desired WT Depth = 85 ± 10 cm ; w/ Feedback ver. of DRAINMOD}					
		SWL = Nominal 65 cm (60-70) Depth					SWL = Nominal 75 cm (70-80) Depth				
1986	133	84	48	72	3	12/23	85	47	72	0	9/22
1987	145	85	60	63	0	11/33	86	59	63	0	7/31
1988	190	103	87	68	2	12/47	104	86	68	0	9/45
1989	177	99	78	71	25	10/37	100	76	70	16	7/35
1990	178	92	86	74	11	12/31	94	84	74	3	9/30
AVG	165	93	72	70	8	11/34	94	70	69	4	8/33
		\multicolumn{5}{c}{Desired WT Depth = 85 ± 10 cm ; w/ BOUSSINESQ-EQ Based Model **_/}					
		SWL = Nominal 65 cm (60-70) Depth					SWL = Nominal 75 cm (70-80) Depth				
1986	133	75	58	66	0	24/26	75	58	66	0	21/23
1987	145	79	66	61	1	19/39	79	66	61	0	14/34
1988	190	97	93	58	0	21/48	98	92	58	0	16/44
1989	177	91	86	62	5	21/42	91	86	62	2	18/39
1990	178	81	97	57	0	19/40	82	96	57	0	14/35
AVG	165	85	80	61	1	21/39	85	80	61	0	17/35

*_/ Drain Spacing = 15 m; Drain Depth = 1.0 m; Feedback adjustment in SI = +15 cm; Crop yields were estimated to be 100% for all runs.
**_/ Predictions at node number 11 for INFIL, RO, ET, AND SEW$_{30}$.

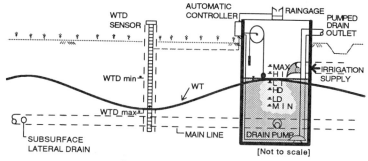

Subirrigation Mode

Figure 1. Schematic of automated outlet water level control structure (sump), shown in subirrigation mode of water table management, where water table depth at the midpoint between drainlines is continuously monitored with an electrical water level sensor system.

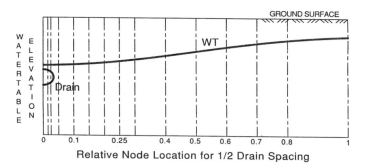

Figure 2. Revised Boussinesq-based model predicts soil-water status and water table depth at 15 Nodes variably spaced between the drainline and the midpoint between drains.

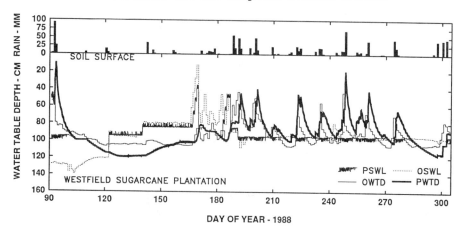

Figure 3. Boussinesq-based model predicted (P...) vs. observed (O...) drainage outlet SWL and midpoint WTD for the 1988 growing season at the Westfield Sugarcane Plantation.

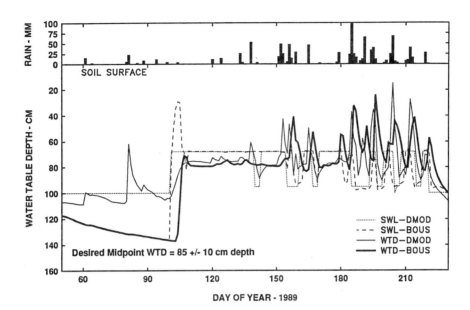

Figure 4. DRAINMOD vs. Boussinesq-based model simulation of automated water table control system; drains at 1.0 m depth and 15 m spacing; nominal SWL = 60-70 cm depth with feedback adjustment in SI = +15 cm.

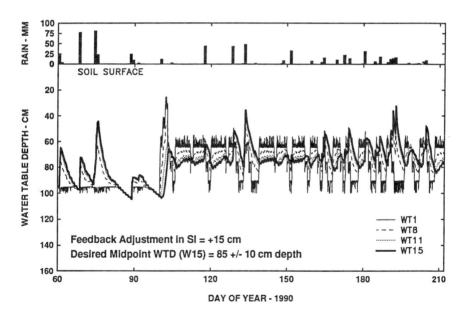

Figure 5. Boussinesq model outputs for a nominal WT1 control range of 55-65 cm depth in SI and 60-70 cm depth in CD.

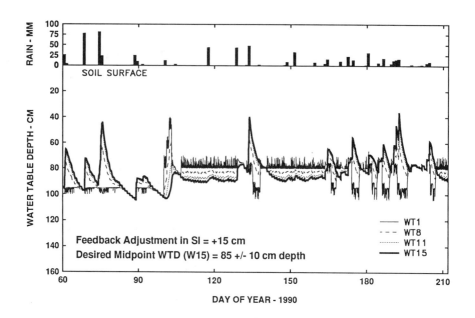

Figure 6. Boussinesq model outputs for a nominal WT1 control range of 70-80 cm depth in SI and 75-85 cm depth in CD.

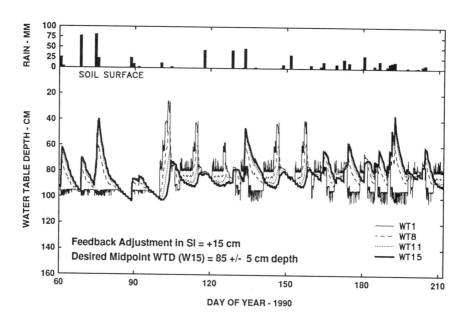

Figure 7. Boussinesq model outputs a nominal WT1 control range of 70-80 cm depth in SI and 75-85 cm depth in CD (with +/- 5 cm desired WTD range).

REFERENCES

Bengtson, R. L., J. L. Fouss, and G. J. McCabe. 1985. Effects of evapotranspiration on modeling in southern Louisiana. ASAE Paper No. 85-2516, ASAE, St. Joseph, MI 49085.

Fouss, J. L. 1985. Simulated feedback-operation of controlled-drainage/ subirrigation systems. TRANSACTIONS of the ASAE 28(3):839-847.

Fouss, J. L., R. L. Bengtson, and C. E. Carter. 1987a. Simulating subsurface drainage in the lower Mississippi Valley with DRAINMOD. TRANSACTIONS of the ASAE 30(6):1679-1688.

Fouss, J. L., R. W. Skaggs, and J. S. Rogers. 1987b. Two-stage weir control of subsurface drainage for water table management. TRANSACTIONS of the ASAE 30:1713-1719.

Fouss, J. L., C. E. Carter, and J. S. Rogers. 1987c. Simulation model validation for automatic water table control system in humid climates. Proc. of the Third International Workshop on Land Drainage. pp. A-55 - A-65. Ohio State Univ., Columbus, Ohio, Dec. 7-11, 1987.

Fouss, J. L. and J. R. Cooper. 1988. Weather forecasts as control input for water table management in coastal areas. TRANSACTIONS of the ASAE 31:161-167.

Fouss, J. L. and J. S. Rogers. 1988. Dynamic simulation model to optimize management/control of dual purpose subdrainage-subirrigation systems. ASCE Irrigation and Drainage Div. Specialty Conf., Planning Now for Irrigation and Drainage in the 21st Century. Lincoln, NE, July 18-21, 1988.

Fouss, J. L., J. S. Rogers, and C. E. Carter. 1989. Sump-controlled water table management predicted with DRAINMOD. TRANSACTIONS of the ASAE 32(4):1303-1308.

Fouss, J. L., R. W. Skaggs, J. E. Ayars, and H. W. Belcher. 1990. Water Table Control and Shallow Groundwater Utilization. ASAE Monograph - Management of Farm Irrigation Systems, Chp. 21:783-824.

Fouss, J. L. and J. S. Rogers. 1992. Adaptive feedback control system for automated watertable management. APPLIED ENGINEERING in AGRICULTURE (pending), ASAE, St. Joseph, MI 49085.

Phene, C. J., J. L. Fouss, T. A. Howell, S. H. Patton, M. W. Fisher, J. O. Bratcher, and J. L. Rose. 1981. Scheduling and monitoring irrigation with the new soil matric potential sensor. ASAE National Irrigation Scheduling Conf. ASAE Publ. 23-81, pp. 91-105.

Rogers, J. S. and J. L. Fouss. 1989. Hydraulic conductivity determination from vertical and horizontal drains in layered soil profiles. TRANSACTIONS of the ASAE 32(2):589-595.

Skaggs, R. W. 1977. Evaluation of drainage-water table control systems using a water management model. Proc. of the 3rd National Drainage Symposium, ASAE Publ. 1-77:61-68.

Skaggs, R. W. 1980. Drainmod-reference report; Methods for design and evaluation of drainage-water management systems for soils with high water tables. USDA-SCS, South National Technical Ctr., Fort Worth, TX. 330 p.

Skaggs, R. W. 1982. Field evaluation of a water management simulation model. TRANSACTIONS of the ASAE 25:666-674.

Skaggs, R. W. 1991. Modeling water table response to subirrigation and drainage. TRANSACTIONS of the ASAE 34(1):169-175.

Smith, M. C. 1983. Subirrigation system control for water use efficiency. M.S. Thesis, North Carolina State University, Raleigh, NC. 182 p.

Smith, M. C., R. W. Skaggs, and J. E. Parsons. 1985. Subirrigation system control for water use efficiency. TRANSACTIONS of the ASAE 28(2):489-496.

DESIGN OF TUBE-WELL DRAINAGE AND WATER MANAGEMENT WITH THE HELP OF A LARGE-SCALE GROUNDWATER MODEL

W. Bogacki, Th. Höddinghaus[*]

ABSTRACT

The situation to be faced in the 190,000 ha Ghotki Fresh Groundwater Irrigation Project in Pakistan is typical for irrigation without adequate drainage. The arid environment of the Lower Indus Basin, intensive irrigation over nearly 30 years and high seepage losses in all parts of the irrigation system taking off from the river Indus have led to a rising groundwater table linked with the hazards of waterlogging and soil salinization. As a standard drainage technique successfully applied in other Salinity Control and Reclamation Projects in Pakistan, more than 1,000 tube-wells were installed along the irrigation canals in order to lower the groundwater table. A second objective of the groundwater extraction is to recover the drainable surplus water stored in the aquifer and to re-use it for irrigation making an additional harvest possible. In order to meet both goals of this dual purpose project, a large-scale finite element groundwater model was developed and extensively utilized to design the tube-well drainage system and to optimize the water management policies.

KEYWORDS: Subsurface drainage, Salinity control, Groundwater model

INTRODUCTION

Pakistan has a long history of irrigated agriculture. With an area of about 14 million ha, Pakistan possesses today the largest contiguous irrigation system in the world. At present the primary and secondary canals of the irrigation network are some 65,000 km long. Tertiary-level canals summarize up-to 1.6 million km, which feed about 90,000 watercourses. In addition, some 250,000 tube-wells exploit fresh groundwater resources to supplement canal supplies (IIMI, 1988).

Two of the most limiting factors in Pakistan's irrigated agriculture are the twin evils of waterlogging and subsequent soil salinization. Long-term and intensive irrigation coupled with high seepage losses in all components of the irrigation system as well as inadequate surface and subsurface drainage has lead to a considerable raising of the groundwater table in many areas. As a result, yields lie far below the expectations and moreover about 25% of the arable land are already fallen out of production.

Since the early 1960s a number of SCARPs[1] have been implemented by WAPDA[2] in order to provide appropriate drainage and to increase agriculture productivity (Awan, 1984). The Ghotki Fresh Groundwater Irrigation Project, financed by the German Financial Cooperation through KfW[3], is one of the recent SCARPs (Peters, 1987). Beside the drainage aspect, this project has the additional objective of re-using the drainage water extracted by tube-wells for irrigation purposes.

---

1) SCARP   Salinity Control And Reclamation Project
2) WAPDA   Water And Power Development Authority of Pakistan
3) KfW   Kreditanstalt für Wiederaufbau

---

[*]   W. Bogacki: Agrar- und Hydrotechnik GmbH (AHT) Consulting Engineers
     Th. Höddinghaus: Institute of Hydraulic Engineering and Water Resources Research,
     Aachen University of Technology

Figure 1. Map of Pakistan with the Project Area

THE PROJECT AREA

Location

The Ghotki Fresh Groundwater Irrigation Project is situated in the shape of narrow belt extending about 100 km north-east from Sukkur on the left bank of the river Indus.

It lies under the command of Ghotki Feeder off-taking from the river Indus at Gudu Barrage. The total culturable command area of Ghotki Feeder is 335,000 ha which is divided into two parts based on the quality of the underlying groundwater. The part having fresh groundwater extends along the river Indus with a culturable command area of 161,100 ha (Fig.2). The part south-east of the fresh groundwater area has highly saline groundwater that cannot be used for irrigation and is therefore excluded from the present project.

The project area is a part of the Indus Valley alluvial plain deposited by the river Indus and its tributaries. The land is virtually flat and slopes gently parallel to the river Indus. Thus, no natural surface drainage is existing.

Climate

The climate of the area can be classified as hot, arid and continental. It is characterized by large diurnal and seasonal fluctuations in temperature. The hottest month is June with a mean monthly temperature of 35.5 °C and a highest monthly temperature of 48.3 °C. January is the coolest month having a mean monthly temperature of 15 °C and a lowest value of 4.4 °C. Precipitation in the area is extremely low and sporadic with a mean annual rainfall at Sukkur meteorological station of 78 mm. Due to the hot and dry climate, evaporation is very high reaching 2000 mm of annual free water evaporation.

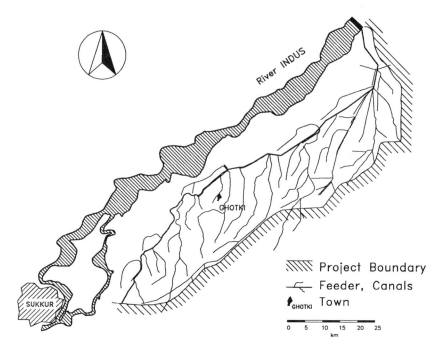

Figure 2. The Project Area

Agriculture

At present the prevailing farming practices are conservative and restrict agricultural production to a subsistence level. The cropping pattern in the project area is typical for non-perennial irrigated areas. Cotton, sorghum and rice are the main Kharif[1] crops while wheat, pulses and oilseeds are the main Rabi[2] crops. The present cropping intensity is approximately 100% with 50% in Kharif and 50% in Rabi (AHT, 1986).

OBJECTIVES OF THE PROJECT

After commission of the Gudu Barrage and the Ghotki Feeder in 1963, groundwater levels in the project area rose steadily. In 1964 the depth to water-table was more than 6 m while by 1985 it was some 1 m or less in many areas showing an annual fluctuation with an amplitude of about 1 m due to the intensive irrigation in Kharif season (AHT, 1986).

The rising of the groundwater table is caused by high seepage losses in the canals, in the watercourses and on-farm. As in any irrigation scheme without sufficient drainage, the hazards of waterlogging and soil salinization have become major problems in the Ghotki area. The implementation of an appropriate subsurface drainage system was therefore the most important objective of the project.

---

1) Kharif Season   (15. April - 15. October)
2) Rabi Season   (15. October - 15. April)

The standard technique successfully applied in several SCARPs for subsurface drainage in the virtually flat area of the Lower Indus Basin is the installation of tube-wells along the irrigation canals. By means of these tube-wells the regional groundwater table is lowered and controlled at a level, at which evaporation from the groundwater table and subsequent soil salinization is prevented.

Beside the drainage effects, an additional aspect of the tube-well operation is the re-use of the extracted groundwater for irrigation. Whereas there is sufficient water in the river Indus during the summer, the improvement of cropping intensity is constrained by a surface water deficit in Rabi season and in early and late Kharif months. Thus the second objective of the project is the provision of additional irrigation water to increase the cropping intensity from todays 100% to anticipated 150%.

It is an interesting approach of the project to pay not too much attention on the reduction of seepage losses, e.g. by the costly lining of canals, but to use the aquifer to store the surplus water in summer and recover it when it is needed.

It becomes obvious, that in order to reach the project goals, a balance has to be maintained between surface water supply, seepage losses in canals and on-farm, evapotranspiration, groundwater extraction, irrigation return flow from re-used groundwater, etc. Furthermore the danger of an upconing and a lateral encroachment of highly saline groundwater exists if the groundwater table is lowered too much, because there is only a limited fresh groundwater lens below the intensive irrigated areas near the river Indus (Bogacki, 1987).

THE GHOTKI FRESH GROUNDWATER DIGITAL MODEL

Numerical models have proven to be excellent planning tools for the groundwater management. The Ghotki Fresh Groundwater Digital Model was developed and extensively utilized to:

- determine aquifer parameters, e.g. transmissivity, storativity, etc.
- determine components of the water balance, e.g. seepage from canals and on farm, evapotranspiration, tube-well extraction, etc.
- develop a better understanding of the groundwater flow regime and the factors affecting it
- evaluate the existing planning, e.g. number, capacity and location of tube-wells, and the effects of several changes to the planning, that had become necessary during the implementation
- investigate the effects of different operation alternatives

Type of Model

The thickness of the Ghotki aquifer varies between some 100 m in the extreme south-west to approximately 270 m in the north-east. The bottom of aquifer is formed by a low permeable layer of silty clay. The alluvial deposit which forms the aquifer consists of fine to medium sands of uniform nature with thin lenses of clay to silty clays. The whole aquifer can be regarded as very homogeneous as numerous grain size distribution analysis show (WAPDA, 1966). Under natural conditions, the groundwater slopes gently parallel to the river Indus with an average gradient of about $0.15\ ^o/oo$. These aquifer characteristics make the Ghotki aquifer to a typical application of a single layer, large-scale, two-dimensional horizontal groundwater model based on the transmissivity approach.

The numerical solution of the governing differential equations of groundwater flow is obtained by the finite element method, as it offers the advantages of convenient handling of all types of boundary conditions, correct representation of tensorial model parameters (i.e. the transmissivity tensor) as well as arbitrary orientated gradients and flow velocities, and an efficient approximation of complex geometries of the model area.

The groundwater model was prepared at the Institute for Hydraulic Engineering and Water Resources Research at the Aachen University of Technology (Bogacki and Pelka, 1987) on behalf of Agrar- und Hydrotechnik Consulting Engineers. Thereafter the model was installed at the WAPDA Computer Centre at Lahore, and groundwater experts of WAPDA were trained in operating the model.

Discretization

The finite element discretization of the project area (Fig. 3) visualizes the high flexibility of triangular finite elements for the modelling of complex project areas. The finite element mesh consists of 2280 nodes and 4428 elements. The model area is larger than the project area to reach well defined boundary conditions as for example the river Indus. It covers a total area of 293,500 ha.

Special care was taken on a correct representation of the irrigation canals by using very small elements, as the seepage from these canals is one important factor of groundwater recharge. To minimize the computational effort, bigger elements were chosen between the canals and outside the main project area, e.g. near the river Indus.

Boundary Conditions

Along the river Indus, the model boundary was described by the transient river water levels. At the north-eastern and south-western borders similar boundary conditions were chosen by prescribing measured groundwater levels. The difficult to define south-eastern border was treated as a gradient dependent flow boundary condition, allowing water to flow into the model area when the water table is lowered inside.

Most components of the water balance, i.e. groundwater recharge from small canals and irrigation return flow from fields and watercourses, groundwater discharge by private and public tube-wells, were incorporated as source respective sink boundary conditions. Seepage from bigger canals was modelled by a leakage boundary condition as seepage will increase if the groundwater table is lowered. Evaporation from the groundwater table was treated similarly, using a depth-evaporation relation depending on the actual soil type and land-use.

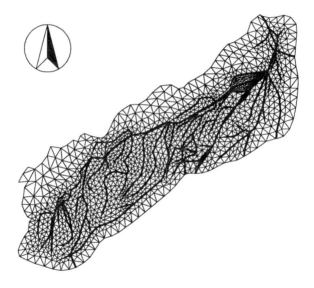

Figure 3. Finite Element Discretization

Model Calibration

One of the most important steps of groundwater model investigations is the model calibration. Obviously, the results of any model will be only as good as the input parameters represent the field conditions. Most of the parameters however, e.g. hydraulic conductivity, storativity, leakage coefficients, recharge or discharge components, etc., are usually badly known. During model calibration the respective parameters are adjusted until the best fit between historic (measured) and calculated groundwater levels is found. Thus, the model calibration offers an efficient way to determine unknown aquifer parameters as well as components of the water balance without time-consuming and costly field measurements.

For the Ghotki Fresh Groundwater Digital Model an automatic calibration algorithm based on the Least Square Method was applied. While the hydraulic conductivity ($7 * 10^{-4}$ m/s) was not calibrated because it was already well determined by numerous pump-tests, several model parameters were successfully estimated by the calibration procedure. A storativity of 30% was found which fits very well with data from adjacent SCARPs. The leakage factors of all bigger irrigation canals lie in the range of 10^{-5} s^{-1}. Seepage losses from fields and watercourses were determined to some 30% of the irrigation water input at the head of the watercources.

At the end of the model calibration procedure, the differences between historic and modelled groundwater levels were reduced to a mean deviation of 0.4 m which has to be regarded as an excellent result taking into account the sparse data available in this area.

RESULTS OF MODEL SIMULATIONS

It can clearly be distinguished between two different simulation phases with the Ghotki Fresh Groundwater Digital Model, which is typical for numerical model studies. The first phase is usually aimed to check the planning and to optimize the design. Often a second phase follows, during which the model is used to find optimal operation policies.

<u>Simulation Phase I</u>

Originally it was planned to install some 720 new tube-wells (56.6, 42.5, and 28.3 L/s) with a total capacity of about 32,848 L/s in addition to 410 tube-wells with a capacity of about 20,954 L/s, that already went in operation in 1978 and 1984. First instationary simulation runs with the model, covering like all following runs the period from 1985 upto the year 2000 showed, that the required drawdown could be achieved in most parts of the project area with an overall pump utilization factor of about 50%. However, these results also indicated the possibility of a significant diminution of the fresh groundwater lens and by this the danger of some planned wells to become saline.

Thus, several model runs were carried out to investigate the effects of global (in the whole project area) and local (near the border to the saline zone) reduction of tube-wells, both in number and in capacity. Subsequent to the simulation with the Ghotki Fresh Groundwater Digital Model computations were carried out with a finite element salt-water model, to predict the movement of the saline water (Höddinghaus, 1990).

The results of these investigations showed, that the diminution of the fresh water lens can be reduced by a global reduction of the tube-well capacity. However, a limited movement of the saline water into the fresh groundwater area cannot be prevented, if the groundwater table is lowered to the required level. Thus, the installation of some tube-wells that were planned near the saline zone had to be cancelled. It is deemed that these model results have saved a lot of money, as these tube-wells would have been most likely become saline in a short time.

Figure 4 shows the predicted depth to the groundwater table for the final well layout in the year 2000 at the end of Rabi season. Finally, 670 new tube-wells with a total capacity of about 28,317 L/s are installed on basis of the preceding model results. As a major change to the first planning, a number of 56.6 L/s wells are substituted by 28.3 L/s wells to minimize the effect of a vertical upconing of saline water. It can be seen, that in large regions of the project area, the groundwater table is located about 3 m below the ground surface. To cover the irrigation water demand according to the anticipated cropping intensity of 150% a high pump utilization factor of about 70% is required during Rabi season. During Kharif a utilization factor of only 30% is sufficient, as most of the irrigation water demand is covered by surface water from the river Indus.

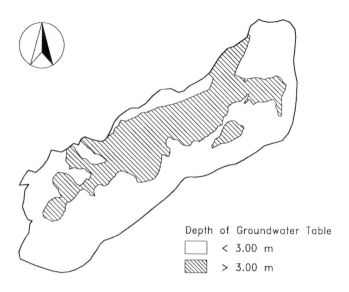

Figure 4: Depth to Groundwater Table for the Final Well Layout
in the Year 2000 at End of Rabi Season

Simulation Phase II

Based on the final well layout, several simulation runs were carried out to evaluate the effects of different water management alternatives on the position of the groundwater table and on the movement of the saline water.

One simulation run was designed to check a situation in which a pump utilization factor of 70% will not be maintained over the whole period of 6 month during Rabi season. A mean utilization factor of 50% was assumed during Rabi which implies a reduction of the groundwater extraction. As it was to be expected, the computation results show the positive effect of a limited encroachment of the saline water. The resulting reduction of the irrigation water supply may be acceptable, as the deduced reduction of the cropping intensity lies in the order of 10%. But the required lowering of the groundwater table is not reached in many parts of the project area, which will have a significant influence on soil salinization and subsequent on agriculture production. Thus the important objective of providing a sufficient drainage cannot be secured under the assumed conditions.

A possible management policy to achieve a fast and sufficient draw-down of the groundwater table is to stop the surface water supply during Rabi season by closing the irrigation canals. This alternative proves to be very efficient concerning the lowering of the groundwater table. However, because of the significant reduction of total water input, the resulting cropping intensity is 106% only, although a pump utilization factor of 60% was assumed during Rabi season. In addition, the saline water would move significantly into the fresh groundwater area, endangering a number of newly drilled tube-wells to become saline. Thus, this management alternative may be used as a short term measure to achieve a fast drawdown of the groundwater table, but is not recommended as a long-term policy.

An optimal operation policy may be found in a compromise of the two extreme alternatives discussed above. The main idea is to allow an acceptable reduction of the anticipated cropping intensity to some 140% and to utilize all free pump capacities to substitute surface water by groundwater. This applies mainly to the Kharif season, where the tube-wells have to be utilized only by 30% because sufficient surface water is available. According to that philosophy, a pump utilization factor of 50% was simulated in Kharif, using the 20% of additionally extracted groundwater to reduce the surface water input from the river Indus while keeping the total irrigation water supply constant. In Rabi the pump utilization factor was taken to 50%, in order to achieve the anticipated cropping intensity.

The results of the simulation of this operation policy are very promising. Both, groundwater table and movement of the saline water are comparable to the simulation of the final well layout (see Fig. 4). Thus, the drainage requirements are achieved and the new tube-wells are most probably secure of becoming saline. The cropping intensity is reduced in the order of 10%, but this depends only on the assumed pump utilization factor of 50% in Rabi, which gives a greater flexibility in case of operation problems. Last but not least, an amount of about 150 million m^3 surface water from the river Indus can be saved and can be utilized in other areas.

CONCLUSION

The Ghotki Fresh Groundwater Irrigation Project is a typical multi-purpose project, where the different objectives are contrary to each other. To achieve the difficile balance necessary to meet the goals of fighting the hazards of waterlogging and soil salinization as well as to increase the cropping intensity and to prevent a salt-water encroachment into the fresh groundwater area, an optimal design and optimal operation policies had to be developed. For this purpose, numerical groundwater models like the Ghotki Fresh Groundwater Digital Model have proved to be excellent planning tools, as once they are calibrated, a large number of alternatives can be simulated very fast and with the best accuracy that is achievable today.

REFERENCES

AHT 1986. Ghotki Fresh Groundwater Irrigation Project. Inception Report.
Agrar- und Hydrotechnik GmbH, Consulting Engineers

AWAN, N.M. 1984. Some Technical and Social Aspects of Water Management in SCARP No. 1, Pakistan. ICID Bulletin, Vol 33, No. 1

BOGACKI, W. 1987. Optimal Management of Fresh/Saltwater Aquifers.
Mitt. Inst. f. Wasserbau und Wasserwirtschaft der RWTH Aachen, Heft 65 (in German)

BOGACKI, W.; W. Pelka .1987. Ghotki Fresh Groundwater Digital Model.
Mitt. Inst. f. Wasserbau und Wasserwirtschaft der RWTH Aachen, Heft 66 (in German)

HÖDDINGHAUS, Th. 1990. Ghotki Salinity Digital Model. Final Report.
Inst. f. Wasserbau und Wasserwirtschaft der RWTH Aachen

IIMI.1988.Annual Report 1987. Diagana Village

PETERS, D. 1987. Ghotki Fresh Groundwater Irrigation Project.
Mitt. Inst. f. Wasserbau und Wasserwirtschaft der RWTH Aachen, Heft 66

WAPDA .1966. Lower Indus Report. Water and Power Development Authority of Pakistan, Lahore

ECONOMIC MODELLING OF SOIL WORKABILITY CRITERIA

K. Eradat Oskoui[*]
Member ASAE

When to start field work in fall and spring is a an economic decision which farmers have to make. The problem is even more important in northern and cooler regions with shorter growing seasons. Working the field too early when the soil is wet can depress crop yield by causing surface and sub-surface soil compaction. Waiting for the soil to dry can cause substantial timeliness penalties due to late planting. Some other economic factors such as energy efficiency and machine and tire damage are also important. In addition to economic aspects, some engineering and practical aspects have to be taken into account when deciding on soil workability. Traditionally, a soil has been defined as workable if a) it has sufficient compressive strength to withstand the weight of the machine, b) it has sufficient shear strength to meet the traction requirement with acceptable wheel slip, and c) a suitable tilth can be produced. A model is developed which predicts the optimum soil working conditions with respect of engineering, economical and biological constraints.

LITERATURE CITED

Bolton et al. (1968), in a study to develop a model for soil work-day prediction, used two measures of soil water levels defined by Broadfoot and Burke (1958), "field maximum" and "field minimum" which correspond to field capacity and wilting point. Field capacity (Rutledge and MacHardy, 1968, Rutledge and Russell, 1971, and Frisby, 1970), available water holding capacity (Morey et al., 1971) water deficit (Smith, 1977), soil water content-rainfall combination (Eliot et al., 1977) rainfall alone (Wendte et al., 1978) and air temperature (Dyer and Baier, 1979) have all been used to define soil workability. Hassan and Broughton (1975), in a study to establish a standard criterion for soil workability reviewed most of the available literature. They summarized the criteria for seed bed preparation for three different soils in terms of both percentage field capacity (FC) and available water holding capacity (AWC).

Of all the above researchers only two make reference to actual tractor and implement type and size when defining workability criteria (Rutledge and MacHardy, 1968; Voorhees and Walker, 1976). Their references to machine configurations have only been from a "traction" point of view. Almost exclusively, none of the research in the available literature has taken compaction and its implications into account when defining soil workability.

---

[*] **Agricultural Engineer, USDA Agricultural Research Service, North Central Soil Conservation Research Laboratory, Morris, MN 56267**

MODEL DEVELOPMENT

A computer model was developed that defines soil workability criteria which encompass several engineering and economical factors. The model takes into account the tractor weight and type (2 or 4 WD etc.), equipment type (soil engaging, surface rolling, or trailed), tire type (low ground pressure or conventional), tire size, etc. Soil type, bulk density, organic matter content, type of growing season (wet or dry and hot or cold), crop type (long or short maturing) and crop variety also have been considered.

The model is developed under the mathematical hypothesis that soil is workable when, traction produced by the drive wheels and tractive efficiency are maximized; wheel slip is optimized and, sinkage (rut depth), rolling resistance, draft required by equipment, tire cost, and yield loss due to compaction are minimized.

In order to estimate the optimum soil water content at which a given farming operation can be carried out with maximum economic return, and also be technically feasible, relationships have to be developed between the above variables and the soil water content. In the following sections a description of these relationships and their development, together with the procedures used for obtaining economic workability criteria are described and finally, some worked examples are presented.

<u>Traction</u>

Traction or mobility of the machine is one of the primary parameters considered when workability decisions are made. For draft equipment, drive wheels of the tractor should produce enough drawbar pull to offset the draft required by the equipment and overcome the rolling resistance of the driven and non-driven wheels. The soil water content at which this is possible is often used as a workability criterion (Rutledge and MacHardy, 1968 and Voorhees and Walker, 1976). This strategy, although appropriate as a technical criterion, may not be a good economic criterion, as this balance (pull vs. draft) could be achieved at almost any soil water content if the farmer is willing to pay for it. Adding more tires (at a cost), increasing weight (at a higher compaction cost), and/or using low ground pressure tires (at a cost) can move the soil water threshold to a considerably higher level. This is often done by farmers to offset potential timeliness penalties, especially in cool and wet regions with heavy soils. Ideally, it is more economical to operate the machine at a soil water level that facilitates maximum pull production with minimum additional costs.

For a given set of soil and machine parameters, coefficient of traction, C_t, (pull/weight) is given by Gee-Clough et al. (1978) as follows:

$$C_t = (C_t)_{max} [1-\exp(-ki)] \qquad (1)$$

where, i, is slip and is calculated:

$$i = 9 + 19 / MN \qquad (2)$$

and $(C_t)_{max}$ is the coefficient of traction at optimum slip, MN is the mobility number and k is rate constant as follows:

$$k(C_t)_{max} = 4.838 + 0.061 \, MN \qquad (3)$$

Mobility number, MN, is calculated from cone index, CI, tire width, w, diameter, d, section height, h, deflection, α, and wheel load, W, using an approach defined by Turnage (1972).

$$MN = CI \times w \times d \times \sqrt{(\alpha/h)} / W \times (1 + w/2d) \qquad (4)$$

Analysis of survey data show that tire deflection under recommended load on a concrete surface as a function of tire width is:

$$\alpha = 0.144 \, w + 0.008402 \qquad (R^2 = 0.92) \qquad (5)$$

Soil cone index (in MPa) is a function of soil water content, bulk density and texture according to Witney et al. (1984).

$$CI = [k_\Theta \times e^{-1 \times \Theta/(1+Cr)} + k_\delta \times \delta_s/(1 + 2 \times Cr)] \times$$
$$e^{\pi/(1 + 2 \times Cr)} \qquad (6)$$

Where Θ is the soil water content (% w/w), δ_s is the soil specific weight (kN/m^3), Cr, is the ratio of clay to silt and sand, and k_Θ and k_δ are empirical constants. Specific weight, δ_s, is a function of soil bulk density, δ.

Tractive efficiency

Tractive efficiency, μ, is a function of wheel slip, coefficient of traction, and coefficient of rolling resistance, C_{rr}, according to Gee-Clough et al (1978):

$$\mu = C_t(1-i)/(C_t + C_{rr}) \qquad (7)$$

Wheel Sinkage (rut depth)

Wheel sinkage affects machine performance and fuel use in two ways. First, increasing sinkage results in increased tire/soil contact area, thus, increased traction which enhances economic returns. Secondly, any increase in sinkage results in increased compaction, rolling resistance and fuel use, thus, lower economic returns. Sinkage ratio (ratio of sinkage over tire diameter), j, is related to mobility number according to data given by Freitag (1965):

$$j = 0.2752 / MN - 0.0138 \qquad (8)$$

Rolling resistance

Rolling resistance, RR, or motion resistance of driven and non-driven wheels is adversely affected by soil water content via mobility number and cone index. The wetter the soil the higher is the rolling resistance in heavy soils. In light soils however, compaction has more effect on rolling resistance than soil water content. According to Gee-Clough et al (1978) coefficient of rolling resistance is a function of wheel mobility number:
$$C_{rr} \text{ or } RR/W = 0.049 + 0.287 / MN \qquad (9)$$

Equipment Draft

Draft required for pulling soil engaging and rolling equipment is affected by soil water content as well as equipment parameters. Draft required, D, in kN, to pull an implement is the sum of soil resistance to cutting and/or the rolling resistance of the implement wheels.

$$D = RR_e + Z \qquad (10)$$

where, Z is the draft required for soil cutting and RR_e is the rolling resistance of equipment wheels. If the equipment is mounted then $RR_e = 0$, or if the equipment is not a soil engaging machine then $Z = 0$. Relationship between RR_e and soil water content (via mobility number) has been given above (equation 9), and the value of Z is calculated:

$$Z = [K_s \, CI + K_d \, \delta_s \, V^2 \, (1 - \cos\beta)/g] \, a_e \, w_e \, n \qquad (11)$$

where a_e and w_e are equipment working depth and width, n, is number of cutting blades, V is travel speed, β, is the lateral directional angle of moldboard tail (angle of the moldboard tail to travel direction) for moldboard plows and angle of approach (angle of cutting blade to the horizontal) for chisel plows and sub-soilers. Other parameters are, g, gravitational acceleration (in m/s^2) and K_s and K_d are static and dynamic coefficients (Eradat Oskoui, 1981). By using equation 11, equipment draft is calculated as a function of soil water content (via cone index).

Fuel cost

With a knowledge of tire and soil parameters, slip, draft and rolling resistance, tractor power, P, is determined for a given implement:

$$P = [(D + RR) V(1-i)] / (\mu \, \mu_t) \qquad (12)$$

Where μ_t, is transmission efficiency of the tractor used and μ and V have been described before.

Calculated power as a function of soil water content for a given machine, wheel size, soil type, compaction level is then converted into $/ha in fuel costs, c_f, by using fuel efficiency, μ_f, in l/kW.h, work rate, μ_r, in ha/h, and unit fuel cost, c_{uf} in $/l:

$$c_f = P \, \mu_f \, c_{uf} \, \mu_r \qquad (13)$$

Tire cost

Tire cost is directly related to tire size, ply rating, tire type (low ground pressure tires or normal) and number of tires (single, duals etc.). It is important to choose a soil water content level which will yield maximum traction using the smallest tire surface area when selecting a workability criterion. Tire dimensions affect the majority of economic parameters mentioned above by affecting the tire mobility number.

A survey of over 1200 common agricultural tires was carried out in the U.S. to obtain a practical relationship between tire parameters and prices. Conventionally, tire sizes are given as "tire width-rim diameter, ply." For calculation of mobility number a knowledge of tire diameter is required. Using the survey data, a relationship between the product of tire width-rim diameter, w (m) x d_r (m) and tire diameter, d, in meters was obtained:

$$d = 0.44669 \, w \, d_r - 0.3247 \quad (R^2 = 0.96) \qquad (14)$$

An attempt was made to obtain a relationship between tire width and tire prices. A significantly high coefficient of determination (R^2) exists between tire width and tire prices for both low ground pressure ($R^2 = 0.84$) and conventional tires ($R^2 = 0.76$). Despite the high R^2, high levels of standard errors indicated the need for other parameters to be included in the price prediction equation. Ply rating was also found to influence tire prices. Similarly, tire and rim diameter had some influence on the tire prices ($R^2 = 0.46$). Tire surface plane was also related to tire prices but goodness of fit was not any better than when the tire width alone was used. Tire surface plane is calculated as the product of tire width and tire circumference. Since the actual tire parameters which will be used for soil compaction modeling (next section) are tire/soil contact area and load carrying capacity, a relationship is obtained between these parameters and tire prices. A new parameter is defined to collectively represent all the tire variables explained here. This parameter is called contact coefficient (CC) and is the product of tire/soil contact area at 75 mm sinkage and ply rating. Use of this parameter considerably reduced the scatter in the data and increased the coefficient of determination while significantly reducing the standard error on the correlation coefficients. The relationship between tire cost, c_{ti}, and contact coefficient is given by:

$$c_{ti} = K_1 \, CC + K_2 \quad (R^2 = 0.91) \tag{15}$$

where K_1 and K_2 are 464.62 and 309 for low ground pressure tires and 308.60 and -106.1 for conventional tires, respectively and contact coefficient is calculated from tire parameters:

$$CC = (0.137 \, d \, w \, \pi + 105.72) \, Pl \tag{16}$$

where, Pl is tire ply rating. The inclusion of ply rating in the cost calculation is justified because a) there is a strong correlation between tire prices and ply rating ($R^2 = 0.66$), b) ply rating dictates the limit (lower limit) on the inflation pressure affecting tire deflection and contact area and, c) ply rating affects tire loading capacity. Tire costs obtained from these equations are in $/tire and are converted into $/ha by using the number and expected life of tire in hours and work rate in ha/h.

Work rate is calculated from equipment width, speed of travel and field efficiency. Calculating the expected life of tire in hours is very difficult if not impossible. Very limited work has been done by the researchers on the expected life of agricultural tires (Zheng, et al. 1987 and A. Andert, 1985). In these works the effect of the torsional vibration in a tractor on energy and tire service life and, influence of the alignment parameters of the tractor front wheel on tire wearing have been investigated. Tire manufacturers have conducted various tests on tire durability and expected life. The results of such tests are extremely variable and are dependent on the terrain, type of work, speed of travel, slip and tire design. According to one test the expected life of a particular tire type varied from 94 hours to 6000 hours depending on the test conditions (Shorter, A.C. 1992). Under normal usage and typical conditions expected tire life for bias ply tires and radial tires can be around 1500 and 2000 hours respectively. For heavy usage in hard surfaces (scraping etc.) a shorter life of 600 and 800 hours could be expected from bias ply and radial tires respectively. Very limited data are available for low ground pressure tires. Life expectancy of these tires are about 10 - 20% less than their radial counterparts.

SOIL COMPACTION

Soil compaction, although one of the most important factors which affects workability criteria, has been widely ignored in the literature due to lack of sufficient knowledge on economics of soil compaction. Soil water content is used as a workability criterion mainly because of its effect on traction. Although, mobility is often the prime concern, and importance of good tilth is also recognized, soil compaction has received little attention.

Working the soil when it is too wet has resulted in considerable yield loss in the subsequent crop. This is due to the increase in final bulk density when the soil is trafficked at higher water contents (Gupta and Larson, 1982); Akram and Kemper, 1979 obtained "S" shaped curves relating final bulk density to soil water content at the time of compaction for a wide range of soil water contents and soil types. This change in bulk density ultimately affects crop yield (Hebblethwaite and McGowan, 1980).

Data supplied by Akram and Kemper, (1979) were analyzed and a series of dimensionless ratios or pi terms were defined. Possible relationships between pi terms and final bulk density after compaction were examined. Of all the possibilities, the following is both statistically most significant ($R_2 = 0.94$) and practically accurate.

$$\delta = c_1 \ln(\epsilon_\Theta) + c_2 \ln(\epsilon_\sigma) + c_3 \ln(\epsilon_\delta) + c_4 \ln(\epsilon_c) + c_5 \qquad (17)$$

where, ϵ_n are pi terms which are defined as, moisture number, $\epsilon_\Theta = \Theta/\Theta_s$; load number, $\epsilon_\sigma = \sigma_z/\sigma_k$; Bulk density Number, $\epsilon_\delta = \delta_i/\delta_k$; and Clay Number (ratio), $\epsilon_c = Cr$. Variables used to obtain these terms are, Θ, soil water at time of compaction, Θ_s, is saturated soil water, δ_i, is initial bulk density, δ_k, is bulk density at 1 kgcm^{-2}, σ_z, is applied stress and σ_k, is reference stress (1 kg cm^{-2}). Using data supplied by Akram and Kemper, (1979) for five soils ranging in clay content 7 to 44% the values of $c_1 - c_5$ are -0.2165, 0.1713, 0.222, -0.113 and 1.366, respectively. A model by Gupta and Larson (1982) was used to calculate bulk density at 1 kgcm^{-2}. Applied stress is calculated from concentration factor (function of cone index), tire/soil contact area (function of sinkage ratio, equ. 8) and wheel load using Boussinesq equation (Soehne, 1953).

Compaction model developed earlier (Eradat Oskoui and Voorhees, 1991) was used to calculate yield loss/gain over trafficked and non trafficked areas for various soil types, tire types, sizes and configurations and water contents at time of compaction.

Overall compaction cost, c_c in $/ha, is given by:

$$c_c = \{Y_o-[c_s\log((1-\delta/\delta_p)-\Theta_p\delta)^2 + c_w\log((1-\delta/\delta_p)-\Theta_p\delta) + c_v]\}p_c \qquad (18)$$

where p_c is crop price ($/Mg), Y_o is uncompacted crop yield (Mg/ha), Θ_p, is soil water content at the time of planting (% w/w), δ is bulk density after compaction and δ_p is particle density of the soil. c_s, c_w and c_v are functions of soil type, weather conditions (growing degree days and annual precipitation) and crop variety (maturity length) according to Eradat Oskoui and Voorhees, (1991).

ECONOMIC SOIL WORKABILITY CRITERIA

The economic workability criterion is defined as the soil water content at which if the soil is worked, cost will be minimum. The cost of an operation is the algebraic sum of compaction cost/benefit, c_c; fuel cost due to draft required and rolling resistance of driven and non-driven wheels, c_f; and, tire cost, c_{ti}, as follows:

$$c = c_c + c_f + c_{ti} \qquad (19)$$

This equation is minimized mathematically with respect to soil water content at the time of the operation and the optimum workability criterion is calculated. To demonstrate the feasibility of the model, three weather scenarios of wet (800 mm), average (600 mm) and dry (350 mm annual precipitation) on a 1000 ha hypothetical farm are assumed and the equation is solved. The results of these solutions are given in Figures 1a-1c. A 4-wheel drive 200 kW tractor weighing 14000 kg was used. It was assumed to pull a 40 tine chisel plow, spaced at 0.3 m and penetrating 0.20 m at an angle of approach to the soil of 45°. Tires used are equal numbers and sizes at the front and rear and assumed to be 20.8 R 38. Other data used were obtained from manufacturers specifications.

In a wet year and heavy soil (clay content over 40%) the least cost soil workability criterion occurred when duals were used at a soil water content of around 18% (Fig. 1A). For higher soil water contents triples should be used to avoid high compaction costs. Using singles on wetter soils, despite reducing fuel cost due to draft, increased the fuel cost due to increased rolling resistance and poorer tractive efficiency. It also increased compaction cost due to poor crop yield. Using duals on wetter soils had lower fuel costs due to reduced rolling resistance (reduced sinkage) and improved tractive efficiency, with much higher compaction costs. At very high soil water contents, duals are not good options in terms of crop yield loss. This is because, the reduction in the spot yield loss (yield loss on wheel tracks) from 6.2 Mg/ha to 5.6 Mg/ha is not large enough to compensate the increase in the area trafficked from around 5% to 9%. Engineering workability criteria, where pull produced was greater than draft required by the equipment and rolling resistance were, at 15%, 22% and 24% soil water contents for single, duals and triples, respectively, which are comparable with the economic criteria.

On an average year and medium soil (20% clay), use of single wheels reduces the economic workability criterion to around 10% soil water content, which is practically un-achievable (Fig. 1B). If duals were used then the economic workability criterion is moved to 15 to 20% which is within practical limits. If it is desired to work at even higher soil water contents (over 25%) it makes economic sense to use triples. This will produce some yield benefit by consolidating the soil and retaining soil water. The fuel cost curve in a medium soil followed almost the same pattern as the one for the heavy soil with slightly less increase due to rolling resistance at higher soil water contents. Fuel cost due to rolling resistance and tractive efficiency was marginally less for duals and triples than for singles. In this case, use of additional wheels increased the soil workability criterion to almost field capacity (20%).

On a light soil (clay content less than 15%) with a dry growing season, least cost criterion was obtained when triples were used at a very high soil water content (Fig. 1c). Although there was no compaction penalty (yield loss) for working at any workability criterion except when using singles at very high soil water contents, by using duals and triples considerable amount of benefits can be gained by spreading the compaction to a wider area and utilizing the extra soil water held in the soil due to compaction. Benefits from the compaction (consolidation) are far more than the extra tire costs, and therefore using duals and triples may be justifiable from compaction point of view.

For a smaller tractor (100 kW engine power), the picture is somewhat different. On a heavy soil the least cost workability criterion is achieved when single wheels are used at a soil water content near field capacity. Only for soil water contents of above 30% the use of duals is justified. Triples are never justified at any soil water contents for this size tractor.

Although, individual segments of the model (traction, draft and compaction) have been validated separately, and the findings of the model is in agreement with some practical

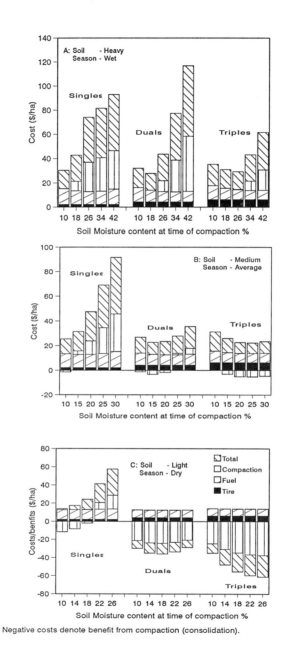

Figure 1. Variations of economic soil workability criteria with soil type, growing season and tire configurations.

results, the overall model may require more validation before it can be used as a decision making tool. The present version of the model can be used with confidence as a decision support, guidance and educational tool.

Conclusions

Use of a single workability criterion such as rainfall or a particular soil water content is not an adequate way of defining soil workability. Soil workability criteria are machine, soil and weather dependent. Use of extended tire width is not always an economical way of increasing soil workability threshold. Also, despite common belief, the benefits of waiting for the soil to dry out may compensate for any revenue loss due to timeliness penalties. The model described here can be used by farmers, researchers, consultants and machinery designers to evaluate optimum soil workability criteria from economic and engineering points of view. This model takes the existing technology one step forward by allowing the user to account for soil, crop, machine and weather effects in defining soil workability criteria.

REFERENCES

1. Akram, M. and W.D. Kemper. 1979. Infiltration of soils as affected by the pressure and water content at the time of compaction. Soil Sci. Soc. Am. J. 43:1080-1086.

2. Andert, A. 1985. Effect of the torsional vibration in a tractor on energy losses and tire service life (Vliv torzniho kmitani u mobilniho prostredku na energeticke ztraty a na zivotnost pneumatik). Zemed. Tech. Cesk. Akad. Zemed-Ustav-Vedeckotech-Inf-Zemed. Praha : Ustav. 31(5):263-278.

3. Boltons, B.; J.B. Penn, F.T. Cooke, and A.M. Heagler. 1968. Days suitable for field work. Mississippi river delta cotton area. Dept. Agric. Econ. & Agribusiness. Louisiana State University. D.A.G. Res. Rep. No 384.

4. Broadfoot, W. M. and H. D. Burke. 1958. Soil water constants and their variations. Southern forest exptl. Sta. Occasional Paper No. 166 USDA-FS.

5. Dyer, J.A. and W. Baier. 1979. Weather based estimation of field workdays in fall. Can. Agric. Eng. 21(2):119-122.

6. Elliot, R.T., W.D. Lambke, and D.R. Hunt. 1977. A simulation model for predicting available days for soil tillage. Trans. ASAE 20(1):4-8.

7. Eradat Oskoui, K. 1981. Agricultural mechanization systems analysis - Tractor power selection for tillage operations. Unpublished Ph.D Thesis, University of Edinburgh.

8. Eradat Oskoui K. and W. B. Voorhees. 1991. Modelling crop response to soil compaction. Am. Soc. Agric. Eng., Pap. No. 91-1556, ASAE, St Joseph, MI, 9 pp.

9. Freitag, D. R. 1965. Dimensional analysis of performance of pneumatic tires on soft soils. Tech. Rept. 3-688. U.S. Army Corps of Engineering Waterways Experiment Station.

10. Frisby, J. C. 1970. Estimating good working days available for tillage in central Missouri. Trans. ASAE 13(5): 641-643.

11. Gee-Clough, D.; M. McAllister, and D.W. Evernden. 1978. The empirical prediction of tractor-implement field performance. J. Terramechanichs, 15(2):81-94

12. Hassan, A. E.. and R. S. Broughton. 1975. Soil moisture criteria for tractability. Can. Agric. Eng. 17(2): 124-127.

13. Hebblethwaite, P.D. and M. McGowan. 1980. The effects of soil compaction on the emergence, growth and yield of sugar beet and peas. J. Sci. Food Agric., 31:1131- 1142.

14. Morey, R. V., R. M. Peart, and G. L. Zachariah. 1971. Optimal harvest policies for corn and soybeans. J. Agric. Engng. Res. 17(1):139-148.

15. Rutledge, P. L. and F. V. McHardy. 1968. The influence of the weather on field tractability in Alberta. Can. Agric. Eng. 10(2):70-73.

16. Rutledge, P. L. and D. G. Russell. 1971. Work day probabilities for tillage operations in Alberta. Res. Bull. 71-1. Dept. of Extension University of Alberta, Edmonton, Alberta.

17. Shorter, A. C. 1992. Agricultural tire engineer, The Goodyear tire and rubber company, Akron, OHIO. Personal communications.

18. Smith, C.V. 1977. Workdays from weather data. The Agric. Eng. 32(4):95-97.

19. Soehne, W. 1958. Fundamentals of pressure distribution and soil compaction under tractor tires. Agric. Eng. 39(5):267-281-290.

20. Turnage, G. W. 1972. Performance of soils under tire loads; application of tests results to tire selection for off-road vehicles. Tech. Rept. No. 3-666. U.S.A. Army Corps of Engineers Waterways Experimental station. CE-Vicksburg Mississippi.

21. Wendte, L.W.; C.J.W. Drablos, and W.D. Lembke. 1978. Timeliness benefit of sub-surface drainage. Trans. ASAE 21(2):484-488.

22. Witney, B. D.; E. B. Elbanna, and K. Eradat Oskoui. 1984. Tractor power selection with compaction constraints. Proc. 8th ISTVS Intnl. Conf. Cambridge, England UK

23. Voorhees, M.L. and P. N. Walker. 1977. Tractionability as a function of soil moisture. Trans. ASAE 20(5):806-809.

24. Zheng, L. Z.; Y.S. Cheng, and S.Q. Deng. 1987. Influence of the alignment parameters of the tractor front wheel on handling lightness and tire wearing. G. Yeh. Chi. Hsieh. Hsueh. Pao. Trans. Chin. Soc. Agric. Mach. Beijing : Chinese Society of Agricultural Machinery. 18(1):1-12.

DRAINAGE FLOW PATTERNS AND MODELING
IN A SEASONALLY WATERLOGGED HEAVY CLAY SOIL

D. Zimmer *

ABSTRACT

Water flow patterns have been determined in a drained seasonally waterlogged heavy clay soil monitored between 1985 and 1990 in western France. During most of the drainage period a perched water table forms above the plow boundary which acts as an impervious barrier. Water flows along the plow boundary before percolating towards the drains through cracks between the trench backfill and the non disturbed soil. The relationship between drain flow rates and mid-spacing hydraulic heads is quadratic. This allows the use of classical drainage formula and modeling in this soil. Consequences are discussed in terms of drainage functioning and design. With conventional subsurface drainage, the water table drawdown cannot be satisfactory in winter and early spring for tillage purposes in this soil.

Key words: drainage, heavy clay soil, flow pattern, modelisation, field experiment, drainage design.

INTRODUCTION

Drainage of heavy clay soils is among the important challenges faced by agricultural water management in the world. Clay soils show adverse properties like (1) high chemical fertility but poor availibility of nutrients for crop roots, or (2) high water retention capacity but poor natural drainage resulting in waterlogging (Lesaffre et al, 1992). To make use of the potentialities of heavy clay soils, drainage has often to be complemented with other improvements like subsoiling, moling or conditioning. However the choice of the proper combination of these techniques remains a headache in many cases (Wesseling, 1988). This is likely due to a great variability of clay materials and to a poor knowledge of the water flow patterns, including depth of the impervious barrier and location and intensity of the water flow. Improvement of this knowledge is hindered by several difficulties:

- the depth to the barrier is not easily observable or measurable in clay soils;
- piezometers do not give accurate information in low permeable materials due to excessively long time responses and borehole smearing problems (Bouma, 1986);
- water flow mainly occurs in large pores or cracks in these soils (Leeds Harrison and Jarvis, 1987; Bouma, 1986).

Therefore drainage design in heavy clay soils is based on theoretical considerations and on experience. As stated by Smedema (1985), this experience is that drainage discharge predominantly takes the form of "shallow drainage"; the discharge to the drains occurs as lateral flow at shallow depth, above the impeding subsoil. However, this assumption already put forward by Flodkvist (in Russel, 1934), Trafford and Rycroft (1973) and many other authors has not been actually verified yet. Neither has the depth of the impeding subsoil clearly identified in most heavy clay soils.

Many clay soils facing seasonal waterlogging have been implemented with subsurface drainage in France during the past recent years. Several field

---

* D. Zimmer, PHD Soil Physics, Head Drainage Division, CEMAGREF, B.P. 121, 92185 ANTONY Cedex, France.

experiments have been monitored and an attempt was made to determine and classify the flow patterns in clay soils (Zimmer and Lesaffre, 1989). This paper presents complementary results regarding water flow and drainage mechanisms in one heavy clay soil. These results are discussed in terms of drainage flow modeling and of drainage design.

FIELD EXPERIMENT AND METHODS

Courcival experimental field is located in western France. It has been monitored since 1984. Its major characteristics are presented in Table 1.

Type of soil	Vertic Haplaquept
Clay content %: - plow layer - below	23-30 35-60
Sand content %: - plow layer - below	50-60 30-50
Dominant clay mineral	Nontronite
CEC of clay fraction (meq/100g)	52
Number of plots	4 (1 ha each)
Treatments and spacings	undrained mole (2-3 m) cum cum pipes (40 m) trencher installation (10 m) trenchless installation (10 m)
Measurements	Rainfall, drain flow rates water pressures
Monitoring time step	1 hour

Table 1. Major characteristics of Courcival field experiment.

The soil develops on a weathered glauconite. Its clay content is rather low and variable in the plow layer which has a sandy clay texture. Below the plow layer the clay content increases until 70-80 cm in depth and decreases underneath where glauconite is less weathered.

The climate is oceanic; the average total annual rainfall depth amounts to 750 mm (standard deviation 143 mm). Waterlogging mainly occurs during winter and early spring. The average water balance (P-ET) amounts to 257 mm during the October-March period.

To determine the water flow pattern, an intensive monitoring was carried out in the plot installed with trencher during 3-4 months of winter seasons 1987-1988 to 1990-1991. This monitoring included : (1) hourly rainfall measurements by use of a tipping bucket recorder; (2) hourly drain flow rates measurements by use of V-notch weirs and ultrasonic sensors; (3) hourly water pressure measurements by use of tensiometers, pressure transducer and scanning device; tensiometers were installed in different locations, inside the trench backfill and at 0.5 m, 1.5 m and 5 m from it; in each location, tensiometers were installed at five depths ranging between 0.1 m to the drain depth i.e. 0.85 m.

EXPERIMENTAL RESULTS

Most of the results presented herein were obtained during the 1989-1990 winter. That winter season was characterized by average rainfall (290 mm) from

November to March, most of the precipitations occurring in December, late January and February.

Hourly sequences of rainfall and drain flow rates (Figure 1) as well as double mass curves (Figure 2) show that two major rainfall events occured during that winter. The corresponding drainage efficiencies amount to 52 % (late January) and nearly 90 % (February). Drainage discharges suddenly ended up in late March.

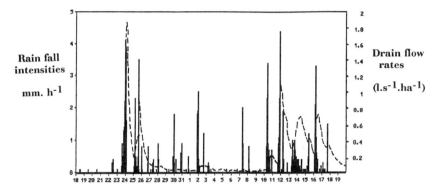

Figure 1. Hourly sequences of rainfall intensities (|▮|) and drainflow rates (---) from January 18th until February 19th 1990.

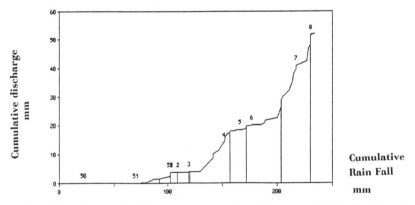

Figure 2. Cumulative discharges versus cumulative rainfall depths from December 11th until March 5th 1990. *Figures indicate the week number.*

The typical total water potential pattern observed in Courcival is presented in Figure 3. During the major part of the winter season, its characteristics are as follows:

- in the saturated zone (i.e. where total potential is greater than elevation), except for the trench, total water potential increases with depth although no arterial aquifer exist in the subsoil; moreover the total water potential values measured in depth can exceed the soil depth; inside the trench the total water potential decreases when depth increases;

- at the drain depth the total water potential does not regularly increase from trench toward drain mid-spacing; the drain pipe does

not significantly influence water potential beyond a distance of 0.50 m;
- in terms of water flow, although some kind of non conventional water table is present (since hydraulic head remains greater than elevation), tensiometer measurements seem to exclude any vertical downward flow below the plow layer except in the trench backfill;
- daily oscillations of the water pressures are observed below the plow layer; their amplitude increases with depth (Figure 4); these daily oscillations are not explained yet; they are related to temperature oscillatures and could be due to the measurement device.

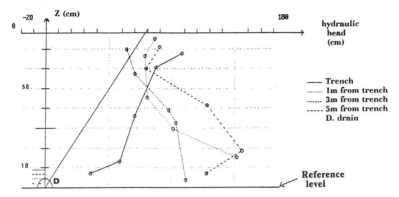

Figure 3. Typical water total potential head profiles observed in Courcival (February 19th 1988).

This surprising potential pattern may be the result of the swelling properties of the clay material. As demonstrated by Philip (1969) and Sposito (1972), the hydraulic head in a swelling material has three components: the first two are the classical gravity and pressure - or matric - components, the third one is an "overburden potential" related to the slope of the shrinkage curve of the material. Accordingly, the hydraulic head at a given depth could reach a theoretical value corresponding to the pressure exerted by the whole above located material (Zimmer and Lesaffre, 1989). This interpretation is not yet fully demonstrated.

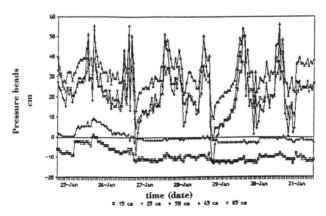

Figure 4. Sequences of water pressure heads at drain mid-spacing from January 25th until 31th 1990.

The water pressures in the plow layer (Figure 4) have a more classical behavior; in particular, their variations are affected by the climatic events and they show that a water table is often present in this more permeable layer. The relationship between water pressures in the plow layer and drain flow rates has therefore been examined (Figure 5).

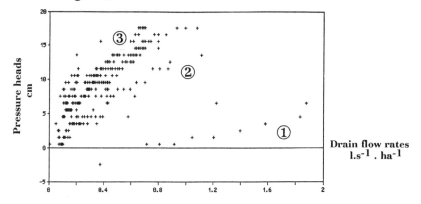

Figure 5. Water pressure heads at the bottom of the plow layer (depth 0.25 m, 3 m from trench) versus drain flow rates from January 24th until February 19th 1990. *(1) first drainage event, first peak flow; (2) first drainage event, second peak flow; (3) second drainage event.*

During most of the drainage season drain flow rates are clearly related to the water pressures inside the plow layer. However the different drainage events and peak flows are characterized by different relationships, thus indicating an evolving behavior of the clay material. In particular during the first peak flow of the first drainage event (1.8 $l.s^{-1}.ha^{-1}$, January 24 th) water pressures and drain flow rates are almost independant; only a slight increase of the water pressure is observed as the flow rate becomes greater than 1 $l.s^{-1}.ha^{-1}$. This behavior is observed in all tensiometer locations and depths. The second peak flow is also part of the first drainage event; it starts less than 24 hours after the first one. Water pressures in the plow layer become suddenly closely related to the drain flow rates, the relation being of a linear type. The second drainage event starts on February 11th and includes three peak flows. During this event the relationship is roughly parallel to that of the second peak flow: for a given drain flow rate the pressure head inside the plow layer has increased by 3 to 4 cm between the two events. The relationship is similar in all locations between trench and drain mid-spacing.

The proposed interpretation of these results is as follows:

- first peak flow: since the rainfall intensity is not higher than for the others, surface runoff can be excluded and preferential bypass flow likely occurs through still open cracks; this could also be the reason why the ratio peak flow rate versus rainfall intensity is relatively higher for this peak flow;

- following peak flows: no more downward bypass flow through cracks occurs; due to the swelling of the material, the cracks have closed up and the bottom of the plow layer becomes impervious;

- second drainage event: the increase in pressure is an indication that the properties of the plow layer - more sandy and very unstable - change, probably as a result of the presence of the water table; the change is either (i) a collapse of the bottom part of the plow layer resulting in an impervious barrier elevated a few centimeters above the previous one and/or (ii) an overall decrease of porosity.

In order to check which type of model is suited to this type of soil pressures were converted into hydraulic heads taking the bottom of the plow layer at the trench location as reference level. The relationship between hydraulic heads at drain mid-spacing and drainflow rates is quadratic - which is typical for shallow impervious barriers - and the hydraulic head corresponding to a nil discharge is zero or slightly above (Figure 6): the field slope is around 0.01 m/m and could explain the difference. This confirms that the bottom of the plow layer acts as impervious barrier.

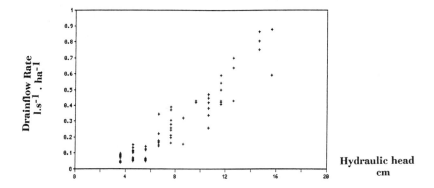

Figure 6. Drainflow rates versus hydraulic heads at drain mid-spacing during the second peak flow from January 25th until 28 1990. *The reference level is the bottom of the plow layer at trench location.*

The water table is rather flat (Figure 7) and the minimum water table elevation does not exactly correspond to the trench location: similar water table heights are measured at the trench and at a distance of 0.50 m from it. The interpretation is likely that water percolates towards the drain pipe through cracks located at the trench sides and not inside the trench backfill itself.

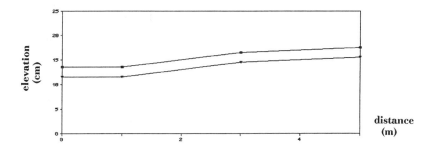

Figure 7. Water table shape during the first drainage event on January 25th 1990; *reference: bottom of plow layer; origin: drainage trench.*

DISCUSSION - CONCLUSIONS

<u>Drainage functioning</u>

In Courcival assuming lateral flow at shallow depth is valid at least during part of the drainage season, as soon as swelling has closed up cracks existing below the plow layer. A classical water table (i.e. where Dupuit-Forchheimer is valid) only forms in the plow layer the bottom boundary of which acts as impervious barrier. Under these conditions, the use of classical models based on Boussinesq's equation and approach (Glover in Dumm, 1954; Guyon, 1961) is allowed in this soil. Preliminary results using Guyon's tail recession equation lead to a value of the saturated hydraulic conductivity of about 2 m/d in the plow layer.

Methodology of water transfer monitoring

Despite possible preferential flows, tensiometers can give accurate and relevant information on water transfers in clay soils. Comparing tensiometer readings with drain flow rates was useful in detecting periods with preferential flow. However two important problems at least remain unsolved in this soil. (i) The meaning of pressures measured in the heavy clay material should be theoretically and practically investigated; is the overburden pressure related to the swelling behavior of the material responsible for the high values measured? What is the meaning of these high values in terms of water transfers? (ii) The daily oscillations of the high pressures are related to temperature oscillations; what is the actual mechanism involved and could they be prevented?

Even though, the experience gained in this experiment is that tensiometer are far more relevant than tubewells. Tubewells have also been tested in 1987-1988 but did not give any information: no relation at all between water table levels in tubewells and drain flow rates or rainfall has been observed.

Drainage design

Whether the use of classical equations and approaches is relevant or not for drain spacing calculation is questionable in this soil. In any case, as observed in the field, the time required to reach satisfactory trafficability is very long as long as the plow layer acts as an impervious barrier. Furthermore plant development is hampered during most of the winter period in this soil; winter cereals are especially affected: humid winters actually ended up with very low yields for winter wheat (less than 3 T/ha). On the contrary maize can give high yields: drainage generally allows a three weeks earlier sowing date and this crop makes good use of the important water storage capacity of this soil. The questions are therefore:

(i) How long does the plow layer act as an impervious barrier during the winter drainage period? This question will be solved with the use of a modeling approach: discharges and water table levels simulated with a model based on the assumption that the plow layer acts as impervious barrier should only fit the measured ones when this assumption is relevant.

(ii) What are the performances of other or complementary drainage techniques ? This question was at the onset of the field experiment. Results regarding crop and drainage efficiencies help classifying the efficiency of the different tested techniques. For hydraulic criteria mole drainage is the most efficient technique; next is trencher installation which gives better results than trenchless installation (Table 2, Auckenthaler and Zimmer, 1991). However the average crop yields are similar in the different drained plots and higher in those plots than in the undrained one. This can be explained by the fairly low impact of drainage on maize (3 crops out of 5 in the 1984-1988 period) in this soil and by the relatively dry climate of the five winter seasons of experimentation.

Conclusion

The existence of shallow drainage as assumed by Flodkvist (in Russel, 1934) and many other authors was verified in one heavy clay soil with high swelling capacity. As long as the impervious barrier remains close to the soil surface, classical subsurface drainage is not more efficient than surface drainage. Mole drainage proves to be more efficient at least from a hydraulic point of view. In any case the quicker removal of excess water allowed by artificial drainage is enough to significantly improve the situation as compared to the undrained situation.

Experimental plot	Mole cum drains	trencher without moles	trenchless without moles	undrained
Mean drainage efficiency %	54	36	23	-
Crop yield index %	102	102	100	88

Table 2. Mean drainage efficiencies and crop yields in Courcival field experiment for 1984-1988 period. Crop yield index 100 corresponds to the yield of the trenchless technique plot.

REFERENCES

1. Auckenthaler J., Zimmer D., 1991 - Comparaison de différents modes de drainage dans les argiles vertes sarthoises. Résultats hydrauliques et agronomiques de l'expérimentation de Courcival. Etudes du CEMAGREF, Hydraulique Agricole, 11,27-46.
2. Bouma J., 1986 - Characterization of flow processes during drainage in some Dutch heavy clay soils. Symposium on Agric. Water Management, Arnhem, The Netherlands, 18-21 June 1985. Balkema, 3-12.
3. Dumm L.D., 1954 - Drain spacing formula. Agric. Eng., 35, 726-730.
4. Guyon G., 1961 - Quelques considérations sur la théorie du drainage et premiers résultats expériementaux. B.T.G.R., 52, 44p. + annexes.
5. Leeds Harrison P.B., Jarvis N.J., 1987 - Soil and water management in drained clay soils. 3rd Int. Workshop on Land Drainage, Columbus, Ohio, USA, 7-11 December 1987. C43-52.
6. Lesaffre B., Favrot J.C., Penel M., Zimmer D., Arlot M.P., 1992 - Diagnosis and identification of problematic soils as related to drainage. Proc. 5th Int. Drainage Workshop, Lahore, Pakistan, February 8-15, 1,13-53.
7. Philip J.R., 1969 - Moisture equilibrium in the vertical of swelling soils. 1. Basic theory. Aust. Jour. Soil Res., 7,99-120.
8. Russel J.L., 1934 - Scientific research in soil drainage. Jour. Agric. Sci., 115(4), 315-320.
9. Smedema L.K., 1985 - Drainage coefficients for heavy land. Jour. of Irr. and Drain. Engng., 111(2),101-112.
10. Sposito G., 1972 - Volume changes in swelling clays. Soil Sci., 115(4),315-320.
11. Trafford B.D., Rycroft D.W., 1973 - Observations on the soil water regimes in a drained clay soil. Jour. of Soil Sci., 24(3), 380-391.
12. Wesseling J., 1988 - The development of drainage in humid temperate regions. Proc. Symp. 25th Internat. Course on Land Drainage, Wageningen, ILRI Public. 42, 14-22.
13. Zimmer D., Lesaffre B., 1989 - Subsurface drainage flow patterns and soil types. ASAE-CSAE Summer meeting. Paper 892139, 17 p.

DRAINAGE SUSTAINABILITY IN THE IMPERIAL VALLEY OF CALIFORNIA

Royal H. Brooks*
Member ASAE

ABSTRACT

The primary function of drains in the Imperial Valley is for salinity control. However, the application of extra irrigation water applied in the Valley for salinity control over a period of time has raised the water table (saturated zone) to profile locations that are near the soil surface that potentially may affect aeration and salinization of the plant root zone.

This paper briefly reviews drainage installations that have occurred in the Imperial Valley over a 50 year period in order to observe trends in design, patterns of system replacement, if any, and the replacement life of subsurface drains The history of drain installations on 200 tracts of land were examined for evidence of needing additional drainage by installing more drains or replacement of existing drains. The magnitude of area drained and the length of drains installed over a 50 year period are shown. The data were reduced to show that drain spacing was substantially reduced over the same period of time to produce crops of lower salt tolerance and increase the export of salt from the Valley. Some evidence of replacement life was obtained from the records examined in this paper by observing installation histories.
Keywords: Drainage, Salt balance, Drain spacing, Permeability, Drain sealing, Replacement life.

INTRODUCTION AND OVERVIEW

The management of water in the soil for purposes of agricultural production can be accomplished through proper methods of irrigation and drainage. Irrigation is a process whereby water is applied by some method so as to increase the water content of the soil. Drainage, on the other hand, is the process whereby water on the soil surface or within the soil profile decreases due to the earth's gravity. The term "drainage" in this paper refers to the decrease of water from the soil profile rather than a decrease in surface water. A decrease of soil-water from a soil profile may be accelerated by a subsurface drainage system consisting of a network subsurface pipes with openings for water entry. Subsurface drainage design has been the subject of intensive research over the past 60 years. This research base has substantially influenced and affected drainage design and the engineer's ability to predict the performance of drains.

The life of drainage systems are often debated in economic studies and for tax purposes. According to principles of engineering economy, when an engineering structure is replaced by another one which is to perform the same service, the new one may be economical for any one of several reasons. In general, there are two economic motives for replacement of structures. The first is increased operation and maintenance costs with age (which includes cost of interruption of service by temporary breakdowns and the hazard of catastrophe), which results from deterioration or wearing out due to both use and the wear-and-tear of the elements.

---

* Royal H. Brooks, PhD, PE., Professor Emeritus, Oregon State University, Corvallis, OR. 97331. Formerly Senior Researcher, Water Research Center, Ministry of Public Works and Water Resources, Cairo Egypt and Professor, Colorado State University, Ft. Collins, CO 80523.

Second, all other motives for replacement that may arise from circumstances which are not related to the wearing out of the structure under consideration. For example, the most common reasons for replacement fall into a class called obsolescence or inadequacy. That is, research or development may have provided improved alternatives or changes in service requirements may demand replacement and make the structure inadequate. When these theoretical principles of "economic life" are considered as they apply to a subsurface drainage structure, the motive for replacement clearly fall into the first category where there may arise increasing costs of operation and maintenance, not due to age, but rather due to elements within its environment.

Historical Perspective

From the beginning of irrigated agriculture in the Imperial Valley when water was first brought into the Valley (1901), the need for subsurface drainage was recognized. However, it took nearly twenty years before this need was translated into action. The Imperial Irrigation District (IID) constructed an open ditch drainage system in the early 1920's in an attempt to improve the salinity and high water table condition that had affected about 25% of the irrigated land at that time. In spite of this initial effort, salt and water continued to accumulate in the soils. It wasn't until 1930 that significant subsurface drainage was initiated in an attempt to control the rise of the water table which was responsible for increasing the quantities of salt in the soil profiles. However, the water table in the Valley continued to rise until about 1940. In 1943, the records showed a trend toward improved water table conditions over the Valley. From a progress report prepared by drainage investigators of USDA, Soil Conservation Service (Fox et al. 1944), the following statement was made with respect to this improvement: "The improvement...(concerning the water table) does not indicate that the water table problem has been solved. It only serves to emphasize that there is still a definite problem in spite of the tremendous amount of effort expended to remedy the situation. It also indicates that getting at the source of the problem by better use of water is one of the most significant factors in the overall drainage picture."

One may conclude from this brief historical overview that there appeared to be about a 40 year build-up of salts from irrigation water imported from the Colorado River before sufficient drainage was provided to bring about improvement in the salt and water balance. It should be pointed out, however, that the All-American Canal was not completed until 1942 and the associated delivery system shortly thereafter. Consequently, large quantities of water from the Colorado began to be imported into the Valley at that time. It was at this time that drainage investigations and research by governmental agencies began in earnest. By 1946, farmers and land owners began to be convinced that the only way to sustain long term agriculture in the valley was by installation of subsurface drains.

Past and Current Design Practices

From the early beginnings of drainage investigations in the Imperial Valley, the ellipse equation for steady flow toward equally spaced drains was used as a basis for drainage design. This equation had been proposed by Hooghoudt, Donnan and others. The equation is written so that spacing of drains is the dependent variable where the permeability or hydraulic conductivity, the drainage requirement and depth to the impermeable layer are the required inputs. This equation was developed for homogeneous - isotropic soils which do not often occur in nature and especially in the Imperial Valley.

The present practice in the Imperial Valley is to still use this equation as the basis of the design of subsurface drainage systems. Based upon experience with correlating permeability measurements with performance of drainage systems, the Soil Conservation Service (SCS) has developed a simplified version of this equation by making certain assumptions that seem to fit within their drainage experience within the Imperial Valley. Their equation is given by:

$$S^2 = 8.70 \, Pd^2$$

where P is the permeability in inches per hour and d is depth of the drain below the soil surface in feet. A nomograph for determining the spacing based upon anticipated depth of drains and an average permeability is used for design purposes. This equation is a simplified version of the ellipse equation and includes constants, conversion factors, expected control depth for the water table, quantity of irrigation water to be drained including the excess needed for leaching and depth to an impermeable layer. The only inputs required by the SCS for determining the design spacing is soil permeability and the depth the drains are to be installed below the soil surface.

Soil Permeability Measurements

Soil permeability is the key factor for determining the design spacing. This is a true statement regardless of the equation used for design. During the early period of drainage investigations and design in the Imperial Valley (1941-1951) in-situ methods for measuring soil permeability were not well advanced. Soil permeability measurements at that time largely consisted of 1) a well pumping test where discharge from a small well was measured along with the steady state draw down curve of the water table, and 2) laboratory tests on disturbed and undisturbed samples taken from the field site. Subsequently, drainage engineers and soil physicists began to question these types of permeability measurements particularly insomuch as data from such measurements were used in engineering design. During the late 1950's and 1960's, considerable research and development was made on field methods of measuring permeability. The auger hole (Kirkham, 1954) method of measuring soil permeability is presently accepted throughout the world as a standard. The SCS and other action agencies presently use this method with consistently good results.

When permeability data obtained in the Imperial Valley by Donnan et al. (1954) are averaged for a large number of samples, the result is 8.71 cm/hr or 3.43 inches per hour. If one assumes that this average permeability would serve as a basis for design, then the spacing obtained from the equation noted above for a drain depth of 5 feet is 265 feet. This spacing is close to the average design spacing used for the Valley at the time (1941-1951). It should be pointed out, however, that these measurements tended to be larger than the actual average permeability of the soil profile. The soil profiles for the Imperial Valley are stratified and the affects of these stratifications cannot be overlooked in design. Stratifications not only affect the average permeability, but the flow path that water takes as it travels to the drain. In general, these early permeability values were too large as evidenced by a trend to reduce spacing as discussed later, and tended to overestimate the drain spacing. Therefore, one might expect that experience would subsequently prevail. and drains would be installed at a reduced spacing.

Installation Practices and Problems

It is not the intention of the author to address in great detail the many installation techniques that have been perfected by the contractors of the Imperial Valley. However, there are some basic concepts followed in the Valley that are used in an attempt to insure adequate performance of the system for a long period of time. The basic concept for most contractors is to place perforated or jointed pipes below the soil surface so that when the water table rises above the pipes (drains) water will enter the perforations or joints freely. This is usually brought about by a combination of properly installed envelope materials and proper backfilling. If the system is installed properly, the water table directly over the drain will be at the top of the drain or slightly above and the water table between drains will tend to be parabolic in shape. Points on the water table midway between two parallel drains will be where the water table is the highest. If the water table does not remain near or at the top of the drainage pipe, it usually means that the permeability of the backfill or the envelope material is too low. This condition may be the result of improper installation or subsequent plugging that may be due to improper installation. In some soils, even though the installation procedure has been properly followed, chemical reactions (oxidation) may cause drain plugging to occur. The most common plugging problem that occurs due to improper installation is siltation. An improper installation may allow fine silt and sand to enter the drain such that is accumulates to a depth that blocks or restricts the flow.

The installation procedure followed by contractors in the Imperial Valley that prevents siltation that is the installation of gravel envelopes. The role of the envelope is to stabilize the soil around the drain so that soil particles will not move. The hydraulic gradient around the drain is usually very large resulting in high water velocities. These high entry water velocities may carry sand, silt and colloids into the drain and cause the drain to collapse or become disaligned, blocking or restricting the flow of water. Most drainage systems installed in the Imperial Valley are installed with pit run gravel as the envelope material. Frequently, pit run gravels are well graded and make an excellent envelope material. However, some contractors find it necessary to mix sand or gravel from two sources to obtain an envelope material having satisfactory size distribution in order to achieve the purpose of the envelope.

In order to ensure long life of installed subsurface drains, the Imperial Valley drainage contractors must also be concerned with proper design grade and its control during installation. Accurate grade control is usually achieved during installation through the use of laser technology. If accurate control cannot be achieved during installation the entire systems may be at risk. For example, a reversal in grade or the grade falling below the design slope may cause siltation which will block or restrict the flow. Recent advances in laser technology has provided the drainage contractor with a tool that insures proper and accurate grade control. High speed trenchers and plows are used in the Imperial Valley and they are usually equipped with laser grade control equipment. These high speed machines require accurate automatic controls. Through a plane of laser light, variation in installation depths of $\pm$ 0.02 feet can be maintained automatically. Such methods have reduced the human eye error in previously used methods and allow for a well designed system that surpasses previously used methods in accuracy.

Maintenance Problems and Practices in the Imperial Valley

Because of soil variability and in spite of careful design and installation, material may move into the subsurface facility or chemical precipitates may occur; all of which may cause a malfunction that will require a regular maintenance program to be developed in order to protect the farmers' investment of the drainage system. The drain openings, joints between clay or concrete pipe sections or holes in plastic pipe, must remain open and clean for the drains to function properly. Some causes of drain malfunction may be due to the following: 1) Mineral deposits or precipitate, 2) Silt, 3) Plant roots.

Each of these may cause sealing of the joints or openings and/or clogging of the pipes. The two terms, sealing and clogging, are defined by Grass et al.. (1975) as: "Sealing is the closing of the drain opening through which water enters and clogging is the blockage of the interior of the drain causing water to back up in the drain or preventing water upstream of the blockage from flowing to the outlet." The result of either sealing or clogging is rise of the water table over the drain in the area upstream of the blockage or sealed section, such that salt begins to accumulate on the soil surface and wet spots begin to appear. These are the normal surface signs of a drain malfunction. In the Imperial Valley, all of the above mentioned causes for drain malfunctioning occur. These causes have been extensively studied in the Valley by USDA - ARS scientists to find improved design and installation methods and cleaning techniques for restoring sealed or clogged system.

Extent of Sealing and Clogging Problems

Grass (1969) reported the results of a survey on tile clogging that appeared to be a major problem in the Valley at that time.. The relative area of the survey, soils, types of drain pipes, age of drainage system, and number of sites were given. The problem area was not restricted to a particular given area but it was widely distributed throughout the Valley. Manganese and iron deposits were the predominate cause of clogging and sealing found in the survey. Grass (1960) stated "the largest quantity of precipitate was found in laterals near their junction with the main line. In severe cases, the laterals were almost filled. The thickness of the deposits gradually decreased with distance away from the base (main) line. The upper portions of the laterals were usually relatively clear of precipitated materials, except in severely affected systems where the entire length of the laterals was affected." Grass et al. (1975) indicated that crystalline gypsum deposits have been found in some drains. These deposits occur mostly as small crystalline beds. Occasionally, large hard crystals may occur. On the other hand, iron deposits most often occur as sludge or soft paste and seldom cause drain clogging even though it is found is clogged drains.

In another study by Grass et al. (1979) they reported that from 100 drain systems inspected, they found 22 drains had silt deposits. They concluded that partial clogging of the drain by silt would tend to increase the likelihood of sealing by manganese of iron deposits. They further concluded that silting problems were mostly confined to clay and bituminous pipes. Bituminous pipes covered with a fiberglass mat are not currently used for agricultural drains in the Imperial Valley.

Cleaning Practices in the Imperial Valley

Cleaning and repairing drains in the Imperial Valley is largely accomplished by six private sector contractors. The contractor often has a fixed clientele but, of course, services all requests for cleaning. The contractor usually maintains a record of cleaning completed for each client as a service to assist in maintenance of the drain system. One contractor (Clayton) estimated that under normal economic conditions, he may clean two to three million feet of drains per year. The average cost for cleaning in 1982 was about 16 cents per foot. Most of the cleaning is accomplished during the summer months during fallow and field crops.

Cleaning of drains in the Valley generally started in 1965 and is fully handled by the private sector. Since 28,507 miles of drains have been installed and assuming that each of the six contractors in the Valley clean two million feet of drain per year and that 90% of the drains are still functioning, then as an average, all drain lines would be cleaned every 11.3 years. Obviously, this is an over simplification of the frequency of cleaning accomplished in the Valley, but it indicates the magnitude of the problem and gives some indication of what one might expect over a long period of time.

IMPERIAL IRRIGATION DISTRICT (IID) TILE DRAINAGE RECORDS

In an attempt to determine the practices that have been or are being developed by the farming industry in the Imperial Valley concerning new installations, additions, and replacements, 200 tile drainage records were selected for analysis over a period of time from 1944 to 1972. The Imperial Irrigation District (IID) began maintaining records on tile drainage systems in 1929. However, it wasn't until 1944 that a numbering system was established for the records. This system, established at that time and used presently, is one in which the drainage plan and design for a tract of land is sequentially assigned a number without regard to installation date or location. The tile drainage (TD) numbers are assigned to the parcel of land being served by the contractor at the time of the initial installation. Subsequent design and installations use the same initial TD number. These TD records are submitted to the IID on a voluntary basis by land owners, tenants, and/or contractors. By the year 1981, 3600 TD numbers had been sequentially assigned since 1944.

In an effort to obtain information concerning replacement activities of drains installed on particular tracts of land, only the first 3000 TD records were considered in this analysis. These records covered the period from 1944 to 1972 (approximately). Because of the large volume of data to analyze, only 200 records were randomly selected and analyzed. It should be noted that by taking a sample of the first 3000 TD records they included installations that occurred as late as 1981 although they are few in number.

Analysis of TD Records

The basic information contained in a typical TD record are: 1) dates of drain installations, 2) location and layout of drains, 3) length and size of each drain line, and 4) type of drainage pipe. The TD records were summarized by tabulating dates of installations, footage of main lines, footage of laterals, and comments pertaining to the system that required a subjective determination or evaluation. The following subjective comments were used in the analysis:

    a. Additional lines installed by splitting the spacing of previously installed drains
    b. Additional area drained by adding additional main line
    c. Double main line installed parallel to old main
    d. Main line abandoned
    e. Extension of existing lines

A total of 200 TD records were analyzed from the 300 TD records randomly selected. The analysis focused primarily upon comment "a" and "d". The conclusions obtained from this analysis are noted below as concerning (1) the rate of installation of drains and changes in design since 1930 and (2) the useful life of subsurface drains in the Imperial Valley.

Historical Installation of Drains and Area Drained

From the analysis mentioned above, about 75% of the drains installed at the request of the farmer after the first installation, split the spacing of the previous installation. A review of the IID records of total length of drains installed and the total area drained as a function of time confirmed that additional requests for drainage were requests to split formerly installed drains spacings.

The IID supplied a summary of the total footage of drains installed since 1929. Included on these records are the annual acreage tiled since 1940. An analysis of these data give some additional insights into the dynamic drainage industry controlled largely by farmers.

The accumulative miles of drains installed in the Imperial Valley since 1930 shows a precipitous rise in the annual footage of drains installed in 1946. This occurred near the end of an era of drainage investigation when the benefits of drainage were becoming well known. This coincided with an upturn in the national economy as indicated by increased sales and profits. The accumulative miles of drains installed continued to accelerate until about 1976. A similar increase in acreage drained also occurred in 1946. However, the rate of the drainage area in the Valley began to decrease in about 1952. The total accumulative miles of drain installed in 1982 was 28,507 miles while the total acreage drained reached 429,537 acres.

The annual rates of change in miles of drains installed and acres drained is shown in Figure 1 where annual miles of drains installed and annual acres drained is plotted for year from 1930 to 1982. The changes in the rate of activity, with respect to quantities of drains installed and rate of acreage drained in the Valley, is more clear when shown as depicted in Figure 1. The rate at which the Valley decreased in acres drained occurred in 1952, while the rate of drain installations did not decrease until 1976. These data indicate that the increased installations were occurring on lands already having some drainage, i.e., these drains are largely additions to the already existing system. When the data shown in Figure 1 are translated into equivalent or average drain spacing for the valley, the results are rather dramatic as shown in Figure 2. These data clearly indicate that most of the new footage installed in the Valley after 1952 were either additional drains that were spaced more closely together than those previously installed in drained areas; or in cases where new acreage was added, more drains were installed per acre in each succeeding year. This is also verified by the analysis of the TD records where 75% of the additional installation were split spacing installations. This clearly indicates that the industry needed or wanted more control over salinity and the water table than previously given to them by earlier designed spacings.

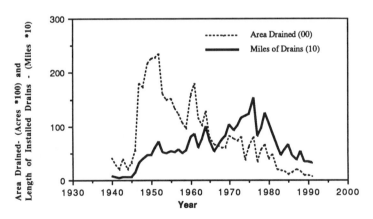

Fig. 1 Annual Rates of Drains Installed and Area Drained from 1940 to Present.

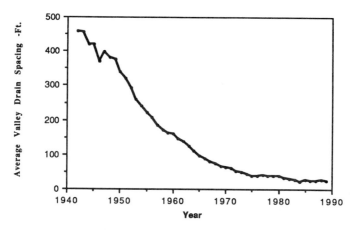

Fig. 2 Average Annual Computed Drain Spacing for the Imperial Valley 1940 to Present.

In a recent publication by Evan et al (1982), salinity increases in the Colorado River at the Imperial Dam were projected by the U.S. Bureau of Reclamation. The forecast calls for an increase to more than 1300 ppm by the year 2000 if no corrective actions occur in the Upper Basin states. In 1952, the salt load of the Colorado River at Imperial Dam was about 879 ppm an 18% increase over the 1952 value. The 1980 annual average was somewhat lower (812). Obviously an increase in the importation of salt into the Valley requires an increase in the leaching fraction of the irrigation water that must pass to the drains. This in turn necessitates increased drainage that must occur at closer spacings if the water table is to remain at a reasonable depth below the soil surface. Data published by IID on salt balance show that over a 25 year period (1965 -1990) more salt has been exported than imported. Due to corrective action in the Upper Basin states since 1982, there has been a general downward trend of salt concentration in the imported irrigation water. Recent data (1990) show that the concentration of imported water was 727 ppm. This trend toward lower concentration of irrigation water and the the trend toward a decrease in drain spacing is essential to sustain agricultural productivity during the next century.

Replacement History of Drains in the Imperial Valley

An analysis of the TD records produced no evidence to support the fact that a replacement practice exists within the Valley. When a replacement occurs it is entirely related to the specific situation or environment in which the drains are placed. A failure may occur over a long period of time due to gradual permanent sealing of the joints. Clogging problems can usually be corrected.

Concerning comment "d" in the TD analysis, it is apparent from the study there is no one single factor or condition which can be used to predict abandonment or the useful life of a drain. There is a lack of data which correlates the effectiveness of the drainage system performance with age. Therefore, a statistical approach was used in order to estimate a drain's useful life.

Given a fixed number (sample) of drainage systems obtained from the total number of drainage systems in the Valley, one would expect the laws of probability to govern the frequency of replacement. That is, one would expect a small number of replacements to occur at a young age and a small number at an old age with the largest number occurring at some median age. The frequency distribution of the actual replacement ages of the total number of drainage systems for the Valley may follow one of several distribution laws. The most common distribution law is known as the standard normal distribution. If it is assumed that the distribution of replacement age in the Valley follows a standard normal distribution, one may obtain an estimate of the drainage system's useful life in the Imperial Valley. Using all the records concerning comment "d" and the standard normal distribution function, the median replacement age is found to be about 33 years. This is very liberal estimate of the useful life of drainage systems for the Valley based upon the above analysis and the assumptions.

A more conservative estimate is to calculate the main line footage replaced as a percentage of the total main line installed on previous occasions. When this percentage is used to reduce the number of previous installation, a more conservative estimate of the useful life, based upon these computations and the above assumptions is 38 years.

REFERENCES AND BIBLIOGRAPHY

1. Donnan, William W Model Tests on a Tile-Spacing Formula Soil Science Society America Proceedings Vol 11, 1946.

2. Donnan, William W. and George B. Bradshaw Drainage Investigations Methods for Irrigated Areas in Western United States ,Technical Bulletin No. 1065, USDA, SCS, Washington DC,September 1952.

3. Donnan, William W..; George B. Bradshaw; and Harry F. Blaney. Drainage Investigations in Imperial Valley, California 1941-51 (A 10 Year Summary) USDA-SCS, SCS-TP-120, Washington 25 DC.September 1954.

4. Evans, Robert G; Wynn R. Walker and Gaylord V. Skogerboe. Defining Cost-Effective Salinity Control Program Journal of Irrigation and Drainage Division, Proceedings of the American Society of Civil Engineers, ASCE Vol 108, No. IR 4, December 1982.

5. Grass, L.B.; A.J. MacKenzie; B.D. Meek; and W.F. Spencer.Manganese and Iron Solubility Changes as a Factor in Tile Drain Clogging: I Observations During Flooding and Drying. Soil Science Society of America Proceedings, Vol 37, No. 1, January-February 1973.

6. Grass, L.B.; A.J. MacKenzie; and L.S. Willardson Inspecting and Cleaning Subsurface Drain Systems , USDA-ARS Farmers Bulletin No. 2258 Washington DC, July 1975.

7. Grass, L.B. and L.S. Willardson High Pressure Water Jet Cleaning of Subsurface Drains Transactions of the American Society of Civil Engineers, Vol 17, No. 5, 1974.

8. Grass, L.B.; L.S. Willardson and R.A. Le Mert Soil Sediment Deposits in Subsurface Drains. Transactions of the American Society of Agricultural Engineers (ASAE) Vol 22, No. 5, 1979.

9. Grass, Luther B Tile Clogging by Iron and Manganese in Imperial Valley, California Journal of Soil and Water Conservation, Vol 24, No. 4, July-August 1969.

10. Grass, Luther B. and A.J. MacKenzie Reclamation of Tile Drains by sulphur Dioxide Treatment The sulphur Institute Journal, Spring, 1970.

11. Imperial County Board of Supervisors Imperial County Agriculture, 1980 Office of the Agricultural Commissioner.

12. Imperial Irrigation District The Soils of Imperial Valley, Community and Special Services, El Centro, California 92244 Fifth Printing 1980.

13. Jensen, M.E. (Editor) Design and Operation of Farm Irrigation Systems. ASAE Monograph No. 3, St. Joseph, Michigan. December 1980.

14. Kirkham, Don Measurement of the Hydraulic Conductivity in Place Symposium on Permeability of Soil American Society for Testing Materials Special Technical Publication 163:80-97.1954.

15. Perrier, E.R.; A.J. MacKenzie; L.B. Grass; and H.H. Shull. Performance of a Tile Drainage System: An Evaluation of a Tile Design and Management Transactions of the ASAE, Vol 5, No. 3, 1972.

16. Sutton, John G Installation of Drain Tile for Subsurface Drainage Journal of the Irrigation and Drainage Division of ASCE Vol 86, No. IR 3, September 1960.

CORN YIELDS, SOIL PROPERTIES, AND DRAINAGE EFFECTS ON CROSBY-KOKOMO SOIL

R. Lal   N. R. Fausey
Member ASAE

ABSTRACT

Corn yields and soil properties were compared at different distances from a subsurface drain on a Crosby-Kokomo soil association near Columbus, Ohio. Subsurface drainage was by means of a 100 mm diameter drainpipe installed at about 90 cm depth. Corn stand and grain yield were monitored for a 5 year period from 1983 through 1987. Combined analysis of the corn grain yield for the 5 year period indicated that the relative grain yield was 100, 98, 97 and 91 for 0, 9, 18 and 27 meter distances from the drain, respectively. Both stand and grain yield were significantly lower at 27 m from the drain than at the drain. The mean daily maximum soil temperature, measured in April, was significantly higher at a distance 27 m away from the drain than in the vicinity of the drainline. Infiltration capacity was not affected by distance from the tile line. Distance from the drain had a significant effect on the soil organic matter content, the soil reaction, and the Bray-1 phosphours concentration. Each of these soil properties was higher at 27 m from the drain compared to near the drain. At 27 m from the drain bulk density was lower, MWD was higher and moisture retention at all suctions was greater then near the drain.

KEYWORDS. Subsurface drainage, Soil temperature, Infiltration capacity, Soil organic matter, pH, Soil structure

INTRODUCTION

In Ohio, over 60% of the cropland is prone to excessive wetness limitation. Excessive wetness and status of drainage can affect soil properties and crop growth and yield. Drainage effects on the magnitude of crop response depend on several factors including species and variety, fertility management, evaporative demand, and antecedent soil properties (Williamson and Kriz, 1970). Field experiments in the U.S. Cornbelt have shown that corn is a particularly sensitive crop even to short periods of inundation and anaerobiosis (Lal and Taylor, 1969; Schwab et al., 1985; Fausey, 1983, 1984; Sipp et al., 1985; Kanwar et al., 1988). Corn yield can also be influenced by tillage methods (Unger, 1984; Sprague and Triplett, 1986; Lal, 1989).

Although the effects of tillage methods on soil properties are well recognized, the effects of drainage on soil physical and chemical properties are not as extensively studied as on crop yields (Hundal, et al, 1976). Steenhuis and Walter (1986) questioned whether drainage increases springtime soil temperatures in cool and humid climates. In contrast, Lieffers and Rothwell (1987) reported higher soil temperatures on a drained rather than on undrained site.

R. Lal, Professor, Agronomy Department, The Ohio State University, Columbus, OH and N. R. Fausey, Soil Scientist, Soil Drainage Research Unit, USDA-ARS, Columbus, OH.

Effects of drainage conditions on soil chemical properties have also not been extensively studied. Calbourn and Harper (1987) reported that drainage limited denitrification losses to about 65% of that from undrained soils. It is partly because of the excessive losses of N that plants grown in poorly drained soils exhibit chlorotic symptoms of nitrogen deficiency (Lal and Taylor, 1969; Cannell, 1979). It may be inferred that because of the lower rate of mineralization all other factors remaining the same, there may be more soil organic matter in poorly drained soils than in well drained soils. Consequently, mineralizable N is generally higher in well drained than poorly drained soils (Rice et al., 1987). For the same reason, the effective CEC may be greater in poorly drained than in well drained soils.

The objective of this experiment was to evaluate the effects of drainage on corn growth and yield and on soil properties for a range of tillage methods. Tillage methods were chosen to create a wide range of conditions of surface drainage, but the effects of tillage are not discussed here (see Fausey and Lal, 1989a, b, and 1992).

MATERIALS AND METHODS

The experimental site is near Columbus in central Ohio (longitude 40°N, latitude 83°W). Soils at the experimental site belong to the Crosby-Kokomo association and are classified as fine, mixed, mesic, Aeric Ochraqualf and fine, mixed, mesic, Typic Argiaquoll, respectively. Crosby silt loam is a nearly level, deep somewhat poorly drained soil. Typically, the surface layer is dark grayish brown, friable silt loam about 23 cm thick. Kokomo silty clay loam is a nearly level, deep, very poorly drained soil. Typically, the surface layer is very dark gray, friable silty clay loam about 23 cm thick.

The subsurface drainage on this site was installed in the late 1950's and consisted of a 100 mm diameter drainpipe installed at about 90 cm depth. Plots were laid out in 1982 at different distances from the subsurface drain (Fig. 1) to create variable conditions of subsurface drainage.

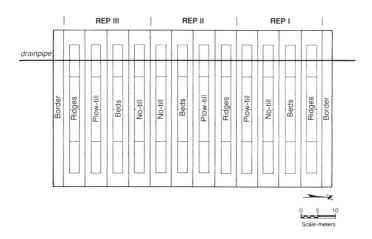

Figure 1: Experimental plot layout illustrating drainage treatment locations.

Depth to the water table was measured during the late winter and early spring in 1983-1985 using small diameter (15mm) plastic wells perforated along their depth. The wells were installed each fall and were read using a blowtube three times each week until the wells were removed for cultivation and planting operations. The wells were not reinstalled for growing season measurments because the water table is mostly seasonal. The wells were located within 1 m of the subsurface drain and at 27 m from the subsurface drain.

Soil temperature measurements using thermistors and an automatic recorder, were made at 5 cm depth in the center of each tillage x drainage treatment plot in replicate 1 once each hour throughout April 1984 and from April 25 through May 8, 1985. The mean daily maximum and mean daily minimum temperatures were calculated from the data and compared according to distance from the drain.

Infiltration measurements were made using the double-ring infiltrometer during July-August, 1987. The inner and outer rings measured 20 cm and 30 cm in diameter, respectively. Both rings were 30 cm deep and were installed 10 cm below and 20 cm above the ground surface. A constant head of 5 cm of water was maintained in the inner ring while free water level was kept in the outer ring. All infiltration tests were run for at least 2 hours. Two measurements were made for each plot -- one on the row and one between rows.

In August, 1987, soil samples were obtained to 50 cm depth using a truck-mounted soil coring tool and cut into 10 cm depth increments. These samples were analyzed by the Research Extension Analytical Laboratory (REAL), Wooster, Ohio. The samples were air dried and ground to pass a 2 mm sieve. Exchangeable Ca, Mg and K were determined using an atomic absorption spectrophotometer. Organic carbon was determined by the method of wet combustion (Allison, 1965). Organic matter content was calculated by multiplying the organic C value by 1.724. Available P was measured by the Bray-1 method (Bray and Kurtz, 1945), and soil pH was measured in 1:1 soil:water suspension.

Soil bulk density was measured on cores 7.5 cm in diameter, 7.5 cm deep and about 300 cm^3 in volume (Blake and Hartge, 1986). These samples were taken with a manually operated tool from each 10 cm depth increment to 50 cm depth near the drain (0 m) and at 27 m from the drain. The same cores were used for assessment of soil moisture retention characteristics. A full range of moisture retention characteristics were measured using a combination of tension table and pressure plate extractors (Klute, 1986). These undisturbed core samples were also used for measuring the saturated hydraulic conductivity.

A bulk sample of soil was taken from each depth and used for aggregate size distribution, measured by the wet sieving technique (Yoder, 1936). Results of wet sieving were expressed in terms of the mean weight diameter (MWD) as described by Youker and McGuiness (1956).

Grain yield was recorded at different average distances from the drain (0, 9, 18 and 27 m). The harvested area was 3 m wide by 9 m in length. Grain yields were expressed at 15.5% moisture content.

Each of the distances from the drain was treated as a separate experiment. Statistical analyses of different parameters were, therefore, done in view of the constraint that the drainage treatments were neither replicated nor randomly assigned.

RESULTS AND DISCUSSION

Water table

The average depth to the water table on each sampling date between 21 March and 15 May in 1983-85 is shown in Fig 2 for near the drain (drained) and 27 m from the drain (undrained). The water table fluctuates in response to precipitation in both areas but is nearer the soil surface much less frequently and for shorter durations near the drain. While we did not measure the redox potential or the oxygen status in the soil, it is clear that an elevated water table for a long duration could have an impact on many aspects of the chemical properties of the soil including the organic matter content.

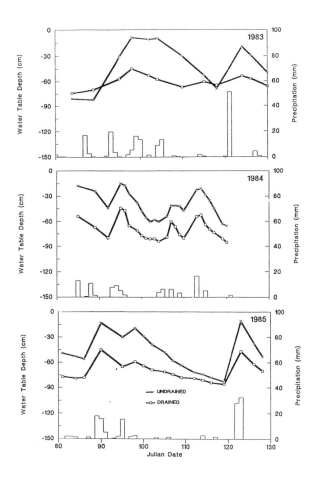

Figure 2: Water table position near the drain (drained) and at 27 m from the drain (undrained) during Spring in 1983, 1984, and 1985. Each data point is an average of six observation wells.

Corn grain yield

Combined analysis of the corn grain yield for the 5 year period from 1983 to 1987 is shown in Table 1. Over the 5 year period, grain yield was significantly affected by distance from the drain. Relative corn grain yield was 100, 98, 97 and 91 for 0, 9, 18 and 27 m distance from the drain, respectively. Crop stand over the 5 year period was also significantly affected by drainage (Table 1). The mean crop stand was significantly lower for 27 m from the drain compared with the other three distances (48,900 versus 54,100).

Soil temperature regime

Soil temperature in early spring was significantly affected by distance from the drain. The mean daily maximum soil temperature generally increased with an increasing distance from the drain (Table 1). The mean daily minimum soil temperature was not affected by distance from the drain. Fig. 3 shows a comparison of hourly temperatures on April 26, 1984 at 5 cm depth above and at 27 m from the drain.

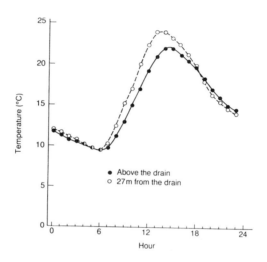

Figure 3: Average hourly soil temperature at 5 cm depth near and at 27 m from the subsurface drain on April 26, 1984. Each data point is an average of three measurments.

Infiltration capacity

Distance from the drain had no effect on infiltration rate.

Soil physical properties

Distance from the drain had no effect on soil bulk density for 0-10 cm and 10-20 cm depths. The overall mean bulk density was 1.44 and 1.43 Mg m^{-3} near and at 27 m from the drain respectively. Furthermore, bulk density was slightly more for 10-20 cm compared with 0-10 cm depth.

Table 1. Effect of distance from the drain on selected crop and soil parameters.

Tillage Method	Distance from drain (m)				t-Test
	0	9	18	27	
Corn yield (t/ha)	6.20	6.06	6.00	5.66	*
Cornstand (1000 plants/ha)	52.8	54.7	54.7	48.9	*
Mean daily max soil temp at 5cm (°C)	13.4	13.6	13.7	14.4	*
Mean daily min soil temp at 5cm (°C)	6.0	5.7	5.7	5.6	NS
Infiltration rate at 120 min (cm/hr)	6.8			6.4	NS
Bulk Density 0-20 cm (Mg/m^3)	1.44			1.43	NS
Mean MWD of aggregates 0-50 cm (mm)	2.49			2.94	+
Water retention at 0 MPa suction (%)					
0-10 cm depth	28.2			38.2	**
0-50 cm depth average	35.6			38.0	NS
Water retention at 0.01 MPa suction (%)					
0-10 cm depth	23.0			31.6	**
0-50 cm depth average	--			--	
Water retention at 0.033 MPa suction (%)					
0-10 cm depth	20.2			29.4	**
0-50 cm depth average	23.4			24.7	NS
Water retention at 1.5 MPa suction (%)					
0-10 cm depth	17.9			27.4	**
0-50 cm depth average	18.1			20.2	*
Organic matter content (%)					
0-10 cm depth	2.7			4.0	*
10-20 cm depth	2.4			3.7	*
0-50 cm depth average	1.9			2.8	**
pH					
0-10 cm depth	6.0			6.4	NS
10-20 cm depth	5.1			5.9	**
20-30 cm depth	5.5			6.3	**
0-50 cm depth average	5.9			6.4	**
Bray-1P (kg/ha)					
0-10 cm depth	91.9			166.6	**
10-20 cm depth	70.0			121.6	*
0-50 cm depth average	41.1			67.7	*

+ Significant at 10% level of probability
* Significant at 5% level of probability
** Significant at 1% level of probability
NS Not significant at 10% level of probability

Structural aggregates were larger and more stable at 27 m from the drain compared with those from near the drain (Table 1). Similar trends existed for aggregates from different depths. The overall MWD of aggregates 27 m from the drain was significantly more (at 10% level of probability) than near the drain (2.94 mm vs 2.49 mm). The highest MWD was observed for the 10-20 cm depth.

At zero suction and for average of all depths, there were no significant differences in moisture retention due to distance from the drain. The overall mean moisture content at zero suction was 35.6% and 38.0% near and at 27 m from the drain, respectively. However, differences in moisture retention were significant for the 0-10 cm depth, with significantly more moisture retained by samples at 27 m from the drain versus near the drain (38.2 vs 28.2%).

Distance from the drain had a significant effect on moisture retention at 0.01 MPa for the 0-10 cm depth. The moisture content for 27 m from the drain was significantly higher than near the drain (31.6 vs 23.0%).

Treatment effects on moisture retention at 0.03 MPa and 1.5 MPa suctions were generally similar (Table 1) with a significant difference due to distance from the drain only at the 0-10 cm sampling depth. Moisture retention at 0-10 cm was 29.4% for 27 m from the drain compared with 20.2% near the drain for 0.033 MPa suction, and 27.4% for 27 m from the drain compared with 17.9% for near the drain for 1.50 MPa suction. The overall mean moisture retention was not affected by distance from the drain at 0.033 MPa suction, but was significantly affected at 1.5 MPa suction. Sampling depth had significant effect on moisture retention for 0.03 MPa suction. At 1.5 MPa suction, however, differences in moisture retention due to sampling depth were significant only at 27 m from the drain.

Soil chemical properties

Organic matter content decreased with depth and this relationship was identical at both distances from the drain. The mean organic matter content over all depths at 27 m from the drain was 2.8 percent. This was significantly greater than the 1.9 percent mean organic matter content near the drain. This seems consistent with the understanding that organic matter is not oxidized as quickly in poorly drained soils as in well drained soils.

The pH in the surface 10 cm of the soil was not affected by distance from the drain, but in the 20-40 cm depth the pH was much lower near the drain than at 27 m from the drain. Overall the pH was 0.5 units lower near the drain. This result is similar to the organic matter content and suggests that good drainage encourages organic matter decomposition which creates acidity of the soil.

Overall Bray-1 phosphorus level was higher in the undrained area than in the drained area. Again, this is consistent with the soil organic matter content differences and indicates that poor drainage favors phosphorus availability.

CONCLUSIONS

Results of this 5 year study indicate that average increase in corn grain yield due to subsurface drainage was about 8% or about 0.5 t ha^{-1}.

Nearness to a subsurface drain did not increase the spring time soil temperature. Mean daily maximum soil temperature in April was highest at 27 m from the subsurface drain.

Infiltration rate was not significantly affected by the nearness to a subsurface drain.

Distance from the drain had a significant effect on the soil organic matter content, the soil reaction, and the Bray-1 phosphorus concentration in the soil. Each of these parameters was higher at 27 m from the drain compared to near the drain. The long term effect of drainage is, then, to reduce the soil organic matter content, lower the pH and make P less available in the Crosby-Kokomo soil association.

The plots at 27 m from the drain had more favorable bulk density (generally lower) and structural characteristis (higher MWD) than the plots near the drain. Moisture retention at all suctions was also generally greater at 27 m from the drain. Favorable soil structure, somewhat higher porosity, and relatively more proportion of macropores are probably due to the higher level of soil organic matter content at 27 m from the drain.

REFERENCES

1. Allison, L. E. 1965. Organic carbon. IN C. A. Black et al., (eds), Methods for Soil Analysis Part 2. ASA Monograph 9:1367-1378.

2. Blake, G. R. and Hartge, K. H. 1986. Bulk density. IN Methods of Soil Analysis, Part I: Physical and Mineralogical Methods, ASA Monograph 9, Madison, WI:363-376.

3. Bray, R. H. and Kurtz, L. T. 1945. Determination of total organic and available forms of phosphorus in soils. Soil Sci. 59:39-41.

4. Cannell, R. Q. 1979. Effects of soil drainage on root growth and crop production. IN R. Lal and D. J. Greenland (eds). Soil Physical Properties and Crop Production in the Tropics. J. Wiley & Sons, Chichester, U.K.:183-197.

5. Colbourn, P. and Harper, I. W. 1987. Denitrification in drained and undrained arable clay soil. J. Soil Sci. 38:531-539.

6. Fausey, N. R. 1983. Shallow sub-surface drain performance in Clermont soil. Trans. of the ASAE 26:782-784.

7. Fausey, N. R. 1984. Drainage-tillage interaction on Clermont soil. Trans. of the ASAE 27:403-406.

8. Fausey, N. R. and Lal, R. 1989a. Drainage-Tillage Effects on Crosby-Kokomo Soil Association in Ohio. I. Effects on Stand and Corn Grain Yield. Soil Technology 2: 359-370.

9. Fausey, N. R. and Lal, R. 1989b. Drainage-tillage effects on Crosby-kokamo soil association in Ohio. II. Soil temperature regime and infiltrability. Soil Technology 2: 371-383.

10. Fausey, N. R. and Lal, R. 1992. Drainage-tillage effects on a Crosby-kokomo soil association in Ohio. III. Organic matter content and chemical properties. Soil Technology 5: 1-12.

11. Hundal, S. S., Schwab, G. O. and Taylor, G. S. 1976. Drainage system effects on physical properties of lakebed clay soil. SSSAJ. 40:300-305.

12. Kanwar, R. S., Baker, J. L. and Mukhtar, S. 1988. Excessive soil-water effects at various stages of development on the growth and yield of corn. Trans. of the ASAE 31:133-141.

13. Klute, A. 1986 Water retention: laboratory methods. IN Methods of Soil Analysis: Part I. Physical and Mineralogical Methods, ASA Monograph 9, Madison, WI:635-662.

14. Lal, R. 1989 Conservation tillage for sustainable agriculture: tropical versus temperate environment. Adv. Agron. 42:85-197.

15. Lal, R. and Taylor, G. S. 1969 Drainage and nutrient effects in a field lysimeter study. I. Corn yield and soil conditions. Proc. Soil Sci. Soc. Am. 33:937-941.

16. Liefers, V. J. and Rothwell, R. L. 1987 Effects of drainage on substrate temperature and phonology of some trees and shrubs in a Alberta peatland. Can J of Forest Res. 17:97-104.

17. Rice, C. W., Grove, J. H. and Smith, M. S. 1987. Estimating soil net nitrogen mineralization as affected by tillage and soil drainage due to topographic position. Canad. J. Soil Sci. 67(3):513-520.

18. Schwab, G. O., Fausey, N. R. and Weaver, C. R. 1985. Tile and surface drainage of clay soils. OARDC Res. Bulletin 1161. Wooster, OH.

19. Sipp, S. K., Lembke, W. D., Boast, C. W., Peverly, J. H., Thorne, M. D. and Walker, P. N. 1985. Water management of corn and soybean on a claypan soil. Trans. of the ASAE 28:780-784.

20. Sprague, M. A. and Triplett, G. B., Jr. 1986. No-tillage and surface tillage agriculture: The Tillage Revolution. J. Wiley & Sons, New York.

21. Steenhuis, T. S. and Walter, M. F. 1986. Will drainage increase spring soil temperature in cool and humid climates? Trans. of the ASAE 29:1641-1645, 1649.

22. Unger, P. W. 1984. Tillage systems for soil and water conservation. FAO Soil Bulletin 54, Rome, Italy.

23. Williamson, R. E. and Kriz, G. J. 1970. Response of agricultural crops to flooding, depth of water-table and soil gaseous composition. Trans. of the ASAE 13:216-220.

24. Yoder, R. E. 1936. A direct method of aggregate anaylsis and a study of the physical nature of the erosion losses. J. Am. Soc. Agron. 28:337-353.

25. Youker, R. E. and McGuiness, J. L. 1956. A short method of obtaining mean weight diameter values of aggregate analyses of soils. Soil Sci. 83:291-294.

Effects of Cultural Practices on Recovery of Permeability in Compacted Subsoil

Norman R. Fausey
Member ASAE

ABSTRACT

In a field where it had previously been demonstrated that a compacted soil layer at 30-50 cm depth was impeding water movement to subsurface drains, plots were cropped as either alfalfa-clover sod or corn/soybean rotation with ridge tillage. Surface irrigation was applied during year 2 and year 7 to induce subsurface drainage. Drain flow rates were measured and the peak rates of discharge were compared to indicate change over time due to cropping practice. The peak rate of discharge was similar in 1985 and 1990 for each cropping practice and the rate was significantly greater for the alfalfa-clover sod than for the corn/soybean, ridge tillage treatment. There was also a very significant drainage by cropping practice interaction, reflecting the gravel backfill effects from a previous experiment. Where there had been no previous enhancement of drainage (by permeable backfill), neither cropping practice increased the peak drain flow rate; but, where permeable backfill had been installed earlier, the peak drain flow rate was always higher for the alfalfa-clover sod compared to the corn/soybean rotation with ridge tillage.
KEYWORDS. Rotation, Macropores, Ridge-till, Subsurface drainage

INTRODUCTION

Soil compaction is a prevalent problem associated with modern agriculture. Practices like annual cash grain production, developed or emerging over time, have caused or contributed to the problem. From a crop production perspective, practices like irrigation and fertilization help diminish the problem because with adequate water and nutrients plants can offset limitations imposed by the compaction (Fausey and Dylla, 1984). From a hydrologic perspective, practices like conservation tillage or no-till planting and use of cover crops and organic residues can help to offset some of the adverse effects of near surface compaction such as increased runoff and less plant water available (Edwards, et al., 1987).
Unfortunately, soil compaction has developed below the usual depth of cultivation and into the subsoil. Subsoil compaction limits rooting depth, causes shallow perched water tables to form more frequently, increases duration of excess soil water and poor aeration, and is very difficult to remediate. Historically the remedial approach used was deep soil loosening or deep plowing. The benefit, if any, was very short lived. This result has been confirmed again recently by Kooistra, et al. (1984) in sandy loam soils in the Netherlands and by Schjonning and Rasmussen (1992) on sandy and sandy loam soils in Denmark. Deep loosening may in fact increase the degree of compaction by subsequent loads as reported by Alblas, et al. (1992). Deep loosening has other negative effects related to plant nutrition and soil conservation, (Hakansson, 1987). The

----

N. R. Fausey, Soil Scientist & Research Leader, Soil Drainage Research Unit, USDA-ARS, Columbus, OH.

data of Edwards, et al. (1987) indicate that no-till cultural practice, including the addition of animal manures, had a dramatic and lasting effect on the hydrologic relationships on a light colored silt loam soil when replacing conventional plow based cultural practice. Runoff was virtually eliminated through the development and preservation of soil macropores. However, no-till on flat, poorly-drained, silty clay and silty clay loam soils has not been successful from a crop production standpoint.

Grower experience also shows that in fields where growers are still using a rotation that includes a sod crop there is less visible evidence of subsoil compaction such as nitrogen deficiency, ponded water above subsurface drain lines, and spring planting delay due to slow soil drying. Hoover and Schwab (1968) reported that the growing season drain discharge averaged 2.7 inches when corn followed oats but was 1.4 inches when corn followed second year meadow. This suggests more complete soil drying by the sod leading to better soil structure and less compactions.

This paper addresses the recovery of permeability in a flat, poorly drained silty clay soil resulting from the use of an alfalfa-clover sod and a ridge-till, controlled traffic, corn/soybean rotation. This is a follow-up to the report contained in the Proceedings of the Fifth National Drainage Symposium in 1987 (Fausey, 1987).

METHODS AND MATERIALS

The research site is in the lake plain physiographic region of Northwest Ohio at the Northwestern Branch of the Ohio Agricultural Research and Development Center. The topography is essentially level and both surface and subsurface drainage are needed for profitable crop production. The soils were formed mainly in fine or moderately fine textured, calcareous, glacial till that was leveled by wave action on lake plains. The soil at the site is Hoytville silty clay loam (fine, illitic, mesic Mollic Ocharaqualfs). The Hoytville series consists of deep, very poorly drained soils (Lal, et al., 1989).

Prior to 1960 the site was undrained and was managed in smaller blocks for agronomic research. In 1960, 127 mm dia. tile drains were installed at a spacing of 12.2 m at a slope of 0.06% and an average depth of approximately 1 m. Surface channels were installed parallel to the tile drains at 73 m spacing and the land was graded to 0.1% slope toward the surface channels.

Between 1960 and 1979, the site was managed as a single unit. The area was initially in corn for two years followed by two years of alfalfa-red clover hay. For the next fifteen years (1965-1979), the area was used to grow a variety of annual crops including corn, soybeans, sugar beets, wheat and oats. Primary tillage was usually by moldboard plowing 0.20-0.25 m deep in the autumn, while secondary tillage normally consisted of field cultivating at a depth of 80 to 100 mm in the spring. Eventually, this annual cultivation resulted in a compacted layer at 0.30 -0.50 m depth that impeded water movement and root penetration.

The work reported here actually began in 1979. Studies were carried out to characterize the existing effective permeability of the soil. These studies involved applying water to a specific area of the field surface at a steady rate until the drain discharge hydrograph became constant. This process was repeated on four different parts of the field. The effective permeability of each area was then calculated from the discharge rate. Peak drain flow rates were calculated from drain discharge measurements using a container of known volume and a stopwatch. The rate is expressed as m^3/day/m-drain length.

During 1980, gravel backfill and shallow mole drains were installed singly or in combination to these field areas, and at various subsequent times water was again applied to the field surface at the same steady rate and the drain discharge was monitored until the hydrograph became constant. The results of that study were published (Fausey, et al., 1986) and the experiment was modified to study the effects of crop and tillage practice on the effective permeability of the soil. A diagram of the plot layout is shown in Figure 1. This experiment involves two treatments: seeding of a leguminous perennial meadow of red clover and alfalfa and adopting a corn/soybean

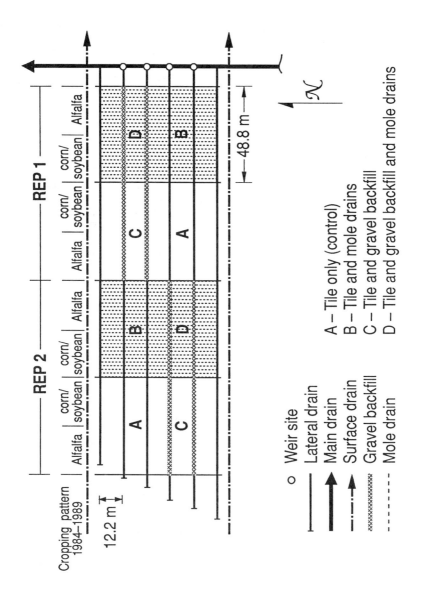

Figure 1. Plot layout on Hoytville silty clay loam.

rotation with ridge tillage and a controlled traffic pattern for all vehicular traffic. The question to be addressed by the meadow treatment is whether, after being compacted, the soil structure and permeability will regenerate under meadow. The use of ridge tillage and controlled traffic is intended to illustrate whether the soil can recover on its own if it is not tilled and driven upon in the process of getting the crop planted and harvested.

The plots in this experiment are 24 m by 24 m. Each plot has subsurface drains spaced 12 m apart and excellent surface drainage. Half the plots have gravel backfill above the drains. Disregarding the ineffective mole drains installed in 1980, there are four replications of each treatment yielding a total of 16 plots. In the autumn, irrigation water is supplied by sprinklers to create steady state flow conditions for both the surface and subsurface drain discharges. Peak discharge rates are measured and compared to evaluate the treatment effects. Crop yields also are measured.

RESULTS AND DISCUSSION

The steady state peak flow rate occurs when a balance has been established between infiltration and runoff. When all the surface microdepressions are full and overflowing and both drain discharge and surface runoff rates are constant, the maximum head causing flow in the soil macropores occurs. The peak drain discharge rate represents the relative amount of macropores in the soil. In 1979 after corn harvest, the average discharge rate for all drainlines was 0.209 m^3/day/m (see Table 1) or approximately 17 mm of surface depth of water per day. This rate was reached about 20 hours after irrigation began and was sustained only as long as the irrigation was continued, after which it dropped off sharply. Flow rates from individual plots varied between 0.11 and 0.31 m^3/day/m.

Table 1. Mean peak drain flow rates from irrigation studies by years (m3/day/m)[1].

1979		1980	1982[2]		1985	1990
0.21						
	without gravel backfill	0.20	0.06			
				ridge-till	0.11	0.20
				alfalfa-clover	0.08	0.22
	with gravel backfill	0.57	0.47			
				ridge-till	0.48	0.38
				alfalfa-clover	1.02	0.67

[1]Each value is the average of measurements made on four drains. Drain length was 24 m during 1979-1982 and 12 m during 1985-1990.

[2]Irrigation studies were conducted in October-November each year except 1982 when the study was conducted in June. Note, no intense drying and cracking of the soil would occur by June.

The plots were plowed in December 1979 after all the irrigations were completed and then seeded to oats in the Spring of 1980 with no additional tillage. Gravel backfill and mole drains were installed in June 1980 (see Fausey, et al., 1986) under very favorable conditions for moling. Dry weather followed and the slit created by the shank of the mole plow opened to the surface. In August 1980 the plots were moldboard plowed, disked twice, cultipacked and cultimulched before reshaping the land surface for surface drainage by 5 passes with a land leveler. Rainfall that occurred 9 weeks after moling resulted in sediment accumulation in the manholes at the weir sites. Visual inspection of the mole channels in November 1980 revealed a small channel (40 mm) with evidence of slaking of the walls.

The irrigation study was repeated in October and November of 1980. The results were an increase in drainflow rate where gravel backfill was installed and a decrease in drainflow rate where mole drains had been installed without a gravel backfill connector to the deeper drain. Flow rate in the control area was up about 15% while flow rate in the moled area was down 25%. With gravel backfill the flow rates increase nearly 200% to 0.57 m^3/day/m.

The plots were plowed again in December 1980 to take advantage of the winter freeze-thaw cycles to create a suitable seedbed condition for planting the next Spring. The 1981 season was very wet, planting was delayed and the yield was poor. Adverse weather did not allow a repeat of the irrigation study in 1981. Because of wet soil conditions, the fall moldboard plowing was initiated at 12 cm depth but only one-third of the area was able to be plowed. The remaining two-thirds of the area was plowed in early Spring 1982 at 20 cm depth.

The irrigation study was repeated in June 1982. Areas in Rep I that were plowed to a shallow depth had quite low flow rates even with the gravel backfill. All plots without gravel backfill had slow flow rates in both replicates. Flow rates in plots with gravel backfill in Rep II were quite high. The average flow rates for the control, mole drained, gravel backfill and gravel backfill plus moles treatments were 0.09, 0.04, 0.54, and 0.39 m^3/day/m respectively. In October 1982, all of Rep II was subsoiled 45 to 50 cm deep, and the entire experiment was moldboard plowed in November 1982.

In the Spring 1983, the plots were split. Half of each plot was planted to corn and the other half was seeded to alfalfa. This was an extremely wet year and both crops failed. The alfalfa plots were plowed up in August in an attempt to reseed them but a suitable seedbed was never achieved. It remained too wet for fall tillage so in February 1984 the alfalfa plots were plowed and ridges were formed in the corn plots using a cultivator and following the existing corn rows. The alfalfa-red clover plots were reseeded in May 1984. This was the beginning of the two cultural practices that were to be evaluated during the next 7 years. No irrigation studies were done in 1983 or 1984 in order to get the new practices installed and functional.

After February of 1984 no primary tillage has been employed. In the ridge-till, controlled traffic, corn/soybean rotation, the top of the ridge is scraped off ahead of the planting shoe and one cultivation is performed in June to reshape the ridge. This is a shallow cultivation about 5 to 8 cm deep in the center between the rows. Typically cracks form in this center between rows during soil drying and a lot of roots are visible along the faces of these cracks.

An irrigation study was performed in October and November of 1985. This and subsequent studies required 8 separate runs rather than the 4 runs required before the plots were split. Average flow rates increased for all treatments except the control treatment where the alfalfa-clover had been established. Increases were modest where no gravel backfill was used and did not reach the 1979 levels for these plots. Increases were substantial (to 1.02 m^3/day/m) where the gravel backfill was used and the alfalfa-clover was planted. No increases were noted (0.48 m^3/day/m) where gravel backfill was used and the ridge-till, controlled traffic rotation was planted.

Weather conditions did not allow another complete irrigation study until August-November 1990. Generally the drain flow rates were greater than in 1985 for plots without gravel backfill and less than in 1985 for plots with gravel backfill. Without gravel backfill the average peak flow rate for the ridge-till and alfalfa-clover plots were 0.20 and 0.22 m^3/day/m respectively. With gravel backfill the average peak rates were 0.38 and 0.67 m^3/day/m respectively.

An analysis of variance was done to test the drainage and cultural practice effects on peak drain flow rates using the 1985 and 1990 data. This analysis indicated that there was no statistical difference due to year. There was a significant difference due to drainage treatment and cultural practice. Differences due to drainage treatment and cultural practices were both significant at the 99% confidence level. The drainage treatments with gravel backfill had a mean peak flow rate of 0.64 m^3/day/m while without gravel backfill the mean peak flow rate was 0.15 m^3/day/m. The mean peak drainflow rate with alfalfa-clover crop was 0.50 m^3/day/m and with ridge-till, controlled traffic, corn/soybean rotation it was 0.29 m^3/day/m.

Average corn (9.07 Mg/ha) and soybean (3.19 Mg/ha) yields remain at high levels in this experiment because the ridge-till, controlled traffic practice allows early planting and provides excellent surface drainage (Fausey, 1990). Hay yield remained good even though the alfalfa was maintained in stand for 7 years. Alfalfa plant counts dropped from 5.5/m^2 in 1985 to 3.6/m^2 in 1990. Clover plant counts dropped from 5.1/m^2 in 1985 to 0 in 1990.

CONCLUSIONS

Fausey, et al.,(1986) summarized the results of this experiment through 1982 and arrived at the following conclusions:

1. The Hoytville silty clay loam soil at the experiment site has a compacted, impaired permeability layer in the profile that impedes water movement through the subsoil and to the subsurface drains.

2. Mole channels installed at approximately 0.50 m depth did not improve water movement to the subsurface drains.

3. Gravel backfill installed over the tile increased the peak drain flow rates; the rates are still small, indicating slow water movement within the cultivated layer.

4. The low crop yields and delays in tillage operations due to wet soil indicated that drainage was still inadequate with all treatments.

5. Excess water remains at the bottom of the plow layer for several days following rainfall and favors compaction and loss of structure.

6. Some technique is required to either re-establish permeability within the subsoil or greatly improve surface drainage.

Additional understanding gained from the studies conducted after the cessation of moldboard plowing and the establishment of a continuous alfalfa-clover sod and a ridge-till, controlled traffic, corn/soybean rotation are summarized here.

1. The increased peak drain flow rates that resulted from installation of gravel backfill over the tile have remained for 10 years.

2. The adoption of a ridge-till, controlled traffic, corn/soybean rotation has eliminated delays in planting, reduced the number of trips over the field for seedbed preparation, and more than doubled crop yields.

3. Where gravel backfill has enhanced drainage, the ridge-till cultural practice treatment did not provide any further increase in peak drain flow rate. No increase in macroporosity.

4. Where gravel backfill has enhanced drainage, the alfalfa-clover sod cultural practice treatment provided a further increase in the peak drain flow rate. The effect was evident in year 2 of the practice and remains at year 7. This suggests an increase in macroporosity or rate of water movement in the cultivated layer. The effect is less in year 7 but the number of alfalfa and clover plants is much less also.

5. The ridge-till cultural practice greatly enhances surface drainage. It provides compacted zones for trafficking and drained ridge tops for planting. Seedling roots are never in an anaerobic or anoxic environment. Roots grow to row middles and grow deeper along cracks.

6. There is an apparent very gradual trend of increase of peak drain flow rates with both cultural practices where the drainage had not been improved by backfill. Unfortunately we will not be able to continue the experiment longer to determine the confirmation of this trend.

REFERENCES

Alblas, J., Wanink, F., Akker, J. v. d., and Werf, H. M. G. v. d. 1992. Impact of traffic-induced compaction of sandy soils on the yield of silage maize in the Netherlands. Soil and Tillage Res. (in press).

Edwards, W. M., Shipitalo, M. J., and Norton, L. D. 1987. Contribution of macroporosity to infiltration into a continuous corn no-tilled watershed: implications for contaminant movement. J. Contaminant Hydrology, Proceeding of AGU Symposium.

Fausey, N. R. 1987. Impact of cultural practices on drainage of clay soils. Proceedings of the Fifth National Drainage Symposium. ASAE, St Joseph, MI.

Fausey, N. R. 1990. Experience with ridge-till on slowly permeable soils in Ohio. Soil and Tillage Res., 18: 195-205.

Fausey, N. R. and Dylla, A. S. 1984. Effects of wheel traffic along one side of corn and soybean rows. Soil and Tillage Res., 4: 147-154.

Fausey, N. R., Taylor, G. S., and Schwab, G. O. 1986. Subsurface drainage studies in a fine-textured soil with impaired permeability. TRANS of the ASAE, 29: 1650-1653.

Hankansson, I., 1987. Long-term effects of modern technology on productivity of arable land. K. Lantbruksakad. Tidskr. 126: 35-40. (in Swedish with English summary).

Hoover, J. R., and Schwab, G. O. 1968. How drain spacing and crop sequence affect tile flow. Ohio Report, 53(5): 73-74.

Kooistra, M. J., Bouma, J., Boersma, O. H., and Jager, A. 1984. Physical and morphological characterization of undisturbed and disturbed ploughpans in a sandy loam soil. Soil and Tillage Res., 4: 405-417.

Lal, R., Logan, T. J., and Fausey, N. R. 1989. Long-term tillage and wheel traffic effect on a poorly drained Mollic Ochraqualf in Northwest Ohio. I. Soil physical properties, root distribution and grain yield of corn and soybean. Soil and Tillage Res., 14: 341-358.

Schjonning, P. and Rasmussen, K. J. 1992. Danish experiments on subsoil compaction by vehicles with high axle load. Soil and Tillage Res. (in press).

DURABILITY OF AGRO-AMELIORATIVE MEASURES ON MINERAL LOWLAND SOILS WITH HETEROGENEOUS SUBSTRATE AND HYDROLOGIC CONDITIONS

L. Müller, U. Schindler[*]

ABSTRACT

Studies to find efficient measures of soil improvement have been conducted on alluvial soils in the valley of the Oder river since 1977. The experiment contains the following treatments: On clay soils Mole loosening (ML) and Deep loosening (DL), on loamy and sandy soils Deep ploughing (DP) and Deep loosening with additional tools for enriching the subsoil with humus from the Ap horizon (DT). Loosening effects were recorded on the basis of repeated digging and measurements of soil profiles, as well as by analyses of soil samples.

On clay soils mole loosening (a special tine having a trailed expander) was the preference measure. The water permeability of the subsoil remained higher for more than 5 years after improvement as compared with the adjacent zone without disturbance. After 10 years there were no significant differences between the average values of the disruption zones and the adjacent undisturbed soil. There is a recompaction of disruption zones, but also a tendency for development of the soil structure parameters of the undisturbed zones in the course of ten years. The measured high values of permeability and air volume are the result of improvement throughout the site. On clay soils monitoring and medium-term control of soil structure and its dependence on water regime are important tasks of further research.

On sandy and loamy soils there was also a recompaction of loosened zones, but as expected no structural development of the non-loosened zones and the areas not ploughed. After ten years, the soil strengths in deep ploughed and loosened zones remained lower. The DP showed no increase in crop yield, but the DT brought an increase of about 0.23 tons per hectare on the average.

KEYWORDS. Lowland soil, Subsoiling, Durability.

INTRODUCTION

The agricultural management of loamy and clayey soils in lowland areas is difficult. Often it is limited by temporally and spatially variable waterlogging or drought. There are certain possibilities to adapt the land use to such conditions. Besides this efficient measures of soil improvement are possible. Adequate ameliorative measures depend on site deficiencies, desired improvement and technical efficiency. To study site deficiencies and to test subsoiling and drainage measures on clay soils some experiments have been conducted (Müller, 1988).

EXPERIMENTS

The following paper refers to a field experiment with differing subsoiling treatments on typical heterogeneous soils containing clayey, loamy, and sandy soils. The effects are studied on soil

---

[*] Dr. agr. L. Müller, Group Leader, Scientist, and Dr. agr. U. Schindler, Scientist, Institute of Hydrology, Centre for Research on Agricultural Landscapes and Land Use (ZALF), Müncheberg, Germany.

parameters and crop yield after ten years. The experimental site is located in the valley of the Oder river about 70 km east of Berlin. Compared with other regions in Germany the site shows relatively low precipitation, the average of 30 years is about 500 mm, and the rainfall distribution over the year is relatively uniform. Potential evapotranspiration amounts to 560 mm on the average (Klimadaten, 1988). Temporary wetness of heavy soils of the site is mainly caused by the nearly steady groundwater recharge from the river Oder.

The experimental field consists of two parts containing different soils. On field part A, heavy soils with water tables of 0.4 to 1.6 m are predominant. On field part B, loamy and sandy soils with water tables of 1.2 to 1.8 m are prevailing (Fig. 1). All soils are Eutric Fluvisols and Eutric Gleysols after FAO classification. The experiment contains the following treatments (Fig. 1, Fig. 2):

On clay soils: * Mole loosening (ML)
                             * Deep loosening (DL)
On loamy and sandy soils: * Deep ploughing (DP)
                             * Deep loosening with additional tools for enriching the subsoil
                               with humus from the Ap horizon (DT)

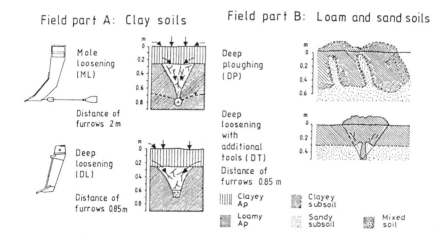

Figure 1. Treatments.

Loosening effects and the performance and durability of mole channels were recorded on the basis of repeated digging and measurements of soil profiles, as well as analyses of soil samples (Schindler and Müller, 1986, Müller, 1988, Heim and Müller, 1988). The soil analyses were done by standard methods (Birecki et al., 1968). Cultivated crops after soil improvement were: alfalfa/grass 1978-1980, winter wheat 1981, 1983, 1984, 1986, 1988, winter barley 1985, 1989, and silo-maize 1982, 1987. The crop yields were recorded and converted into GE (1 GE = 0.1 tons of grains) to have a comparable basis for averaging the crop yield over the years. The yields were analysed by three-factorial variance analyses. Variance components are treatments, years and blocks.

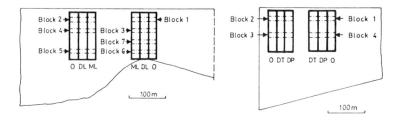

Figure 2. Plan of the Plots.

RESULTS

Durability of Disruption Zones and Mole Channels on Clay Soils

As expected, loosening zones recompact after several years. On clay soils after about three years recompaction was found, but the differences of all parameters were still significant (Table 1).

After ten years water permeability and air volume showed a marked recompaction of the loosened zones to have taken place, but the penetrometer resistance remained lower in loosened zones. There are no clear differences between the quality of the loosened zones of ML and DL treatments, but differences between parameters of different years are very marked. There is a tendency for development of the soil structure parameters of the undisturbed zones in the course of ten years. The measured high values of permeability and air volume (Table 1) are the result of improved soil conditions throughout the site. Such soil structure developing processes are typical for weakly drained clayey soils, but not for more preconsolidated soils.

ML treatment consists of loosened zones combined with mole channels. The durability of the channels depends on stresses caused by the water regime and loads of heavy machinery. Mole channels mostly disintegrate within five years. Disintegration of the mole channels leads to a decreased hydraulic performance. Badly disintegrated channels also provide sufficient water regulation (Schindler and Balla, 1989).

Mole loosening leads to an increase in yield of 2.5 GE per hectare as related to non-loosened plots on the average of ten years (Table 2).

The large number of replications allows to state the relatively small increase in yield of 2.5 GE/ha to be significant. The variance analysis showed the factor 'loosening' ranking only third.

The crop yield of clay soils is influenced by

(1) Years (containing weathering, farming and other factors),
    probability after F-test < 0.0001,

Table 1. Durability of Disruption Zones on Clay Soils.

Parameter	Treatment	Years After Improvement					
		0.5		3		10	
		Disruption Zone	Adjacent Zone	Disruption Zone	Adjacent Zone	Disruption Zone	Adjacent Zone
Saturated Conductivity k_f (m/d)	ML	0.64	0.02*	0.51	0.14*	1.65	1.09
	DL	n.m.	n.m.	0.88	0.01	0.82	0.44
Infiltration (m/d)	ML	n.m.	n.m.	8.79	2.23*	2.68	2.26
	DL	n.m.	n.m.	5.15	1.03*	3.25	2.54
Air Volume (%)	ML	12.2	5.0*	10.2	5.4*	6.9	8.2
	DL	16.0	10.4*	9.7	5.2*	7.7	7.6
Penetrometer Resistance (N/mm^2)	ML	0.7	1.7*	0.6	1.2*	1.1	1.3*
	DL	0.5	2.1*	0.5	1.1*	1.0	1.2*

ML = Mole Loosening, DL = Deep Loosening, n.m. = not measured
* = Significant at level 0.05 (Wilcoxon's test)

Table 2. Crop Yields of Mole and Deep Loosening Measures (Average of 10 Years * 7 Blocks * 3 Replications).

Loosening Treatment	Crop Yield GE/ha	Significance (& = 5 %)
Zero	48.2	
Mole Loosening	50.7	*
Deep Loosening	49.3	
LSD 5 %	1.9	

(2) Drainage states (containing water table, soil structure, micro-relief and other factors), probability after F-test < 0.001, and
(3) Mole loosening, probability after F-test 0.01 - 0.05.

This means that weathering, adapted farming and medium-term water regime control have a larger influence on crop yield than loosening measures. On the other hand, the mole loosening as combined with drainage markedly changes the soil drainage status (Müller, 1988). The layout of the experiment considered here, not containing such a combination also shows a site development. It is a task of further research to study the course and reversibility of such and similar soil development processes and their dependence on water regime.

Durability of Subsoiling on Sandy and Loamy Soils

On sandy and loamy soils there also was a recompaction of loosened zones, but as expected no structural development of the adjacent zones, as well as the non-ploughed soils (Table 3). Due to recompaction, the dry bulk density reaches the value of the adjacent soil after 1.5 to 2.5 years, or can increase to values more than the initial level of 1.56 g/cm^3. The air content of the

DP treatment decreases rapidly, and after 1.5 to 2.5 years the value is similar to that of the adjacent soil. The decrease in air content of the DT treatment occurs more slowly. After ten years differences are no longer recognizable.

The parameter 'penetrometer resistance' seems to be very sensitive to differences in soil disturbance in the field. After ten years the penetrometer resistance in deep ploughed and loosened zones remained lower.

DP showed no increase in crop yield, but DT brought an increase of about 0.23 tons per hectare on the average (Table 4). It is typical that subsoiling effects are markedly in the first years, but decrease thereafter.

Table 3. Soil Physical Parameters After 2 and 10 Years at a Depth of 30 to 50 cm.

Parameter	Treatment	Values after 2 Years	Values after 10 Years
Dry Bulk Density (g/cm^3)	Zero	1.56	1.55
	DP	1.58	1.56
	DT	1.62	1.61
LSD 5 %	DP: Zero	0.05	0.06
	DT: Zero	0.17	0.05
Air Volume (%)	Zero	13.8	11.4
	DP	14.9	13.7
	DT	20.9*	10.7
LSD 5 %	DP: Zero	3.1	4.8
	DT: Zero	6.1	4.2
Penetrometer Resistance (N/mm^2)	Zero	2.6	2.1
	DP	2.1**	1.8**
	DT	1.1***	2.1
LSD 5 %	DP: Zero	0.36	0.18
	DT: Zero	0.26	0.11

Table 4. Crop Yields After Subsoiling on Soils of Light and Medium Texture (Average of 10 Years * 4 Blocks * 3 Replications).

Treatment	Crop yield GE/ha
Zero	47.7
DP	46.2
DT	50.0*
LSD 5 %	2.1

DISCUSSION

The main effect of subsoiling consists in an increase in the rootable soil depth including improved soil water conditions (Marks and Soane, 1987, Werner, 1988, Zajjdelman, 1989). The speed of recompaction partly depends on the geometry of loosened zones and the soil management after subsoiling (Bechtle, 1985, Werner, 1988, Heim and Müller, 1989). These conditions can be considered as to have been made, if the subsoiling process causes the initiation of a secondary soil structural development in the loosened zone or of the whole site. That seems to be relevant on the field part A containing clay soils. Werner (1988) also observed stable structure developments in the deeper subsoil of loosened zones on loam soils.

Concerning the values of crop yields, our results are in general accordance with those of Kuntze and Bartels (1980) who observed increases of 0.3 tons cereals per hectare after subsoiling of marshy soils. In field experiments commonly not all the parameters show a uniform behaviour. The meaning of a single parameter should never be over-emphasized. This holds especially for the factor 'crop yield' in single years because of this factor depends on unknown other factors and boundary conditions (Bechtle, 1985).

CONCLUSIONS

If soil loosening is required on clay soils, mole loosening (ML) is the measure preferable to adopt. Using this measure is possible preferably combined with ditches and backfilled subdrains and also in controlled drainage-subirrigation systems (Müller et al., 1989). Using mole loosening without direct hydraulic connection with drains or ditches as considered here is also possible, if the site conditions are similar. Because of the relatively small increase in yield, the cost - yield relation is more decisive.

On loamy and sandy soils deep ploughing (DP) did not lead to sustainable effects and is not recommended for comparable sites.

Deep loosening with additional tools (DT) is practicable, but there are difficulties to avoid recompactions, and it is necessary to decide whether the significant but small increase in yield provides an economic return.

REFERENCES

1. Bechtle, W. 1985. Erfahrungen und Ergebnisse aus Tieflockerung in Baden-Württemberg. In: Die Gefügemelioration durch Tieflockerung - Bisherige Erfahrungen und Ergebnisse, Schriftenreihe des Deutschen Verbandes für Wasserwirtschaft und Kulturbau e. V. (DVWK), Heft 70, 37-73.

2. Birecki, M., A. Kullmann, I.B. Revut and A.A. Rode. 1968. Untersuchungsmethoden des Bodenstrukturzustandes. Edt. by the International Soil Science Society (ISSS). VEB Deutscher Landwirtschaftsverlag Berlin, 1. Auflage, 504 p.

3. Heim, H. and L. Müller. 1988. Field studies on the structure of alluvial clay soils as a precondition for the determination of the drainage situation. Archives of Agronomy and Soil Science, Berlin, 32, 3, 141-151 (in German, Engl. summary).

4. Heim, H. and L. Müller. 1989. Structure of natural backfills of drain trenches of different ages in alluvial clay soils of the Oderbruch region. Archives of Agronomy and Soil Science, Berlin, 33, 6, 335-342, (in German, Engl. summary).

5. Heim, H. and L. Müller. 1990. Durability of soil loosening zones on an alluvial clay site. Archives of Agronomy and Soil Science, Berlin, 34, 12, 819-827 (in German, Engl. summary).

6. Klimadaten der Deutschen Demokratischen Republik. Ein Handbuch für die Praxis. Reihe B, Band 6, Verdunstung, 59 p., Reihe B, Band 14, Klimatologische Normalwerte 1951/80, 148 p., Potsdam, 1988.

7. Kuntze, H. and R. Bartels. 1980. Unterbodenmelioration in der Marsch. Zeitschr. für Kulturtechnik und Flurbereinigung, Berlin und Hamburg, 21, 27-37.

8. Marks, M.J. and G.C. Soane. 1987. Crop and soil response to subsoil loosening, deep incorporation of phosphorus and potassium fertilizer and subsequent soil management on a range of soil types. Part 1: Response of arable crops. In: Soil Use and Management, 3, 115-123.

9. Müller, L. 1988. Efficiency of subsoiling and subsurface drainage in heavy alluvial soils of the G.D.R. Soil & Tillage Research, Amsterdam, 12, 121-134.

10. Müller, L., R. Mittelstedt, U. Schindler and S. Heim. 1989. Construction and durability of mole channels in heavy alluvial soils. Journal of Rural Engineering and Development, Hamburg and Berlin, 30, 394-403, (in German, Engl. summary).

11. Schindler, U. and L. Müller. 1986. Studies in the change of soil physical parameters of alluvial sites after mechanical soil loosening with tine-shaped tools. Archives of Agronomy and Soil Science, Berlin, 30, 8, 475-484 (in German, Engl. summary).

12. Schindler, U. and D. Balla. 1989. Hydraulic efficiency of mole drains in alluvial clay soils. Archives of Agronomy and Soil Science, Berlin 33, 4, 209-215 (in German, Engl. summary).

13. Werner, D. 1988. Einfluß des physikalischen Bodenzustandes auf den Wasserentzug landwirtschaftlicher Fruchtarten. Arch. Acker- Pflanzenbau Bodenkd. Berlin, 32, 5, 303-310.

14. Zajjdelman, F.R. 1989. Glubokoe rykhlenie i krotovanie poverkhnostno pereuvlazhnennykh tjazhelykh pochv. Ehffektivnost' i limitirujushhie uslovija primenenija. (in Russian). Papers, International Scientific Conference 'Soil Melioration'. Sofia, Bulgaria, Sept. 12-16, 102-107.

ENVIRONMENTAL SUSTAINABILITY OF DRAINAGE PROJECTS

Chandra A. Madramootoo[*]
Member ASAE

ABSTRACT

Significant investments are being made in drainage of irrigated lands for salinity and waterlogging control. It is imperative that these investments be protected, and that systems are sustainable in the long-term. Environmental sustainability will, therefore, be pivotal in future designs of irrigation and drainage systems. Major concerns revolve around the management, reuse and disposal of drainage water from irrigated areas. The technical and policy issues which ought to be considered are presented. A scaling checklist and interaction matrix were developed, to assist in environmental impact assessments and scoping studies.

INTRODUCTION

In arid and semi-arid regions, irrigation is necessary in order to meet crop evapotranspiration requirements. Two consequences of irrigation in these regions are secondary soil salinization/alkalization and waterlogging. Salinization is the build-up of water soluble compounds containing the following cations: Na, Ca, Mg, and K, and the following anions: Cl, SO_4, HCO_3, CO_3, and NO_3. Continual irrigation of soils containing these chemical constituents, under high rates of evaporation lead to saline conditions in the plant root zone. As relatively purer water evaporates from the soil surface, precipitated salts accumulate in the soil profile. Saline ground waters, and intrusion of sea water are also sources of salinization. Alkalization (a specific form of salinization) is the accumulation of sodium in the soil due to the precipitation of calcium sulphate, calcium carbonate and magnesium carbonate. Irrigation water with an EC greater than 3 dS/m restricts the growth of most crops. Similarly, soil with an EC greater than 4 dS/m in the saturation extract is characterized as saline. Plants are unable to extract soil-water under saline conditions. Waterlogging is the gradual accretion of the watertable, due to continual irrigation over time. Seepage from irrigation canals is also a major source of waterlogging in several regions. There are over 250 million ha of irrigated land worldwide, of which nearly 30 million ha are affected by salinity and waterlogging. It is estimated that nearly 1.5 million ha of cropland are being destroyed by salinization each year. Yields of major crops have been reduced on 16 million ha of irrigated land in Pakistan, 8 million ha in India, and 3.5 million ha in Egypt.

Subsurface drainage, either through horizontal pipe systems, or vertical tubewells, is a technological practice for the reclamation of saline and waterlogged soils. It is, therefore, regarded as a complementary activity to irrigation. Drainage lowers the watertable to some critical depth, thereby preventing salts from rising to the root zone. Furthermore, with additional

---

[*]Chandra A. Madramootoo, Associate Professor, Agricultural Engineering Dept., McGill University, Macdonald Campus, Ste. Anne de Bellevue, Quebec, Canada.

water applications, salts can be leached from the soil profile and the effluent collected by the drains.

Future water mangement practices should be environmentally sustainable. This implies that projects should not have a deleterious effect on biodiversity within the ecosystem, create human health hazards, reduce soil and crop productivity, impair quality of life in the long-term, or degrade water quality. The purpose of this paper is to highlight the major environmental issues related to the drainage of irrigated lands, and discuss the relative merits of mitigative measures. Environmental impact assessment tools are also proposed, to aid project planners and environmentalists in decision making.

AN OVERVIEW OF DRAINAGE AND IRRIGATION WITHIN THE ECOSYSTEM

Irrigation and drainage impact, both positively and negatively, on the water, soil, plant, animal, and atmospheric constituents of the ecosystem within a project area. The impacts could be either direct, or indirect, and affect the entire food chain. The interactions of irrigation and drainage processes within the ecosystem are shown in Fig. 1. The mechanisms causing change within

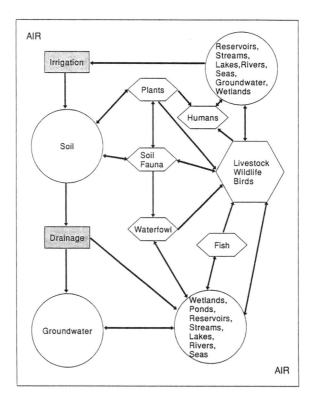

Figure 1. Interactions of Irrigation and Drainage in the Ecosystem

the physical, chemical, and biological components of the ecosystem are complex. Some components are inter-related. This is demonstrated by the soil-plant-humans-livestock components in Fig. 1. Irrigation with water containing salts and toxic elements changes soil structure and chemical constituents in the soil. This is not only harmful to plant growth, but could also be detrimental to humans and livestock consuming the plant products. Furthermore, the changes in soil structure and chemical properties affect groundwater recharge and quality. The complexity of the system can be shown by the public health impacts. Subsurface drainage of waterlogged lands reduces the incidence of malaria in surrounding villages. However, if the disposal of this drainage water is to ponds and marshes, then malaria could be spread to neighbouring regions. This also demonstrates that the solution to a problem could lead to the creation of other environmental problems. Another significant factor is the irreversibility of ecological damage. This is best illustrated by the toxic effects of water quality on aquatic biota.

Madramootoo (1992) indicated that the major environmental issues surrounding irrigation and drainage projects are: salinization, waterlogging, water quality, spread of waterborne diseases, public health, ecological changes due to reservoir and canal systems, proliferation of aquatic weeds, destruction of wetlands, oases, marine ecosystems, waterfowl and wildlife, pathogen contamination of humans and food products due to wastewater irrigation, disposal of drainage water, modification of the hydrologic regime, and changes in biodiversity.

ENVIRONMENTAL CONSIDERATIONS IN THE DRAINAGE OF IRRIGATED LANDS

The focus of this paper is on the problems related to the management, reuse, and disposal of drainage water in irrigation projects. However, it is recognized that the environmental issues stated above, must also be considered in the overall sustainability of irrigation and drainage projects.

Subsurface drainage remains the best available technology for lowering the water table and keeping salts out of the root zone. Irrigation water in excess of crop evapotranspiration requirements is applied, to leach salts from the root zone, in irrigated arid and semi-arid regions. The subsurface drains intercept the leachate, and maintain the water table below the root zone. Salts are thereby prevented from rising to the root zone by capillary action. In Pakistan, vertical tubewells are used not only to supply irrigation water, but also to control the groundwater table. The flow of water, salts and chemicals in an irrigation and drainage system is depicted in Fig. 2. However, the disposal of drainage water requires special consideration. There are several options, revolving primarily around outlet availability and water quality. These are summarized in Table 1.

The volume of water to be removed by the drains should be minimized. This necessitates precise calculation of the irrigation and leaching requirements. Irrigation efficiencies will have to be improved and irrigation return flows reduced. The volume and quality of water collected by subsurface drains is also a function of lateral drain depth and spacing. Deeper drain installations will initially produce a larger volume of drainage water. Higher salt concentrations may be found in drainage effluent, as leaching occurs within a deeper soil profile. Therefore, minimum drain depth for optimum leaching and water table control is advantageous. An added benefit to shallow drain depths is that subirrigation would be feasible for some soils and crop conditions, once reclamation is achieved. Wider drain spacings also reduce drain flow. It should, however, be pointed out that smaller drainage volumes will result in higher concentrations of salts and chemicals. Shallower drains installed at wider spacings reduce installation and system costs. As reclamation occurs with time, the salinity of the drainage water should decrease.

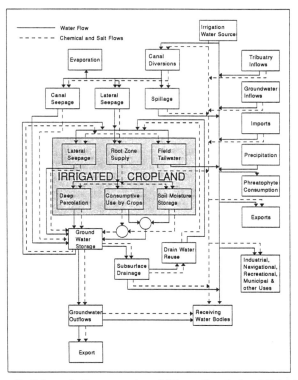

Figure 2. Schematic Representation of Water, Chemical and Salt Flows in an Irrigation Project (Adapted from Tanji, 1990).

In some regions, drainage water can be outlet directly to rivers and watercourses. This will increase the salinity of these water bodies, and affect freshwater fish species, and other aquatic biota. Where there are outlet restrictions, saline drainage effluent is disposed in evaporation ponds. However, there are several environmental problems with these ponds (Mian and van Remmen, 1991). The saline effluent seeps out from the ponds, and increases the salinity and waterlogging problem of adjoining areas. There could also be salt movement downwards to the aquifer. These conditions are counterproductive to the function of the subsurface drains. In addition, the drainage water contains fertilizer, pesticides, trace elements and other toxic chemicals. The accumulation of these chemicals, salt and toxic substances poses a severe environmental threat to aquatic organisms, wildlife, plants, and waterfowl. The National Research Council (1989) has fully documented the effects of selenium in drainage water on waterfowl and fish in the Kesterson National Wildlife Refuge in California's San Joaquin Valley. Selenium occurs naturally in soils, and is leached by the drainage water. Human and animal health could also be affected if drainage effluent contaminates drinking water, or water from the evaporation ponds is directly consumed. The problem is further compounded if sewage and other contaminants enter the pond from surface runoff. Use of evaporation ponds for fish farming must be studied carefully, given the presence of toxic chemicals in the water. Human consumption of fish containing chemicals leads to illness.

The Australian experience with evaporation ponds indicates that N and P in the drainage effluent increases algae growth, and due to anaerobic conditions, there is an unpleasant odour (Mian and van Remmen, 1991). As water evaporates from the pond, the toxicity of the chemicals remaining is very concentrated. The accumulated salt in the pond would, therefore, not be suitable for human use. Due to chemical contamination, the area served by the pond could not be used for other purposes. Barber (1985) recommended against the use of evaporation ponds in northern India for several reasons. He noted that an area of 10 to 30% of the total project area would have to be set aside for evaporation ponds, which would rapidly accumulate salt, and the water being evaporated would approach the concentration of brine. As salt concentration increases, the rate of water evaporation decreases. This would further limit the efficiency of the pond.

Table 1: Options to be considered for the management, reuse, and disposal of drainage water

1. SOURCE CONTROL FOR DRAINAGE WATER QUALITY AND QUANTITY

Water Management; irrigation and leaching efficiencies
Drainage Management; depth and spacing of lateral drains
Crop Management
Alternate Land Use

2. REUSE

Agriculture
Industrial Utilization
Cooling Water for Power Plants
Solar Ponds for Electrical Energy
Evaporation Ponds for Salt Production
Aquaculture
Agroforestry

3. DISPOSAL

River Discharge With and Without Dilution
Evaporate Drainage Waters in Ponds
Deep Well Injection
Physical, Chemical, and Biological Treatment Processes

4. HUMAN, AQUATIC AND WILDLIFE ECOLOGY

Protection, Restoration and Enhancement; removal of toxic chemicals
Provision of adequate freshwater

5. INSTITUTIONAL CHANGES

Water Pricing and Marketing
Modify or Eliminate Irrigation Subsidies
Modify Tax Structures and Rates
Impose Drainage Effluent Fees
Reallocate Water from Agriculture to Other Uses

(Adapted from Tanji, 1990)

Reuse of drainage water is another possibility for disposal (Fig. 2), which should be actively promoted. It is advantageous for the following reasons. Firstly, large volumes of effluent are disposed of relatively cheaply. An expensive treatment process is not required. Secondly, if good quality water is scarce for irrigation and reclamation, drainage water can be supplemented. Rhoades (1990) advocated the reuse of drainage water and showed how it can be used beneficially for reclamation without significantly affecting crop yields. However, the SAR of the irrigation water and soil infiltration characteristics need to be evaluated. It is certainly possible to blend saline water with fresh water (if available), before irrigating.

There are two methods of reusing drainage water for irrigation and reclamation. The **cyclic** strategy involves the use of relatively good quality irrigation water, or less saline drainage water for pre-plant and early irrigation of crops, and the application of more saline drainage water during later stages of crop growth, when the plant is more salt tolerant. This strategy should ensure that soil properties, such as permeability are sustained in the long-term. An essential methodology to be developed is an irrigation and reclamation plan which takes crop sensitivity to salt, time of growth, and salt concentrations of irrigation and reuse water into account.

The other strategy for the reuse of drainage water is **blending**. This involves mixing water of two different salinities, to obtain irrigation water of suitable quality. The waters can be blended in the soil, or irrigation canal. A major disadvantage of blending is that good quality water becomes unusable when it is blended with highly saline drainage water. Other disadvantages of blending are that highly salt-sensitive crops cannot be included in the rotation, and a blending facility is required. For these reasons the cyclic strategy is preferred. It is more flexible and soil salinity can be lowered at critical times. While there may be some benefits to drain water reuse, there are several unanswered questions. For example, the effects of drainage water and soils variability on leaching efficiency, and on crop productivity are not fully known.

Drainage water may have to be stored in ponds or reservoirs, prior to reuse. The environmental problems associated with these structures were previously mentioned. It is necessary to ensure that the concentrations of salts and toxic compounds in drainage water are not harmful to humans, plants, animals, and waterfowl. The salt tolerance levels for several crops and trees are given in Maas (1990). The viability of evaporation ponds will be improved, if fish farming and salt mining are included in their operation. Municipal and industrial effluent should not be disposed in the evaporation ponds. It may be advantageous to allow some surface runoff into the ponds, to reduce salt and chemical concentrations. In cases where ponds have to be installed on coarse textured soils, a geotextile liner should be placed on the bottom and sides, to eliminate seepage to ground water.

Deep well injection of drainage water is possible, provided that the geologic conditions permit deep drilling, and groundwater contamination problems do not arise. This is especially important when the receiving aquifer is a source of drinking water. There is the possibility of treating drainage water by reverse osmosis (RO) to remove salts and trace minerals. However, large evaporation ponds may be required to store the by-products, viz. brine. While the technology is well known (Trompeter and Suemoto, 1984), there is a need to design small-scale RO treatment facilities for developing countries. Biological reactors and chemical treatment processes using ferrous hydroxide do not appear to be economic or technically feasible at this stage. Although, (Squires et al., 1989) stated that economic viability could improve if the quantity of drainage water was reduced, and other wastes could be treated. Likewise, there are currently technical and economic limitations with solar ponds for both water treatment and power generation.

There is scope for drainage and irrigation projects to be more closely allied to the agroforestry sector. Barber (1985) recommended afforestation as an intermediate solution to lowering the water table on saline, waterlogged soils. The trees need to be planted before the soil is heavily salinized, thereby ensuring their establishment. He estimated that a closed canopy of trees could remove more than 1.0 m of water from the soil profile in a year, and up to 600 mm over a dry season. Most salt tolerant trees could also be used for wood or fuel. However, this is not an ultimate solution, because in the long term, it will become necessary to evacuate accumulated salts from the profile. Nonetheless, drainage water could be used to irrigate tree crops which have a high salt tolerance.

Methods of improving on-farm water management and reducing canal seepage must also be an intergal part of the solution. Over-irrigation due to poor water management results in high quantities of drainage water. This modifies the hydrologic regime of rivers and watercourses, or requires a large evaporation pond. Lining of irrigation canals and the use of pipelines for conveying irrigation water are recommended. An environmental benefit of pipelines is the reduction of schistosomiasis, and other waterborne diseases. Water pricing and pumping of water may force farmers to be better users of irrigation water. Smaller scale irrigation schemes may also enable irrigation authorities to better predict irrigation requirements and deliver more precise water quantities at the farm level.

The imposition of drainage effluent fees may force farmers not to over-irrigate, and also partially cover water treatment costs. There are also management problems which should not be overlooked. A single drainage outlet serves several farmers, growing a variety of crops at any one time. Water management conditions and drainage requirements, therefore, vary considerably within a single drainage system. This imposes system constraints and makes effective use of drainage water extremely difficult. Stricter monitoring of water applications, cropping sequences, and salinity levels in soil and water on blocks of farm units will be required.

ENVIRONMENTAL IMPACT ASSESSMENT TOOLS

In order to properly assess the environmental consequences of irrigation and drainage projects, planners and environmentalists are required to perform environmental impact assessments (EIAs). These assessments may be in the form of scoping studies, or detailed predictive modeling studies. A scaling checklist was, therefore, developed (Table 2). A total of 30 major ecological indictors is identified and listed. The nature of the likely impacts, and their possible significance, or degree of complexity can be indicated. Application of this checklist to a specific project provides a strong indication of whether to proceed with the development. If the decision is to proceed, the checklist identifies impacts of concern which require further detailed study. The checklist should be applied separately, upstream of the project area, within the project area, and downstream of the project area, as well as to major project components such as reservoirs, dams, etc. Table 2 should be used as the basis for project planning and implementation.

Furthermore, an interaction matrix (Table 3) was developed, to help project planners focus on critical environmental issues, which have been previously neglected. Municipal and industrial effluent is included in the matrix, because it is sometimes dumped in watercourses and canals, thereby affecting both drainage and irrigation water quality. Similarly agricultural practices, including chemical usage, interact with water flow processes during irrigation and drainage to affect the ecosystem. The matrix can be used simply to indicate an interaction, by entering a check-mark in the appropriate box. Or, the magnitude and importance of impacts can be ascribed using a grading system similar to that of Table 2.

CONCLUSIONS AND RECOMMENDATIONS

Irrigated lands which have been degraded by salinity and waterlogging are now being reclaimed with subsurface drainage technology. However, steps must be taken to ensure that drainage outlets and drainage water do not adversely affect the environment.

Environmental sustainability is, therefore, a key element in project planning. The ecological variables to be protected include: biological diversity, wetlands, fisheries, coastal and marine habitats, wildlife, waterfowl, human health, crop yields, soil physical and chemical properties, quality and quantity of water resources, and river basin hydrology.

In order to improve the design of projects in the future, more information on the effects of mitigative measures is required. It is, therefore, recommended that ecological variables be monitored over the long-term on some projects. This will also assist with future environmental impact assessment studies. Apart from technological inteventions, there are policy and institutional issues which must be addressed. Improved on-farm water management requires new policy initiatives on water availability to farmers and water pricing.

Finally, a scaling checklist of environmental issues and interaction matrix were developed. These EIA tools should give a clear indication of whether the beneficial impacts significantly outweigh the adverse impacts, during project planning. They also help to identify those environmental issues which must be studied in more detail, prior to project implementation.

TABLE 2: SCALING CHECKLIST OF ENVIRONMENTAL IMPACTS OF IRRIGATION, DRAINAGE, AND FLOOD CONTROL PROJECTS

ITEMS	Nature of Likely Impacts	
	Adverse ST LT R IR L W	Beneficial ST LT SI N
Aquatic ecosystems		
Fisheries (shell fish, fin fish, crustaceans)		
Mangroves and other swamp/coastal ecosystems		
Reservoir, estuarine, and marine habitats		
Rare and endangered species		
Wildlife		
Waterfowl and bird habitats		
Forests		
Wetlands		
Oases		
Surface water hydrology		
Groundwater hydrology		
Surface water quality		
Groundwater quality		
Navigation		
Waterlogging		
Soil salinization/alkalinization		
Soil physical and chemical properties		
Sedimentation		
Irrigation water quality		
Irrigation return flows		
Human settlements		
Human health		
Incidence and transmission of disease		
Bacteria and contaminants in soil, water, crops, animals		
Human nutrition & income levels		
Incidence of aquatic weeds, crop pests, animal diseases		
Industrial, agricultural and urban development		
Tourism and Recreation		
Aesthetic and historical sites		

ST = Short-term   LT = Long-term   R = Reversible   IR = Irreversible   L = Local
W = Wide   SI = Significant N = Normal   * = Negligible

TABLE 3. INTERACTION MATRIX TO IDENTIFY ECOLOGICAL IMPACTS OF SPECIFIC IRRIGATION AND DRAINAGE SYSTEM COMPONENTS, EFFLUENT DISPOSAL, AND AGRICULTURAL PRACTICES.

ACTIONS	ECOLOGICAL INDICATORS																		
	Soil Properties		Air	Groundwater		Surface water		Biota				Wetlands	Forests	Marine and Coastal Habitats	Oases	Crop Yields	Human Health	Food Products	Pasture lands
	Physical	Chemical		Quality	Levels and Quantity	Quality	Quantity	Plant	Animal	Aquatic	Avian								
Drainage Outlets																			
Drainage Water Reuse																			
Drainage Water Disposal																			
Evaporation Ponds																			
Irrigation Return Flows																			
Wastewater Irrigation																			
Municipal and Industrial Effluent																			
Farming Systems and Techniques																			
Pesticide and Fertilizer Usage																			

ACKNOWLEDGEMENTS

The financial support of the Canadian International Development Agency (CIDA) is gratefully appreciated. Messrs. Aly Shady of CIDA, and Walter Ochs, Ashok Subramanian, Lambert Smedema, and Tom Brabben of the World Bank are also thanked for their inputs.

REFERENCES

1. Barber, W. 1985. The Rising Water Table and Development of Water Logging in Northwestern India. A Drainage Sub-sector Report. South Asia Projects Department, Irrigation II Division, The World Bank, Washington, D.C.

2. Maas, E.V. 1990. Crop Salt Tolerance. **In:** K.K. Tanji (ed.). Salinity Assessment and Management, pp. 262-304. American Society of Civil Engineers, New York, N.Y. 10017

3. Madramootoo, C.A. 1992. Environmental Sustainability of Irrigation and Drainage Projects: An Analysis of the Issues and Future Needs. A Report to the Natural Resources Division, Canadian International Development Agency, Hull, Quebec, Canada.

4. Mian, M.A. and T. van Remmen. 1991. Evaporation Ponds For Disposal of Saline Drainage Effluent in Pakistan: Feasibility Study (Part1, Data Collection). NRAP report No. 18, Wageningen, The Netherlands.

5. National Research Council. 1989. Irrigation-induced Water Quality Problems. Published by the National Academy Press, Washington, D.C.

6. Rhoades, J.D. 1990. Overview: Diagnosis of Salinity Problems and Selection of Control Practices. K.K. Tanji (ed.). **In:** Agricultural Salinity Assessment and Management, pp. 18-41. American Society of Civil Engineers, New York, N.Y. 10017.

7. Squires, R.C., G.R. Groves and W.R. Johnston. 1989. Economics of Selenium Removal from Drainage Water. Journal of Irrigation and Drainage Engineering, 115(1): 48-57.

8. Tanji, K.K. 1990. Nature and Extent of Salinity. **In:** K.K. Tanji (ed.). Agricultural Salinity Assessment and Management, pp. 1-17. American Society of Civil Engineers, New York, N.Y. 10017.

9. Trompeter, K.M. and S.H. Suemoto. 1984. Desalting by Reverse Osmosis at Yuma Desalting Plant, pp. 427-437. **In:** R.H. French (ed.). Salinity in Watercourses and Reservoirs: Proceedings of the 1983 International Symposium on State-of-the-Art Control of Salinity, July 13-15, 1983, Salt Lake City, Utah.

SUBIRRIGATION AND CONTROLLED DRAINAGE: MANAGEMENT TOOLS FOR REDUCING ENVIRONMENTAL IMPACTS OF NONPOINT SOURCE POLLUTION

P. K. Kalita [*] R. S. Kanwar[*] S. W. Melvin[*]
Member, ASAE Member, ASAE Member, ASAE

ABSTRACT

A dual pipe subirrigation-drainage system was installed in 1988 on an area of 0.85 ha at Iowa State University Ankeny Research Farm, Iowa. Field experiments were conducted to evaluate water-table management (WTM) effects on ground water quality by maintaining variable water table depths in the field through subirrigation and controlled drainage during the corn growing seasons of 1989-91. Nitrogen fertilizer and herbicides were applied at planting. Shallow groundwater samples were collected by installing piezometers and solute suction tubes at various depths. Water samples were collected biweekly in 1989 and monthly in 1990-91 to analyze for nitrate-N atrazine, and alachlor concentrations in groundwater. Three years of data reveal that nitrate-N, atrazine, and alachlor concentrations in groundwater were decreased by maintaining shallow water table depths (< 0.6 m) in the field. Results of these experiments clearly suggest that subirrigation and controlled drainage practices have the ability to improve groundwater quality in agricultural areas.

Key Words: Subirrigation, Drainage, Water quality, Water table management.

INTRODUCTION

At present, ground water contamination from agricultural nonpoint sources has become a major environmental concern. Hallberg (1986) reported an almost linear increase in ground water nitrate-N concentration over the last 20 years. Logan et al. (1980) reported Nitrate-N concentration ranging from 0.5 to 120 mg/L in tile drainage water under corn in Iowa, Minnesota, and Ohio. In Iowa, Baker and Johnson (1981) observed nitrate-N levels of 10 to 70 mg/L in subsurface water samples from tile lines under corn rotated with oats or soybean. Hubbard and Sheridan (1989) documented that in many agricultural areas, nitrate-N levels in drinking water were significantly higher than the maximum contaminant level of 10 mg/L nitrate-N set by the U.S. Environmental Protection Agency. Also the detection of pesticides in groundwater has led to a serious concern about the potential water quality problems associated with the increasing use of organic chemicals. Pesticide usage in the United States totaled nearly 300 million kilograms in 1982, and 85 percent of this usage is in the corn belt (Hallberg, 1986). Widely used herbicides such as alachlor, atrazine, metolachlor, and cyanazine have been detected in groundwater systems of several states.

Reviews of several studies on the effects of agricultural practices on subsurface water quality are presented by Baker and Johnson (1977) and Hallberg (1986). Hallberg (1986) has suggested that infiltration recharge may be the primary delivery mechanism of agriculture related contaminants to the ground water. Recently, management practices have received much attention as potential measures to reduce pollution hazards to groundwater systems. Water table management (WTM) using subirrigation and controlled drainage has been rapidly accepted in the Midwest and Southwest Regions of the USA (Skaggs, 1987). The use of subsurface drainage systems for both drainage and irrigation has become a popular water management technique where conditions are favorable for its use. It is particularly

---

[*]P.K. Kalita, Post-Doctoral Research Fellow, Agricultural Engineering Dept., Washington State University, Pullman, WA 99164, and R.S. Kanwar and S.W. Melvin, Professors, Agricultural and Biosystems Engineering Dept., Iowa State University, Ames, IA 50011.

attractive as an irrigation alternative where parallel drainage systems have been previously installed in a flat, poorly drained soils (Melvin et al., 1990). Subirrigation and controlled drainage systems have been successful in providing excellent crop production as well as an excellent technique for management of shallow groundwater quality (Evans et. al., 1987; Kalita, 1992). The concentrations of nitrate-N, atrazine, and alachlor in shallow groundwater could be reduced and corn yield could be increased with WTM practices (Kalita, 1992; Kalita and Kanwar 1992a). Bengtson et al. (1989) reported that atrazine and metolachlor losses were reduced by 55% and 51% respectively, by subsurface drainage. Results of these studies have accelerated research interests on the use of WTM practices for water quality benefits under different soils, crops, and climatic conditions (Kanwar, 1990; Thomas et al., 1991; Bengtson et al., 1991; Belcher and Merva, 1991; Skaggs et al., 1991; Fausey et al., 1991). This paper reports the effects of subirrigation and controlled drainage on shallow groundwater quality from three years of field experiments with corn in Iowa.

METHODS AND MATERIALS

Site Description: Experiments with subirrigation and controlled drainage were conducted during 1989-91 at Iowa State University's research center in Ankeny, Iowa. The soils at this site are Nicollet silt-loam in the Clarion-Nicollet-Webster soil association. Some of the physical properties of these soils are presented by Kalita and Kanwar (1992b). A dual-pipe subirrigation system was installed at this site in 1988 on a 0.85-ha area with natural surface slope of 2.5 percent. The principle of this system is the management of a shallow water table in a field with the use of parallel subsurface pipes, one set at shallow depth and used for subirrigation, and other set at deeper depth, which is used as drainage pipes installed midway between irrigation lines and connected to drain head control structures. The basic concept of the dual-pipe subirrigation system is illustrated in Figure 1. Shallow irrigation pipes were installed at a depth of 0.5 to 0.6 m parallel to and midway between drainage pipes, which were installed at a depth of 1.2 m. The natural slope along the length of the field allowed water tables to be maintained at various depths by controlling the subsurface drainage outflows in the head control structure and by supplying irrigation water through the subirrigation pipes. An integral part of this system was a reservoir used to store excess drainage water during wet periods. This water was used as a major irrigation source during periods of deficit moisture.

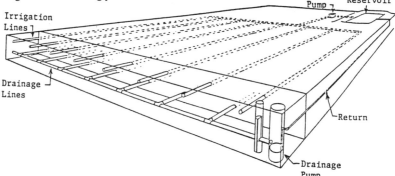

Figure 1. A schematic sketch of the dual-pipe subirrigation system

Cultural management and water-table treatment: Corn 'Pioneer 3379' was planted on 23 May in 1989, 8 May in 1990, and 27 May in 1991 at the Ankeny site. Harvesting dates were 31 October in 1989, 16 October in 1990, and 10 October in 1991. The plant population was 66,600 per ha with a row-to-row distance of 0.75 m and a seed-to-seed distance of 0.2 m every year. Urea nitrogen fertilizer was applied at planting every year at the rate of 200 kg-N/ha. Herbicides Atrazine and Alachlor (common name Lasso) were applied at the rate of 2.2 kg/ha in 1989 and 1991. Crop growth parameters on plant height and photosynthesis were monitored biweekly and corp yield data were collected at harvest every year. Water-table depths in the subirrigation field ranged from 0.03 to 1.25 m during the growing seasons. The average water-table depths at five locations A, B, C, D, and E (where monitoring devices were installed) were maintained at 0.2, 0.3, 0.6, 0.9, and 1.1 m, respectively. Water table position at each location, however, fluctuated during and after heavy rainfall events. A maximum water-table depth of 1.25 m was observed at location E in the beginning of the season (before subirrigation started), and a minimum water-table depth of 0.03 m was observed at location A once during the growing season due to heavy rainfall in 1990. The average water-table depths were maintained through subirrigation from 53 to 96 days after planting (DAP) in 1989 and 1990 and from 45 to 97 DAP in 1991.

Subsurface water sampling: Both solute suction tubes and piezometers were used for collecting shallow ground water samples. Solute suction tubes were made by coupling a 2-bar porous ceramic cup at the end of each 38 mm diameter pvc pipe of 0.9, 1.2, 1.5, and 2.1 m lengths. Two 6 mm dia polyvinyl tubes were inserted into the pipe (one long enough to collect water from the bottom of the pipe, and a short one to pump air out of the pipe) through a rubber stopper which was used to seal at the top of the pvc pipe. Solute suction tubes were installed at 0.9, 1.2, 1.5, and 2.1 m depths at three locations (B, C, and D) with three replications at each location. Piezometers were made with 25 mm diameter pvc pipes and installed at 1.2, 1.8, and 2.4 m depths at locations A, C, and E with three replications at each location. Observation wells (25 mm diameter and 1.5 m long pvc pipes) were installed to monitor water table positions at each location of the subirrigation field. Water samples were collected from piezometers and solute suction tubes biweekly in 1989 and monthly in 1990 and 1991 for nitrate-N and herbicide analysis. For collecting samples from the solute suction tube, air was taken out of the tube with a vacuum pump to create suction one day before sampling. Piezometers were also pumped out one day before sampling. Water samples were collected from each location and depth on the following day, and preserved in a cold chamber at $4^\circ C$ for analysis.

RESULTS AND DISCUSSION

The concentrations of nitrate-N, atrazine, and alachlor in groundwater changed with different WTM practices at different soil depths and at different times of the growing season every year. Amount of rainfall during the growing seasons also caused variation of chemical concentrations from year to year. Rainfall data were collected within 100 m from the experimental field. The 1989 season was relatively dry with May to October rainfall 494 mm at the Ankeny site. The 1990 season was very wet in comparison to the previous year with May to October rainfall of 775 mm at the same site. Seasonal rainfall at Ankeny in 1991 was 535 mm.

Nitrate-N concentrations in ground water: The average nitrate-N concentrations in ground water from the piezometer samples in the subirrigation field are shown in Figures 2, 3, and 4. Figure 2 shows variations in nitrate-N concentration as a function of depth on different times during the growing season in 1989. The average water-table depths in the field at sampling time are shown in the same figure. Water-table depths at three locations where piezometers were installed fluctuated between 1.6 and 1.0, 1.1 and 0.35, and 0.8 and 0.12 m during the growing season and are referred to as deep, medium, and shallow water-table depths. Nitrate-N concentrations in ground water under shallow water-table depths were always less than those with medium and deep water-table depths, and the lowest nitrate-N concentrations were observed on 84 DAP with shallow water-table depth. Nitrate-N concentrations in ground water decreased with increased soil depth under all three water-table conditions. At 1.2 m soil depth, concentration of nitrate-N in ground water varied from 7 to 2.5, 15 to 8.0, and 20 to 17 mg/l under shallow, medium and deep water-table depths, respectively. Similarly, variations in nitrate-N concentration were observed from 4 to 2.5, 15 to 8, and 20 to 14 mg/l at 1.8 m depth, and from 4 to 2, 10 to 5, and 18 to 11 mg/l at 2.4 m depth for shallow, medium, and deep water-table conditions, respectively. These results indicate that nitrate-N concentrations in ground water can be lowered by maintaining shallow water-table depths. In 1990, water samples were collected monthly to reduce the number of samples and analytical costs. Figure 3 shows the concentrations of nitrate-N as a function of water-table depth for 1990. A nitrate-N concentration of as high as 67 mg/l was observed at 1.2 m depth under the deep water-table condition. Nitrate-N concentrations during the early part of the growing season of 1990 were higher than those in 1989. These values varied from 42 to 18, 17 to 13, and 6 to 7 mg/l at 1.8 m depth, and from 36 to 18, 2 to 1.5, and 4 to 1 mg/l at 2.4 m depth under deep, medium, and shallow water-table conditions, respectively. The major rainfall events occurring in the late spring and early part of the growing season of 1990 perhaps moved a considerable amount of

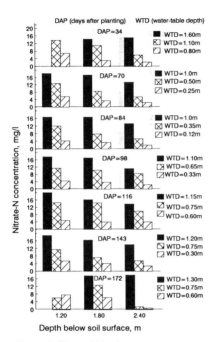

Figure 2. Nitrate-N in piezometer water samples in 1989

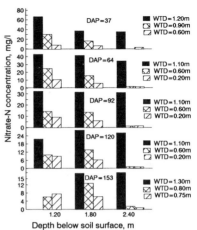

Figure 3. Nitrate-N in piezometer water samples in 1990

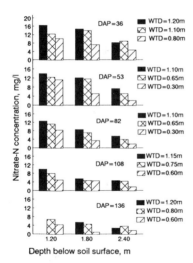

Figure 4. Nitrate-N in piezometer water samples in 1991

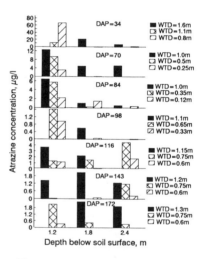

Figure 5. Atrazine conc. in piezometer water samples in 1989

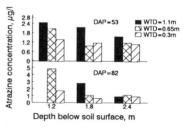

Figure 6. Atrazine conc. in piezometer water samples in 1991

nitrate-N from surface layer to the deeper depths immediately after N application. In 1990, plant growth was also restricted at the Ankeny site because of weed damage resulting from a herbicide application error. Nitrate-N concentrations, however, were lower under shallow water-tables than under deeper water-tables. The year 1991 received high rainfall during spring, but low rainfall during the rest of the growing season. Figure 4 shows nitrate-N concentrations at the Ankeny site for 1991. Water tables were almost maintained at 1.1, 0.65, and 0.3 m depths at the piezometer locations between 45 and 97 DAP. Nitrate-N concentrations observed in 1991 at 1.2, 1.8, and 2.4 m depths were very similar to those observed in 1989 under similar water-table conditions.

Concentrations of nitrate-N in ground water from suction tube water samples as a function of water-table depth at the Ankeny site are presented in Table 1. Ground water samples at 0.9, 1.2, 1.5, and 2.1 m depths showed trends of nitrate-N concentrations similar to the piezometer samples under shallow, medium, and deep water-table depths. These results also indicate that nitrate-N concentrations in ground water can be reduced by maintaining shallow water-table depth. Nitrate-N concentrations at 0.9 m depth were reduced from 21 to 6, 49 to 8, and 31 to 3 mg/l by maintaining shallow, medium, and deep water-table depths through subirrigation practice in 1989. At 1.2, 1.5, and 2.1 m soil depths, nitrate-N concentrations were always lower under shallow water-table depths than those under medium and deep water-table depths. The concentrations of nitrate-N in ground water were reduced to 2.3, 1.9, and 6.2 mg/l at 2.1 m depth at the end of the 1989 growing season under shallow, medium, and deep water-table depths. Table 1 shows that average nitrate-N concentrations in ground water in 1990 were much higher than those in the previous year. These differences were also due to higher rainfall amount in 1990 than in 1989 in addition to poor plant growth (and reduced N uptake) resulting from a herbicide application error in 1990. The trends of nitrate-N concentrations in ground water at 0.9, 1.2, 1.5, and 2.1 m soil depths were similar to those observed at 1.2, 1.8, and 2.4 m depths (Figure 3). In 1991, nitrate-N concentrations in ground water at 0.9, 1.5, and 2.1 m depths were less than 20 mg/l (Table 1). Nitrate-N concentrations were much lower when shallow water-table depths were maintained during the growing season.

The concentrations of nitrate-N in ground water in the unsaturated zone were higher than those in the saturated zone. Data in Figures 2 through 4 and Table 1 show that, by maintaining water table at shallow depths, nitrate-N concentrations in ground water could be reduced. The reduction of nitrate-N concentration at shallow water-table depth was possibly enhanced by increased denitrification. In the saturated zone where soil-air was replaced by water, the bacterial reduction of nitrate-N to dinitrogen was possibly greatly enhanced. Soil physical and chemical factors such as oxygen, organic C, pH, and would affect soil denitrification. Myrold and Tiedje (1985) reported that active denitrifier biomass was directly related to moisture content and organic carbon, whereas pH had no consistent effect on denitrification, but, under water saturated conditions, denitrification capacity could be significantly increased. These reports support the results of our WTM experiments in which nitrate-N concentrations in groundwater were greatly reduced by maintaining shallow water-table depth through subirrigation and controlled drainage.

Groundwater atrazine concentrations: Atrazine concentrations in groundwater samples taken from the piezometers at the Ankeny site during 1989 growing season are shown in Figure 5. Atrazine concentrations varied from 0 to 67 ug/L. The highest concentration of atrazine was observed at 1.2 m depth in groundwater samples taken on 34 DAP (before the start of subirrigation practice). With subirrigation, atrazine concentrations in groundwater were reduced at this site. In most of the groundwater samples taken in 1989, atrazine concentration decreased with shallow water-table conditions and decreased with soil depth with few exceptions. The fluctuations of water table might have caused some irregularities in this trend. Atrazine concentrations in groundwater decreased with time. Groundwater samples collected at the end of the growing season showed that under shallow water-table condition, atrazine concentration at 1.2 m soil depth was only 0.34 ug/L, and no atrazine was detected at 1.8 and 2.4 m soil depths. Figure 6 shows atrazine concentrations in piezometer water samples collected on 53 and 82 DAP in 1991. Samples collected on other dates are not yet analyzed and results could not be presented in this report. However, results of the pesticide analysis for 1991 samples showed a definite influence of WTM practices on pesticide concentrations in groundwater.

Table 1. NO_3-N concentrations in suction tube water samples for the Ankeny site in 1989, 1990, and 1991

Year	DAP	WTD, m	NO_3-N, mg/l				Year	DAP	WTD, m	NO_3-N, mg/l			
			0.9m	1.2m	1.5m	2.1m				0.9m	1.2m	1.5m	2.1m
1989	34	1.40	30.8	-	19.5	15.3	1990	92	0.90	30.4	29.9	31.9	31.0
		1.10	48.6	17.0	15.4	6.2			0.60	26.6	23.0	19.7	13.7
		1.00	21.0	16.1	13.7	1.8			0.30	19.0	18.3	8.5	5.3
	70	0.90	25.0	-	13.5	11.2		120	0.90	26.7	18.0	20.3	12.7
		0.50	14.9	13.6	11.4	5.6			0.60	16.6	17.4	15.5	8.1
		0.35	14.4	11.0	8.0	2.4			0.30	14.4	14.2	4.5	6.4
	84	0.80	20.2	16.7	16.8	7.4		153	1.20	-	-	14.0	17.9
		0.35	13.4	12.7	11.8	3.7			1.00	-	15.0	12.9	8.3
		0.20	12.0	9.8	7.5	1.4			0.90	13.8	12.3	9.1	5.3
	98	1.00	7.0	13.5	10.2	10.2							
		0.65	11.3	10.3	8.7	4.2	1991	36	1.20	-	-	14.7	12.7
		0.50	10.0	9.3	5.7	1.3			1.10	-	-	14.4	13.2
	116	1.10	3.0	9.1	8.1	7.9			0.90	14.8	-	12.0	9.3
		0.75	8.1	7.8	6.5	3.1		53	0.90	-	-	14.2	12.3
		0.60	6.3	5.8	5.0	2.1			0.65	19.9	-	13.9	13.2
	143	1.10	6.9	10.4	10.1	-			0.45	15.4	-	11.1	9.4
		0.75	7.5	6.6	4.1	1.4		82	0.90	14.8	-	12.1	12.7
		0.70	6.8	6.3	3.7	2.4			0.65	10.8	-	12.8	9.1
	172	1.15	-	9.9	8.2	6.2			0.45	10.7	-	8.8	4.3
		0.90	8.0	6.9	3.6	1.9		108	1.00	10.5	-	12.1	8.1
		0.80	7.5	5.9	2.9	2.3			0.75	-	-	7.2	5.5
									0.65	5.4	-	4.1	2.5
1990	37	1.20	35.5	34.9	28.1	26.7		136	1.10	-	-	10.7	6.1
		0.90	23.9	21.5	18.8	10.4			0.80	-	-	5.5	5.2
		0.75	16.5	11.8	13.8	4.0			0.70	4.2	-	2.7	2.5
	64	0.90	32.1	31.5	30.4	29.1							
		0.60	28.1	22.4	19.5	11.5							
		0.30	18.0	16.6	13.9	3.2							

DAP - days after planting; WTD - water-table depth.

Table 2. Atrazine and alachlor concentrations in suction tube water samples for the Ankeny site in 1989 and 1991

Year	DAP	WTD, m	Atrazine, µg/l				Year	DAP	WTD, m	Alachlor, µg/l			
			0.9m	1.2m	1.5m	2.1m				0.9m	1.2m	1.5m	2.1m
1989	34	1.45	4.4	-	8.4	23.1	1989	34	1.45	27.5		8.7	1.0
		1.10	-	20.6	9.0	7.6			1.10		1.3	0.0	0.0
		1.00	22.2	8.6	7.4	7.2			1.00	2.4	0.0	0.0	0.0
	70	0.90	13.7	-	-	-		70	0.90	0.0			
		0.50	-	-	2.2	-			0.50			0.0	
		0.35	4.2	-	4.3	-			0.35	6.2	0.0	0.0	
	84	0.80	12.9	8.8	6.7	7.8		98	1.00			5.0	0.0
		0.35	0.2	9.1	1.0	0.9			0.65	0.0	0.0	0.0	0.0
		0.20	0.8	1.8	1.8	1.2			0.50	0.0	0.0	0.0	0.0
	98	1.00	-	-	0.7	3.7							
		0.65	13.6	7.3	0.6	0.3	1991	53	0.90			1.0	0.2
		0.50	3.8	2.4	1.0	1.7			0.65	0.2		0.0	0.0
	116	1.10	8.4	2.3	2.6	3.4			0.45	0.0		0.0	0.0
		0.75	6.1	4.4	0.3	3.4		82	0.90	0.1		0.7	0.0
		0.60	6.8	0.4	0.3	0.4			0.65	0.4		0.0	0.0
	143	1.10	0.5	5.9	1.3	5.0			0.45	0.0		0.0	0.0
		0.75	2.2	4.8	1.0	0.4							
		0.70	2.6	0.8	0.6	0.8							
	172	1.15	-	-	0.4	0.8							
		0.90	1.2	1.2	0.2	0.2							
		0.80	2.1	0.9	0.4	1.3							
1991	53	0.90	-	-	4.6	0.6							
		0.65	1.7	-	1.3	0.4							
		0.45	1.9	-	0.7	0.3							
	82	0.90	8.9	-	4.8	0.5							
		0.65	3.8	-	1.1	0.3							
		0.45	3.0	-	0.6	0.3							

DAP - days after planting; WTD - water-table depth

Atrazine concentrations were lower under shallow than under deep water-table conditions, and these concentrations decreased with increased soil depth. In 1991, shallow, medium, and deep water table depths were maintained at 0.3, 0.65 and 1.1 m., and were almost constant during subirrigation period. This might be the reason for a better atrazine concentration trend with WTD in 1991 than in 1989.

Atrazine and alachlor concentrations in groundwater samples collected from the suction tubes at the Ankeny sites in 1989 and 1991 are shown in Table 2. In 1989, the highest atrazine concentration of 23 ug/L was observed at 2.1 m depth before subirrigation was started. Table 2 shows that shallow WTD reduced atrazine concentration at 0.9, 1.2, 1.5, and 2.1 m soil depths in 1989. The fluctuation of water-table during the growing season at each location caused some irregularities. At the end of the growing season in 1989, atrazine concentrations in groundwater reduced to 2.1, 0.9, 0.4 and 1.3 ug/L at 0.9, 1.2. 1.5, and 2.1 m soil depths, respectively, where water table was maintained at shallow depth. Water samples from suction tube also showed higher atrazine concentration at 2.1 m depth than at 1.5 m depth towards the end of the growing season when subirrigation was cut off. In 1991, water samples were collected from 0.9, 1.5, and 2.1 m soil depths. Table 2 shows that atrazine concentrations decreased with increased depth on DAP 82 in 1991. It was also observed that atrazine concentrations in groundwater was lower with shallow than with deep WTD at all soil depths. Piezometer and suction tube water samples gave identical results. These results showed a positive influence of WTM in reducing atrazine concentrations in shallow groundwater.

Groundwater alachlor concentrations: Concentrations of alachlor in suction tube water samples at the Ankeny site in 1989 and 1991 are shown in Table 2. Alachlor was not detected in many samples collected on different dates in 1989. Table 2 shows only the dates when alachlor was detected in groundwater samples in 1989 and 1991. In 1989, alachlor was detected only once in piezometer water samples at Ankeny. In suction tube water samples, alachlor was detected at shallow depths. Alachlor is less persistent in soil than atrazine. This study shows that alachlor was detected even on 98 days after its application with a concentration of 5.0 ug/L at 1.5 m soil depth in 1989. There was not enough data on alachlor concentrations and not many samples are yet analyzed from our WTM experiments to find trends of alachlor concentration in groundwater with WTM practices.

CONCLUSIONS

Analyses of data on nitrate-N and pesticide concentrations in shallow groundwater from three years of WTM experiments at Ankeny, Iowa reveal that subirrigation and controlled drainage practices are useful in reducing environmental impacts of non-point pollution. Nitrate-N concentrations in ground water could be reduced to less than EPA health standard of 10 mg/L by maintaining shallow water-table depths of 0.3 to 0.6 m during the growing season. With the increase in soil depth below any water-table position, nitrate-N concentrations decreased. Water samples collected from suction tubes and piezometers gave consistent and similar results in all three years. Both atrazine and alachlor concentrations in shallow groundwater were affected by WTM practices. With subirrigation, pesticide concentrations were minimum under shallow water-table conditions. Atrazine concentrations were reduced from 67 to 0 ug/L by maintaining shallow WTD during the growing season. Atrazine concentrations in suction tube and piezometer water samples were almost similar in 1989 and 1991. When compared to nitrate-N concentrations under similar WTM practices, atrazine concentrations followed similar trends. Lower nitrate-N and pesticide concentrations were observed under shallow WTD conditions and vice-versa. Therefore, appropriate use of WTM with subirrigation and controlled drainage is recommended for humid regions to reduce groundwater quality degradation from agricultural chemical use.

REFERENCES

Baker, J.L. and H.P. Johnson. 1977. Impact of subsurface drainage on water quality. Proc. 3rd National Drainage Symp., pp.91-98, ASAE, St. Joseph, MI.

Baker, J.L. and H.P. Johnson. 1981. Nitrate-nitrogen in tile drainage as affected by fertilization. J. Environ. Qual. 10:519-522.

Belcher, H.W. and G.E. Merva. 1991. Water table management at Michigan State University. ASAE Paper No. 91-2025, ASAE, St. Joseph, MI.

Bengtson, R.L., L.M. Southwick, G.H. Willis, and C.E. Carter. 1989. The influence of subsurface drainage practices on herbicide losses. ASAE Paper No. 89-2130, ASAE, St.Joseph, MI.

Bengtson, R.L., C.E. Carter and J.L. Fouss. 1991. Water management research in Louisiana. ASAE Paper No. 91-2021, ASAE, St. Joseph, MI.

Evans, R.O., J.W. Gilliam, R.W. Skaggs, and W.L. Lamke. 1987. Effects of agricultural water management on drainage quality. In Proc. 5th Nat. Drain. Symp., pp.210-219, ASAE, St. Joseph, MI.

Fausey, N.R., A.D. Ward and L.R. Brown. 1991. Water table management and water quality research in Ohio. ASAE Paper No. 91-2024, ASAE, St. Joseph, MI.

Hallberg, G.R. 1986. Overview of Agricultural chemicals in ground water. In Agricultural Impacts on Ground Water. National water well Association, Warthington, OH. pp.1-67.

Hubbard, R.K. and J.M. Sheridan. 1989. Nitrate movement to ground water in the Southeastern Coastal Plain. J. Soil Water Conserv. 44(1): 20-27.

Kalita, P.K. 1992. Measurement and Simulation of Water Table Management Effects on Groundwater Quality. Unpublished Ph.D. Thesis, Iowa State University, Ames, IA.

Kalita, P.K. and R.S. Kanwar. 1992a. Shallow water table effects on photosynthesis and corn yield. TRANSACTIONS of the ASAE 35(1), pp.97-104.

Kalita, P.K. and R.S. Kanwar. 1992b. Energy balance concept in the evaluation of water table management effects on corn growth: experimental investigation. Water Resour. Res.(in press).

Kanwar, R.S. 1990. Water table management and ground water quality research at Iowa State University. ASAE Paper No. 90-2065, ASAE, St. Joseph, MI.

Kladivko, E.J., G.E. Van Scoyoc, E.C. Monke, K.M. Oates, and W. Pash. 1991. Pesticide and nutrient movement into subsurface tile drains on a silt loam soil in Indiana. J. Environ. Qual. Vol.20: 264-270.

Logan, T.J., G.W. Randall and D.R. Timmons. 1980. Nutrient content of the tile drainage from cropland in the North Central Region. NC Regional publ. 268. Res. Bull. 119. OARDC, Wooster, OH.

Melvin, S.W., R.S. Kanwar, and D.G. Baker. 1990. Evaluation of a dual level subirrigation system. In Proc. 3rd National Irrigation Symposium, pp.204-210, ASAE, St. Joseph, MI.

Myrold, D.D. and J.M. Tiedje. 1985. Establishment of denitrification capacity in soil: effects of carbon, nitrate and moisture. Soil Biol. Biochem. 17(6): 819-822.

Skaggs, R.W. 1987. Design and management of drainage system. In Proc. 5th National Drainage Symposium, pp.1-12, ASAE, St. Joseph, MI.

Skaggs, R.W., R.O. Evans, J.W. Gilliam, J.E. Parsons and E.J. McCarthy. 1991. Water management research in North Carolina. ASAE Paper No. 91-2023, ASAE, St. Joseph, MI.

Thomas, D.L., M.C. Smith, G. Vellidis, C.D. Perry and B.W. Maw. 1991. Status of water table management research in Georgia. ASAE Paper No. 91-2022, ASAE, St. Joseph, MI.

ASSESSMENT OF SOIL WATER REGIME REQUIREMENTS OF WILD PLANT SPECIES

G. SPOOR, J. M. CHAPMAN, P. B. LEEDS-HARRISON*

ABSTRACT

Little quantitative data is available on the water requirements of wild plant species. This paper describes an approach for determining this in a form usable by engineers and habitat managers. Use is made of data obtainable from established wildlife habitats where little change has occurred in water regime over a considerable time. These habitats in moist and wetland areas usually have some form of water control measures, and the water regime varies across the area due to local changes in elevation and to differences in position relative to the boundary water source or sinks. Detailed botanical analyses are made at specific locations and the past water regime estimated at each position. Species/water regime relationships can then be determined, by combining the two. The water regime estimates are made using recently developed Silsoe College models capable of determining the regime at any point in an area for ditch arrays of various geometries.

To extend the quantified results to a much wider range of species, use is being made of a very comprehensive but qualitative wetness classification developed by Ellenberg for European species. Water regime limits can be placed on the Ellenberg classes on a basis of the specific requirements of the species investigated.

INTRODUCTION

Within the agricultural sphere, the need for sound quantitative information on the water and aeration needs of crops has long been recognised. Such information has been considered an essential pre-requisite within agriculture for efficient soil and water management for crop production. Where existing soil water regimes were found wanting, appropriate drainage or irrigation measures were instituted to rectify any problems. This has not been the case in many natural environment situations where at best, any quantification of need has been largely subjective and assessed in general terms of wetness or dryness. In addition, rather than positively managing water, things have often been left to nature and the habitats which developed accepted for what they were.

Requirements within the natural environment are changing rapidly as a result of new demands and the subjective assessments of wetness and dryness available to predict change and requirement are becoming increasingly inadequate. Environmental impact assessments are mandatory within all river, coastal and flood protection schemes likely to have an impact on the environment. Significant deterioration has occurred in a number of habitats as a result of water regime change, stimulating interest in restoration

---

* G SPOOR, Professor, J M CHAPMAN, Senior Research Officer, and P B LEEDS-HARRISON, Senior Lecturer. Silsoe College, Cranfield Institute of Technology, Silsoe, Bedford MK45 4DT. UK.

work. With expansion of the agricultural set-aside scheme in Europe, there is considerable interest in possibilities for converting some of the land back to habitat. Interest in modifying conditions within an existing habitat to improve the environment for specific species, such as breeding wader birds is also increasing. All these demands and requirements can be best met and most efficiently satisfied with sound quantitative information on the water regime requirements of particular species and communities.

The quantification of the water regime requirements of natural species could be approached in a number of ways. If, however, their aeration and water needs could be defined in a similar way to those for agricultural crops, all water control theories and models currently available and in use in agriculture, could be readily applied in these natural environment situations. Recent work at Silsoe College has been examining this possibility and assessing the feasibility of applying such data in field situations.

APPROACH

Two possible approaches to defining the water regime requirements of wild plant species were considered. The first was based on following lysimeter studies with individual species and the second making the assessment from existing habitats. Although individual lysimeter studies offer possibilities for considerable control and apparent accuracy, considering the number of species involved, the scale of the problem is vast, as would be the time required. The approach adopted, therefore, utilises where possible known hydrological and ecological information for existing habitats and uses models to link plant presence and response to past soil water regime.

Within what is often perceived by the ecologist and others to be a fairly uniform wetland site and habitat containing a diversity of species, closer examination reveals significant local variations in both topography and species composition. The wetland areas also often contain skeletal ditch systems of varying geometric shape which induce different phreatic surface levels across the area. These variations in surface elevation and spatial phreatic surface levels combine to produce quite distinct differences in local water regime. The water regime differences tend to correspond with areas of different species composition, thus providing the past water regime is known or can be predicted, the requirements for the species and community growing in those local areas can be defined. The apparent uniform field is in effect a collection of natural lysimeters which have been subjected to different water regime treatments.

The in field methodology adopted to define water regime requirements can be summarised as follows:

a. identify sites having available as much species diversity and hydrological information as possible and where water level behaviour in surrounding watercourses has been fairly constant for some years

b. identify species composition and distribution in specific locations using quadrat analysis

c. determine surface elevation and spatial location of quadrats relative to the surrounding ditch system

d. determine past water regime at each quadrat location

e. relate water regime data to the species present at each quadrat.

RESEARCH METHODOLOGY

Techniques are well established for habitat surveys and the analysis of hydrological data. Methods for determining past soil water regimes within fields and quantifying them in a usable form for engineers and habitat managers have, however, until recently been unavailable. Although many habitat surveys have been completed over long time periods, the ecologist rarely, if ever, monitored anything associated with soil water. Determination of past water regime has, therefore, usually to be achieved by prediction using whatever soil, hydrology and climatic information is available for the site. Methods have recently been developed (Youngs et al., 1989, and Youngs, 1992) to predict temporal and spatial water table position and movement within fields using existing or readily obtainable channel stage data, soil and climatic information. These water table estimates can be extended where necessary to quantify the soil moisture status in the unsaturated zone.

Water Table Movement Model

The model used to estimate water table movement with time within fields is that developed by Youngs et al. (1989). This model predicts seasonal water table movement in flat low-lying lands intersected by a network of ditches. It has been derived from land drainage theory and non-steady water tables are assumed to behave as a continuous succession of steady states. The flux through the water table is given by the sum of the components due to rainfall and evaporation through the soil surface and due to the water released or taken up by the unsaturated soil above the water table. A simple steady state drainage equation is used to determine the relationship between water table height and flux and soil specific yield is assumed to have a constant value for all water table positions.

The model requires data on soil hydraulic conductivity and specific yield, a knowledge of the geometric configuration of the flow region, boundary water levels and meteorological records. Output from the model is shown in Figure 1 for the water table movement at one site in 1987. It illustrates the seasonal change of water table height and shows good agreement with dipwell observations.

Specific yield and hydraulic conductivity assessments can be made from a knowledge of soil structure and texture or by direct measurement in the case of saturated hydraulic conductivity. It is important, however, to note that the model is quite sensitive to specific yield, so great care has to be exercised in choosing an appropriate value, particularly in layered soils.

Estimates of potential evaporation demand are made from climatic data. Actual evaporation is taken to be the potential value except when the soil condition limits capillary rise i.e. with deep water tables. When this situation arises a conductivity factor dependant on soil type is used. Little difference in simulated water table response occurs between using either daily values for rainfall and evaporation or those averaged over a week with weekly ditch water level readings. Even with a time increment of one month, the simulated response does not vary very much.

Assessment of Water Table Variation over whole Area

The actual water table position or temporal variation at any particular location in a field is assessed using the seepage analysis model of Youngs (1992). This model allows the development of contour maps such as that illustrated in Figure 2 which shows the water table variation over an area surrounded by water filled ditches. The shape factor values W assigned to the map contours, indicate the water table height at that location, as a fraction of the mid drain water table height that would occur for the same water flux, if the field was drained with a parallel drainage system at a spacing equal to that between the closer ditches. (Example: for an estimated mid drain water table height above ditch

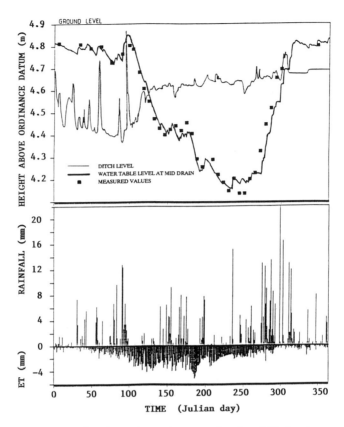

Figure 1. Comparison of measured and modelled water table heights in the centre of a rectangular field surrounded by water-filled ditches.

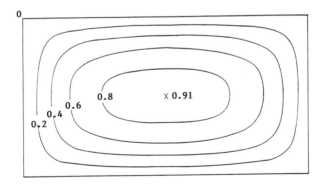

Figure 2. Contours of dimensionless shape factor values W in a rectangular field surrounded by water-filled ditches.

level of 0.5 m with a parallel ditch system, the water table height at the centre of the area in Figure 2 would be 0.455 m). These contours of W also indicate locations where there is the same seasonal range of water table rise and fall. This model has been developed for application in rectangular fields bounded by ditches on three or four sides, and in triangular, trapezoidal and elliptical shaped fields. Such a range of field boundary conditions allows the model, with appropriate approximations, to be applied to a wide range of practical situations.

Quantifying Soil Moisture Status in the Unsaturated Zone

The water table movement model allows estimates to be made of the lower boundary of the unsaturated soil i.e. the water table. Accounting for the capillary fringe, capillary rise or infiltration as influenced by soil type and climate, predictions of soil moisture tension in the unsaturated zone can be linked to soil moisture release data, to indicate likely moisture stress levels in the root zone.

WATER REGIME/SPECIES RELATIONSHIPS

Within the agricultural sphere, crop water regime requirements are conveniently defined in terms of parameters such as water table depth or moisture deficit. To be useful, similar or equivalent parameters must be identified to define water regime/wild species relationships. This requires consolidating the within year and between year variations in water regime into representative parameters. A number of approaches are currently being investigated to achieve this.

Parameters

Parameters currently being considered and examined for their simplicity, ease of determination and discriminating qualities between species include the following:

   a. mean winter water table depth, when soils in UK are at field capacity and aeration stress is likely to be the major problem

   b. mean summer water table depth, with soil usually below field capacity, when lower water tables and water deficits are likely to prevail

   c. mean length and timing of flooded periods

   d. mean maximum soil moisture deficit

   e. variation, using the sum of the exceeded water table or moisture deficit level (SEW approach), relative to a steady state water table position or given deficit level respectively.

To determine the water table depth and deficit values, the elevation and position of the botanical survey quadrat locations are determined. The water table elevations at these same locations are predicted using the water table movement and variations models. Water table depth below the surface and/or moisture deficit at any particular quadrat location can then be determined and related to the species present.

Species Classification and Quantification

Due to the large number of species to be handled, it would be more convenient to group plants with similar water regime requirements into classes and to define the requirements for the class rather than for each individual species. This approach also opens up the possibility of making best use of appropriate existing subjective classifications.

One of the most comprehensive and relevant classifications for the European flora is that developed by Ellenberg (1988) for Central European conditions. This classification assigns to a very comprehensive range of species, detailed subjective information on tolerance levels for growth factors such as wetness, soil nitrogen status, pH and temperature. In the wetness category, Ellenberg divides the species into 12 classes whose growth conditions range from continuous submergence to very dry.

Field results show that some species are not particularly sensitive to water regime, being found growing over a wide range of water table depths. Other species are only found within narrower water regime ranges and these are currently being used as indicator species to correlate the Ellenberg class or number with quantitative measurements of the chosen parameters. Some success has been achieved in assigning mean depths to water table during winter and summer to the Ellenberg classes, but work is continuing with the other parameters with the objective of achieving greater sensitivity and discrimination. A very tentative relationship between water table regime and Ellenberg number for the species observed is identified as follows for conditions in southern England:

Mean depth to water table (mm)		Ellenberg F value	Drainage Condition
Winter	Summer		
0-100	200-300	9, 10	Badly drained with some flooding
100-200	300-400	7, 8	Badly/moderately drained
250-350	450-550	4, 5, 6	Well drained

It must be stressed the above relationship is very tentative and work is continuing, to significantly improve confidence levels.

EXTENSION OF CLASSIFICATION TO OTHER SPECIES

Whilst the ideal situation is one where the requirements of each species are examined in the field individually, time constraints and the lack of available hydrological data for many habitats are likely to preclude this. Providing, however, there is a high level of confidence in the ranking of species within existing subjective classifications, the boundary water regime parameters can be applied to these other species. Two approaches have been used successfully to assess this level of confidence, these being based on studies on ditch banks and in wetland field areas. (In the UK context, these studies were made to validate the suitability of the Ellenberg Classification, developed in Central Europe for UK conditions).

Considerable botanical data was available from ditch bank studies on the range of heights above ditch water level at which specific plant species were found. Such sites provide a gradation in water regime from ditch water level to bank top and into the field. Figure 3 shows the very clear relationship derived from observations in 24 peat ditches between plant position on the bank and its Ellenberg moisture value. The drier loving plants, corresponding with the lower Ellenberg values, being mainly found higher up the bank. Analysis of plant distribution in ridge and furrow areas also provides similar information.

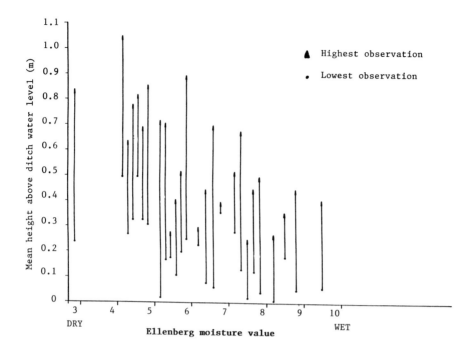

Figure 3. Distribution of plant species on ditch banks in relation to mean height above water level and Ellenberg moisture values (24 sites)

In wetland field areas, measured or predicted information on depth to water table and plant composition at that location can also be used to assess ranking confidence. Statistical programmes such as CANOCO (Ter Braak, 1988), commonly used by ecologists, are available for assessing how a range of species respond to external factors. Figure 4 shows an ordination diagram based on this approach, indicating the position of different species relative to mean annual depth to water table and fluctuation around this depth. The Ellenberg values are superimposed on the Figure and show good agreement with the axis representing mean annual depth to water table.

CONCLUSION

The techniques identified for assessing the soil water regime requirements of wild plant species, show considerable promise for providing engineers and others with sound quantitative data for assessing the impact of water management measures on wildlife habitats.

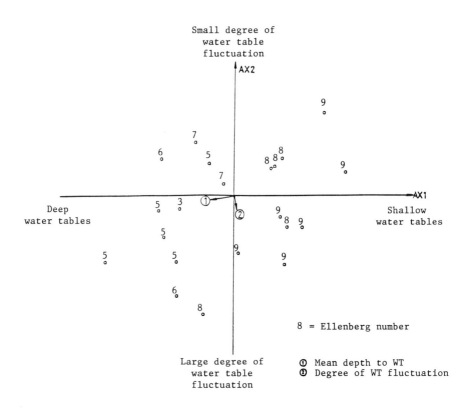

Figure 4. Ordination diagram with selected species (o) and environmental variables (arrows)

ACKNOWLEDGEMENTS

The authors are grateful to the River and Coastal Engineering Group of the Ministry of Agriculture, Fisheries and Food, UK, for their support of this work and to inputs from the Institute of Terrestrial Ecology, Monks Wood, Cambridgeshire.

REFERENCES

1. Ellenberg, H. (1988). Vegetation ecology of Central Europe. 4th ed. Cambridge University Press. (First published in German as *Vegetation Mitteleuropas mit den Alpen*).

2. Ter Braak, C.J.F. (1988). CANOCO - a FORTRAN program for canonical community ordination by [partial] [detrended] [canonical] correspondence analysis, principal components analysis and redundancy analysis (version 2.1). Technical Report LWA-88-02, Agricultural Mathematics Group, Box 100, 6700 AC Wageningen, The Netherlands.

3. Youngs, E.G., Leeds-Harrison, P.B., Chapman, J.M. (1989). Modelling water table movement in flat low-lying lands. Hydrological Processes, 3, 301-315.

4. Youngs, E.G. (1992). Patterns of steady groundwater movement in bounded unconfined aquifers. Journal of Hydrology, 131, 239-253.

ENVIRONMENTAL CONSEQUENCES OF MAJOR IRRIGATION DEVELOPMENT

A CASE STUDY FROM SRI LANKA

Nalini Amarasekara[1] and Khalid Mohtadullah[2]

ABSTRACT

The importance of environmental impacts of irrigation development has been increasingly recognized during the past three decades. Most of the irrigation development projects however, have come to attract a lot of criticism from environmentalists for a variety of reasons. Any large scale irrigation projects are bound to have environmental impacts. If these impacts are properly identified in the planning stage, it would help in minimizing the adverse effects and in maximizing the advantages of the projects. Recently there has been an curiosity to contemplate and evaluate major irrigation projects for their impacts on the environment. This paper discuss the environmental impacts of the most recently developed major irrigation project in Sri Lanka with particular reference to adverse effects like salinity, waterlogging, health hazards, loss of land fertility, displacement of people, change in river regime and changes of ecological balance of flora and fauna. Reformatory measures to minimize such adverse effects of the project are outlined.

Key words: Environmental Impacts, Irrigation, Kirindi Oya, Sri Lanka.

INTRODUCTION

The past 25 to 30 years have been a period of unprecedented irrigation development in many countries of the developed and developing world. A number of storage reservoirs in the world with a capacity larger than 100 million m3 has increased, and most of the major river valleys developed rapidly during the years 1951-85(Biswass, 1989). Although, irrigation developments of all types i;e, major and small scale, are progressing at a rapid rate and many beneficial effects are being recorded, the environmental impacts of some irrigation projects have not been what water resources planners had expected. Serious problems emerging from time to time have drawn the attention of all concerned to the fact that irrigation projects in both developed and developing countries, may yield a mixture of desirable and undesirable effects. The introduction of <u>Garrision</u> irrigation project in North <u>Dakota</u> on the <u>Mississippi-Missouri</u> river has caused serious environmental impacts and the loss of a sizable area of cultivated land (Davis,1983).

---

[1] Environmental Scientist

[2] Director for Research, IIMI, Colombo, Sri Lanka

As pointed out in *Our Common-Future* - the recent report of the World Commission on the Environment and Development, the so-called Brundtland Commission - good development will protect and enhance the environment. Attention paid to environment concerns will strengthen developmental progress and prospects. To achieve these goals, in terms of irrigation management, must be planned, implemented and operated in an "environmentally sound" way. What is needed is development and management of water resources such that the resources base is maintained and enhanced over the long term.

ENVIRONMENTAL SIGNIFICANCE IN SRI LANKA

Since the late 1970s, with the introduction of the new economic development policy in Sri Lanka, a large number of development projects have been lunched both by the public and private sector. Such development projects have often paid little or no attention to associated environmental consequences. It was soon realized in serious adverse environmental impacts such as soil erosion flooding, loss of valuable fauna and flora, and loss of historical and cultural resources. Consequently, the need for integrating environmental economic and social considerations with the planning and decision making process was realized and the government of Sri Lanka decided to introduce the Environmental Impact Assessment (EIA) of all development projects as an aid to the decision making process through which possible adverse environmental impacts of proposed development projects could be minimized or prevented.

Accordingly, it was made mandatary that all development projects be subject to an EIA from January 1984. Even though, over the years enough awareness has been created among the water resources planners to consider EIA as a part of their project planning, due to lack of guidelines or understanding, the application of EIA for irrigation projects has not been received wide attention even in large irrigation systems in Sri Lanka. So far, among the water resources development projects, EIA had carried only Accelerated <u>Mahaweli</u> Development project(TAMS, 1990). Most of the major river basins have already developed for multipurpose programs, and a large sum of money has been spent so far, environmental consideration had not take into account (Table 1).

Table 1.   Investment of major irrigation   1950 - 1982 (Rs '000).

Year	Major works	River basin development	Total
1950-54	171.9	84.7	256.6
1955-59	133.8	35.3	169.1
1960-64	153.6	15.3	168.9
1965-69	245.3	20.4	265.7
1970-74	175.0	280.7	455.7
1975-79	362.0	1,654.2	2,016.2
1980-82	1,200.3	7,100.0	8,300.3
Total	2,441.9	9,190.6	11,632.5

<u>Source</u> - IIMI (1990)

MAJOR IRRIGATION SYSTEMS

In Sri Lanka, major irrigation schemes are classified as those with command areas ranging from 200 to 400 hectares (ha) to about 10,000 ha (500 to 1,000 acres to about 25,000 acres). Currently, major irrigation systems are governed under the Irrigation Ordinance. Their design and construction, and operation and maintenance are undertaken by the Irrigation Department, and the management of selected projects is entrusted to the Irrigation Management Division established in 1984.

Many of the major irrigation reservoirs and their irrigation systems have been in operation for over 5,000 years. Throughout the long history of Sri Lanka, irrigation has been of vital importance. In ancient days, it formed the basis on which the Sinhala civilization flourished. The extensive network of canals and reservoirs constructed by the kings of Anuradhapura and Polonnaruwa brought prosperity to their people. However, the irrigation works were vulnerable targets in times of war. Their destruction often led to acute crises resulting in famine and disease, and finally to a general decline of irrigation. Since Sri Lanka (then Ceylon) gained independence in 1948, restoration work has been undertaken since most of these systems, which had been in an abandoned stage. Major irrigation projects were implemented recently at Mahaweli, Gal Oya, Kirindi Oya, and Walawe Ganga.

PROJECT AREA

The Kirindi Oya Irrigation and Settlement Project (KOISP), is the largest irrigation project in the south of Sri Lanka and is situated in the Hambantota District. KOISP, projected to irrigate 21,500 acres of agricultural land, affecting about 4,000 resettled families in several hamlets and villages.

The specific objectives of the project were:

1. To cheat a reservoir across the Kirindi Oya at Lunugamwehera to store 230 mcm water

2. To enable to irrigate 4100 ha of new agricultural land obtained by felling an equal area of forest and scrub etc

3. To provide improved irrigation facilities to 4584 ha of existing paddy land by augmenting existing tanks and reservoirs

4. By the end of both phases to settle about 8325 farm families in the area, establishing 28 hamlets and 5 village centers

THE CHALLENGE OF THE PROJECT

The project has a great deal of criticism on the basis that no proper Environmental Impact Assessment(EIA) was done prior to implementation, feasible alternative sites for the construction of the reservoir were not adequately considered. The inflow of Kirindi Oya has been a subject of debate ever since the project was initiated. The project has been in operation for the past five years and during this period continuous shortage of water has been experienced. There have been number of studies made about the water potential of Kirindi Oya at the Lunugamvehera reservoir and the area which can be irrigated.

It should be noted that there was another selection for alternative site for the construction of the reservoir. But due to political and other influence the present side, Lunugamwehera, was selected for construction. Only justification for taking up Lunugamwehera for construction then, was that there would be delay if alternative site, Huratgamuva was to be taken up for investigation. In the event of various reasons construction of Lunugamwehera did not begin till late 1978. The estimated time for construction at that time was 4 years. However, for various reasons construction actually took double that time and was completed only in 1986. Meanwhile cost of construction also escalated widely, and it was said that the original estimate ultimately increase by a factor of four when all claims were add up after completion. And the Huratgamuwa site was never investigated (Mendis,1990)

Even though, the objectives of the project appear to be useful, the project has drawn great deal of criticisms since its inception.

The original project proposal has been contained to grow 60% of Other Field Crops and 40% irrigated rice(ADB 1982). But the canal net work has been constructed only for irrigated rice only. Unfortunately, Lunugamwehera reservoir failed to provide expected water for the irrigable area.

ENVIRONMENTAL IMPACTS

Positive Impacts Intended: *Flood Control:* A considerable hydrological effect of construction of the Lunugamwehera Reservoir is the disappearance of flood peaks. The complete suppression of floods had led to the disappearance of floodlands which have dried up. There is much benefit for economic development from Lunugamwehera reservoir which was constructed across the Kirindi Oya. Among such benefits, are control of flood, annual, seasonal, weekly and diurnal runoff redistribution and creation of water surface for irrigation purposes, transformation of the hydrological regime for rational land-use, recovery of non productive land for accumulation of water, attenuation or complete elimination of natural phenomena such as floods, mud-flows, siltation of canals, etc, improvements of natural conditions in the neighboring area; milder climate, construction of a special network of water bodies, etc.

Increase of agricultural Production and Employment generation: The stated aims of Phase 1 of the project are "increased agricultural production particularly of paddy; employment generation; foreign exchange savings; and land settlement" Under the Phase II, the overall objectives are reiterated but they are to be achieved through, in addition to irrigation and settlement,forestry and livestock development and crop diversification.

Land Ownership: With the settlement scheme, most of farmers migrated the area with the legal settlement title. These people haven't had any legal provision for the land.

Negative Impacts Expected: *Changes of Hydrological Regime:* The construction of Lunugamwehera Reservoir and its operation involve a number of undesirable environmental changes. The most significant of these are (a) inundation of lands (b) transformation of reservoir bed and shores (c) raising of groundwater levels. Downstream of the reservoir the river valley landscape also has changed, especially in the case of the seasonal or long-term runoff regulation. Significant environmental changes are observed downstream of the reservoir along the river regulated flow all though these are not visible. Changes of flood regime, spreading of vector-borne diseases due to stagnant water, changes of quality of water are some of these. Traditional flood plain cultivation has been abandoned automatically with the construction of the reservoir. Changes in stream flow and water releases from the dam have affected irrigation systems in the lower basin.

There are considerable impacts on reduction of flood. Two negative impacts have been identified. One is a lot of fertility soil from the upstream flowing towards the downstream with the flood water is deposited on the paddy fields. With out floods now no fertility soil deposits on the field in order to increase production. Second is reduction of floods is harmful to aquatic ecology in Bundala lagoon. Aquatic water ecosystem, aquatic life rare and valuable aquatic fauna and fisheries development have been reduced significantly in the lagoon. The low flow hydrological regime of the Kirindi Oya has changed substantially by more than 20 per cent. Due to this change, adverse effects have occurred in the old area such as salinity, increase of malaria, decrease of crop production, and loss of fertile soil. The rotational water distribution from the reservoir is not adequate to maintain pre-diversion ecological balances in the downstream.

Kirindi Oya is developed as a "one-off" river basin development concept, i.e., one big reservoir which collects all the runoff from the catchment area and from which reservoir water is supplied to the whole irrigation system. To achieve this "one-off" system, the already existing small reservoirs under this new reservoir (except the Ellagala system) were broken and developed as new irrigable areas. The one-off system has the advantage, from hydraulic engineering point of view, that one reservoir covers a comparatively smaller area for the volume stored. Thus, the evapotranspiration losses will be less. These tanks have created an attractive set up for the area as well as a sanctuary for birds.

Watershed Conditions and Changes of Land Use:

The natural vegetation of the area has already been seriously affected by long-term extensive loot land use, shifting cultivation as well as illicit felling. Development of the irrigation service area depended on the clearing of nearby 6,000 hectares of secondary forest area of primarily low scrub, under uncontrolled slash and burn cultivation, which results in soil losses due to erosion. Forest clearance and agricultural activities have already had a drastic detrimental impact on the hydrological function of the catchment of the main reservoir and other five tanks in the project. Upstream of Kirindi Oya, chena cultivation is the most practiced form of traditional agriculture. Since the project was initiated, very little attention has paid to the current problems of chena cultivation.

Losses of Forestry, Wildlife and Birds: The KIOSP area is an existing bird sanctuary. The Bundala Lagoon and the existing five major tanks in the project service area are the prime importance focal points for water birds including many migrants pelican, flamingo, heron, duck, cormorant, stork, plover, teal, among others. It also contains three endangered turtle species. Bundala is probably the most threatened area.

The terms of reference of the Feasibility Study requested an evaluation of the environmental impact of the project, and possible "methods and feasibility level designs of measures required, if any, to eliminate or minimize undesirable environmental effects." The feasibility study concluded that drainage into Bundala bird sanctuary had to be prevented because of its brackish nature, and its special value for bird life.

The present drainage into Bundala has led to serious problems within Hambantota district because it appears that about 300 shrimp fishermen were dependent on shrimp cultivation in the lagoons of the sanctuary. These shrimp cultures require brackish conditions, and are very vulnerable to pollution. Thus, they would be severely threatened by the mass inflow of polluted drainage water.

In the project proposal (ADB 1982), it has been mentioned that a half-mile buffer zone should be established to separate wildlife from inhabited or cultivated areas. The project authority has planned to develop additional wildlife corridors around the project area with the ultimate objective of providing a linked forest reserve system throughout the country. But the proposed wildlife corridors and the buffer zone could not be developed as planned. The ultimate result is that many elephants are roaming the project area. In the dry season many elephants are forced to migrate outside the Yala national park due to luck of water (Table 2). The increasing elephant man conflicts have come about in the last 10 years or so. Raids on chana farms used to be limited by careful watching and were not a necessity for elephants. The change has come because of the massive expansion of farming especially under the settlement scheme.

Table 2

Estimation of elephant population in the parks towards the project

Location	Approximate number of elephants	
	Minimum	Maxium
Ruhuna National Park, Blocks 1,11,111 1V, V and Yala South National Park including Palawatte	350	400
Hambntota District outside Ruhuna National Park(between Walawe Ganga and Kirindi Oya)	150	160
Uda Walwe National Park and environs	150	200

Elephants have been driven back from the former ranges of Lunugamwehera and Bundala sanctuary. They are squeezed into remaining jungle pockets and some forest plantations and chena plots at night. The hope that they might find sanctuary in the 3 parks including new Lunugamwehera Park, has not been fulfilled. In part, they are trapped and can not follow their former migration paths to reliable dry season feeding: but also the parks do not have sufficient fodder and water.

Socioeconomic Impacts: Most of the settlers lacked the skills and background to successfully colonize the resettlement site. The main reason for this is the failure to take into account the needs of concerned people which led to conflicts between new settlers and the local groups. The most difficult problems the settlers faced were in obtaining domestic water and transport services. The life in general of the project has described on the living conditions. Most of these settlers are from adjoin districts and they have started the life here from the beginning of the project. This settlement scheme has been a boon to many poor farmers in adjoining districts but it has also been characterized by irrigation inefficiency, resulting in low agricultural yield and poor rate of return on national investment, inequitable distribution of water, lack of cooperation among settlers and inability of settlers to maintain their system.

Salinity: Since KOISP was initiated, in newly developed areas, drainage facilities have been neglected mainly because of capital investment. In old area also there is no proper drainage canals due to encroachment of drainage canals. Existing un proper drainage system caused secondary salinity. The extent of the problem however in Kirindi Oya is not known. The water management study done by Agrar-und Hydrotechnik in 1987 refers to about 110 ha as salt land that can not be reclaimed. It also cites 500 ha in the coastal part of the project and 200 ha in the Badagiriya system that could be reclaimed.

Surface water quality: The most important physical quality of surface water is turbidity. The turbidity of kirindi Oya is expected to vary over a wide range, due to possible variations of flow. The chemical quality of the selected sources could be expected to have high variations during the year. The irrigation canals flow through highly agricultural areas and the surface runoff will carry large quantities of insecticides and weedicides which include compounds of Arsenic, Copper, Phosphorus and synthetic organic materials. The existing irrigation tanks are bound to contain them in harmful quantities.

The lower reaches of Kirindi Oya flows through the dry zone and is a major source for bathing and washing for the inhabitants of its surrounding area. Thus faecal pollution is inevitable due to the human and animal waste discharges. Further more, the surface runoff will contain excreta from human, cattle etc. and also many harmful bacteria from top soils. This faecal pollution is more aggravated by the minimum flow during drought periods, due to less dilution and salt discharges from impounding.

Channel Maintenance Problems: Due to un proper maintenance, serious loss of conveyance capacity was occurring but a problem of quantification arises. The effects of sediment and weeds on channel performance cannot adequately be judged although it appears that the sediment loads in channels are not yet high. The main problem of canal maintenance is inadequate funds as well as farmers participation.

Sedimentation: As other reservoirs, Lunugamwehera storage reservoir also retains virtually all of the sediment inflow until the reservoir capacity is so depleted its efficiency is reduced. This reservoir has a large capacity in relation to the annual flow of the Kirindi Oya and it has a long useful life. Several studies have indicated that even 100 years sedimentation may not be a problem. Inadequate data however, in this nature reduced the analytical capacity in KOISP. It is apparent that most studies have tended to assess the sediment problem in a similar and rather simple manner by comparison of different sediment yield estimates ranging from: field measurements, regional recommended values, universal values. Design aspects considering actual deposition (sediment profile) which depends on sediment constituency, reservoir size, site topography, stream flow regime and reservoir operation has been given little attention in KOISP project.

Health: Malaria is endemic in the project area. It is revealed 11 species of anopheline and 27 of culicine found from the KOISP. Among these are a number of species known to be vector or potential vectors of malaria, filaria and arboviral diseases. The most vector-borne diseases in KOISP are malaria, filariasis, dengue and Japanese encephalitis. A vast network of irrigation canals, not maintained properly and even not lined properly have created a good breeding ground in KOISP.

REFERENCES

Biswas, Asit K: (1989) Use Knowledge where it counts. *In International Irrigation Management Institute Review. Vol. 3 No. 1. IIMI, Sri Lanka.*

Davis, C: (1983) Garrision Irrigation Project. *A talk delivered on 14 September 1983 at Department of Environment, Government of India.*

TAMS: Environmental Impact Assessment, Accelerated Mahaweli Program. (1981) *Tippetts-Abbett-Mc-Carthy-Stratton Engineers, Architects and Planners. U. S. Agency for International Development. New York.*

Mendis, D.L.O; (1990) *Irrigation Development and Underdevelopment in Southern Sri Lanka. In Economic Review. Published by Peoples' Bank. Sri Lanka.*

ADB, (1982) Appraisal Reformulated Kirindi Oya Irrigation Settlement Project (Phase 1) in Sri Lanka. Asian Development Bank, Report no. SRI: AP. 28, November 1982.

Hydraulic Effects of Upstream Drainage Reconstruction on Existing Downstream Floodplains for Small Watersheds in Delaware

Richard T. Smith P.E., Ronald F. Gronwald P.E., and John O. Kelley[1]
ASAE Member   ASAE Member

ABSTRACT

Reconstruction of drainage channels in Delaware has evolved during recent years to minimize construction in the downstream reaches where channels have contiguous wooded flood plains containing freshwater wetlands. These flood plains serve to contain flood waters during storm events greater than the drainage design frequency. The effect of upstream channel reconstruction must be evaluated in order to minimize or eliminate the construction in these downstream wetland areas. There is potential for damage in these lower reaches if higher flood flows occur. The offsetting effect of reduced runoff due to increased infiltration in the upstream reaches is evaluated and compared to the increased hydraulic efficiency of the reconstructed channels. Watershed models are developed using the SCS TR-20 computer program. These models will evaluate pre-construction and post construction flood flows taking into account the soil and channel alterations. This information is used to determine where full channel reconstruction must begin to provide an outlet for the upstream drainage needs while minimizing the amount of construction performed in the downstream flood plain outlet.

---

[1]The authors are Richard T. Smith P.E., Drainage Program Administrator, Delaware Department of Natural Resources and Environmental Control, Division of Soil and Water Conservation, Georgetown DE, and Ronald F. Gronwald P.E., State Conservation Engineer, John O. Kelley, Civil Engineering Technician, USDA Soil Conservation Service, Dover, DE.

BACTERIAL DENSITY CHANGES UPSTREAM AND DOWNSTREAM OF A SLUDGE DISPOSAL FACILITY

Ling Cheng, Brett Buras, D. M. Griffin, Jr., and James D. Nelson[1]

ABSTRACT

Fecal coliform and fecal streptococcus densities were determined daily for 92 days (August-October 1991) at five points upstream and downstream of the Ruston landfarm. All data sets were found to follow a lognormal distribution with negligible autocorrelation. In addition, the fc/fs ratio was computed daily and found to be unreliable as an indicator of the source of contamination in this natural system.
Keywords: coliforms, water quality, sludge, bacteria, sludge disposal.

Impetus for This Study and the Experimental Design

This study was initiated as a result of the rather unique design of the 55-acre (22.2 ha) sludge disposal facility in Ruston, Louisiana, hereafter referred to as the "landfarm." Waste activated sludge is pumped from the Ruston sewage treatment plant to a storage lagoon at the landfarm. Periodically, sludge is sprayed on the ground surface, using one of 28 nozzles located uniformly over the area. Hay is grown on-site, harvested periodically and sold. The topography of the site may be described as a large drainage

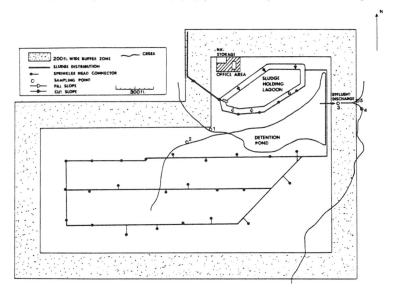

Figure 1. Sampling Locations.

---

[1]Ling Cheng and Brett Buras are graduate students in civil engineering and D. M. Griffin, Jr., and James D. Nelson are professors of civil engineering at Louisiana Tech University, Ruston, Louisiana.

swale bordered on two sides by rolling terrain having an average slope of 4.0%. Smaller swales enter the site from the northwest and west respectively, merging into the larger one which is dammed at the east end. A valved discharge line extends beneath the dam, providing the only outlet from the site. Releases from the site enter Choudrant Creek, which eventually flows into Lake D'Arbonne. Any accumulated drainage can be stored behind the dam at the south end of the site and pumped back to the storage lagoon or discharged. A concrete-lined interception ditch surrounds that portion of the site where uncontrolled runoff would occur, resulting in a closed catchment.

Direct runoff from spraying operations is avoided if possible. The quantity of runoff leaving the site is monitored, using a recording and totalizing flow meter. Water can get onto the site only through drainage swales, direct precipitation or irrigation, and can leave only through the discharge at the dam. This allows a rather unique opportunity to examine changes in drainage quality across the site. The purpose of this study was to compare the bacteriological quality of drainage entering the site with that leaving the site and to examine the validity of bacteriological parameters used to indicate the source of bacterial contamination.

The specific parameters chosen for measurement were fecal coliform and fecal streptococcus bacteria. The first was chosen because it is the standard parameter used to assess bacterial contamination from warm-blooded organisms. The second was chosen in order to allow computation of the fecal coliform to fecal streptococcus ratio (fc/fs). This ratio is sometimes cited as a way of distinguishing between bacterial contamination originating from animal sources and that of human sources. A ratio greater than 4 indicates contamination of human origin while a ratio less than 1 suggests animal origin (Metcalf and Eddy, 1991). However, to the authors' knowledge, the validity of these ratios as indicators in a natural system has not been well studied.

Samples were collected from a total of five sampling points, two upstream and three downstream of the site, for 92 consecutive days, from August 1 to November 1, 1991. Samples were also collected directly from the sludge storage lagoon on an aperiodic basis. Daily volumes of sludge applied and daily rainfall amounts were also measured. Daily estimates of discharge from the site were recorded by city personnel but later were found to be in error and were not used in this analysis.

As shown in Fig. 1, Sampling Point 1 was located up-gradient of the site in a drainage swale entering the site from the northeast. It was assumed that water quality at this point was indicative of that coming onto the site. Sampling Point 2 picks up runoff from precipitation falling onto the site as well as any runoff which might occur from irrigation. At one point during the study, heavy precipitation precluded sample collection from this point for about ten days.

The other three sampling points were located downstream of the site. Point 3 was on the downstream side of the discharge weir. Point 4 was in Choudrant Creek, upstream of the discharge. Data from Point 4 were assumed to represent background conditions in the creek. Point 5, also in Choudrant Creek, was located downstream from the discharge, approximately 30 yards from the weir.

Plating, incubation and enumeration procedures were those for the membrane filter methods described in the 15th edition of *Standard Methods, 1989*, and technical material published by Millipore (1984). Using appropriate dilutions, determined before sampling began, three replicates of each test were plated out from each sample. The measured concentration of each organism for that day was taken as the average of the three replicates. All bacterial concentrations were reported as colonies per 100 ml.

Data Analyses

Example data sets from Sampling Points 1 and 3 are shown in Figs. 2 and 3. Similar plots for fecal coliforms, fecal streptococcus and the fc/fs ratio were developed for all sampling points. Rainfall amounts and sludge volumes applied were collected daily during the study. However, analysis of the data indicated no quantifiable interactions between these variables and effluent bacterial densities.

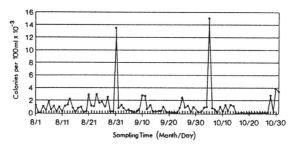

Figure 2. Fecal Coliform Concentration at Sampling Point 1.

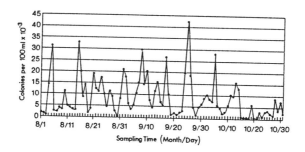

Figure 3. Fecal Coliform Concentration at Sampling Point 3.

Many statistical tests assume that the underlying structure of the data set being analyzed is normal. However, environmental data often exhibit a lognormal distribution, meaning that the logarithms of the data are normally distributed rather than the data themselves. A simple technique for determining if a data set fits a lognormal distribution is to plot the data on log-probability paper as described by Gilbert (1987). Figure 4, obtained by plotting the fecal coliform data from Sampling Point 1 on log-probability paper, shows that these data fit such a distribution quite well. Once again, all data sets exhibited this property. Thus, we may use those statistical tools which have been developed for analyzing normal distributions once the data have been transformed according to the equation $y = \ln X$.

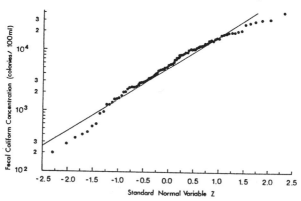

Figure 4. Log-Probability Plot of Data from Sampling Point 1.

Another common problem when attempting to analyze environmental data closely spaced in time is autocorrelation. If the samples are autocorrelated, any sample statistic will contain less information than the same statistic computed using independent data. The assumption of independence is common in many statistical tests, and violating it often renders the results invalid or increases the number of observations required to obtain a specified degree of statistical validity (Gilbert, 1987). Therefore, prior to analyzing these data, it was necessary to determine if autocorrelation existed. This was done by computing autocorrelation coefficients over all possible lag periods. Box and Jenkins (1976) suggest that the most satisfactory method of estimating the ith lag autocorrelation, ρ_i, is

$$\hat{\rho}_i = \frac{\sum_{t=1}^{n-i}(x_t-\overline{x})*(x_{t+i}-\overline{x})}{\sum_{t=1}^{N}(x_t-\overline{x})^2} \qquad (1)$$

where $\hat{\rho}_i$ is the sample estimate of the ith lag autocorrelation coefficient, x_t is the individual observation value, and $\overline{x}$ is the mean of n observations.

Autocorrelation coefficients based on transformed data for each sampling point were computed using lag periods ranging from 1 to 20 days. Results for Sampling Point 1 are shown in Fig. 5. These results are similar to those obtained at other sampling points and indicate little if any autocorrelation. Thus, even though the sampling interval is relatively short, the data points may be considered independent.

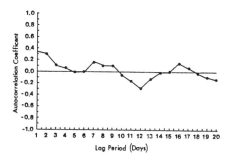

Figure 5. Probability Plots for Sampling Points 1 and 3.

The upper one-sided 100(1-α)% confidence interval for the mean is given by Equation 2 (Gilbert, 1987).

$$UL_{1-\alpha} = \exp(\overline{y} + 0.5*s_y^2 + \frac{s_y*H_{1-\alpha}}{\sqrt{n-1}}) \qquad (2)$$

where α is the specified confidence interval; $H_{1-\alpha}$ is obtained from tables provided by Land (1975); n number of observations and $\overline{y}$ and $(s_y)^2$ are the sample mean and variance based on the log transformed data.

Computed values of the geometric mean, μ_g, for each data set, as well as the upper one-sided 95% confidence limits on the mean, are listed in Table 1.

Table 1.   Geometric Mean and Upper 95% Confidence Limit for Four Sampling Points and Sludge Lagoon

Organism/Sampling Point	μ_g	Upper 95% Confidence Limit
Fecal Coliform (colonies/100 ml)		
background	1,190	1,618
discharge weir	9,931	13,219
downstream of discharge	11,721	16,350
upstream of discharge	2,852	3,824
storage lagoon[a]	98,459	N.C.
Fecal Streptococcus (colonies/100 ml)		
background	1,740	2,480
discharge weir	8,550	11,261
downstream of discharge	10,421	14,704
upstream of discharge	4,003	5,043
storage lagoon[a]	19,385	N.C.
Ratio		
background	1.57	2.13
discharge weir	2.23	3.06
downstream of discharge	2.14	2.63
upstream of discharge	0.93	1.19
storage lagoon[a]	5.08	N.C.

[a]computed using 20 sample values, upper C.L. not computed because data set is incomplete
N.C. = not computed

Interpretation of Results

It was somewhat surprising that all data sets exhibited little if any autocorrelation. Other investigators (Sanders, et al., 1983) have indicated that this can be problem when attempting to analyze environmental data. Based on the data collected, the underlying structure of both the fecal coliform and fecal streptococcus populations is lognormal. This knowledge can be quite useful because it means that although effluent (runoff) quality is not predictable in a deterministic sense, statistically valid inferences can be made. Since this study was conducted from August to October, no inferences concerning seasonality of the data can be made. A summary of the probability of occurrence of various coliform concentrations and fc/fs ratios is provided in Tables 2 and 3.

This study also demonstrated the effectiveness of the landfarm in reducing the coliform concentration in the applied sludge. The ratio of the average fecal coliform concentration in the storage lagoon to that at the discharge weir is 9.91, while that for fecal streptococcus is 2.26. Thus, the coliform concentration is reduced by 90% while the streptococcus concentration is reduced by only 55%. These results appear to contradict other work (e.g., USEPA, 1983) which suggests that, relative to coliforms, fecal streptococci have limited survivability in natural systems.

In the opinion of the writers, the most significant result of this study concerns the reliability of the fc/fs ratio as an indicator of the source of contamination. Probability plots of the fc/fs ratio at the upstream swale and the discharge are presented in Fig. 6. As stated earlier, this ratio has been suggested as a means of determining the origin of bacterial contamination—values less than 2 indicating animal origin, values between 2 and 4 inconclusive, and values greater than 4 indicating human origin. The average fc/fs measured in the storage lagoon was 5.08. Specific sources of contamination upstream of the site were not evident. Therefore, given the nature of the system being studied, it seems reasonable to expect that the fc/fs downstream would be substantially greater than that upstream. Fig. 6 shows that the two distributions are nearly identical, suggesting that the origin of the bacterial contamination upstream and downstream of the landfarm is the same, which is clearly not the case. The authors' interpretation of these results is that the fc/fs ratio does not appear to be a reliable indicator of the source of bacterial contamination in natural systems. Others have suggested that due to the limited survivability of fecal streptococci in a natural environment, samples be obtained close to the source of contamination in order for the ratio to remain valid; however, these results suggest that even when this advice is followed, misleading conclusions can be obtained when using this indicator.

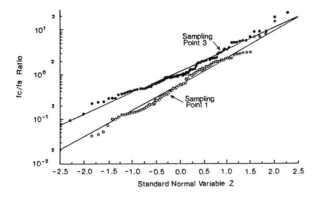

Figure 6. Correlogram Computed for Sampling Point 1.

Table 2. Probability of Occurrence of Fecal Coliform Concentration Less Than or Equal to the Indicated Value (%)

Sampling Point	1000 colonies /100 ml	2000 colonies /100 ml	4000 colonies /100 ml
background	62.0	85.9	96.7
discharge weir	9.6	18.2	38.7
upstream of discharge	25.7	34.3	77.1
downstream of discharge	13.1	24.1	32.9

Table 3. Probability of Occurrence of FC/FS Ratio Less Than or Equal to the Indicated Value (%)

Sampling Point	FC/FS ≤ 2.0	2.0 < FC/FS ≤ 4.0	FC/FS > 4.0
background	81	14	6
discharge weir	68	18	14
upstream of discharge	90	9	1
downstream of discharge	69	17	15

Conclusions

1. Results obtained from this study suggest that the fc/fs ratio is not a reliable indicator of the source of bacterial contamination in natural systems, even when samples are collected close to the source of contamination.

2. Fecal coliform and fecal streptococcus data collected from several points above and below the Ruston landfarm exhibited little, if any, autocorrelation.

3. Based on the data collected, the underlying populations of fecal coliforms and fecal streptococci are lognormally distributed. Thus statistically valid inferences can be made regarding the bacterial quality of the influent to and effluent from the landfarm.

4. Results of this study showed that coliform reduction across the landfarm (90%) was substantially greater than streptococcus (55%). This seems to conflict with statements by others (USEPA, 1983) regarding the relative survivability of fecal streptococcus bacteria.

BIBLIOGRAPHY

1. Aitchison, J., and J. A. Brown, 1969, *The Lognormal Distribution*, Cambridge University Press., Cambridge, Mass.

2. Box, G. E. P., and G. M. Jenkins, 1976, *Time Series Analysis: Forecasting and Control*, 2nd ed., Holden-Day, San Francisco.

3. Gilbert, R. O., 1987, *Statistical Methods for Environment Pollution Monitoring*, 1st ed., Van Nostrand Reinhold, New York.

4. Land, C. E., 1975, "Tables of confidence limits for linear functions of the normal mean and variance," in *Selected Tables in Mathematical Statistics*, Vol III, American Mathematical Society, Providence, R.I., pp. 385-419.

5. Metcalf and Eddy, Inc., 1991, *Wastewater Treatment, Disposal and Reuse*, 3rd ed., McGraw-Hill, New York.

6. Millipore, 1984, *Water Microbiology/Lab and Field Procedures*, Millipore Corporation, 1984, Massachusetts.

7. Sanders, T. G., R. C. Ward, J. C. Loftis, T. D. Steele, D. D. Adrian, and V. Yevjevich, 1983, *Design of Networks for Monitoring Water Quality*, Water Resources Publications, Littleton, Colorado.

8. _____. *Standard Methods for the Examination of Water and Wastewater*, 15th ed., 1980.

9. U. S. Environmental Protection Agency, *Process Design Manual for Land Application of Municipal Sludge*, Office of Research and Development, 1983, EPA-625/1-83-016.

PESTICIDES IN STORM WATER RUNOFF
FROM AGRICULTURAL CHEMICAL FACILITIES

E. O. Ackerman*   A. G. Taylor*

ABSTRACT

During the late 1970's and early 1980's, it became increasingly evident that the quality of surface water runoff discharging from agricultural chemical and fertilizer retail facilities in Illinois was having an adverse effect on water quality. This was illustrated by fish kill incidents, pesticide/fertilizer spills, and drainage areas void of vegetation immediately downstream from such operations. There were an estimated 1500 such facilities in Illinois providing service to the agricultural community by supplying fertilizers and pesticides for production agriculture. The facilities handled large volumes of agricultural pesticides and fertilizers in both the liquid and dry forms. Management practices at these operations resulted in numerous spills and discharges of pesticide and fertilizer products. Some of the spills have been sizeable; however, routine, incidental spillage combined with a lack of secondary and operational area containment structures was also obviously contributing to the poor quality of natural surface drainage from the sites. As a result of these conditions, the Illinois Environmental Protection Agency undertook a sampling project in 1985 and 1986 to document the quality of water discharging from agricultural chemical and fertilizer operations.

An additional factor prompting the initiation of the sampling project in 1985 was the pending National Pollutant Discharge Elimination System (NPDES) point source discharge program. These types of facilities were being considered by the USEPA for inclusion in the NPDES storm water permit regulations. Therefore, IEPA-DWPC Agricultural Specialists were requested to sample storm water discharges from agricultural chemical retail outlets during the 1985 spray season. Samples were collected either during or immediately following a precipitation event which yielded surface runoff. The 1985 project involved 8 agricultural chemical facilities.

In 1986, the project was expanded to include 31 sites. This was done to provide a larger base of laboratory data in order to better describe the quality of water discharged during storm events. The 1986 study also provided a limited opportunity to examine the quality of ponded or puddled water on site and the effect of the storm water discharge on the receiving stream.

---

*E. O. Ackerman, Agricultural Engineer, Division of Water Pollution Control/Field Operations Section, Illinois Environmental Protection Agency (IEPA), Peoria, Illinois. A. G. Taylor, Agricultural Advisor, IEPA, Springfield, Illinois

The results of the 1985/1986 sampling project revealed the presence of herbicides and nutrients prevalent in the discharges, on-site puddled water and receiving streams. Water quality violations were noted as stream concentrations exceeded Illinois environmental standards and regulations.

PROCEDURES

During the project period, 8 agrichemical facilities were selected in 1985 and 31 facilities in 1986 in Illinois in order to characterize the concentration of contaminants discharging from such operations during and immediately following precipitation events. Sites were chosen based on several factors including the manner of storage and handling of pesticides and the proximity of the site to a receiving stream. The extent of environmental stewardship and secondary containment and spill containment structures utilized at the facilities varied greatly.

Grab samples from the discharge were collected during precipitation events. In some instances, samples were collected from puddled water on-site and from the receiving stream as well. A typical surface water discharge sampling station is shown in photograph 1. The samples were placed in locked coolers and promptly shipped via UPS to established laboratories for analyses. In 1985, all samples were analyzed by IEPA laboratories. During 1986, the Agency contracted with Southern Illinois University-Carbondale/Plant and Soil Science Laboratory to perform pesticide analyses. Inorganic analyses were conducted at the IEPA Laboratory in Champaign. Laboratory analysis for pesticide determination and quantification were performed by gas chromatography with electron capture detector according to USEPA Method 608.

The 1985 sampling project included analysis for alachlor, atrazine, butylate, cyanazine, metalochlor, metribuzin, trifluralin, ammonia-nitrogen and phosphorus. During the 1986 project, samples were analyzed for the same compounds as 1985 plus the following additional parameters: 2,4-D, Ester, pendimethalin, chloramben, dicamba, linuron, bentazon, ethalfluralin, simazine and nitrates. The principal pesticides discussed in this report are: alachlor, atrazine, cyanazine, metolachlor, metribuzin and trifluralin.

RESULTS AND DISCUSSION

The project was not set up to statistically compare the analytical results from each of the facilities. The primary objective was to identify the principal contaminants and their relative concentrations in runoff/discharges from agricultural chemical facilities during precipitation events. The procedure simulated conditions the Agency could expect to find during routine field investigations.

The data collected during the project period included approximately 100 samples from more than 30 facilities in Illinois. Table 1 provides a portion of the laboratory analysis showing the concentration of herbicides at the point of discharge from agrichemical facilities. Difficulties at the laboratory required the combining of analytical results for alachlor and metribuzin for some 1986 samples as shown. Figure 1 illustrates the concentration of atrazine in the runoff water from several facilities during 1986.

Table 1. Herbicide Concentration in Storm Water Discharges.

(Units = mg/L)

Date	Facility	Alachlor	Atrazine	Cyanazine	Metolachlor	Metribuzin	Trifluralin
4/30/86	3D (c)	0.95	2.51	(b)	1.07	0.03	0.004
4/30/86	3D (c)	3.71	20.32	4.289	1.69	0.113	0.002
5/06/86	4B	3.0337	130.72	24.3378	2.0456	(a)	0.233
5/07/86	5I	2.2087	10.606	(b)	1.51	(b)	0.0021
4/30/86	3A	6.9471	2.9113	12.092	1.5285	(a)	0.0012
5/16/86	5M	4.112	9.391	1.08	0.273	(a)	0.0088
5/15/86	5K	5.954	48.148	(b)	3.129	(b)	0.0088
5/15/86	3F	5.534	252.482	0.2431	(b)	4.0345	0.6999
5/15/86	3E	6.14	160.52	0.6915	3.7055	(b)	(b)
5/27/86	3C	2.9826	2.4862	5.8273	0.3459	(b)	0.0102
5/28/86	6A	.6069	0.0825	4.177	0.497	(a)	0.0014
5/28/86	6C	1.099	0.3798	9.9598	0.122	(b)	0.0035
6/05/86	3H	1.0416	0.27	1.55	0.431	(a)	0.0045
6/04/86	3G	6.4478	24.1437	3.8273	4.643	0.0507	0.0047
6/05/86	5A	0.159	0.361	3.4527	0.0235	0.0141	0.0028
6/09/86	6C	0.2124	0.0927	5.968	0.0725	0.022	0.0049
6/09/86	6B	1.3751	0.4346	35.5859	0.3124	(a)	0.0034
6/10/86	3C	1.5579	0.5475	4.6451	0.2225	(b)	0.0033
6/10/86	3B	1.4937	0.201	36.2476	1.3726	0.0216	0.0027
7/02/86	7B	(b)	(b)	32.892	0.0351	0.0243	0.0045
7/11/86	5E	0.0366	(b)	27.068	(b)	(b)	(b)
7/11/86	5F	0.0649	0.0471	77.58	0.026	0.0186	0.0049
7/11/86	5J	0.1014	0.0417	75.392	0.0202	0.0136	(b)

a - Indicates alachlor + metribuzin.
b - Indicates trace amounts or below detection limits.
c - Samples collected at two different discharge points from the same facility.

Photograph 1. A Typical Storm Water Runoff Sampling Station at an Agricultural Chemical and Fertilizer Facility is Shown.

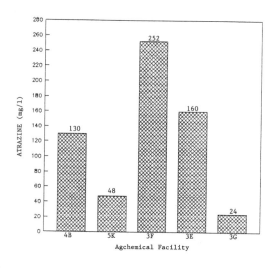

Figure 1. Atrazine Concentration in the Storm Water Runoff from Agricultural Chemical Facilities (1986).

Table 2 reports the nutrient concentration in 26 storm water discharge samples from 12 different agricultural chemical facilities.

Table 2. Nutrient Concentrations In Storm Water Discharges From Agricultural Chemical Facilities in Illinois.

		(Units = mg/L)		
Date	Facility	Ammonia	Nitrates	Phosphorus
4/30/86	3A	180		34
4/30/86	3D	87		15
5/15/86	3F	210		69
5/15/86	3E	2100		120
5/27/86	3C	220		2.2
6/10/86	3B	81		12
6/10/86	3C	80		1
6/05/86	4A	9.7		3.9
6/30/86	4A	110		12
5/06/86	4B	1500		670
6/05/86	4B	1800		820
6/30/86	4B	420		30
7/10/86	4B	580		100
6/30/86	4B	27		28
5/06/86	4C	96		530
5/06/86	4C	690		120
6/05/86	4C	3700		75
6/05/86	4C	4300		18
6/30/86	4C	230		10.1
6/30/86	4C	1040		65
7/10/86	4C	410		33
7/30/86	4C	740		31
4/30/86	6B	1070	350	170
5/28/86	6C	110	140	20
6/09/86	6C	180	91	58
5/15/86	7A	91		5.3

Figure 2 demonstrates the ammonia-nitrogen concentration in the surface water discharge from several facilities.

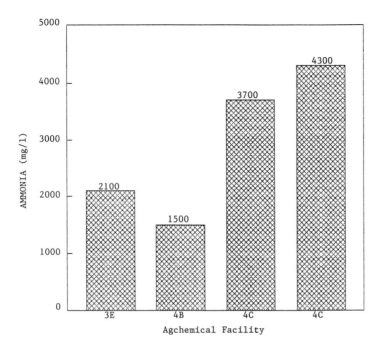

Figure 2. Ammonia Concentrations in Storm Water Runoff at 3 Agricultural Chemical Facilities in Illinois (1986).

Tables 3 and 4 characterize the quality of surface water puddled or ponded on-site at agricultural chemical facilities.

Table 3. Herbicide Concentration in On-site Ponded Surface Water at Agrichemical Facilities in Illinois.

Date	ID	Alachlor	Atrazine	Cyanazine	Metolachlor	Metribuzin	Trifluralin
			(Units	= mg/L)			
05/07/86	5C	2.155	1.232	0.767	(b)	1.668	0.0025
05/07/86	5H	2.294	4.447	(b)	1.6799	0.206	0.0057
05/07/86	5I	4.438	159.955	(b)	2.712	(b)	0.0016
05/09/86	5H	2.661	7.032	0.862	0.189	0.0733	0.0064
05/15/86	5K	8.9839	(b)	(b)	5.062	(b)	0.0379
05/16/86	5L	0.1	41.486	0.49	1.333	(b)	(b)
05/16/86	5F	7.04	166.191	1.418	0.5544	(b)	2.3717
05/29/86	5E	0.3499	0.188	8.9487	0.125	(b)	0.0038
05/30/86	3B	1.625	0.8227	4.5524	1.2276	0.0203	0.0028
06/30/86	6B	9.598	330.1716	27.741	5.1348	(a)	1.0733
06/30/86	6B	10.455	249.2868	49.2927	5.5141	(a)	0.8081
07/11/86	5E	0.0366	(b)	27.068	(b)	(a)	(b)
07/11/86	5F	1.0834	5.9717	35.475	0.1405	0.0058	0.0058
07/11/86	5J	1.712	1.5105	41.945	0.1337	0.0035	0.0035

a - Indicates alachlor and metribuzin.
b - Indicates trace amounts or below detection limits.

Table 4. Nutrient Concentration in On-site Ponded Surface Water at Agricultural Chemical Facilities in Illinois.

Date	Facility	(Units = mg/L) Ammonia	Nitrates	Phosphorus
04/03/86	4D	53000		
05/19/86	5I	330	360	66
05/27/86	5B	240		34
05/27/86	5C	212		18
05/07/86	5C	27	31	11
05/09/86	5C	26	28	16
05/12/86	5D	100	84	33
05/29/86	5E	19	100	4.8
05/16/86	5F	1180	710	380
07/11/86	5F	80	220	15
05/09/86	5H	290	450	11
05/19/86	5H	210	430	5.5
05/27/86	5H	310		5
05/07/86	5H	140	550	27
05/19/86	5I	84	150	14
06/30/86	6B	4370	1230	1500
06/30/86	6B	5220	1420	1600

One of the more graphic illustrations of the effect of storm water runoff from agricultural chemical facilities is shown in the case of facility 3B during a 1985 sampling event. A sample collected from a discharge at the facility during a storm event revealed an ammonia-nitrogen concentration of 210 milligrams per liter. The discharge travelled approximately 400 meters in a roadside ditch before entering a perennial stream. An upstream sample collected from the receiving stream showed 0.12 mg/L ammonia. A sample collected in the receiving stream 10 meters downstream from the confluence with the agricultural chemical facility runoff revealed an ammonia content of 51 mg/L in the stream. This concentration of ammonia violates Illinois water quality standards which allow for 1.5 mg/L of ammonia. Figure 3 provides a graph of the water quality impact for this incident. The impact of phosphorus was also examined during this incident with the laboratory analysis provided in Table 5.

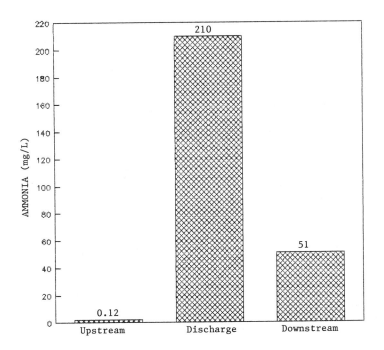

Figure 3. Ammonia-nitrogen Impact on Receiving Stream Due to Storm Water Discharge from Agricultural Chemical Facility (May-1985).

Table 5. Nutrient Impact on Receiving Stream During Storm Event at an Agricultural Chemical Facility (May-1985).

Parameter (mg/L)	(Sampling Station)		
	Upstream	Discharge	Downstream
Ammonia-nitrogen	0.12	210	51
Phosphorus	0.04	80	18

OBSERVATIONS

The results of this two year storm water sampling project indicate that herbicides and nutrients are discharged from agricultural chemical and fertilizer facilities during rainfall events. The single highest concentration in a discharge was atrazine with greater than 250 mg/L being discharged during a May 15, 1986 sampling event. Cyanazine was also prevalent in the storm water runoff with several samples showing greater than 30 mg/L in the discharge. Alachlor, metolachlor and trifluralin were also present in the samples. Elevated concentrations of ammonia, nitrates and phosphorus were noted in the storm water. The ammonia-nitrogen content was greater than 2,000 mg/L in several effluent samples.

Samples collected from on-site puddled water also demonstrated the presence of herbicides and nutrients. One sample collected at a facility in Central Illinois had an ammonia-nitrogen concentration of 53,000 mg/L. An atrazine concentration greater than 300 mg/L was noted in a puddle of liquid on-site at a facility in Southern Illinois during a June 30, 1986 sampling event. The same sample contained 27.7 mg/L of cyanazine. Several puddled samples exceeded 200 mg/L of nitrates with one sample at a Southern Illinois facility containing 1420 mg/L. The Illinois drinking water standard for nitrate-nitrogen is set at 10 mg/L. The existence of such contaminated and uncontained liquid at agricultural chemical facilities generates a concern that soil infiltration and groundwater contamination problems may occur.

Although only limited data were available for receiving stream observations, the study did show a potential impact on receiving streams due to storm water runoff at agricultural chemical facilities. A review of the herbicide data shows several apparent water quality violations for the herbicide alachlor. During the study period, the water quality standard for alachlor as set by the Illinois Pollution Control Board Rules and Regulations Title 35, Subtitle C, was 0.28 mg/L. This was based on one-tenth of the 96-hour median tolerance limit. Off-site water quality violations were also noted for atrazine, cyanazine and metolachlor. Storm water discharges from at least two sites resulted in violations of ammonia-nitrogen water quality standards. In the case of facility 3C, the storm water discharge caused an elevated alachlor concentration in the receiving stream. Additional data, not detailed in this report, showed dilution effects of high intensity storms and effluent mixing with uncontaminated runoff.

The scope of this project was not sufficient to demonstrate the complete and combined impact of herbicides on water quality. It is recommended that further studies be conducted to document the water quality impacts in receiving streams located near agricultural chemical plants. Such studies should include an assessment of ecological impacts and effects on stream biota.

In 1990, the State of Illinois adopted the Title 8: Chapter I, Part 255 Containment Rules for Agrichemical Facilities. These rules provide design criteria for the construction and installation of spill containment and secondary containment structures. The adoption of these rules should serve to minimize the extent of contaminated storm water discharges from agricultural chemical sites although there may be continued adverse impacts on surface and groundwater due to residual contamination.

DRAINAGE EFFECTS IN MARSH SOILS
I. EFFECTS ON WATER TABLE AND DRAINFLOW

Ibrahim.S.M[*]

SUMMARY

Drainage effects in marsh soils caused by ditches and pipes are shown by results of long term field experiments (1982/90) in the Nordkehdingen brackish marshes at Elbe river coastal region of Northern Germany. Three fields were selected for this study to demonstrate the effects of drainage on the ground water table and drainflow comparing ditch drainage and pipe drainage with 7 m, 14 m or 28 m drain distances and 0.9 m, 1.0 m or 1.2 m drain depths.
The results indicate, that without pipe subdrainage the ground water table remains near soil surface during heavy rainfall periods. Under favorable hydraulic and topographic condition in the investigated region, a systematic pipe drainage with a moderate drain intensity with open main outfalls (ditches) is recommended.
The results also show that with closer drain distances higher drainflow and deeper ground water table result. The calculated drain spacing of 14 m is not necessary. With more advantageous a drain depth of 1.0 m and an optimum drain distance of 13 - 20 m was determined. If the outlet ditch is deep enough and no foreign water influence exists, one can propose 28 m drain spacing for this site.
Instead of grassland and traditional ditch drainage, the arable use with subdrainage system has become possible.

1. Introduction

A great portion (30 %) of soils in Northern Germany is influenced by high ground water level and generally high rainfall. Severe problems with high ground water tables especially exist in the marsh regions at the Elbe river. This coastal region in Northern Federal Republic of Germany due to large average amounts of annual precipitation (750-800 mm) and rather low average annual evapotranspiration (500-550 mm) suffers under an extended period of wet soil conditions. The annual precipitation exceeds evapotranspiration by 200 to 300 mm. A portion of this excess water infiltrates the soil and frequently causes the ground water table to rise near the soil surface during extended wet periods in winter and spring. This high water table causes difficult farming conditions.In many cases large areas of land became impassable to heavy machinery. In the past the region had been drained with a closely meshed network of shallow bed ditches with cambered beds, and the soil had been used only as grassland.
In 1976 for the region of Nordkehdingen (5000 ha) the buildup of a new Elbe dike was completed and a land drainage project could be started. Instead of the traditional grassland, arable cropping should become possible.
Several experimental fields have been installed at the site to obtain better information about the most suitable and economical drainage system in this area. Three fields were selected for this study to evaluate the effect of different drainage intensities on

---
[*] Dr. Shaban M. Ibrahim, Soils Department, Faculty of Agriculture
Kafr El-Sheik/Egypt

soil water table and drainflow. Further work will concentrate on the effects of drainage on soil properties and grain yield (s. also IBRAHIM, 1991) .

2. Study procedures

The experimental site near Freiburg at Elbe river coastal region of northern Germany (topographical map 2120 and 2121) lies about 2.0 m above main sea level and has nearly level topography. North Germany belogs to the humid climate zone of Europe. Perennial winds from west and southwest cause a maritime climate with mild winter and relatively cool summer. Table 1 shows the main climate data for the region of Nordkehdingen (Hydrologischer Atlas der Bundesrepublik Deutschland, 1978). The soils at the experimental site belong to the brackish marshes.

Table 1: Mean data of climate for Nordkehdingen region

	air-temp. oC	rain-fall mm	evapotrans-piration mm	discharge mm
year	8-9	750-800	500-550	200-300
Nov. - Apr.	3-4	300-350	100-150	100-200
May - Oct.	14-15	450	400	< 100

The influence of drainage on the water table and drainflow was studied on three fields. In field Nr. 1 , 7.65 ha, the following 4 variants were installed in August 1982
- without pipe subdrainage
- without pipe subdrainage combined with 60 cm deep ploughing
- pipe subdrainage without deep ploughing
- subdrainage with 60 cm deep ploughing

The subsurface drainage on this field was installed at 1.0 m depth with drain distance of 14 m. The soil without subdrainage remained drained by bed ditches 20 m distance.
The second field , 6.87 ha, was installed in August 1982 with two drain distances - 7 m and 14 m - and two drain depths - 0.9 m and 1.2 m.
The third field , 10.0 ha, was installed in 1986 with three drain distances - 7 m, 14 m and 28 m - and drain depth of 1.0 m. One half of this field at the same time was 60 cm deeply ploughed.
The subsurface drainage of the three fields consisted of 50 mm diameter corrugated PVC-drain tubes, grade 0.2 %. The drain pipes were installed by a cutter-chain drainage machine on the third field without filter and on the first and second field with Filtan(R) filter. - Filtan consists of cellulose wool, acryl and fibrous peat -. The drains empty into outlet surface ditches about 1.6 m depth with 120 m distances.
The soil characteristics of the three fields are nearly similar. Only in small parts of the second and third fields, near the Elbe dike, the soil to a depth of 60 cm has a higher sand and a lower clay content. Particles > 0.60 mm were almost not found. The relevant soil characteristics of two soil profiles of the first and third field are given in Table 2.
The soil profile consisted usually of surface grayish brown Ap horizon to a depth of about 30 cm, underlain by a stratified gray, brown mottled Go/r horizon to a depth of about 120 - 150 cm. The

Table 2: General soil analytical data

field	depth (cm)	part.size distr. (% W/W) <2μm	2-60μm	>60μm	O.M. %	dB (g/cm3)	kf1 (cm/d)	pore space (% V/V) tps2	Dp3	Ap4
1	10-20	34.7	63.9	1.4	8.1	1.05	98	55.1	12.2	7.0
	30-40	30.9	64.7	4.4	7.0	1.17	72	54.2	13.9	8.9
	50-60	25.1	64.2	10.7	5.0	ni	ni	49.7	17.3	7.1
	90-100	31.4	66.0	2.6	6.8	1.04	236	65.5	10.9	4.9
3	10-20	16.8	62.0	21.2	6.6	1.19	10	49.4	17.0	3.8
	30-40	15.3	76.6	8.1	3.0	1.33	33	48.2	16.9	6.9
	50-60	16.7	56.1	27.2	2.3	1.35	81	45.6	19.5	9.1
	90-100	28.2	54.5	17.3	3.4	1.14	152	56.6	13.8	3.9

1: kf   saturated hydraulic conductivity, core method
2: Tps   total pore space
3: Dp   drainage porosity, >10 μm = Tps - pF 2.5
4: Ap   air porosity, > 50 μm = Tps - pF 1.8
5: ni   not investigated

third horizon, depth > 150 cm, is very dark blue (Gr horizon). The soil is usually clayey silt to silty clay, especially in drain depth the soil in all tested profiles is almost homogeneous with a clay content usually > 20 %. The pH of the soil is higher than 7 (7,1 - 7,7) and the O.M. content ranges between 1 - 9 %.

The saturated hydraulic conductivity was measured
a) in the field by the auger hole method acc. Hooghoudt-Ernst (BEERS, 1970), the water permeability ranges between 2 and 75 cm/d with an avarage permeability of 39 cm/d,

b) in the laboratory, using an undisturbed core method (HARTGE, 1966), geometrical main value was 114 cm/d and 72 cm/d for first and third field respectively (horizontically sampled cores). The ground water table fluctuation in between 2 drains was measured in 15 cm diamter observation wells, drilled down to 3 m depth - and fitted with water level recorders. 30 days rotation time gives a continuous record of the water table amplitude. Drainflow was measured with drainflowmeter type Cambridge (Rycroft, 1972) with 7 or 30 days rotation time.

3. Results and discussion

In the past the soils of the Nordkehdingen marshes had been drained with a closely meshed network of shallow trenches or deep ditches with cambered beds, 20 m width. Deficient main outfall on a flat field, danger of silting and ochre clogging and a closely meshed network of ditches usually indicate only individual branch drains. This drainage system is ineffective, the water table rises frequently up to soil surface for extended periods of time. This high water table avoids arable use.

Under favorable hydraulic and topographic conditions in the region investigated, after the construction of the Elbe dike and deep opened main outfalls, a systematic pipe drainage could be installed with opened main outfalls (ditches) as main drains. Without pumping station the water drained can directly flow to the Elbe river interupted by sluices at high tides. The water table

dates presented and the drainflow dates for three fields give a good approximation of the average situation for the long years tested period.
Subdrainage resulted in a clear drop of ground water table. Without subdrainage the ground water table remains near the soil surface for extremely wet periods (Fig. 1). As a consequence greater air volume, rooting depth and farming intensity can develop on the pipe-drained plots. Subsurface drainage is needed

in the investigated region also to provide better trafficable conditions for seedbed preparation and planting in the spring, to insure a suitable environment for plant growth, and to provide trafficable conditions for harvest operations in the autumn.

Water table levels obtained on the second field on both 7 m and 14 m drain distance variants from November 1984 until August 1987 are given in fig. 2. Generally, a similar water table course on the 7 m and 14 m was measured, but on the 14 m the water table distance to the soil surface was usually 1 till 40 cm smaller. The groundwater measurements in summer (May to October) showed that there was little difference between the water table levels in the plots drained with 7 m drain spacing and the plots drained with 14 m drain spacing, and the water table was actually often below drain depth.

In wet winter periods (November to April) higher differences of ground water level between 7 m and 14 m drain spacing variants were measured. Comparing the ground water level of 1.2 m and 0.9 m drain depth plots shown that with 1.2 m depth approximately 20 - 30 cm deeper ground water level can be found than for the 0.9 m drain depth plots.

Drainflow for the 7 m and 14 m drain spacing, drain depths 0.9 m and 1.2 m, is presented in table 3.

There was a higher discharge by 7 m than 14 m drain spacing. A wide range of values was observed for all plots, depending primarily on both antecedent moisture content and the amount and intensity of rainfall. The discharge increased from 33.2 - 49 % of incident rainfall on 14 m drain spacing variants to 46.4 - 70.5 % on 7 m drain spacing variants. The drainflow of 1.2 m drain depth plots was something higher than of the 0.9 m drain depth plots.

The ground water measurements on the third field (see fig. 3) show that the water table on the 7 m drain spacing plot was near drain depth, but on 28 m drain spacing plot - south - it was about 40 cm higher. Except extremely wet periods here the groundwater depth was deep enough for agricultural operations however for shorter times than at 14 m and 7 m drain distance. The water table level for the 28 m drain spacing - north - was higher than 28 m drain spacing - south - due to foreign water influence.
As shown in figure 4, the ground water table level is more often too high on 28 m drain spacing - north, nearby the Elbe dike - especially during high mean tidal water level of the Elbe. This means that percolation through and below the dike can happen during river high water, so that the soil up to 150 m, away from the dike remain under foreign water influence. Another comparison between drain discharges at three drain spacing (7 m, 14 m and 28 m) and drain depth 1.0 m is given in table 4. There was usually higher discharge for 7 m than 14 m and 28 m resp. The closer the

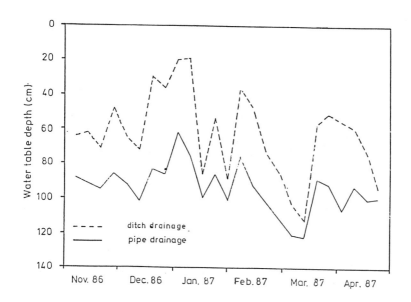

Fig. 1 : Ground watertable levels measured for pipe-drained and ditch-drained plots from Nov.1986 till Apr.1987

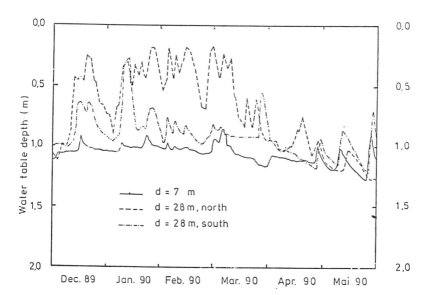

Fig. 3 : Ground watertable levels and different drain distances (d)

Fig.2: Ground water level fluctuations influenced by drain distances (d)

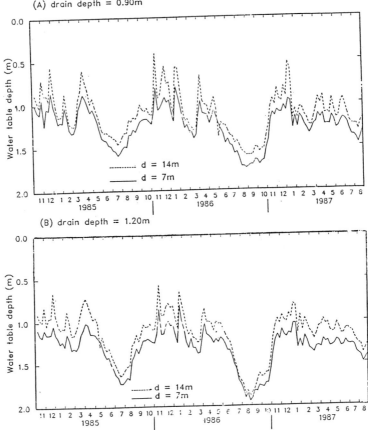

(A) drain depth = 0.90m

(B) drain depth = 1.20m

Table 3: Monthly precipitation (P) (mm) and drainflow (F) (mm) in the winter period 1985-1987 with different drain depths and spacings

		Nov.		Dec.		Jan.		Feb.		March		April		sum			
m depth		0.9	1.2	0.9	1.2	0.9	1.2	0.9	1.2	0.9	1.2	0.9	1.2	0.9		1.2	
spacing (m)														mm	% of P	mm	% of P
1985	7 F	31.5	64.4	44.9	44.1	20.2	24.0	23.7	22.3	33.2	28.7	48.0	50.0	201.5	66.0	215.5	70.
	14 F	20.1	28.3	16.5	19.0	14.4	16.8	14.9	14.0	23.7	19.5	30.2	34.0	119.8	39.2	131.6	43.
	P	133.7		75.3		92.3		3.1		58.0		25.5		387.9			
1986	7 F	54.4	55.3	61.6	67.8	67.3	77.4	1.5	2.1	36.4	39.7	10.4	8.2	231.6	59.7	250.5	64.
	14 F	38.9	42.8	48.1	57.6	41.1	52.1	1.0	0.3	28.8	31.5	7.5	5.9	165.7	42.7	190.2	49.
	P	73.0		130.4		55.8		35.3		31.6		50.2		367.3			
1987	7 F	23.0	20.1	74.7	78.1	43.2	45.3	9.8	10.0	10.5	12.6	13.3	17.2	176.5	46.4	182.8	48.
	14 F	23.3	16.9	49.6	58.6	27.1	26.1	6.2	6.8	9.4	11.6	9.5	12.3	125.1	33.2	132.3	35.

drain distance the higher the quantity and the rate of drainflow will be. The drainflow increased from 36.9 % of incident rainfall for 28 m drain spacing to 47.2 % for 14 m and to 63.2 % for 7 m. During the observed periods, the average drainflow rate was 1.72, 2.34 and 3.19 mm/d for 28 m, 14 m and 7 m drain distance respectively.

The results also show that without pipe drainage the ground water level was too high in spring and in autumn and thereupon very little field working days resulted. Because the ground water differences between 7 m and 14 m distance remained very small, the 7 m drain space should not recommend in the investigated site because this narrow distance increases the installation costs up to 100 % without obvious advantages in comparison to a 14 m drain spacing. The calculated optimum drain spacing, by drain depth of 1.0 m, in the investiged region was determined between 13 - 20 m. Under favorable topographic and hydraulic conditions - deep outlet ditches and no foreign water influence - one can propose to 28 m drain spacing.
Drainage depth at the beginning of a branch drain should be sufficent after the beds have been partially leveled and the bed ditches filled up. An average drain depth of 1,0 m or greater is more advantageous than shollow ones with small spacing.
The silting and ochre clogging in the investigated site till today are very small. Therefore drain pipes can be installed without filter. A moderate drainage intensity is effective enough. By this the drainage costs in the site investigated remain suitable.

Conclusions

Clayey silt marsh soils in humid regions need artificial drainage for modern farming on these sites. After a soil survey with measurement of the hydraulic conductivity by the auger hole method and additional laboratory tests a pipe drain distance of 14 m at a depth of 1 m was calculated.
Farmers of a new diked polder in the river Elbe marsh region in Northwest Germany meant, that this drainage intensity could be insufficient with respect to an early and long enough, trafficabilty (field working days) for their heavy mechanery.
Others feared the too high costs increasing with smaller pipes and were looking for a lower cost drainage.
Several long termed fild trials therefore were started in 1981 to demonstrate under different weather conditions how drain distance and depth together with or instead of deep ditches can improve the ground water level depth.
No advantages for farming on these silty soils could be found by drain pipe spacing less than 14 m. Except fields near the dike (< 150 m) with additional tidal water influences in the average of 8 years the soils have been sufficiently drained even with 28 m distance of pipe drains at an average depth of 1.0 m.
A supporting drainage effect is given by the 1.6 m deep main outlet ditches. No problems of pipe silting or ochre clogging occur. Totally filter wrapped pipes therefore are not recommended.
Following these results the amelioration costs for these difficult silty soils could be reduced by a distance of 20 m down to 60 % of the installation costs and without filtering additionally down to 50 % of the material costs. For a 5000 ha project originally average drainage costs of 2000 DM/ha were calculated = 10 Mio DM in the total. Reduced to 6 Mio DM the costs of field trials over a 10 years period amout only 10 % of the costs spared.

Table 4 : Drainflow (mm) during discharge periods in 1987-1989 in relation to precipitation (p)

Discharge period	31.12.87- 11.1.88	14.1.88- 10.4.88	6.10.88- 10.10.88	14.2.89- 23.2.89	24.4.89- 2.5.89	x	x
precipitation (mm)	91.4	225.8	70.0	29.0	26.0	% of P	Flow (mm/d)
7 m drain distance	59.5	216.0	25.4	18.3	14.5	63.2	3.19
14 m "	53.5	207.6	15.3	4.6	12.4	47.2	2.34
28 m "	41.3	155.3	8.6	7.5	8.4	36.9	1.72

Fig. 4 : Mean tidal waterlevel and ground water level, third field- Jan.1987 till Nov.1989 -

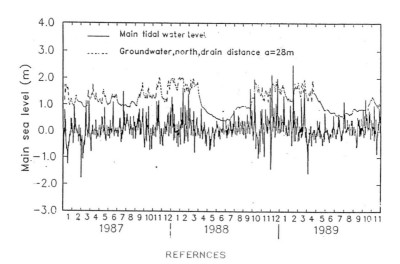

REFERNCES

BEERS, W.F.J. VAN, 1970: The auger-hole method. - Int. Inst. for land reclamation and improvement/ILRI, Bull. 1. Wageningen, The Netherlands.

HARTGE, K.H., 1966: Ein Haubenpermeameter zum schnellen Durchmessen zahlreicher Stechzylinderproben. Z.f.Kulturtechnik und Flurbereinigung 7, 155-163, Berlin, Hamburg.

HYDROLOGISCHER ATLAS DER BUNDESREPUBLIK DEUTSCHLAND, 1978: Gesamtleitung Reiner Keller, herausg. Deutsche Forschungsgemeinschaft, Bonn-Bad Godesberg.

IBRAHIM, S.M., 1991: The need and possibility of drainage of silty brackish marsh soils- studied in model and long term field experiments.- Diss. Univ. Göttingen/F.R.Germany (in German with summary in Englisch) .

RYCROFT, D.W., 1972: Drainflow hydrographs. Field Drainage Experimental Unit. Ann. Rep. 1971, S. 47-55, Cambridge/UK.

CONTRACTOR PANEL MEMBER REMARKS

Cy Schwieterman

I have chosen to put my remarks in a question/answer format.

Question 1: Since many areas of the United States and Canada do not have a subsurface drainage contractor, could ASAE members teach a school for training future contractors in the installation of subsurface drainage systems and other conservation practices?

ANSWER 1: Yes, I believe that an ASAE member would be well qualified to teach not only the installation of subsurface drainage, but many other soil conservation practices that are in demand from land owners and farmers in the United States and Canada.

Subsurface drainage is a good business for 6 to 7 months per year, but in order to fill the balance of the year the contractor will have to learn to build waterways, clean open ditches, clean fence rows, do earth moving projects that require backhoe loaders and bulldozers, etc. Also, home sewage systems and plowing snow in the winter time are a source of extra income.

Contractors should become a member of and participate in the community.

Above all, the contractor will have to stay friendly with the banker.

Summing up -- The contractor will become a good member of the community. He/she will most probably marry and raise a family which in time will bring more people into the community.

QUESTION 2: Could ASAE members do research on the quality and nutritional value of food from crops grown on subsurface drained soil as opposed to crops or food that is grown on soil that has no subsurface drainage?

ANSWER 2: Yes, I believe that ASAE members have the ability to do such research if they so desire. Stop and think that if the crops grown on subsurface drained soil contain more food and nutritional value people most certainly would rush to purchase such a product. I am sure that the American farmer would certainly begin to produce more corps with subsurface drained land.

Cy Schwieterman, Drainage Contractor, St. Henry, Ohio

SUBSURFACE DRAINAGE DESIGN

Owen R. Row
Member ASAE

Design of subsurface drains is presently based on full pipe flow, no pressure, slope, "n", and drainage coefficient. There are a lot of subsurface main drains installed in the early 1900's. Most are in need of replacement. A number of County Surveyors/Engineers will not allow the use of plastic drainage material of the same size as the existing mainly because of "n" values.

In practice we all know the drains carry more when there is pressure flow. Another consideration is the ability of the soil to allow more or less than the drainage coefficient to enter the drain. Pressure flow is more prevalent today due to more surface inlets and/or french drains being installed low areas that have compaction problems. Deep compaction is occurring causing lower rates of infiltration throughout the fields.

Two projects I feel how ASAE can help the design engineer and drainage contractor are:

1. A simplified capacity chart to give flow rates based on head, slope, length of drain, "n", and permeability of the soil.

2. Based on the above chart develop a flow routing table to see how long it will take to remove different rain falls.

With this information we can compare the two materials, concrete and plastic, on flow rates and time. Basing only on "n" values of .013 for concrete and .020 for large plastic, it should take 50% longer for water to get away. In practice it does not appear to happen that way.

Why this concern for getting plastic accepted? I have seem new hare tile installed. Spaces between tile occur. Crayfish are known to exit through these spaces. Pressure flow will occur in these mains. This can lead to undermining the bearing surface of the tile. Presently that has been the major cause of failure in old systems, pressure flow and crayfish.

Plastic systems are continuous, no spaces. Pressure flow will occur but no undermining.

Cost of installed projects are about the same for different materials. Concrete more labor, less materials costs. Plastic less labor, more material costs.

If nothing else, this project will bring designs closer to what happens in the field.

Owen R. Row, P.E., Ag Engineer, Drainage Contractor, Monroeville, IN.

CONTRACTORS COMMENTS

FRED GALEHOUSE
DRAINAGE CONTRACTOR

I have been a subsurface drainage installer since 1948 and my father was a drainage contractor. I have been a member of American Society of Agricultural Engineers (ASAE) since 1971 and have used the information garnered at meetings and in papers to improve my knowledge of subsurface drainage and do qulaity work. I have been active in attending meetings and have been a ASAE subcommittee chairman, as well as helping to rewrite and approve several standards. The relationship with ASAE I feel has been profitable for both ASAE and myself.

I have always considered myself a conservationist and a steward of the soil doing a much needed service and making this country a better place in which to live and do business. But in the last few years subsurface drainage has been under attack as a destroyer of wetlands and a polluter of water. I do not feel either of these labels are justified to the extent they are labeled. As a result my work is considered by many as something which is bad, this is a bitter pill which has taken 3/4 of my business. We have to do a better job of communicating, in a persuasive way, the benefits of subsurface drainage and water table control, to the legislators and the general public. I feel remise in not proposing a solution so am asking for help. Research is one of the first steps in collecting the information, but how to communicate the results to the public is the part I do not know how to accomplish. The research papers are there but the results have had little publicity except to those interested in supporting drainage and water table control.

Since the 1985 farm bill "Food Security Act" (FSA) the swampbuster provisions have told farmers to stop draining land as it might be a wetland. Subsurface drainage is perceived as draining swampland when in reality the water table, subsurface drainage is trying to control is a perched water table of short duration and not a wildlife wetland. The farm economy has not been good and the drought years have not helped the economy and as a result the farm drainage industry has suffered a severe depression. This has forced diversification among contractors and loss of many contractors.

The farm bill is still trying to define wetland in usable terms so the potential of loss of benefits is still an unresolved problem. After seven years of FSA, I cannot tell a farmer whether his field is or is not a wetland with any degree of certainty.

ASAE has provided a meeting ground for discussion of drainage contractors problems and has written standards for design and installation. Yet only a few contractors have participated or used the standards. Most did not wish to spend the time or money to become informed and use the standards. The work or design is not inspected except in the case of cost share work which is a very small portion of the work. The farmer/landowner has little knowledge of the standards and does not specify material or installation according to the standards. The work is priced by the foot so speed is money in the pocket and quality is left by the wayside.

Subsurface drainage is a very durable practice and I have found and sometimes repaired systems that are 50 or more years old and still functioning very well. Maintenance is a small yearly item and most often neglected. In repairing or maintaining existing systems, there are three places to investigate: 1. Outlet for missing or damaged outlet pipe without an animal guard or the ditch is filled up; 2. Holes in the field where chipped connections, broken tile, or utility crossings were not properly repaired; 3. Tree roots plugging the line. All these problems could have been avoided if standards had been followed or available. If a truck or tractor or other equipment was purchased with the same amount of money as spent on a subsurface drainage system it would be seen, insured, protected and talked about but because subsurface drainage is out of sight except for the outlet and crop response, it is mostly out of mind and neglected, with little maintenance. Records of systems have not been well kept except in a few places and most people looking at the farm or the field do not see evidence of the existence of a subsurface drainage system. I will have to admit my own records are not as good as they should or could have been if I had spent a little more time on keeping records.

The list of subsurface drainage benefits is long but all are permissive, by allowing beneficial things to happen. Earlier planting dates, later harvesting dates, better growing conditions during the year, better soil structure, less compaction, less soil erosion, better water quality. Some of these items do not have enough research to verify the extent of the benefits attributed to subsurface drainage.

I see a COMMUNICATION problem of great magnitude in convincing, farmers, environmentalist, legislators, and others, of the benefits of water management and subsurface drainage. There needs to be a coordinated effort to incorporate the research into the design to be used by contractors and farm planners and the benefits that accrue to the land owner/farmer the environment and the general public.

The 1985 farm bill was a golden opportunity to promote the benefits of drainage but instead it has been a platform to condemn all forms of drainage. I have been seated beside an activist environmentalist for a whole day of meeting and because I was a drainage person anything I said was not heard because he was an anti drainage person.

I commend ASAE for the progress and information it has produced and I wish it to continue. I have many papers from ASAE covering many years on drainage subjects but I will have to admit they are not general knowledge used by designers, contractors, environmentalist, legislators or planners. I am looking for ways to get some of this information included into the teaching and public sector information.

I have consistently asked the writers of the Universal Soil Loss Equation (USLE) and the researchers for the WEPP erosion prediction equations for some inclusion of the benefits of subsurface drainage. They have all agreed there is a relationship but the problem is quantifying the amount. The computer models EPIC, CREAMS, and GLEAMS have components that include subsurface and surface drainage. An example of erosion control is the grass waterway. Nearly all grass waterways have a subsurface drainage system to compliment the waterway, just why is that done? The answer is to control the water table to permit the growth of grass and stop the erosion which will occur if it is not done: ie.permissive.

There is a need for subsurface drainage and water table management to protect the wetlands and to promote the most efficient use of the now existing cropland. How to get this understanding is a real communication problem and needs to be worked out.

I have seen many fields in the dry years of 1988 and 1991 where the fields with subsurface drainage had better crops than similar fields without subsurface drainage. This is a little difficult to understand at first but the bottom line is the crop had a better environment early in the season, had a better established root system and was better able to use a larger volume of soil to withdraw moisture and nutrients and had less stress resulting in a better crop. How many of the general public have seen this? I dare venture very few, as the crop was not as good as in a normal year. I have recounted several times about driving to an adjacent county to the conservation district office and was greeted by the remark that there was not much need for drainage. I explained they had not looked at the problem in the same way I had on the way up, I was looking for benefits and there was an abundance of evidence but because the subsurface drainage system was not visible it was not given credit. These fields were not wetlands. They would not grow cattails, yet the environmentalist through the FSA has largely stopped subsurface drainage by not having a usable definition of an environmental wetland.

I thank you for letting me ramble on and present my view. I am not getting any younger but am willing to try to help some of these situations but cannot carry the ball alone.

THE FUTURE OF DRAINAGE CONTRACTORS

Ronald D. Cornwell
Member ASAE

What is the future of the Drainage Contractors? This would all depend upon the future of the need for subsurface drainage. As long as there is a need there will be contractors. But, will this be true?

As long as the farmers continue to farm and they receive an overabundant amount of rainfall, drainage will be needed. But, there are not always the abundance of contractors to do the work. I see politics, economics, and attitude as the three problems facing contractors.

First politics. Many laws are made about wetlands, clean air, clean water, and so on with little regard to any scientific research. Political issues must be pursued at all times. Organizations must be maintained and used to lobby our views. Research must be ongoing to prove our views, but who will pay for this work.

This brings me to: second economics. The ultimate consumer to drainage is the farmer. And there is no question that there is a need for drainage. But it is not always economically possible to do the required amount. The farmer is being hard pressed. He must make business decisions and to drain is not always possible. Commodity prices are low. This is trickle down affect to contractors. Many times the contractors do not have the resources to belong to the organizations to lobby lawmakers, and contribute to the research needed.

Third attitude. The attitude of the drainage contractors has a false sense of optimism. Many times the contractor is more concerned with the fellow contractor that is "just down the road" competing for the same work, than trying to help his industry grow. This same contractor does not get involved because he hopes his fellow contractors will get involved. Any why do anything, drainage has always been around and it will always be in the future!

I see our industry get weaker. If we look at the support industries and suppliers for the contractors and farmers we can see a definite trend. This tile manufacturers, machine manufacturers and the laser manufacturers have all started with the drainage business. If they are still in business today, they had to learn to diversity their business. If anyone would check the percent of drainage business now compared to what it was when the company was formed, it is extremely low. Like manufacturers the drainage contractor had to diverse into other areas. For many contractors, the drainage business is a low percent of their gross business.

Will the manufacturers and contractors continue to support an industry that they receive continually lower percent of their business? The people I have talked to enjoy the drainage business. It is a satisfying line of work for everyone. Maybe certification of contractors will help. I do not have any answers. The drainage business had lost many good people over the past several years. I hope we do not continue this loss.

Ronald D. Cornwell, President, Cornwell and Sons, Inc., Drainage Contractors, Arlington, OH.

LONG TERM PERFORMANCE OF GEOTEXTILES ON HORIZONTAL SUBSURFACE DRAINAGE SYSTEMS

R.B.Bonnell[1]   R.S.Broughton[1]   J. Mlynarek[2]
Member CSAE,ASAE   Member CSAE,ASAE   Member IGS,CGS

This paper presents results of a field and laboratory examination of nine distinct fabrics which had been installed four to fifteen years ago as envelopes on drain pipes, in silt and sand soils. All but one fabric continued to perform their design functions very well. No measurable amount of sediment deposition was noted in any drain. The fabrics exhibited inconsequential amounts of internal clogging by soil particles and were not torn or damaged from washing or drying in the laboratory. The one exception was a fibreglass fabric of an unknown age (greater than 15 years) which was easily torn and was brittle upon drying. Microscopic examination of the soil to fabric interface established the existence of a network of soil arched and of a large quantity of macro-pores at this interface.
Keywords. Geotextiles, Durability, Drain pipe envelopes.

Introduction

In Canada and in Europe, thin synthetic geotextiles have been used successfully for filtration and soil stabilization since the early 1970's (Rollin et al., 1987; Martinek, 1986). A few authors extracted such geotextiles from the ground two to ten years after their installation, and have reported favourably on their continued good performance as drain envelopes (Sotton et al., 1983; Sotton, 1984; Bolduc, 1988; Saathoff, 1988; and deu Hoedt, 1989; and Mlynarek et al., 1990). Yet the long term durability of synthetic envelopes as drain pipe filters has been questioned. At a joint seminar (Rilem 1988) held by The International Union of Testing and Research Laboratories for Material and Structures, the International Colleges of Building Science and the International Geotextile Society (IGS), a major conference recommendation was that "researchers need to form an international data bank, monitor geotextiles in use, develop suitable test procedures and identify factors

---

[1]R.B.Bonnell, Research Associate and R.S.Broughton, Professor, Agricultural Engineering Dept., McGill University, Montreal, Canada.

[2]J.Mlynarek, Research Associate, Dept. of Civil Eng.,Polytechnique, Montreal.

which will provide indicators to the aging process".

It is accepted that many large subsurface drainage development projects are needed in the arid and semi-arid regions of the world. The costs involved in such projects necessitate a firm confidence in geotextiles as functional synthetic drain envelopes before they will be chosen over the relatively costly bulk gravel envelopes. This research project was initiated to check the condition of synthetic fabric envelopes which had been installed on subsurface drainage systems up to 15 years ago.

Materials and Methods

A field investigation was executed in the autumn of 1991 on two sites to inspect the condition of various synthetic drain envelopes which had been installed on drainage systems. The first drainage system; located near Ormstown, Quebec, on the farm of Mr. P. Finlayson, consisted of 30 parallel lateral drains of 220 m length with six replicates of five envelope treatments in a silt-loam soil. Installed October 1983, the treatments included four fabrics (Mirafi, Silt-sock, Texel F200 and Alidrain) and one control drain with no fabric (Bolduc, 1988). Another lateral, installed in this field in 1987 with a Sand-sock, was also sampled.

The second drainage system, located near Notre Dame du Bon Conseil, Drummond County, Quebec on the farm of Mr. Valois, consisted of 21 laterals, 134 m in length, with four replicates of 5 treatments plus one lateral with a gravel envelope installed July 1976 in a field of loamy-sand soil (Gibson, 1978). The treatments included four fabrics (Typar, Cerex, Reemay and a Nylon-sock) and a control with no fabric. At both locations, the pipes with no fabric became clogged with sediment and were replaced with fabric covered drain pipes within two years. Table 1 gives a description of the fabrics.

Table 1 : Description of Fabrics on Finlayson and Valois Farms.

NAME	DESCRIPTION
Mirafi	Woven slit-film polypropylene, needle punched stable fibres, side sewn
Silt sock	Knitted polyester with pile on one side
Texel	Calendered, needle punched, non-woven polyester, side sewn
Alidrain	Wet laid bonded, non-woven polyester, side sewn
Sand sock	Knitted polyester
Fibreglass	Non-woven fibreglass
Nylon sock	Knitted nylon yarn stocking
Typar	Spunbonded polypropylene, glue jointed
Reemay	Spunbonded polyester, glue jointed
Cerex	Spunbonded nylon, glue jointed

Sampling conditions were in the dry at the Ormstown location but at the Valois farm the water table was at the top of the drain pipes due to controlled drainage being practised at this location. Controlled drainage is practised at the Valois site in order to irrigate the crop, and in an effort to maintain an anaerobic condition at drain level to discourage iron ochre formation and deposition on the outside and inside of drains. A 50 mm by 50 mm undisturbed fabric plus soil sample was retrieved from each drain. This was accomplished by first isolating the sample by carefully flaking away the surrounding soil, leaving a covering of less than 5 mm. This was then soaked in a potting compound and allowed to dry and harden before removal. Laboratory sample preparation consisted of cutting the samples across the soil to fabric interface, polishing the surface and photographing it under a microscope. Under the dry conditions at the Ormstown location, it was possible to flake off soil patches from the fabric. By this means, an undisturbed soil surface was obtained and photographed in the laboratory. Also, at each sampling site, a one meter length sample of drain pipe plus fabric was removed and two soil samples were taken. The first sample was extracted from the soil profile at a horizontal distance of 150 to 200 mm from the drain, and the second sample was collected from the soil layer within 5 mm of the drain pipe. Finally, soil profile characteristics, amount of sediment within the drains and the general condition of the pipe and fabric were noted.

The one meter long fabric samples were allowed to dry in the laboratory and were visually examined. Then the loose soil adhering to the surface of the fabric was shaken off and the fabric was weighed. Each fabric sample was then hand washed with warm water, dried in an oven for 20 minutes to air dryness and weighed again.

Results and Discussion

Soil conditions:

In the field, it was noted that the soil in immediate contact with the geotextiles exhibited an abundance of macro-pores, suggesting that the soil fines had been removed from a zone of soil measuring 2 to 4 mm in thickness. Photograph 1 shows what the soil surface, in contact with the fabric, looked like. Evident, is an abundance of macro-pores. This pattern of macro-pores was not present when clods of the soil located away from the drains were broken apart. The macro-pores in the photograph are a result of the proximity of the soil to the fabric.

Table 2 and Figure 1 present the textural analyses of the soil samples obtained using the hydrometer method. From the textural analyses, it is evident that for all but one case (the Silt- sock) the soil samples adjacent to the fabric contained more clay then the original base soil as represented by the samples taken at a distance of 150 mm from the fabric surface. It is thought that water movement towards the drain carries some soil fines to the fabric-soil interface. The data means, adjacent and at 150 mm from the drains, at the two locations were calculated. A Student's

Photograph 1: Soil surface which was in contact with the Mirafi fabric at the Ormstown location (actual size).

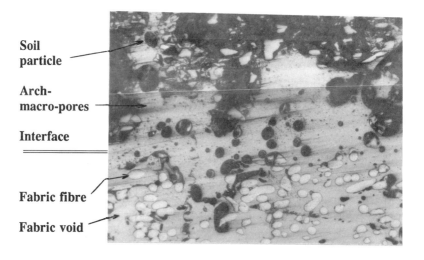

Photograph 2: 100 times magnification of the soil surface in contact with the Texel fabric. Note that very little of the intra-fabric pore space is occupied by soil particles. Also note the arch-bridging of the soil, resulting in macro-pores adjacent to the fabric.

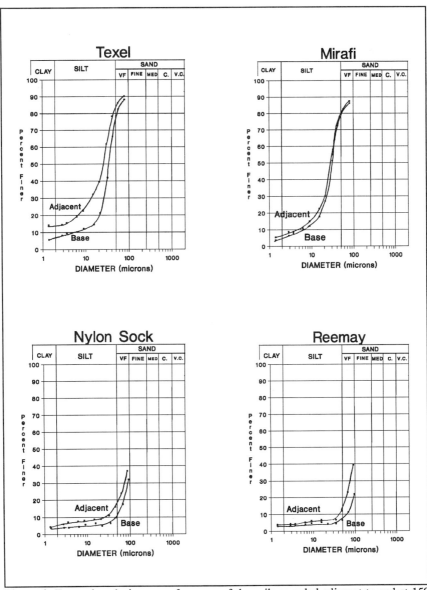

Figure 1: Textural analysis curves for some of the soils sampled adjacent to and at 150 mm from the fabric. Mirafi and Texel from Ormstown and Nylon sock and Reemay from Valois location.

"t" test (Steele and Torrie, 1960) of these means showed no significant difference, at the 5% confidence level, between clay content of samples taken adjacent to or at 150 mm from the drains. The data suggests a trend of more clay adjacent to the fabric, but more replicates would be needed in an effort to establish a statistically valid, significant difference. The visual evidence of a network of macro-pores, noted within the soil adjacent to the fabric, cannot be caused by removal of only fines from the soil. The macro-pores are thought to be as a result of all soil particle sizes moving into the fabric with some of the coarser ones being trapped within the fabric and the finer continuing through into the drain. Thus the formation of a more porous soil, a "natural soil filter" in the immediate vicinity of the fabric. This feature was common to all sites at both the Ormstown and Valois locations.

In a similar study carried out in the Netherlands, Stuyt (1992) took core samples of soils surrounding drain tiles wrapped with synthetic filter materials installed in weakly cohesive soils. Most of the drains were observed to have "clogged" soil near them, that is fine particles (≤ 30 µm) moved towards the drains and filled the pores among the larger particles. The author concluded that the filter materials around the drains appeared to have little or no effect on this process. The result of this soil clogging was that hydraulic conductivities were greatly reduced in the vicinity of the drains, and water flow was slowed considerably.

Stuyt continues, by noting that only in a small percentage of the cases observed was the "washing" out of fine particles near drains, and the formation of a natural soil filter seen. It was concluded that rapid water flow occurs at only a few places along a drain length, while most of the soil surrounding drains is hydraulic inert, and the type of envelope wrapped around drains has little effect on this phenomenon.

Condition and performance of the fabric:

With regard to the filtering performance of the fabrics, none of the plastic drains contained more than a trace of sediment. At the Valois location the drain pipes were one-half to two-thirds full of an ochre gel. Upon drying, the thickness of this gel was reduced to 5-15 mm. All of the fabrics at this site were stained a red colour. At the Ormstown location, which was in the dry, careful flaking of the soil away from the fabric revealed very clean fabric. That is, the soil did not adhere strongly to the fabric. Visually, the fabrics differed from the virgin samples from the manufacturers in colour only. The fibres were not crushed or cracked. The easy separation of soil from the envelope was most particularly noticeable on the Texel fabric. Minor amounts of small roots on, within and under the fabric occurred at all of the Ormstown sites, but at none of the Valois sites. The lack of roots at the Valois site is likely due to the high water table which is maintained artificially at this site.

Table 2 : Percent Sand, Silt and Clay in Soil Near the Drain Pipes.

Sample site	Soil*	Sand	Silt	Clay	Fabric Name
F1	A	21	73	6	Mirafi
F1	B	21	74	5	Mirafi
F2	A	21	75	4	Silt sock
F2	B	18	75	7	Silt sock
F3	A	26	70	14	Texel
F3	B	24	69	7	Texel
F4	A	18	75	7	Alidrain
F4	B	33	63	4	Alidrain
F5	A	17	70	13	Sand sock
F5	B	21	69	10	Sand sock
V2	A	84	12	4.7	Nylon sock
V2	B	90	6	4	Nylon sock
V3	A	81	13.3	5.7	Typar
V3	B	93	4.0	3.0	Typar
V4	A	88	7.5	4.5	Reemay
V4	B	94	2.5	3.5	Reemay
V5	A	81	15.5	3.5	Cerex
V5	B	93	3.5	3.5	Cerex

* Note: A and B represent samples adjacent to and 150mm from the soil-- fabric interface, respectively. Some complete particle size distribution curves are given in Fig.1.

Table 3 presents some of the results of the sampled fabrics. The pre-washed weights are those recorded after shaking any loose soil off the fabric. Then the fabrics were hand washed, dried and weighed again. Hand washing removed all soil adhering to the fabric surface. It is realized that this washing action may have forced more soil into the fabric and/or some soil particles out from within the fabric. But the post-washed weight can give some idea of the amount of soil clogging to expect for each fabric type. Column "B" of Table 3 presents numbers representing the percent weight of soil left within the fabric after washing, based upon weight of the original fabric. The lower the number, the less clogging has occurred.

Weights of the fibreglass fabric "V1" were not recorded because this fabric was found to be in a very fragile state. It cannot be recommended for use as a drain envelope because, upon handling, it easily tore and once dry it became brittle. This fabric was found covering the top half of a 100 mm clay tile lateral at the Valois location, of unknown origin and of unknown age, but it was certainly older than the plastic drain pipes excavated. The clay tile was full of a sandy sediment, but this may not be attributed to poor performance of the fabric, which covered only the top half of the drain. It is currently recognized that sediment can enter a drain from below, as well as through the top of a drain pipe.

Table 3: Fabric Parameters

Sample	Name	Years in soil	Pre-washed g/m^2	Washed g/m^2	Virgin g/m^2	A	B
F1	Mirafi	8	284	188	171	66	10
F2	Silt-sock	8	572	241	230	148	5
F3	Texel F200	8	276	227	200	38	14
F4	Alidrain	8	158	94	119	33	-20
F5	Sand-sock	4	318	101	95	235	6
V1	Fibreglass	15					
V2	Nylon-sock	15	210	110	119	76	-7
V3	Typar	15	205	176	120	71	47
V4	Reemay	15	43	29	30	43	0
V5	Cerex	15	47	30	20	135	50

"A" : Percent weight of soil on fabric before washing, based upon weight of original fabric. ex. $((284 - 171)/171)*100 = 66$

"B" : Percent weight of soil left within fabric after washing, based upon weight of original fabric. "Virgin": weight of clean fabric as quoted by the manufacturer.

Photograph 2 is a cross-section taken of the soil to Texel fabric interface. Note that the bulk of the fabric pore space is free of soil particles. Also note the arching of the soil and macro-pores at the soil to fabric interface.

Conclusions

After eight to fifteen years in the soil, all fabrics investigated, continued to perform their designed functions. The fabrics were not clogged, the drains were clear of sediment deposition and the fabrics had no tendency to tear while being sampled or hand washed. The outflow from the drains has been satisfactory. The farmers are satisfied that the drainage of these fields has been adequately rapid ever since the drains were installed. Measurements at the Finlayson farm show peak drainage rates as high as 15 mm/day. The design drainage rate was 9 mm/day. The one exception was a fibreglass fabric of unknown origin, over fifteen years in the soil, which tore easily and was brittle upon drying.

A "natural soil filter", the occurrence of macro-pores in the soil at the soil to fabric interface, is not the result of fines being removed from the soil immediately adjacent to the fabric. In fact, the soil adjacent to the fabric tended to contain more clay sized particles than the soil located 150 mm away from the drain although this difference was shown to be not significant at the 5% confidence level. The occurrence of macro-pores at the soil to fabric interface is a result of washing all particle sizes into the drains and leaving arches in the remaining soil. Minor amounts of soil particles were trapped within the fabrics.

References

Bolduc, G.F. 1988. Développement de géotextiles pour utilisation en drainage souterrain dans les sols limoneux. Ph.D. Thesis, Université de Montréal, Ecole Polytechnique, Montréal.

deu Hoedt, G. 1989. Durability experience in the Netherlands, 1958-1988. In; "Durability and Aging of Geosynthetics". Publ. Elsevier Applied Science, New York, pp. 82-94.

Gibson, W. 1978. A field evaluation of some subsurface drain envelopes. M.Sc. Thesis, Department of Agricultural Engineering, McGill University, Macdonald Campus, Montreal, Quebec.

Leflaive, E. 1988. The different aspects of long-term behaviour of geotextiles. In "Durability of Geotextiles", Rilem, Publ. Chapman and Hall, New York NY.

Martinek, K. 1986. Geotextiles used by the German Federal Railway - experiences and specifications. Geotextiles and Geomembranes, Vol. 3:175-200.

Mlynarek, J., J. Lafleur and J.B. Lewanolowski. 1990. Field study on long term geotextile filter performance. Proc. of Fourth Int. Conf. on Geotextiles, The Hague, PP. 259-262.

Rollin, A.L., R.S. Broughton and G.F. Bolduc. 1987. Thin synthetic envelope materials for subsurface drainage tubes. Geotextiles and Geomembranes, Vol. 5:99-122.

Rilem. 1988. "Durability of Geotextiles". Rilem (The Int. Union of Testing and Research Laboratories for Material and Structures) and the Int. Colleges of Building Science with the support of the Int. Geotextile Society, in Saint-Rémy-les-Chevreuse, France. Publ. Chapman and Hall Ltd., 29 West 35 Street, New York, NY 10001.

Saathoff, F. 1988. Examinations of long-term filtering behaviour of geotextiles. In; "Durability of Geotextiles", Rilem, Publ. Chapman and Hall, New York, NY, pp. 86-114.

Sotton, M. 1984. Durability of Geotextiles. 23rd Inter. Man-Made Fibres Congress, Dornbirn.

Sotton, M., B. Leclerq, J.L. Paute and D. Fayoux. 1983. Quelques élements de réponse au probleme de la durabilité des geotextiles. Proc. Second Int. Conf. on Geotextiles, Las Vegas. IFAI Publ., St. Paul, Minnesota, USA. pp 553-558.

Steele, R.G.D. and J.H. Torrie. 1960. "Principles and Procedures of Statistics". McGraw-Hill Book Co. Ltd., Toronto.

Stuyt, L.C.P.M. 1992. The water acceptance of wrapped subsurface drains. Ph.D. Thesis, Wageningen Agricultural University, Publ. Winand Staring Centre for Integrated Land, Soil and Water Research, Wageningen, the Netherlands. 305 pp. + 7 plates.

Acknowledgements

The authors gratefully acknowledge the financial support of the Canadian International Development Agency which made these investigations possible. We acknowledge the cooperation of the farmers who allowed us to dig up and investigate the drain pipes and fabric envelopes. We also acknowledge the assistance of Dr. G. Bolduc with the field investigation at Mr. Valois's farm.

DRAINAGE NEEDS OF THE SMALL INLAND VALLEY SWAMPS OF WEST AFRICA

Y.M. Mohamoud[*]

ABSTRACT

Inland Valley Swamps (IVSs) are small streams that are subject to floods and high water table during the rainy season. Because of these water management problems, crops other than wetland rice require enormous drainage operations and may not be suitable for most of the IVSs. However, IVSs are suitable for wetland rice cultivation if flooding and submergence of the rice crop are avoided by surface drainage. At the end of the rainy season, evapotranspiration and base flow deplete the water table. The water table recedes slowly and farmers plant cassava and sweet potatoes on raised beds known as mounds or ridges. The use of mounds and ridges is to prevent crop stress from excess moisture by creating artificial rooting depth without lowering the water table by drainage.

KEYWORDS. Drainage, Inland valley, Hydrology, Geomorphology, Wetland rice

INTRODUCTION

IVSs can be geomorphologically defined as small order streams that have a shallow water table and become inundated during the rainy season. These small streams are known as dambos in east and central Africa (Balek and Perry, 1973) as fadamas in northern Nigeria (Ipinmidun, 1970 and Turner, 1977) and as bas-fonds in French-speaking west Africa (Kilian and Teissier, 1973). The total area of IVSs in tropical Africa is estimated to be 340,000 km^2 (Balek, 1977). Hekstra et al (1985b) estimated the inland valley area in west Africa to range between 100,000 km^2 to 200,000 km^2. Andriesse (1985) reported that the landforms of west Africa consist of coastal plains, interior plains, plateaux and highlands which constitute about 5, 50, 30 and 15 percent of the moist savanna and humid forest zones of west Africa. Each of these landforms has inland valley swamps that are geomorphologically related to the major landform.

The major water management problems that restrict the utilization of inland valleys for crop production are flooding and high water table which can be eliminated by selection of an appropriate water management system. Inland valleys that are located in traditionally wetland rice growing areas are developed and cultivated during the rainy season. In some areas, particularly in the humid forest zone of Cameroon, where wetland rice is not traditionally grown, inland valleys are uncultivated during the rainy season. However, farmers cultivate inland valleys during the dry season when water management problems are minimum. Drainage operations require capital, labor and management skills that are scarcely available to the small scale farmers of west Africa. The choice of the crop determines the drainage needs of the inland valleys and therefore, to reduce drainage needs, farmers grow wetland rice at the valley bottom and non-rice crops at the valley-side.
The objectives of this paper are (i) to present some geomorphologic and hydrologic conditions of inland valleys and, (ii) to recommend drainage systems and appropriate cropping patterns for these geomorphologic and hydrologic conditions.

---

[*] Y.M. Mohamoud, Agricultural Engineer, International Institute of Tropical Agriculture (IITA), Ibadan, Nigeria.

MATERIALS AND METHODS

Hydrologic data were collected on a small inland valley experimental watershed in Central Nigeria. The IVS watershed has an area of 8.7 km^2 and locates near Gbako river floodplain which is a tributary of the Niger river. The inland valley watershed has sandy loam soil classified as Typic Tropaquents (Smaling et al. 1985b) with a flow impeding lateritic layer at some depth below the soil surface and a sedimentary geological formation.

Daily rainfall amounts were recorded using manual and recording rain gauges. Stream flow was measured with 130° V-notch weir and water table depth was monitored along a cross-sectional transect of the valley using perforated PVC pipes. Water table depth and stream discharge vary with the physiography, soil, depth of the impeding layer and the rainfall of the inland valley swamp. Therefore, the hydrologic data shown in this paper is not representative for all inland valley swamps of west Africa even though inland valleys that are located in the moist savanna and humid forest zone have relatively similar stream discharge and water table responses to rainfall.

Quantitative geomorphologic parameters were also estimated from several drainage basins of west Africa using 1:50,000 scale topographic maps. Third order stream watersheds were selected as the largest unit of an inland valley watershed. The geomorphological parameters measured from the third order stream watersheds are stream channel slope, valley-side slope, stream frequency, drainage density and valley bottom width.

RESULTS AND DISCUSSION

Geomorphology of the IVSs

Table 1 shows some of the geomorphologic characteristics of the IVSs that influence land drainage. Inland valleys at the interior plains have gentle stream channel slopes and may have high water table and risk of rice submergence. Conversely, IVSs that are located in the plateaus and highlands have steep stream channel slopes and less drainage problems. However, inland valleys that are located in the plateaus and the highlands may have risk of soil erosion and flooding because valley-side slopes influence the response of the watershed to rainstorm since hydraulic gradient increases with an increase of the valley-side slope assuming similar soil hydraulic properties and predominance of subsurface runoff components.

Figure 1 shows IVSs from highland, plateau and interior plain landforms of Cameroon, west Africa. The geomorphologic characteristics of the IVSs that influence drainage are the stream channel slopes, valley-side slopes and width of the valley bottom. IVSs in the interior plains have relatively flat valley-side slopes when compared with inland valley swamps that are found in the plateaus and highlands (Table 1). A gentle valley-side slope indicates that farmers can diversify the cropping pattern by cultivating wetland rice at the valley bottom and other crops at the valley-side. Conversely, the inland valley swamps that are found in the highland landforms have high valley-side slopes and therefore the upland may not be cultivable without appropriate management practices (e.g; terracing).

Crop diversification in the valley bottom and thus drainage becomes important when non-rice crops cannot be cultivated in the valley-side due to steep valley-side slopes. Also, crop diversification in the valley bottom area is important when dry season crops cannot be cultivated in the valley bottom during the dry season because of limited residual moisture.

Table 1. Geomorphologic Parameters of Third Order Stream Watersheds from Selected Areas in West Africa.

Country	Area km^2	Stream frequency km^{-2}	drainage density km^{-1}	stream. channel slope (min.)%	Valley side slope %	Valley bottom width	Landform
Cameroon							
Bafoussam	4.5	1.77	1.6	0.93	17.0	wide	Highland
Ngaoundere	5.5	2.18	1.7	1.10	20.0	wide	Highland
Betare-oya	12.0	1.40	1.5	0.60	8.0	narrow	Plateau
Garoua	4.5	1.77	2.1	1.30	5.0	wide	Plains
Mbalmayo	3.8	2.63	1.7	0.80	6.6	narrow	Plains
Sierra Leone							
Kabala	2.2	2.72	1.5	1.70	10.0	wide	Highland
Panguma	3.9	1.54	1.4	1.50	13.0	wide	Highland
Nigeria							
Ilesha	8.2	0.73	1.1	1.0	6.0	narrow	Plateau
Zaria	9.4	0.53	0.5	0.9	3.0	narrow	Plains
Bida	60.2	0.06	0.4	0.4	2.0	narrow	plains

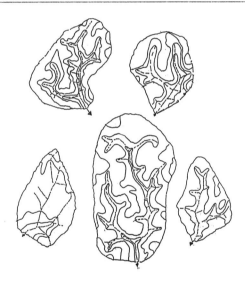

Figure 1. Third Order Stream IVSs Watersheds from Selected Areas in Cameroon.

IVSs at the interior plains have fewer drainage needs for crop diversification purposes since non-rice crops can be grown at the valley-side during the rainy season. In areas where farmers need to diversify their cropping pattern in the valley bottom (e.g; central Benin), non-rice crops such as corn are grown at the valley bottom before the water table rises to the surface. Corn is harvested when the water table rises and wetland rice is planted as a second crop. However, the possibility of growing corn depends on the response of the watershed to rainfall and the resulting water table depth at the valley bottom. When the water table rises very early in the rainy season, farmers would need drainage to grow crops other than wetland rice.

Hydrologic Regimes of an IVS: A Bida Case Study

The hydrologic regimes of inland valleys depend on the climate, soil, geology and the topography of the watershed. The Bida area has soils with high permeability and subsurface flow is the dominant runoff component. Rainfall infiltrates into the soil and later reaches the valley bottom by subsurface runoff or base flow. Early rains fill the available storage capacity of the soil until hydraulic gradient develops between the valley-side and the valley bottom. Subsurface flow coming from the valley-side rises the water table at the valley bottom (Balek and Perry, 1973). Surface runoff from the valley-side is not commonly observed, however, saturation flow occurs at the valley bottom when the water table rises above the soil surface.

Figure 2 shows water table depth measured along a cross-sectional transect of the Gara valley in Bida at different periods during the year. The first period coincides with the beginning of the rainy season and illustrates a deep water table. The second period coincides with the peak of the rainy season when adequate moisture is stored in the soil and consequently the water table rises above the soil surface. The third period coincides with the end of the rainy season when evapotranspiration and base flow started to deplete the water table. Figure 2 shows that only hundred meters of the valley bottom width has a shallow water table that is suitable for wetland rice cultivation. The water table at the valley-side is below the soil surface and therefore other crops can be cultivated without drainage.

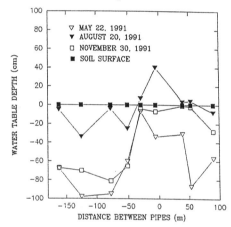

Figure 2. Water Table Depth Measured at Different Times During the Rainy Season Along a Transect at Gara Valley Bida, Nigeria.

Figure 3 shows the daily water table fluctuations at the valley bottom as affected by rainfall amount and frequency during the rainy season. The water table rises after a heavy storm or sequences of storms. However, due to evapotranspiration and base flow, the water table drops after few non-rainy days. A maximum water table depth of 40 cm above the soil surface has been recorded in August 1991 when consecutive heavy rainstorms coincided with a period of limited moisture storage capacity of the soil. Appropriate selection of the rice transplanting period and central drain maintenance generally protect the rice crop from submergence due to heavy floods.

Figure 3 shows the daily stream discharge as influenced by rainfall amount and water table depth during the rainy season. Base flow is generally observed

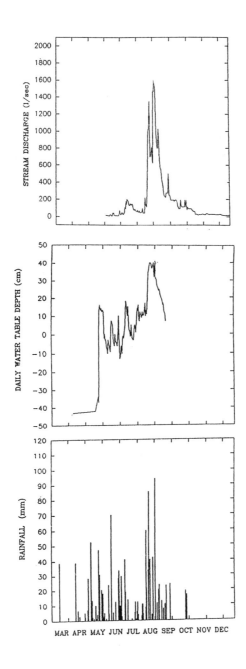

Figure 3. Daily rainfall, Water Table Depth and Stream Discharge Data Measured at an Inland Valley Near Bida, Nigeria.

until the end of February or early March depending on the available storage from last year's rainfall. Stream discharge begins after several rainstorms raise the water table, however, peak stream discharge can be observed only when the water table is above the soil surface. The period with the peak stream discharge coincides with the period when water table is high and drainage is required to avoid flooding and submergence of the rice crop.

IVSs Drainage Needs During the Rainy Season

Most of the inland valleys in the humid forest zone of west Africa are cultivated particularly where wetland rice is a commonly grown crop. Inland valley development in the humid forest zone of west Africa requires considerable land clearing operations. These land clearing operations can only be justified where farmers are traditionally wetland rice growers (e.g; Liberia and Sierra Leone) since other crops would require enormous drainage operations. In areas where wetland rice is not grown, inland valleys have high water table which renders impossible to grow other crops without land drainage. Also, valley bottoms are narrow and farmers use the valley-side to grow a variety of crops during the rainy season. Therefore, it is not necessary to drain the valley bottom for non-rice crops when enough land is available for crop production at the valley-side. Small scale farmers in west Africa leave their inland valley farms fallow during the rainy season or otherwise grow wetland rice and reduce drainage needs that would have been required if non-rice crops are grown.

Owing to the topography and the hydrologic regimes of IVSs, surface drainage is required to avoid rice submergence and flooding at the valley bottom. It is difficult to devise a standard water management system for all IVSs, however, the components of an IVS water management system may consist of a central drain, peripheral canal and head-dike or any combination of these components.

The central drain which is often the natural stream channel is the most important inland valley water management system. A well designed and maintained central drain system can eliminate high water table and risk of submergence during the peak of the rainy season. In areas where flash floods occur, head-dikes are used to raise the water level in the stream so as to divert stream flow into the peripheral canal and to attenuate floods. The purpose of the peripheral canal is to convey water from the valley bottom to the valley-side and thereby increase the valley-side area under wetland rice cultivation through irrigation.

IVSs Drainage Needs During The Dry Season

In the Bida area, dry season crops are grown at the inland valley bottoms using the residual soil moisture from the water table. However, water table is high at wetland rice harvest and some type of water management practices are needed to grow cassava, sweet potato or vegetables during the dry season. Mounds and ridges are commonly used water table management practices for cassava and sweet potato cultivation in the valley bottoms. Farmers grow these crops on mounds or ridges to prevent plant stress from excess moisture by creating an artificial rooting depth without lowering the water table. Water table is the source of soil residual moisture during the dry season and therefore any attempt to drain the water will create moisture deficiency unless irrigation is used. In some areas, inland valley swamps are irrigated during the dry season by pumping water from shallow water tables. However, irrigation may only be profitable in areas where farmers grow vegetables for urban consumers in nearby cities.

Drainage/subirrigation seem to be a potential dry season water management system for IVSs. However, drainage/subirrigation may not be appropriate for most of the inland valleys because these small scale farmers can neither afford the cost nor the proper management of the system. Also, inland valley bottoms are narrow and represent a very small portion of the watershed with a non-flat geometry.

CONCLUSION

Drainage needs of inland valley swamps depend on the crop choice, geomorphology and the hydrologic regimes of the valley. Drainage needs are reduced if wetland rice is cultivated at the valley bottom during the rainy season. The appropriate drainage for wetland rice cultivation is the central drain system. This water management system protects flash floods and submergence of the rice crop during the peak of the rainy season. Valley bottom drainage for non-rice crops is not feasible because of the small valley bottom area and the topography of the valley bottom which is prone to flash floods and high water table. Also, drainage of the valley bottom for non-rice crops is not justifiable when farmers can grow non-rice crops at the valley-side.

REFERENCES

1. Andriesse, W. 1985. Wetlands in subsaharan Africa area and distribution. In: The Wetlands and Rice in subsaharan Africa. Eds. Juo, A.S.R. Lowe, J.A. Proceedings of the International conference on Wetland utilization for rice production in subsaharan Africa. 4-8 November 1985. IITA, Ibadan, Nigeria.

2. Balek, J. (1977). Hydrology and water resources in tropical Africa. Elsevier, Amsterdam 1-208

3. Balek, J. and Perry, J.E. 1973. Hydrology of seasonally inundated African headwater swamps. J. of Hydrology. Vol. 19, 227-249.

4. Hekstra, p. and Andriesse, W. (1983). Wetland utilization research project, west Africa. Phase I, The Inventory. Vol. IV The maps. Wageningen. Netherlands soil survey Institute.

5. Ipinmidun, W.B. 1970. The agricultural development of fadama with particular reference to Bomo fadama. Nigerian Agricultural Journal, 7/2 152-163.

6. Kilian, J. and Tiessier, J. 1973. Method d'investigation pour l'analyse a la classement des bas-fonds dans quelques regions de l'Afrique de l'ouest: proposition de classification d'aptitude des terres a la riziculture. Agronome Tropical 28 (2): 156-171.

7. Samling, E.M.A., Kiestra, E. and Andriesse, W. 1985b. Detailed soil survey and qualitative land evaluation in the echin woye and kunko benchmark sites, Bida area, Nigeria. Wageningen, Netherlands Soil Survey Institute.

8. Turner, B. 1977. Fadama lands of central northern Nigeria: Their classification, spatial variation, present and potential use. PhD. Thesis, University of london.

ELIMINATING DOWNSTREAM SEDIMENT TRANSPORT BY PUMPING BASE FLOWS AROUND CONSTRUCTION AREAS [1]

Richard T. Smith [2]

ABSTRACT

A 4000 feet reach of perennial channel was restored to its original depth and 40 feet bottom width while base flow was diverted around the construction area by using a dredge booster pump. The base flow was easily handled for the 8.4 square mile agricultural watershed. This method was chosen so no sediment transport downstream into a sensitive tidal creek would occur during construction. Additionally an existing tree canopy was retained and construction was performed by a long-boom hydraulic excavator reaching through the standing trees. As a result the town of Dagsboro, Delaware received flood control for a 50-year storm and the construction did not require filling wetlands. Costs and techniques are reviewed and compared to conventional construction techniques.

---

[1]Contribution from Department of Natural Resources and Environmental Control, Division of Soil and Water Conservation, Georgetown, Delaware.

[2]The author is a Drainage Program Administrator, P.O. Box 567, Georgetown, DE 19947.

Palustrine Wetland Retention, Restoration and Creation as Part of Delaware's Tax Ditch Program

Richard T. Smith, P.E., Thomas G. Barthelmeh, and Sally L. Griffith-Kepfer[1]
Member ASAE

Introduction

In Delaware over 50 miles (80.5 km) per year of tax ditches are reconstructed to serve as outlets for on-farm conservation practices. Typically these projects are developed for tax ditch organizations in the coastal plain region having 10 to 15 miles (16 to 24 km) of channels to be reconstructed. Urban areas and small towns also rely on these channels for flood protection and drainage. Since 1951 over 2000 miles (3220 km) have been reconstructed, and it is estimated by the authors that at least another 500 miles (805 km) of channel will be petitioned for reconstruction.

In the last decade the importance of wetlands for many functions such as water quality and wildlife habitat has been well documented (Tiner, 1985). Recent channel reconstruction efforts in Delaware have demonstrated that natural resource impacts from this reconstruction can and should be minimized. This has resulted in the re-evaluation of how channels are planned for reconstruction, including weighing wetland impacts against the drainage benefits prior to recommending reconstruction of the deteriorated channels.

To effect these changes several steps have been added to the planning process for channel reconstruction. Taken together these steps are referred to as "mitigation." Individually the steps consist first of avoidance of impacts to wetlands to the maximum extent possible, then minimizing the impacts where they cannot be avoided, and finally, compensation for unavoidable impacts by restoring previously drained wetlands, enhancing existing low value wetlands, and creating wetlands in upland areas.

Project Description

Impacts to freshwater wetlands within tax ditch watersheds normally occur due to fill placed on wetlands when it is side cast during channel reconstruction. These restored channels can also lower surface and subsurface water levels. Our experience in Delaware indicates that subsurface drainage of wetlands adjacent to open channels is typically very limited. This is due to the predominance of silty soil lenses near the surface preventing the surface water in these wetlands from infiltrating downward.

The reduction of the hydraulic capacity of the channels is normally caused by sediment deposition in the low gradient channels. Hydrologic conditions sufficient to support hydrophytic vegetation develop, or in other words, the areas adjacent to these channels revert to freshwater wetlands.

Freshwater wetlands account for 130,000 acres (52,610 ha) or 10 percent of the State. The impacted wetlands consist primarily of palustrine forested deciduous wetlands with "J" and "A"

---

[1]The authors are Richard T. Smith P.E., Drainage Program Administrator, Thomas G. Barthelmeh, Drainage Program Manager, Delaware Department of Natural Resources and Environmental Control, Division of Soil and Water Conservation, Georgetown DE and Dover DE, respectively, and Sally Griffith-Kepfer, State Resource Conservationist, USDA Soil Conservation Service Dover DE.

water regimes (PF01J, PF01A, respectively)[2]. Fifty-four percent of the freshwater wetlands in Delaware are of these types (Tiner, 1985, Saveikis, 1992). Impacts to these palustrine forested deciduous wetlands are primarily due to the construction of the outlet portion of the channel system, *not* from the conversion of these areas for agricultural production. Impacts to palustrine forested deciduous wetlands with "C" and "E" water regime (PF01C and PF01E) account for a very small portion of the wetlands impacted. These types of wetlands are usually avoided during the channel reconstruction.

Federal Mitigation Policies

Drainage projects performed under the USDA-Soil Conservation Service (SCS) PL-566 small watershed program have re-established many previously existing agricultural drainage outlets. Since 1979 SCS policies have required mitigation for most wetland impacts. These impacts usually consist of filling and/or drainage of wetlands. We are compensating for these impacts on a 1:1 ratio based on habitat evaluation.

These compensations are performed within the same watershed whenever possible. When this is not possible, compensation sites are established where diversity can be added near an existing wetland, such as adding a shallow emergent wetland area near a wooded wetland. Fortunately Delaware has a very high percentage of freshwater wetlands that are palustrine forested. Adding sites with emergent properties provides diversity and is practicable in the high water table soils. Many of these locations were previously drained for agricultural production during the last 200 years. Re-establishing these wetlands has been done as a component of on farm conservation plans in cooperation with the owners. Other sites were on State property that had been drained for agricultural fields prior to the State's ownership. The State Division of Fish and Wildlife desires to re-establish these wetlands for water quality and wildlife management benefits.

Wetland fill associated with drainage for *existing* agricultural crop production is specifically exempt from the U. S. Army Corps of Engineers' Section 404 regulations. Portions of the projects, such as residential drainage, are not exempt. To date these portions have been small enough to fall within the range of Nationwide Permits, eliminating the need for lengthy individual permit review. Compensation for these small impacts has not been required by the Corps of Engineers.

State Mitigation Policies

In May of 1988 Governor Castle issued Executive Order 56 which established a wetland review policy for projects utilizing State funding or technical assistance. Simply stated this policy requires the review of State projects by the Department of Natural Resources and Environmental Control for wetland impacts prior to the initiation of the project by a state agency. If no agreement for avoidance of impacts or mitigation can be reached between the agencies then the Cabinet Secretaries would notify the Governor for resolution of the disparity. Political pressure is very effective in preventing such an impasse from being forwarded to the Governor. This policy was similar to planning policies already established (at a lower administrative level) within the Department of Natural Resources and Environmental Control for the planning of channel reconstruction projects to minimize wetland impacts. Once the Executive Order was issued all State agencies had to comply.

---

[2]Water regime and wetland designations are those used by the USFWS National Wetland Inventory.
J = Intermittently flooded
A = Temporarily flooded
C = Seasonally flooded
E = Seasonally saturated
F = Semipermanent

During the initial tax ditch planning process the lead agency (normally the Division of Soil and Water Conservation, Drainage Section) would request assistance from the Wetland Section for evaluation of the wetland resources in the project area. Additional assistance is also provided from the SCS State Biologist in evaluating the project impacts and determining the potential for violation of USDA Food Security Act provisions for individual farms and properties. This interagency cooperation is consistently excellent and the evaluation team has (to date) shown remarkable good common sense in recommending areas to be avoided and developing compromises to effect the necessary drainage and provide resource protection. This team has also developed recommended compensation ratios for unavoidable wetland impacts for these projects. Additionally, ratios to apply compensation credit for wetland restoration, enhancement and creation efforts performed by the Tax Ditch organization were recommended. (Fig. 1)

IMPACT	REQUIRED COMPENSATION RATIO	COMPENSATION CREDIT RATIO
Filling or draining wetland with a "J" (intermittently flooded) water regime	¼ to 1	
Filling or draining wetland with an "A" (temporarily flooded) water regime	1 to 1	
Filling or draining wetland with a "C" (seasonally flooded) water regime	1¼ to 1	
Filling or draining a wetland with an "E" (seasonally saturated) or "F" (semipermanent) water regime	2 to 1	
Restoration of previously drained wetland		1 to 1
Creation of wetlands in uplands		1 to 1
Enhancement of existing wetland by introducing a wetter water regime. (i.e., increasing an existing A water regime to a C water regime)		½ to 1
Farmed Wetland Abandonment		½ to 1
Farmed Wetland Enhancement		¾ to 1

Figure 1. Required wetland compensation ratios and mitigation credit ratios for tax ditch drainage projects and related wetland restoration and creation.

The planning review team first reviews the proposed projects in the office using aerial photography, soils maps, and National Wetland Inventory maps. From these data sources the Wetlands Section staff and SCS staff advise the tax ditch planner of areas that appear to contain significant wetland resources. Obtaining this advice before landowner contacts are made prevents the planner from promising channel reconstruction to landowners and then later having to renege on that commitment because of unacceptable wetland impacts. Once initial contact with landowners is made the detailed field evaluation of probable wetland impacts is done. Also detailed engineering data for the channel designs is completed. This aids the final evaluation of potential wetland impacts. It also provides guidance in determining which wetland areas can be avoided, retained, or restored as part of the drainage project. And finally this data allows the determination of which channels can be reconstructed with minimal wetland or other environmental impacts and which will require on site evaluation from the planning team.

Once it is determined that channel reconstruction is required through a wetland area an on site evaluation for minimizing wetland impacts is performed. This field evaluation will consider the applicability of:

1. Relocation of the channel away from the wetland.
2. Installing reverse berms to prevent surface water flow from the wetland.
3. Setting pipes through the built up channel berms to retain the surface water in the wetland at or above historical water levels using biological benchmarks.
4. One sided construction to minimize disturbance. (Fig. 2)
5. Off sided construction to place a small berm between the channel and the wetland on the minor construction side. This will retain wetland surface water adjacent to the channel. (Fig 3)
6. Blockage of drainage channels which serve only wooded areas.
7. Placement of spoil only on old berms if they exist from previous reconstructions.
8. Compliance of affected landowners with USDA Food Security Act provisions where drainage channel reconstruction occurs.

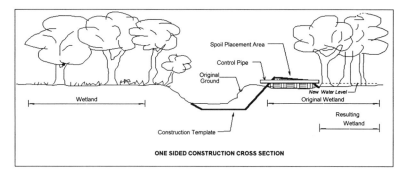

Figure 2. One sided construction cross section. Note that spoil placement reduces the wetland area, but also serves as a barrier to surface water flow into channel. Restoration or expansion of the wetland area behind barrier is achieved by adjusting control pipe elevation.

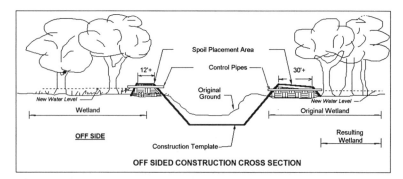

Fig. 3. Off Sided Construction Cross Section. Similar to One sided (Fig. 2) but a narrow construction area is established on the off side (minor construction side) to prevent draw down of the surface water into the reconstructed channel. Surface water elevation behind the berms is controlled by the elevation of the control pipes on each side. The off side will not be mowed during future channel maintenance.

Once the planning is complete the same concern for wetland impacts is retained during the design stages of the project. The hydraulic design is analyzed for the following:

1. Maximize the use of clearing and snagging techniques as an alternative to full construction in the downstream channel reaches and through sensitive areas when possible.
2. Minimize outlet construction (start construction of the outlet channel as far upstream as possible).
3. Minimize clearing limits and place fill on as narrow a "foot print" as possible to minimize fill area through wetlands.
4. Installation of water control structures to retain water in channels. This can help to retain water in adjacent wetlands by decreasing the lateral hydraulic gradient between the water in the channel and the adjacent wetland during the majority of the year when full drainage is not required.

In order to assure that our efforts for wetland protection are successful we continue our protection efforts into the construction phase. Project reviews are continuously held between planners, designers and the construction team. The construction equipment operators and inspectors are the critical members of the construction team. Obtaining the desired results for minimizing construction impacts may require an entirely different construction technique. In one instance wooden mats were used to span the channel top width in a wooded wetland area, then the large hydraulic excavator removed the sediment deposits from the channel while suspended over the channel on the mats. This requires extraordinary skill on the part of the operator.

Achieving understanding of the need to protect these resources within the project team involves training. In 1991 and 1992 wetland training sessions were held for the equipment operators, construction managers, tax ditch planners, surveyors, construction inspectors, and design engineers and technicians involved in these projects. These sessions have significantly improved the understanding of the project staff on the construction impacts to these wetlands and why we need to conserve and manage them. The participants in these sessions are directly involved in developing construction methods that will avoid or minimize wetland impacts.

Avoiding and Minimizing Wetland Impacts

Avoiding negative impacts to wetlands is the primary goal of wetland management within these projects which is not always consistent with the water management goals of the project. These conflicts are resolved on a case by case basis. Often there are technical reasons why avoidance cannot be achieved, such as having too deep of an excavation in the relocation route or the wetland area is too extensive to go around. Just as often there are political reasons why relocation from a wetland area cannot be done. Typically a landowner will not want the channel re-routed through their property when it has historically been located on the property boundary.

Using minimal clearing widths and narrowing the "foot print" of the spoil and debris disposal areas reduces negative wetland impacts. One sided and off sided construction techniques are the best examples of this technique (Figs.2,3). This requires consideration of the construction equipment to be used and the amount of debris and spoil expected from the reconstruction. Modern hydraulic excavators can work in very narrow confines. The minimum construction width that can be used is about 25 feet (8 m) if there is minimal spoil and clearing debris. Typically the cleared area along the channel is 30 to 35 feet (9 to 11 m) wide for the hydraulic excavator compared to 45 to 60 feet (14 to 18 m) wide to accommodate draglines, which have not been used on our projects for about ten years. This has the added advantage of lowering the total construction costs, however, if the width becomes too narrow the excavation operation becomes very inefficient because of maneuvering around timber and cleared debris.

Constructing downstream only far enough to obtain the hydraulic outlet for the drainage problems leaves existing flood plain wetlands intact. This can substantially reduce total construction costs. It requires reliable data on the existing downstream channel conditions and the outlet conditions required for the upstream watershed drainage. Before the downstream limit of construction is determined the potential for changes in outlet conditions used for the design must be considered. Of special consideration is the potential for the area below the start of construction to become blocked with silt and debris. This may alter the design assumptions for downstream hydraulic elevations at the construction limit. If routine maintenance is required it may be advisable to perform full construction or develop access routes outside the wetland area if possible.

Maximizing the use of clearing and snagging diminishes the debris and spoil disposal impacts to wetlands adjacent to the outlet channels. This alternative is less expensive than full construction, however gaining access to remote areas with hydraulic excavators can require substantial time and cost. This can negate cost savings if significant or persistent future maintenance is necessary.

Elevated channel berms control surface water flow from adjacent wetlands into the channel. This control is effective and inexpensive to obtain since the channel berm must be shaped as part of normal construction operations. Inexpensive control pipes (typically 12 inch (0.3 m) diameter) are installed through the channel berm at the grade necessary to re-establish water elevations in the adjacent wetlands to historical surface water levels. Benchmarks determined from staining on trees and other biological indicators are used to determine what these levels should be. The water must rise to this level before the control pipe allows the water to flow into the channel. If desired an elbow can be attached to the inlet and swivelled to adjust the water height behind the berm.

Water control structures allow manipulation of the ground water elevations in adjacent wetlands. The use of water control structures to raise the water surface elevation in channels when full drainage draw down is not needed can reduce lateral ground water flow from the wetland into the channel extending the hydro period of the wetland. These structures range in price from $2,500 to over $15,000 and have the added benefit of providing for sediment retention and nutrient reduction (Evans, et al, 1989). The cost effectiveness reduces rapidly if the drainage area exceeds 350 acres (142 ha) due to the large culvert sizes necessary to carry the design flows at a low hydraulic head.

Wetland Compensation

Restoration of previously drained forested wetlands is accomplished by blocking old channels in wooded areas. This effectively prevents the surface water from running into the outlet channel. Control of the water level in these areas is achieved using control pipes through this low berm. Forestry management impacts must be evaluated and landowner consent must be acquired before this can be done. For surficially recharged wetlands this is a very reliable method of compensation.

Initial mitigation sites were constructed as traditional wildlife ponds consisting of 4 to 1 side slopes with a uniform bottom elevation. Typical constructed depths were 3 feet (1 m). As a result of reviews and monitoring of the response at these sites these standards have been revised. Design criteria now includes flatter, longer, more gradual side slopes, variable bottom depths and in some instances scrub shrub planting in order to create shallow water wetland complexes. (Fig. 4) These areas are excavated to achieve the desired side slopes (10:1 or flatter). Each excavated area is planned to have approximately 15 to 20% of its area excavated to hold 3 feet (1 m) of water. The remaining area will have water depths varying from 1 inch to 24 inches (3 to 60 cm) with gradual side slopes between these areas. Some areas will have "humps" in the bottom and will include island and peninsula shapes to further diversify the complex. A cover crop is seeded to stabilize the soil until natural vegetation prevails.

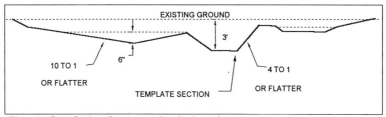

Figure 4. Cross Section of an excavated wetland complex.

Critical to the establishment of a freshwater wetland is the determination of a bottom elevation range that will maintain hydrophytic conditions in the created wetland bottom, thereby promoting wetland plant development. Monitoring of proposed compensation construction sites for periods up to a year has been done to determine optimum pond bottom elevations to achieve hydrophytic conditions. If the pond is too shallow the essential saturated conditions will not be realized for a period long enough to establish wetland vegetation. If the pond is too deep the result becomes a dugout pond with water too deep to promote emergent vegetation.

Nine mitigation areas have been constructed using shallow pond construction or shallow wetland complex construction. (Fig. 5) These ponds range in age from 4 years to less than 1 year old and all are providing wetland functions. Recent inspections to three of the sites by SCS and DNREC biologists revealed that the sites are functioning as designed. In all cases the sites' hydrology was established and then the sites were allowed to develop naturally. Some test planting of wetland plants was done on the Barthelmeh mitigation area using wetland plants available commercially. It is too soon to determine if this will be successful. All of the sites will be monitored periodically to determine how their establishment and development progresses.

PROJECT	MIT. ACRES	TOTAL EXCAV. C.Y.	REHANDLE QUANTITY C.Y.	TOTAL COST	COST PER C.Y.	COST PER ACRE OF MITIGATION (per hectare)
Delaware Pond	2.0	8500	1900	$25,500	$2.45	$12,750 ($31,500)
Spaceship Pond	1.3	4800	1800	$14,400	$2.18	$11,077 ($27,370)
Site P2, PB	0.4	1155	1000	$3,465	$1.61	$8,662 ($21,400)
Site P2, PD	0.9	4235	1400	$12,705	$2.25	$14,117 ($34,880)
Site H6, PA	0.2	860	440	$2.580	$1.98	$12,900 ($31,880)
Site H6, PB	0.5	1810	945	$5,430	$1.97	$10,860 ($26,840)
Barthelmeh Mit Area	1.6			$19,410		$12,131 ($29,980)
Babola Mit. Area	1.4			$7,690		$5,492 ($13,570)
Blackiston Mit Area	7.7			$19,000		$2,468[a] ($6,100)

Figure 5. Wetland creation costs.
[a]Includes a small embankment which creates a majority of the wetland acreage.

Summary

Innovative planning of a project can result in effective techniques for avoidance and restoration of wetlands. Minor changes in construction techniques have returned large dividends in wetland avoidance and retention. Cooperation of project staff in seeking a mutually acceptable solution is required from initial planning to final construction. This plan must be sensible and sensitive to both the water management project's goals and it's impact on the adjacent wetland resources.

More intensive design efforts using hydrologic computer models often result in a reduction of outlet reconstruction required. Clearing and snagging can be effectively used to restore downstream hydraulic efficiency while reducing impacts to adjacent wetlands. These techniques may result in higher future maintenance costs to the tax ditch organization.

The development of water management projects has become very complex. Projects which 10 years ago would have been planned and implemented by 2 people now require up to 12 technical and professional staff reviewing potential impacts on natural resources within the project areas.

We do not thoroughly evaluate all the economic impacts of wetland management done for these projects. Scientific research indicates that wetland functions are a valuable component in a productive environment, so they are included in the project development. Effective project planning with implementation of wetland management measures insures continuation of these water management projects for multiple public benefits.

References

1. Tiner, Ralph W., 1985. Wetlands of Delaware. US Fish and Wildlife Service National Wetlands invetory, Newton Corner MA. and Delaware Department of Natural Resources and Environmental Control, Wetlands Section, Dover DE. Cooperative Publication

2. Saveikis, David E. 1992. Personal Communication

3. Evans, R. O., J.W. Gilliam, and R. W. Skaggs. 1989. Managing Water Table Management Systems for Water Quality. N.C. State University, Raleigh NC.

LOW COST CHANNEL SIDE SLOPE SCARIFICATION AND SEEDING [1]

Richard T. Smith   Ronald F. Gronwald   John O. Kelley [2]

ABSTRACT

Over 50 miles of drainage channels with 1:1 side slopes are constructed each year in Delaware. Side slope vegetation for erosion control has been difficult to establish. For the past 2 years all side slope have been scarified horizontally with protrusions welded onto the trapezoidal buckets of the hydraulic excavators. The resulting horizontal grooves in the side slopes have helped to hold the seed which is applied within 24 hours of excavation. Results have been generally good to excellent. The cost of applying the grooves and seed is low when compared to conventional seeding methods and the results are superior. Costs of equipment modifications and seed application are presented along with guidelines for making the equipment modifications.

---

[1] Contributions from both Department of Natural Resources and Environmental Control, Division of Soil and Water Conservation, Georgetown, Delaware and USDA - Soil Conservation Service, Dover, Delaware.

[2] The authors: Drainage Program Administrator, PO Box 567, Georgetown, Delaware 19947; State Conservation Engineer and Civil Engineer Technician, 210 Treadway Towers, 9 E. Loockerman Street, Dover, Delaware 19901, respectively.

STATUS OF THE WATER TABLE MANAGEMENT-WATER QUALITY RESEARCH PROJECT IN THE LOWER MISSISSIPPI VALLEY

Guye H. Willis, James L. Fouss, James S. Rogers, Cade E. Carter,
and Lloyd M. Southwick*

ABSTRACT

The rate of agrochemical transport through soils may be significantly affected by various soil-water/water table management methods. Corrugated plastic drain tubing (102-mm diameter) was installed at a 1-m depth and 15-m spacing in each of 16 bordered plots (35 by 61 m). Each plot was surrounded by a subsurface 2-m vertical plastic barrier and was equipped with appropriate sump, pumps and instrumentation for microprocessor control of water table depth via regulated subdrainage/subirrigation flows. Water table measuring pipes with depth sensors, soil water (matric potential) sensors, soil-water pressure sensors, tensiometer-pressure transducers with ceramic cups, water table sampling tubes, piezometers, and soil temperature sensors were installed at various depths in the root/vadose/water table zones and at appropriate distances from the center drainline in each plot. Initial water management treatments include four replications each of (I) surface drainage only, (II) conventional subsurface drainage at a 1-m depth, (III) controlled water table at 45 $\pm$ 5 cm depth, and (IV) controlled water table at 75 $\pm$ 5 cm depth. Initial agrochemical applications will be in accordance with currently recommended practices. Plot installation was near completion in October, 1992.

Keywords. Leaching, Runoff, Subdrainage, Subirrigation, Controlled water table.

INTRODUCTION

Pesticide, fertilizer, and other chemical contamination of ground water is a national problem that needs timely and rational solution. For economic reasons continued use of pesticides and fertilizers is expected for the foreseeable future in U.S. agriculture, and there is potential for extensive ground water contamination from continued, long-term agrochemical use. Since ground water provides drinking water for about half of the U.S. population (Pye et al., 1983), prudence suggests that steps be taken to rectify this potential problem.

About 25%, i.e., 40 million hectares, of the total U.S. cropland needs drainage (U.S. Dept. Agric., 1987). Much of this land is usually flat, highly fertile, and has no serious erosion problems. These potentially productive wet soils are primarily located in the prairie and level uplands of the Midwest, the bottom lands of the Mississippi Valley, the bottom lands in the Piedmont and hill areas of the South, the coastal plains of the East and South, and irrigated areas of the West (Schwab et al., 1981). During most or part of the year these soils have shallow water tables that are potential sinks for agrochemicals that leach below the root zone. Reports of pesticide and nitrate contamination in ground waters in the lower Mississippi River Valley (LMV) have been made (Williams et al., 1988; Calhoun, 1988; Cavalier and Lavy, 1987; Acrement et al., 1989; Whitfield, 1975). Water levels in the LMV generally are less than 9 m below land surface, and are much closer to the surface (0 to 2 m) in southern areas. These shallow water tables fluctuate considerably and respond mainly to rainfall (1150-1500 mm annually), the major source of recharge for the LMV ground water (Whitfield, 1975; Morgan, 1961; Poole, 1961; Dial and Kilburn, 1980).

---

*Guye H. Willis, Research Leader, U. S. Department of Agriculture, Agricultural Research Service, Soil & Water Research Unit, P.O. Box 2507, Baton Rouge, LA 70894 and Biological and Agricultural Engineering Dept., Louisiana State University Agricultural Center Baton Rouge, 70803, USA.

Concepts of Water Table Management

The "optimal" management of soil-water for agricultural cropland in humid areas of the U.S. via control of water table depth in the soil profile involves complex daily operational/management decisions because of the erratic spatial and temporal distribution of rainfall. Periods of excess and deficit soil-water conditions in the active root-zone often occur within the same growing season. Thus, controlling water table depth within a desired range relative to the root-zone requires facilities for regulating both subsurface drainage flow from and irrigation into the soil profile. The management process is further complicated because soil-water management must be integrated with improved fertilizer and pesticide application practices. Integrated methodology is needed to manage soil, water, ground cover, and agrochemical applications in such a way that pesticides and fertilizers are contained within their "action zones", thus reducing the risk of surface and ground water pollution. Improved soil-water management technology may reduce the amount of pesticides and fertilizer used, thereby increasing crop production efficiency and farmer profitability, while reducing pollution. Gilliam and associates developed controlled-drainage practices for reducing nitrogen and phosphorus levels in surface/subsurface effluent from agricultural lands (Gilliam et al., 1979; Gilliam and Skaggs, 1985; Deal et al., 1986). These practices are being used in eastern and southern coastal plains soils.

Research Objectives

The research objectives of this study are to:
1. Identify and characterize chemical and physical factors and processes that affect the rate and mode of pesticide and plant nutrient transport in surface runoff and in the root, vadose, and saturated zones of shallow water table soils.
2. Characterize and quantify the effects of water-soluble organic matter on pesticide transport in soil.
3. Determine the effects of water table management on losses of pesticides and plant nutrients via surface runoff, subsurface drainage outflow, and leaching to ground water.
4. Develop models needed to devise water table management strategies that will avoid and/or alleviate ground water contamination by pesticides and fertilizers.

MATERIALS AND METHODS

General Plot Layout and Site Characterization

The study will be conducted on 16 bordered, 0.21-ha plots (located on the LSU Ben Hur Research Farm near Baton Rouge, LA) instrumented for automatic, microprocessor-controlled measurement and sampling of surface runoff and subsurface drain outflow, and water table management. The plots are on a Commerce silt loam soil which consists of layers of silt and clay mingled with sand lenses that were deposited by past Mississippi River overflows. The top 45 cm of the soil profile is relatively high in clay (up to 34%). Consequently, the hydraulic conductivity is relatively low and the soil is easily compacted by wheel traffic (Lund and Loftin, 1960). At depths from 45 to 90 cm the clay content decreases to about 22% while the silt and sand contents range up to 47%. Hydraulic conductivity values determined from auger holes in the soil surrounding the plots were 15 mm/h for 60-cm deep holes, 23 mm/h for 90-cm deep holes, 38 mm/h for 120-cm deep holes, 44 mm/h for 150-cm deep holes, and 30 mm/h for hole depths between 180 and 240 cm (Rogers et al., 1991). Water table depths averaged between 30 and 50 cm below the soil surface. Hydraulic conductivities of the soil profile below 240 cm have not been reported. However, the soil clay content below 150 cm increases somewhat and the hydraulic conductivity should decrease accordingly.

Each plot, 35 by 61 m, has a 20-cm high dike around the perimeter, a 0.15 mm polyethylene subsurface barrier installed 30 cm below the soil surface and extending down 2 m, three subsurface drain lines (102-mm diameter corrugated plastic tubing) installed 15 m apart and 1.0 m below the soil surface, a 1.2- by 1.2- by 3.0-m steel sump to control drainline outlet water levels, and an H-flume at the surface runoff outlet (Fig. 1). Each plot is precision-graded to a 0.2% slope with 0.2% cross-slope. The drainlines next to the longitudinal borders control border effects between adjacent plots. The area centered over the middle drainline (15 by 61 m) is assumed to be representative of an area in a larger field with the same drain spacing. Surface runoff and subsurface drain outflow are measured, sampled, and directed, via a 300-mm diameter PVC subsurface "main", to a collector sump for diversion into a surface drainage ditch.

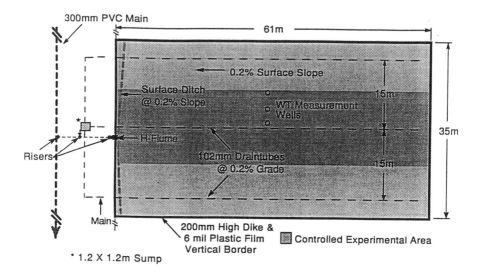

Figure 1. Plan view of a controlled-watertable research plot with surrounding plastic-film subsurface barrier.

A National Weather Service Class-A automated weather station is located 250 m from the plots. Rainfall, relative humidity, pan evaporation, wind speed and direction, total radiation, and air temperature data obtained from the weather station's automatic data-logger are used in conjunction with the study. Evapotranspiration is estimated from pan evaporation and the modified Penman equation (Jensen, 1974).

The treatments, imposed in a randomized complete block design, consist of four replications each of (I) surface drainage only [the subsurface drain lines will be plugged], (II) conventional subsurface drainage [water table kept at the level of the drainline or below], (III) controlled water table at 45 ± 5 cm depth, and (IV) controlled water table at 75 ± 5 cm depth.

Treatment I is the control and will be used to characterize pesticide and nutrient transport in surface runoff and soil leachate in the absence of subsurface drainage and water table control. Chemical movement in this treatment should be representative of that under farming practices typical to millions of hectares of alluvial soil. Treatment II provides subsurface drainage but no water table control: pesticide and nutrient transport in surface runoff should decrease (Southwick et al., 1989; Bengston et al., 1988), but the amounts leached below the plant root zone and potentially into the shallow water table may increase (Gilliam et al., 1979). In this treatment the water table will not be allowed to rise closer than 1 m (nominally) below the soil surface (during wet periods), but will be allowed to fall below that depth through normal water table decline during periods of low rainfall. The year-round presence of an unsaturated soil surface zone at least 1 m deep will enhance rainfall infiltration and reduce runoff volume. Greater infiltration will cause larger fractions of surface-applied chemicals to enter the plant root zone and reduce losses in surface runoff. Depending on rainfall amount and distribution patterns, and the chemical's persistence and water solubility characteristics, a greater chance exists for the chemical to move into the shallow water table zone. Treatments III and IV provide nearly constant water tables 45 cm and 75 cm below the soil surface within tolerances of ± 5 cm. These depths will provide less storage for infiltrated rainfall than treatment II, but will greatly reduce the potential for chemical leaching below the 1-m-deep drainlines. These treatments will encourage rainfall infiltration (treatment IV more so than treatment III) and concomitantly reduce chemical losses in surface runoff. The continuous presence of a water table above the drainlines will prevent downward chemical movement except for those times when water is pumped from the drainlines to lower the water table to the prescribed elevation. Thus, except during controlled-drainage periods, the infiltrated chemicals will be kept in the soil profile above the drainlines where they will remain for extended periods during which utilization and/or modification/degradation can occur. Plant growth and yield data and pesticide/fertilizer use efficiency for the 45- and 75-cm water table depths will be compared as the initial step in developing a management program of controlled variable-depth water tables designed to provide an optimized combination of profitable crop production, efficient pesticide and fertilizer use, and decreased chemical transport in surface runoff and leaching to ground water.

Subsurface Drainage/Subirrigation Systems

The dual purpose controlled-drainage and subirrigation system was installed in each plot as shown in the plan view of Fig. 1. The drainline spacing was selected so that the water table depth could be accurately controlled in the experimental area of each plot for all three modes of system operation: subsurface drainage, controlled-drainage, and subirrigation. The 15-m drain spacing selected was based upon computer simulation results (DRAINMOD, and a Boussinesq-equation based model) for various system designs and operational modes. Simulated results for the 15-m spacing indicated that even with the water level at the drain outlet held constant at a 60-cm depth during the growing season, the excess soil-water in the root-zone (30 cm depth, expressed as SEW-30) was less than 150 cm-days, which is considered to be an acceptable level of performance.

Because of the relatively flat, low-lying land at the experimental site and the large amounts of surface runoff that often fill the surface drainage ditches in the area, sump outlets for the subsurface drains are necessary. A relatively large sump (1.2- by 1.2- by 3-m) was installed at each plot. All drainage effluent enters the sump and is pumped into a 300-mm diameter main buried close to each sump. Irrigation water is supplied from a pressurized water line into each sump. This sump is large enough to house outlet water level control, flow measuring, and flow sampling equipment, and provide room for a technician to enter the sump for servicing the instrumentation.

Water table Control System

Each drainline outlets into the sump separately, and the water levels at the outlets of the draintubes are controlled in 300-mm diameter plastic pipe risers mounted inside the sump (Fig. 2). The water levels in the risers are controlled by pumping out of the risers with small electric pumps [for drainage or controlled-drainage (CD)], or adding water from a pressurized water supply line [for subirrigation (SI)]. The water level in the riser pipe for the experimental (center) drainline in each sump is automatically controlled with an electronic data-logger/controller system which operates the small electric drainage pump and the electrical solenoid valve on the irrigation supply line. The water level control in the other riser pipe for the buffer drainlines is "slaved" to the experimental drain riser pipe in the same sump.

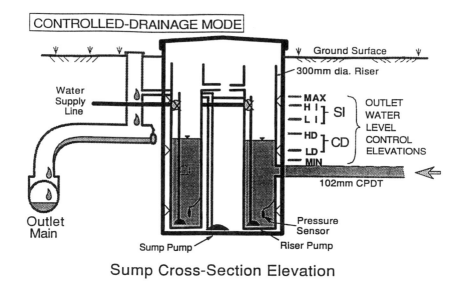

Figure 2. Schematic drawing of the automated control for a single research plot (middle and buffer drainlines controlled by right and left risers, respectively.

All water pumped from the risers is discharged into a 300-mm diameter plastic gravity flow drainage conduit (outlet main) buried nearby. The operation of the experiment is continuously monitored and controlled by four microprocessor-based, data-logger/controller systems (Campbell Scientific, Inc., Model No. CR7X). Each data-logger/controller unit monitors and operates all four plots in one experimental replication. The microprocessor in the data-logger is programmed to record (every 5 seconds for precise water table control) the water level in the riser pipe for the experimental drainline in each sump and the water table depth (midway between drainlines) in each field plot, and to independently control the drainage pumps and irrigation solenoid valves on the research site. All data-logger-acquired data are automatically transferred hourly to a personal computer via a coax-cable network.

The water level in the outlet-water-level (OWL) risers for each drainline is measured with a strain-gage pressure transducer (Druck model 950, 2.5 psi range) mounted in each riser. The field water table depth (WTD) is continuously monitored with a linear-resistor type water level sensor (Metritape, Inc., Type LA) housed within a 50-mm diameter, perforated, plastic pipe installed to a depth of 1.5 m. The recorded riser water levels and field water table depth data are summarized (averages and standard deviations) by the data-logger on an hourly and a daily basis. The field water table depth is also recorded (as a backup) with a standard FW-1 water stage recorder and float in a 150-mm diameter WT measurement well (Bengtson et al., 1983).

A cross-sectional schematic of the experimental drain outlet riser pipe and the riser water-level-elevation sensor mechanism for the automatically-controlled plot treatments is shown in the controlled-drainage mode in Fig. 2; for the subirrigation mode of operation (not shown) the water level in the riser is typically maintained at a higher elevation than for controlled drainage.

The water table management system operation is automatically switched as needed from the controlled-drainage to subirrigation, and vice versa, to maintain the OWL between the maximum (MAX) and minimum (MIN) water elevations specified. Feedback of the measured field WTD midway between drains is an optional control parameter implemented during the growing season to help maintain the WTD within the desired range.

With no feedback control, the OWL in the riser is maintained between programmed high (HD) and low (LD) elevations (about 10 cm apart) for controlled drainage and between preset high (HI) and low (LI) elevations (also about 10 cm apart) for subirrigation (Fig. 2). With feedback control, the measured field WTD is used to automatically adjust the water level control threshold elevations up or down as a function of the deviation of the measured WTD from the desired depth range.

A more detailed description of the data-logger/controller instrumentation, microprocessor program, optional automated control modes, and remote telecommunications control features are presented by Fouss et al. (1987, 1990).

For the conventional subsurface drainage (treatment II) the drainage effluent collected in the outlet riser pipe is pumped out so that the OWL always remains below the drainpipe outlet into the riser. For automated water table management (treatments III and IV) the systems are operated (1) in a conventional subsurface drainage mode after pesticide concentrations in drain effluent decrease to zero (typically January through March), (2) in the controlled-drainage mode with the OWL maintained about 10 to 30 cm above the outlets during April and again from mid-September through December, and (3) in the subirrigation mode from May through mid-September with the feedback option activated to maintain the field WTD within the desired range, except as overridden by the rainfall amount/time threshold, or by high water table conditions which are governed by cumulative rainfall amounts. The system control is occasionally switched from the subirrigation to controlled-drainage mode, via remote computer communications, in advance of predicted significant rainfall (Fouss et al., 1987, 1990), or to adjust the water table depth prior to application of fertilizer or pesticide.

Soil-Water, Runoff, and Temperature

Soil water measurements are made near the center drainline in each plot at 1/2, 1/4, 1/8 of the distance between drains. Measurements are made with a neutron scattering device and soil matric potential sensors (current generating diode type) located at depths of 30, 60, and 90 cm. Neutron scatter readings are made weekly, and more often during intense study periods where knowledge of water content is important to assessing the status of applied chemicals and crop response. The data-logger controlled matric potential sensors are normally read hourly, but can be read every 5 to 10 minutes, for example, to determine when an upward-moving wetting front approaches the sensor location.

Two sets of porous-cup soil water samplers are installed midway between drains in each plot. A set contains four cups located at depths of 30, 60, 90, and 120 cm. These cups are used to sample water for both unsaturated and saturated conditions. Additionally, 2.5-cm diameter piezometer wells are installed to depths of 150, 180, and 210 cm for collecting water samples under saturated conditions.

A 0.46 m (1.5 ft) H-flume with a FW-1 chart recorder is used to measure surface runoff rate and volume. A multi-turn potentiometric transducer on the FW-1 float wheel permits electronic recording of flow data by the data-logger. The runoff measuring system is designed to handle a 25-year frequency, 24-hour duration rain storm (254 mm).

Soil temperatures are obtained with resistance thermometers (Campbell Scientific, Inc. Model No.

108B) at depths of 10, 20, and 30 cm in each plot. Resistance sensors are read every 5 sec and hourly average and daily maximum and minimum temperatures are recorded by the data-logger.

Sampling Techniques and Frequency

ISCO refrigerated samplers are used to collect flow-proportioned water samples throughout each surface runoff event. After collection the samples are taken to the laboratory for analysis.

Drainline effluent samples for the analysis of pesticide and nutrient content are collected as water is pumped from the outlet water level control riser pipes for each center drainline. A composite effluent sample is collected for each storm event; a minimum 6-hour period with no more than 2.5 mm of rainfall defines the start of a new storm event. A pulse-type flow meter provides electrical signals for data-logger recording of cumulative flow versus time. An orifice-type sampling device is used on the discharge pipe of the riser pump to collect an effluent sample that represents about 0.5% of the total flow volume. The data-logger/controller system advances the sample collection containers between storm events occurring on the same day. The data-logger/controller programmed functions also permit activation of an alternative orifice for flow sampling so that the sample volume collected can be decreased in advance of predicted large storm events (e.g., 0.25% for a hurricane).

Specific sampling frequencies for water from saturated and unsaturated soil depend on rainfall occurrence and pesticide persistence. In general, water is collected from the porous-cup soil water samplers and the piezometer wells weekly. Sampling is sometimes more frequent following rains and for several weeks following pesticide application to the plots. Conversely, sampling is sometimes less frequent during periods of little rain or several months after pesticide application.

The surface soil (top 2.5 cm) is sampled before pesticide application, 0 and 2 days, 1 and 2 weeks, and 1, 2, 4, 6, 9, and 12 months after application. Soil samples are also collected in 15-cm increments with a 7-cm diameter auger down to the water table. Following sample collection, all augured holes are back-filled with soil (collected from between the plots) to prevent serious disruption of normal water flow paths in the soil profile. Sample sites will be recorded to prevent subsequent sampling from the same spot. Samples are taken before pesticide application and at 2 and 6 weeks and 3, 6, 9, and 12 months after application.

SUMMARY AND COMMENTS

The paper describes a controlled-water table experiment designed to (i) help characterize pesticide and nutrient leaching in soil and (ii) devise a management strategy for minimizing pesticide and nutrient leaching in soil. The description of a system of surface- and subsurface-drained, bordered plots equipped for microprocessor control of water table depths follows brief reviews of potential ground water contamination in the shallow-water table soils of the lower Mississippi River valley and the concepts of water table control. Sampling frequencies and techniques for surface runoff, subsurface water, and soil are also presented.

Installation of the described system was near completion in October, 1992. The described system has the potential for integrating methods for crop production efficiency and pollution reduction. The study should lead to a strategy for managing water table depth to enhance plant fertilizer use efficiency, thereby reducing fertilizer needs and potential pollution by fertilizers. Further, there should be an opportunity to reduce pollution by pesticides through management of pesticide applications and water table depths as dictated by prevailing and predicted weather conditions.

REFERENCES

1. Acrement, G.J., L.J. Dantin, C.R. Garrison, and C.G. Stuart. 1989. Water resources data for Louisiana, water year 1988. U.S. Geol. Survey. Baton Rouge, LA. Rept. No. USGS/WRD/HD-89/262.

2. Bengtson, R.L., C.E. Carter, H.F. Morris, and J.G. Kowalczuk. 1983. Subsurface drainage effectiveness on alluvial soil. Transactions of the ASAE 26:423-425.

3. Bengtson, R.L., C.E. Carter, H.F. Morris, and S.A. Bartkiewicz. 1988. The influence of subsurface drainage practices on nitrogen and phosphorus losses in a warm, humid climate. Transactions of the ASAE 31:729-733.

4. Calhoun, H.F. 1988. Survey of Louisiana groundwater for pesticides. Louisiana Dept. Agric. Forestry. Baton Rouge, LA.

5. Cavalier, T.C., and T.L. Lavy. 1987. Eastern Arkansas groundwater tested for pesticides. Ark. Farm Res. 36:11.

6. Deal, S.C., J.W. Gilliam, R.W. Skaggs, and K.D. Konyha. 1986. Prediction of nitrogen and phosphorus losses as related to agricultural drainage system design. Agric. Ecosys. Environ. 18:37-51.

7. Dial, D.C., and C. Kilburn. 1980. Groundwater resources of the Gramercy area, Louisiana. Louisiana Dept. Transpor. Develop., Water Res. Tech. Rept. No. 24.

8. Fouss, J.L., C.E. Carter, and J.S. Rogers. 1987. Simulation model validation for automatic water table control systems in humid climates. Proc. Third Internat. Workshop on Land Drainage. Ohio State Univ., Columbus, OH. pp. A-55 to A-65.

9. Fouss, J.L., R.W. Skaggs, J.E. Ayars, and H.W. Belcher. 1990. Water table control and shallow groundwater utilization. ASAE Monograph Management of Farm Irrigation Systems. pp. 783-824.

10. Gilliam, J.W., and R.W. Skaggs. 1985. Use of drainage control to minimize potential detrimental effects of improved drainage systems. In Development and management aspects of irrigation and drainage systems. Amer. Soc. Civil Engrs., New York, NY. pp. 352-362.

11. Gilliam, J.W., R.W. Skaggs, and S.B. Weed. 1979. Drainage control to diminish nitrate loss from agricultural fields. J. Environ. Qual. 8:137-142.

12. Jensen, M.E. 1974. Consumptive use of water and irrigation water requirements. Irrigation and Drainage Division, Amer. Soc. Civil Engrs., New York, NY.

13. Lund, Z.F., and L.L. Loftin. 1960. Physical characteristics of some representative Louisiana soils. U.S. Dept. Agric., Agric. Res. Ser. ARS 41-333. 83 p.

15. Morgan, C.O. 1961. Groundwater conditions in the Baton Rouge area, 1954-1959. Louisiana Geol. Survey, and Louisiana Dept. Public Works, Baton Rouge, LA. Water Res. Bull. No. 2.

16. Poole, J.L. 1961. Groundwater resources of East Carroll and West Carroll parishes, Louisiana. Louisiana Dept. Public Works, Baton Rouge, LA.

17. Pye, V.I., R. Patrick, and J. Quarles. 1983. Groundwater contamination in the UnitedStates. Univ. Pennsylvania Press, Philadelphia. p. 38.

18. Rogers, J.S., H.M. Selim, C.E. Carter, and J.L. Fouss. 1991. Variability of auger hole hydraulic conductivity values for a Commerce silt loam. Transactions of the ASAE 34:876-882.

19. Schwab, G.O., R.K. Frevert, T.W. Edminster, and K.K. Barnes. 1981. Soil and water conservation engineering. John Wiley and Sons, New York, NY.

20. Southwick, L.M., G.H. Willis, R.L. Bengtson, and T.J. Lormand. 1990. Effect of subsurface drainage on runoff losses of atrazine and metolachlor in southern Louisiana. Bull. Environ. Contam. Toxicol. 45:113-119.

21. U.S. Dept. Agric. 1987. Farm drainage in the United States: history, status, and prospects. G.A. Pavelis (Ed.). Misc. Pub. No. 1456.

22. Whitfield, Jr., M.S. 1975. Geohydrology and water quality of the Mississippi River alluvial aquifer, northeastern Louisiana. Louisiana Dept. Public Works, Baton Rouge, LA. Water Res. Tech. Rept. No. 10.

23. Williams, W.M., P.W. Holden, D.W. Parsons, and M.N. Lorger. 1988. Pesticides in groundwater data base: 1988 interim report. U.S. Environ. Protection Agency, Washington, DC.

IN-SITU HYDRAULIC HEAD INDICATOR SYSTEM
FOR MONITORING FIELD WATER TABLE ELEVATION [1]

James L. Fouss	Matthew Mahler	Nicola A. Ellis [2]
Member ASAE	Stud. Mem. ASAE	Stud. Mem. ASAE

ABSTRACT

For water table management (via controlled-drainage and subirrigation), the monitoring of field water table elevation (or depth) mid-way between drainlines is important as a feedback parameter to properly adjust the drainage outlet water level to control the field water table elevation within prescribed limits. Electrical instrumentation wire to remotely "read" a water level sensor in a measurement "well" can be buried below tillage depth, however, the exposed pipe at the ground surface is still an obstruction and a hinderance to farming operations. In this paper a system is illustrated and described that uses a perforated, horizontal section of plastic drain pipe buried at the midpoint between subsurface drainlines and connected via small-diameter tubing to a standpipe observation well near the drainage outlet or other suitable off-site location. This water table sensor functions much like a piezometer device which measures hydraulic head above the elevation of the perforated pipe section buried in the soil. This device provides a means of monitoring the water table elevation and its fluctuations from a remote location without having above-ground obstructions in the field.

A theoretical analysis and laboratory tests were conducted with a prototype system to evaluate time-response characteristics to changes in water table elevation. Good agreement was found when comparing the theoretically predicted and laboratory measured response characteristics in the remote riser piper. From the results it was concluded that, for an in-situ water table sensor located 150-300 m from a 50-mm diameter observation riser pipe, a 10-mm minimum diameter tubing should be used to connect the sensor and the riser. For connection tubing smaller than 10 mm in diameter significant time delays occur between fluctuations of field water table elevation and the observed changes in water level in the remote riser pipe.

---

[1] Contribution from the Soil and Water Research Unit, USDA-ARS, Baton Rouge, LA in cooperation with the Louisiana Agricultural Experiment Station, Louisiana State University Agricultural Center, Baton Rouge, LA.

[2] The authors are Agricultural Engineer, USDA-ARS, Soil and Water Research Unit, and Students of Biological & Agricultural Engineering, Louisiana State Univ., Baton Rouge, LA. A portion of this research was conducted as a Senior Design Project by the engineering students.

IN-SITU WATER TABLE MONITORING CONCEPT

The concept of the in-situ monitoring of field water table for a sump-controlled water table management system is illustrated in Fig. 1. The water table management system (Fouss et al., 1990) is shown in a subirrigation mode of operation, where water is pumped from the well into the sump to raise the water level in the outlet sump (Fig. 1). In the subsurface drainage or controlled-drainage modes of operation, water is pumped out of the sump into surface drainage channels. The water table elevation or depth (WTD) sensor is invisioned as a one-meter long section of perforated, plastic drainage pipe, which is covered with a synthetic envelope to prevent sediment entry. The WTD sensor is buried deeper than drain depth at the midpoint between drainlines in the field, which is typically a significant distance (e.g., 150-300 m) from an off-site drainage outlet water level control structure (e.g., a sump). The WTD sensor does not have a vertical riser pipe section which extends above the soil surface, and could hinder farm field operations. Thus, the WTD sensor can remain in the ground permanently, and does not need to be removed during farming operations. The remote riser pipe, or observation "well", is shown mounted in the outlet water level control sump. Modern instrumentation can be provided for the measurement of both the field water table elevation in this riser pipe and the water level in the sump. These monitored water levels can be used as inputs to a microprocessor system to operate drainage/irrigation pumps or valves to automatically control the outlet water level, and thus maintain the field water table within desired elevation limits.

Performance Requirements: The rate that the in-situ WTD sensor and remote monitoring system can indicate an actual change in field water table elevation (depth) is important for timely automated control of water table management systems. For example, if the indicated water table level significantly lags behind actual changes in water table elevation in the field, excessive soil-water events could go undetected long enough to seriously delay needed corrective (feedback) action by the drainage flow control (pump operation) in the sump. To prevent over pumping from or into the sump (based upon the senior author's experience), the field water table elevation indicated in the sump riser pipe should not lag a rapid change in field water table elevation (e.g., a 300 mm change in elevation in 15 minutes) more than 15 to 20 minutes. For a slow but constant rate of change of field water table elevation (e.g., 4 mm/min elevation change), the indicated water table in the sump riser should not display a time-lag greater than about 10 minutes.

THEORETICAL ANALYSIS and LABORATORY TESTING

A theoretical analysis was conducted to determine the range of tube diameters that should be tested for the small diameter tubing connecting the field WTD sensor and the riser pipe in the sump (Fig. 1). The analysis also permitted the evaluation of various pipe diameters for the sump riser. A schematic of the laboratory set-up to test the various components of the the water table sensor system, and to compare results with theory, is shown in Fig. 2. A large-diameter field riser was used to represent the soil-water surrounding the field water table sensor (Fig. 1). The field riser in this set-up was at least four times larger in diameter than the sump riser pipe, thus a small change of head (h_f) in the field riser would provide the flow volume through the small-diameter tubing to cause the head in the sump riser (h_s) to equilibrate with the head in the field riser. A small-diameter sump riser was desired to minimimize the volume of water displaced for the two risers to equilibrate in head.

The hydraulic flow characteristics of the laboratory test system (Fig. 2) were described theoretically with the energy formula (Roberson et al., 1990),

$$\frac{p_f}{\gamma} + \alpha_f \left[\frac{V^2}{2g}\right] + h_f = \frac{p_s}{\gamma} + \alpha_s \left[\frac{V^2}{2g}\right] + h_s + \left[\frac{32\mu LV}{\gamma d^2}\right] \quad (1)$$

where,

p_f = pressure at water surface in field riser, (N/m^2)
p_s = pressure at water surface in sump riser, (N/m^2)
α_f = orifice coefficient for small tube connection at field riser
α_s = orifice coefficient for small tube connection at sump riser
g = gravitational acceleration constant, (m/s^2)
γ = specific gravity of water, (N/m^3)
μ = absolute viscosity (water), (N-s/m^2)
h_s = hydraulic head in sump riser pipe, (m)
h_f = hydraulic head in field riser pipe, (m)
V = velocity of flow in small diameter connecting tube, (m/s)
L = length of small diameter connecting tube, (m)
d = diameter of small diameter connecting tube, (m).

Since both the field and the sump riser pipes were open to the atmosphere, p_f and p_s cancel out of the calculations. Laminar flow conditions governed because flow velocity in the small diameter tubing (by system design) was very low, and thus the [$V^2/2g$] term was negligible. The last term in Eq. (1) represented the head loss in the small diameter tubing due to frictional resistance to water flow in the tube. Therefore, Eq. (1) was simplified, as follows, to solve for flow velocity (V) in the small diameter tubing with a known hydraulic head difference between the field and sump riser ($h_f - h_s$).

$$V = \frac{\gamma d^2 (h_f - h_s)}{32 \mu L} \quad (2)$$

Eq. (2) was iteratively solved for the falling/rising head case by assuming that the flow velocity (V) was constant over a <u>one</u> second time-step. The volume of water displaced from one riser to the other during each time step was used to adjust the hydraulic heads in the risers before the next iteration.

System Components Evaluated: Two diameters for the sump riser pipe were theoretically evaluated; 50- and 75-mm diameter pipe. For an assumed length of 150-300 m for the small diameter connecting tubing between risers, we evaluated three tubing diameters; 6.35-, 9.53-, and 12.7-mm diameter tubing. The initial head differentials considered were 50, 100, and 150 mm. Additionally, two rates of constantly rising water level in the field riser (e.g., a rise of 4 to 10 mm/min) were imposed to determine the steady-state time lag characteristic of the water level in the sump riser.

Laboratory Instrumentation: The water levels in the field and sump riser pipes (Fig. 2) were measured with an electrical resistor-type water-level sensor (MetriTape™, Inc., Type LA).[3] Signals from the water-level sensors were continuously monitored with an electronic data-logger system (Campbell Scientific, Inc., Model 21X), which was also used to compute and store summary data for later processing. The riser water levels were monitored every 5 seconds with the data-logger, and electronically transferred to a PC via an RS-232C serial cable. Campbell Scientific, Inc. Program, PC-208, was run on the PC to allow the continuous downloading and PC-monitor display of acquired water level data during a test run. Data stored in the data-logger for later processing included "average" riser water levels at 5 minute intervals. Approximately five 5-second readings, taken just before of end of each interval, were averaged to represent the riser water levels at that time during the test.

Theoretical vs. Lab Test Results: In general, the theoretically predicted water level in the sump riser pipe followed changes in the field riser level quite well. Only one example of results is presented here for a 10-mm diameter connection tubing, as shown in Fig. 3 for a slow constant rate of rise of the water level (about 4 mm/min) in the field riser pipe. The theoretically predicted head changes in the risers as a function of time closely approximated the laboratory measured (observed) time-related changes in water levels. The steady-state time-delay between the water level in the sump riser and the field riser for this test was only 2.5 min., well within the performance requirements stated above.

Tests with the 6.34-mm diameter connection tubing indicated an excessive time-delay in response (greater than 30 min.), and tests with the 12.7-mm diameter tubing indicated via the speed of response (less than 2 min.) that the tubing diameter was larger than necessary. The time-delay of response for the 75-mm diameter sump riser pipe was more than twice that for the 50-mm diameter sump riser (directly related to the cross-sectional area of the riser pipe). Thus, we concluded from the theoretical and laboratory testing, that for a 150-300 m distance between the field sensor and the sump riser pipe, a mimimum 10-mm diameter connection tubing between risers, and a maximum 50-mm diameter sump riser pipe (results shown in Fig. 3), should be used in the in-situ water table monitoring system to insure time-response characteristics as specified.

---

[3] Trade and company names are included in this paper for the benefit of the reader and do not imply endorsement or preferential treatment of the product(s) listed by USDA or cooperators.

PROTOTYPE IN-SITU WTD SENSORS

Two configurations of a prototype WTD Sensor were assembled and tested in a laboratory sand box as illustrated in Fig. 4. Both WT Sensors were made from perforated, 100-mm diameter, corrugated-wall polyethylene plastic drain pipe, and each was about 1.0 m in length. The sensors were covered with a synethic fabric ("Bean Sock^{TM}") to prevent sand entry. WTD Sensor #1 had a vertical riser section at one end (Fig. 4), which extended to about 15 cm below the soil (sand) surface; this riser section was not perforated. The top end of the corrugated pipe riser section on Sensor #1 was capped, but the cap did have a few holes drilled in it to allow air exchange with the surrounding soil (sand). WTD Sensor #2 included only the 1-m long horizontal section of corrugated tubing (Fig. 4).

Laboratory Test Procedure: The prototype WTD Sensors were installed in a laboratory sand box about 0.5 x 1.0 m by 0.75 m deep. The Sensors were placed at about the 0.5 m depth below the sand surface. This laboratory set-up did not represent the entire soil profile, but just the soil media surrounding the buried WTD Senor. A 5-cm diameter perforated plastic drain pipe was installed at the bottom-center of the sand box, and was equipped with a valve at the outlet end to control drainage flow when lowering the water table in the sand box. Water was applied on the surface of the sand via an irrigation drip-tube when it was desired to raise the water table in the sand box; only minor ponding was allowed to occur. All tests were conducted at night, when the air temperature was relatively constant, because the 150 m lengths of 10-mm diameter tubing were laid outside of the laboratory and exposed to air temperature variations. Only the 10-mm diameter connection tubing between the WTD Sensors and the 50-mm diameter riser pipes was used in these tests. The remote riser pipes used were 50 mm in diameter.

A separate water table measurement "well" was placed in the box to monitor the water table in the box during a test (see "field" water level sensor in Fig. 4). The MetriTape^{TM} water level sensors were used in all three exposed riser pipes, namely the "field" sensor, the riser for Sensor #1, and the riser for Sensor #2 (Fig. 4). During a test the measured water level in each riser was observed on the PC Monitor with the same instrumentation and PC software discussed above for the inital laboratory testing. The data-logger was programmed to store the measured water level in each riser on one-minute intervals (no readings were averaged), which provided data for plotting water table and responses in the remote risers. Tests were conducted for one and two ramp-step changes in "field" water table.

Prototype Test Results: Example results from tests conducted with these prototype WTD Sensors and connected 50-mm diameter risers are given in Figs. 5 to 8. The WTD Sensor-Riser Pipe monitored responses to a 380 mm rise in the "field" water table in about 15 minutes is shown in Fig. 5, and to a 400 mm recession in "field" water table in about 20 minutes is shown in Fig. 6.

For the rising water table case (Fig. 5), the water levels in the remote riser pipes lagged behind the changing water table in the "field" by about only 2-3 minutes (this is a steady-state time-lag), which is very satisfactory. The steady-state response monitored by Sensor #1 was slightly more sluggish than Sensor #2; that is, the lag-time for Sensor #1 was about 3 minutes, and about 2 minutes for Sensor #2 (Fig. 5). After the field water table became constant again (i.e., following the ramp-step), about 5-6 minutes was required for the water levels in the remote riser pipes to stabilize and equlibrate with the field water table elevation, which is very acceptable performance. For the receding water table case (Fig. 6), the monitored responses by both Sensor #1 and #2 were essentially the same, and lagged behind the field water table by a maximum of 2 minutes; very satisfactory performance.

Where the "field" water table was raised or lowered in two ramp-steps, initiated a short time-interval apart (Figs. 7 and 8), the monitored responses with Sensors #1 and #2 were similar to those found for the single ramp-step changes in water table elevation. When the field water table was rapidly raised in step-changes of 225 mm and 150 mm about 10 minutes apart (Fig. 7), the time required after the step for equilibrium in the remote riser water levels and the "field" WT was about 10 minutes. During the rapid two-step receding of the field water table (Fig. 8), the monitored riser water levels closely tracked the actual field water table.

DISCUSSION AND COMMENTS

This in-situ WTD Sensor system, with remote off-site monitoring riser pipe, offers some noteworthy potential for automating water table control systems without the need for exposed water table monitoring pipes in the field that hinder farming operations. Actual field testing of this system is recommended to more fully evaluate it for future applications.[4] The possible future application of this in-situ WTD Sensor system should provide the means for all water table management system to be equipped with automated feedback controls. In the past water table management with feedback has not often been attempted because of the hinderence to field operations and other difficulties encountered with exposed, above-ground water table monitoring "wells" in the field.

REFERENCES

1. Fouss, J. L., R. W. Skaggs, J. E. Ayars, H.W. Belche. 1990. Water Table Control and Shallow Groundwater Utilization In Management of Farm Irrigation Systems. ASAE Monograph, <u>Management of Farm Irrigation Systems</u>, Chapter 21:783-824.

2. Roberson and Crowe, 1990. <u>Engineering Fluid Mechanics</u>. Houghton Mifflin Co., Boston. MA. pp. 270-497.

---

[4] The authors intend to install the prototype system for field testing on a site in Southern Louisiana which has an automated water table control system in operation. In-field water table monitoring "wells", equipped with the MetriTape™ water level sensors, are currently used at the site for feedback control of field water table elevation. Comparison of both water table sensor-monitoring systems will be compared.

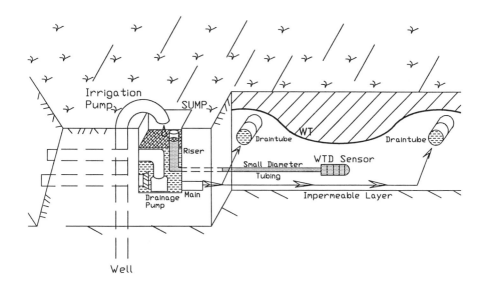

Figure 1. Sump-controlled water table management system with water table depth sensor, located mid-way between drainlines, connected via small diameter tubing to a remote observation riser pipe (within sump).

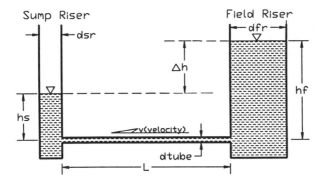

Figure 2. Schematic of laboratory test set-up to evaluate sump riser pipe diameter and connecting small diameter tubing for water table sensor system; water in field riser represents soil-water surrounding field water table sensor.

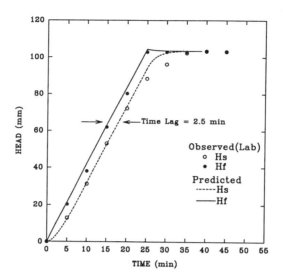

Figure 3. Laboratory test results (observed) vs. predicted response of water level rise in sump riser pipe (Hs) where water level in field riser pipe (Hf) was raised at a constant rate; sump riser dia. = 50 mm; connection tubing dia. = 10 mm and length = 150 m.

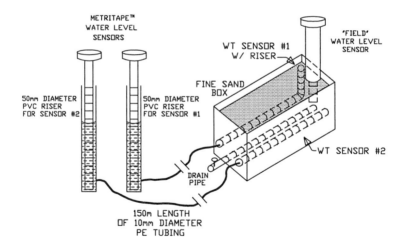

Figure 4. Laboratory test set-up illustrated for evaluation of response characteristics for water table sensor with riser section (sensor #1) and without riser (sensor #2) that are installed in fine sand; the water table in the sand ("field") and riser pipes measured with Metritape water level sensors.

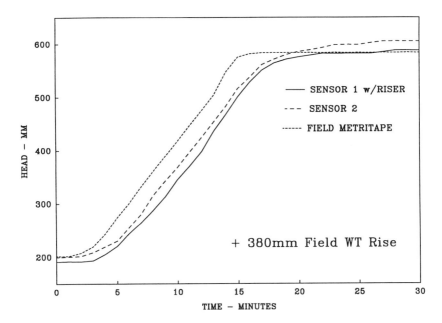

Figure 5. Laboratory observed responses of water level risers for water table sensor systems #1 and #2 in sand where the "field" water table was raised 380 mm in approximately 15 minutes.

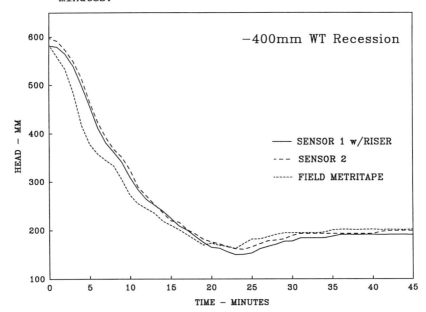

Figure 6. Laboratory observed responses of water level risers for water table sensor systems #1 and #2 in sand where the "field" water table was lowered 400 mm in approximately 22 minutes.

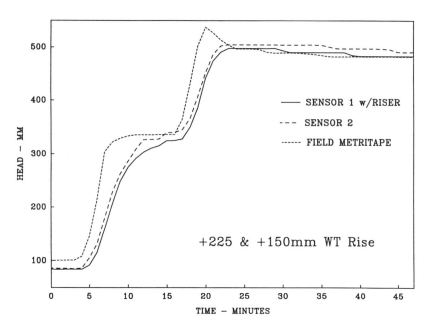

Figure 7. Laboratory observed responses of water level risers for water table sensor systems #1 and #2 in sand where the "field" water table was raised in two steps of 225 mm and 150 mm in approximately 25 minutes.

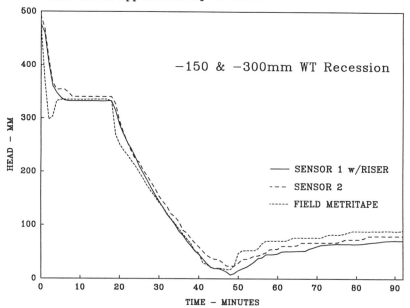

Figure 8. Laboratory observed responses of water level risers for water table sensor systems #1 and #2 in sand where the "field" water table was lowered in two steps of 150 mm and 300 mm in approximately 50 minutes.

A WATER TABLE - WATER QUALITY MANAGEMENT RESEARCH FACILITY FOR THE SOUTHEASTERN COASTAL PLAIN: PROGRESS REPORT

C. R. Camp   K. C. Stone   P. G. Hunt[*]
Member, ASAE   Member, ASAE

ABSTRACT

A research facility to investigate automation of controlled drainage/subirrigation (CDSI) systems for soils in the southeastern Coastal Plain has been constructed at the Coastal Plains Soil and Water Conservation Research Center, Florence, SC. This facility contains four separate systems, each consisting of three drain lines connected to individual control tanks. Water is added to or removed from the control tanks to regulate field water table depths. Each system is managed by a datalogger/controller that activates relays and solenoids to adjust the control tank water elevation to the position required to maintain target water table depths in the field. Surface runoff volume will also be measured. Water samples collected from control tanks, drain discharge lines, wells, and surface runoff will be analyzed to determine chemical concentrations. Future work will include the development of more sophisticated automatic control, management of the water table to reduce off-site water quality degradation, and modelling of water and chemical movement through the soil profile.

Keywords. Water table, Controlled drainage, Subirrigation, and Water quality.

INTRODUCTION

Systems that reduce and/or control wet soil conditions, store excess water, and supply water for crop requirements during drought periods would conserve water and increase crop productivity in many areas of the southeastern Coastal Plain. Large areas of the region have water tables within 1.5 meters of the soil surface during a significant part of the year. The development of DRAINMOD, a model that allows the evaluation of the drainage-water table control systems for a range of soil and climatic conditions during both drainage and subirrigation, significantly aided the design and evaluation of these systems for a wide range of soils and climates (Skaggs 1978, 1981). However, design and management criteria for water table management (WTM) or controlled drainage/subirrigation (CDSI) systems on these soils have not been fully developed (Shirmohammadi et al., 1992). These criteria must include techniques for reducing contaminants in surface and ground waters (Thomas et al., 1992).

Although investigators have shown that CDSI provides most of the water needed for crop production (Doty and Parsons, 1979), implementation of current and improved management criteria is often limited by lack of automated and/or remotely-operated control structures.

---

[*]C. R. Camp and K. C. Stone are Agricultural Engineers and P. G. Hunt is a Soil Scientist, Coastal Plains Soil and Water Conservation Research Center, USDA-ARS, P. O. Box 3039, Florence, SC 29502-3039

Manual adjustment of control structures for controlled-drainage and CDSI systems is difficult and is often not accomplished because of conflicts in the work schedule. Recent developments have resulted in prototypes of systems to automate this process and to link it to weather forecasts and computer data bases (Fouss, 1985 and Fouss and Cooper, 1988).

The impact of WTM or CDSI on water quality has been studied on a limited basis. Fields with conventional subsurface drains lose more nitrogen than fields with improved subsurface drainage (Jacobs and Gilliam, 1985). In another study, about 10 times more nitrate was lost from fields with good subsurface drainage than from fields with primarily surface drainage (Gilliam and Skaggs, 1986). However, reductions of about 50 percent in nitrate movement into drainage outlets from controlled drainage systems were reported by Gilliam et al. (1979). Evans et al. (1989) reported that average nitrate-nitrogen concentrations for controlled-drainage systems remained below 10 mg/L in 11 of 13 studies in the southeastern Coastal Plain. Thomas et al. (1992) reported limited data regarding phosphorus and pesticide losses from conventional subsurface drainage systems and no pesticide-transport data from controlled drainage and CDSI systems. They concluded that additional research is needed, particularly with respect to pesticide losses through these systems.

There is less annual drainage discharge with controlled drainage or CDSI than with conventional subsurface drainage (Evans et al. 1989). Consequently, more water is available for evapotranspiration and vertical seepage. Additionally, the higher water table increases the system sensitivity to events such as rainfall and chemical applications. The objectives of this paper are to describe a research project and facility that is directed toward development of an automated management system for CDSI and to report initial progress. This facility will also be used to investigate the movement of agricultural chemicals in water table systems and to develop management criteria that will minimize movement of agricultural chemicals out of the field, into either surface outlets or the ground water.

SYSTEM DESCRIPTION

A water table management research facility has been installed in a 1-ha Carolina Bay at the Coastal Plains Soil and Water Conservation Research Center, USDA-ARS, Florence, SC. The soils in the area are Coxville loam (*Clayey, kaolinitic, thermic Typic Paleaquults*) and Dunbar loamy fine sand (*Clayey, kaolinitic, thermic Aeric Paleaquults*). The facility consists of four separate systems, each with a sump outlet. Within each system, three subsurface drain lines, which enter the sump separately, are spaced 15 meters apart. Each drain line is connected to a separate control tank so that each drain line can be controlled independently. The systems are positioned in pairs, such that two are located immediately adjacent to each other with respective exterior drain lines spaced 15 meters apart; thus, each pair can also be operated as a combined system consisting of six drain lines with each drain line controlled by a separate tank. Schematic diagrams showing the soil boundaries, system locations, and water table control system are included as Figs. 1 and 2.

The water table elevation in the soil adjacent to the drain line can be adjusted by changing the water elevation in the control tanks, either by energizing a pump to remove excess water or by opening a solenoid valve to add water from a pressurized water supply. The system was initially managed using float controls in the early stages of development, but the first phase of an automatic control system is now operational. This control system consists of a central

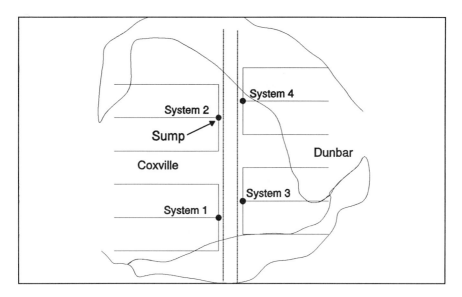

Figure 1. Schematic diagram of four water table management systems on two southeastern Coastal Plains soils, Coxville loam, and Dunbar loamy fine sand.

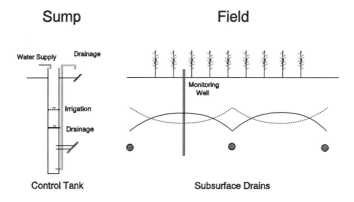

Figure 2. Schematic diagram of water table management system including drain lines, control tank, and monitoring wells. Each sump contains three control tanks, one for each drain line.

datalogger/controller (Campbell Scientific Inc. CR-7X[**]) to measure and record sensor values

---

[**]Mention of trademark, proprietary product, or vendor does not constitute a guarantee or warranty of the product by the U. S. Dept. of Agr. and does not imply its approval to the exclusion of other products or vendors that may also be suitable.

and to energize switches (relays, solenoids, etc.) as directed by a control algorithm stored in the datalogger/controller. Water level sensors presently include pressure transducers (Druck model PDCR 950, 2.5 psig) and linear resistors (Metritape Aquatape type AGS). Water level sensors are located in each control tank and in wells. The wells are located in each plot at selected distances from the drain lines (drain line, quarter-spacing, and half-spacing). Currently, water flow into and out of the control tanks is measured with positive displacement flow meters with manual readout, but pulse-output flow meters will be installed on both the supply and discharge lines for each control tank in the near future.

Communication between the field datalogger/controller and a personal computer in the laboratory is accomplished using a radio-frequency (RF) telemetry system manufactured by Campbell Scientific, Inc. The telemetry system consists of a base station connected to the computer, communication software, UHF portable transmit/receive radios at both the base station and the field station, antennae, and modems. Memory in the datalogger/controller is adequate to store the program and data for several days and will be interrogated each day when fully implemented. The system status or value of any control point or sensor can be determined at any time. Also, the datalogger/control algorithm can be edited and the values of all parameters can be adjusted remotely.

The current automatic control algorithm includes a single set of control parameters, one for the irrigation cycle and one for the drainage cycle, for each control tank. These parameter values can be altered via the remote computer or on site via keyboard entry. When fully implemented, the automatic control program will operate with feedback from water table measurements in each of the field plots. As more is learned about the system, soils, and control dynamics, it may be possible to simplify the system by eliminating the feedback portion of the control program and relying predominantly on long-term weather records or soil properties and weather forecasts.

Refrigerated pump samplers (Isco model 3700) will be installed adjacent to the control tanks in each system to collect water samples for quality analysis. Plans are now being completed for the fabrication and installation of surface runoff collection and measurement equipment for each system.

OPERATIONS

Because of the delay in installation of the water supply and an early season drought, it was impossible to achieve adequate water table control in any system in 1991, especially in Systems 3 and 4, where a higher fraction of the coarse-textured soil is located. In fact, corn suffered from drought stress at the mid-point between drains even on the fine-textured soil. The primary reason for this, especially on the Coxville loam, is that much of the water in the root zone was extracted by the crop before water and controls were available for subirrigation. By the time that water was available, the hydraulic conductivity of the soil was so low (because much of the soil profile was not saturated) that it was not possible to raise the water table.

In 1992, water table control was intermittent during the late winter and early spring months because of conversion from an unreliable, temporary water supply to a permanent one and because the control system was being converted from float control to automatic control using the datalogger/controller and sensors. Water table depths at three locations relative to the drain lines for Systems 1 and 2 during a 4-week period during summer are shown in Fig. 3. The water table elevations at the above-drain location are very similar to the water elevations in the control tank. In these two systems, the water table at the mid-point between drain lines was generally responsive to the control tank water elevation. This was not true in the other two systems.

Although 2-3 times more water was pumped into System 4 as into Systems 1 or 2, a water table

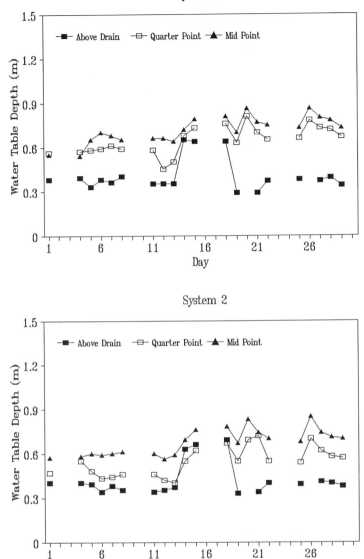

Figure 3. Water table depths during a 4-week period in 1992 at three locations relative to subsurface drains for two water table management systems on a Coxville loam soil in the southeastern Coastal Plain.

did not exist at the drain-line depth in System 4. A water table was measured at the quarter-point and occasionally at the mid-point between drains in System 3. Based on these preliminary data it may not be possible to successfully maintain water tables in the Dunbar portion of these two systems.

Although the system was not fully operational and the water supply was not installed until mid-season, a corn crop (*Zea mays* c.v. Hybrid Pioneer 3165) was planted in 1991 to evaluate soil variation. It was suspected that there would be considerable variation in soil pH and nutrient level (confirmed by measurement) and other chemical properties. Lime was applied based on pH measurements and soil type. However, soon after corn emergence, it became evident that there were significant corn growth differences. After further sampling, it was determined that there was significant variation in soil pH in this area. Addition of more lime in both fall 1991 and spring 1992 was required to significantly alter soil pH. Barley was grown during the winter of 1991-1992 to help diagnose problem areas using crop growth differences caused by soil pH variation. Sorghum was planted in 1992, to evaluate both soil variation and water table management system performance. The later planting date for sorghum allowed necessary construction in the immediate area to continue, accommodated delays caused by rainfall during the late winter and early spring, and allowed the completion of soil surface smoothing in the experimental area. A final adjustment in soil pH will probably be needed in fall 1992, following harvest of the sorghum crop.

FUTURE PLANS

Simulations of water table elevations, chemical concentrations, and movement of agricultural chemicals through the soil profile will be accomplished using DRAINMOD, CREAMS, a linked version of DRAINMOD and CREAMS, and other appropriate models. Models or their components will be evaluated, modified, and/or developed as necessary to describe these processes in Coastal Plain soils. Initial studies will include determination of nitrogen and phosphorus losses from the system, but later studies will include selected pesticides. Samples collected from the control tanks, surface runoff, and wells will be analyzed to determine concentrations of nutrients and pesticides for establishing baseline levels for selected chemicals and to evaluate water and/or soil management practices studied in the future.

REFERENCES

1. Doty, C. W., and J. E. Parsons. 1979. Water requirements and water table variations for a controlled and reversible drainage system. *Trans. ASAE* 22(3):532-536, 539.

2. Evans, R. O., J. W. Gilliam, and R. W. Skaggs. 1989. Managing water table management systems for water quality. ASAE Paper No. 89-2129. St. Joseph, MI: ASAE.

3. Fouss, J. L. 1985. Simulated feedback-operation of controlled drainage-subirrigation systems. *Trans. ASAE* 28(3):839-847.

4. Fouss, J. L., and J. R. Cooper. 1988. Weather forecasts as control input for water table management in coastal areas. *Trans. ASAE* 31(1):161-167.

5. Gilliam, J. W. and R. W. Skaggs. 1986. Controlled agricultural drainage to maintain water quality. *J. Irrig. and Drain. Eng.* 112(3):254-263.

6. Gilliam, J. W., R. W. Skaggs, and S. B. Weed. 1979. Drainage control to diminish nitrate loss from agricultural fields. *J. Environ. Quality* 8(1):137-142.

7. Jacob, T. C. and J. W. Gilliam. 1985. Riparian losses of nitrate from agricultural drainage waters. *J. Environ. Quality* 14(4):472-478.

8. Shirmohammadi, A., C. R. Camp, and D. L. Thomas. 1992. Water table management for field-sized areas in the Atlantic Coastal Plain. *J. Soil and Water Cons.* 47(1):52-57.

9. Skaggs, R. W. 1978. A water table management model for shallow water table soils. Water Resources Res. Inst., Univ. N. Car., Raleigh. Rpt. No. 134: pp. 178.

10. Skaggs, R. W. 1981. Water movement factors important to the design and operation of subirrigation systems. *Trans. ASAE* 24(6):1553-1561.

11. Thomas, D. L., P. G. Hunt, and J. W. Gilliam. 1992. Water table management for water quality improvement. *J. Soil and Water Cons.* 47(1):65-70.

FIELD ESTIMATION OF DRAINABLE PORE SPACE AND APPLICATION IN DRAINAGE DESIGN

S.K. GUPTA[*]

ABSTRACT

Drainable pore space is often treated as a constant although its variation with depth to water table is known. For making appropriate field estimates of this parameter, Glover and Dumm equation was integrated to develop formulae which would allow calculation of the drainable pore space utilizing field measured hydraulic heads and drain discharges. These relations were used to work out drainable pore space for alluvial soils at two stations in India. The analyzed data revealed that drainable pore space at one station increased linearly with increasing depth to which water table was lowered. At the other station, drainable pore space did not vary with hydraulic head. Test results at the first station reveal that a better match with observed data is obtained with increasing than with constant drainable pore space. It is further observed that estimates of drainable pore space in this way allowed a larger lateral drain spacing and it was 50 percent more than the design spacing calculated with the assumption that drainable pore space is constant at this station.(Keywords: Hydraulic conductivity, drainable pore space, drainage design)

INTRODUCTION

The drainable pore space, also called specific yield, invariably appears in drainage design equations derived on the basis of non-steady state theory. Because of difficulties in its estimation, in the application of these theories, drainable pore space is assumed to be a constant. Although concern for assigning a constant value to the drainable pore space has been expressed (Taylor, 1960; Zhang, 1986), the assumption continues to find favour with designers as well as practical users. Field investigations, on the other hand, have shown that drainable pore space varies with depth to water table or the hydraulic head

---

[*] Head, Division of Drainage and Water Management, Central Soil Salinity Research Institute, Karnal- Haryana (India)

(Bishay et al., 1978; Taylor, 1960). When other conditions do not vary, the drainable pore space increases with increase in depth to water table. Inverse techniques have been used to find out hydraulic conductivity (K), drainable pore space (f) or a ratio KD/f in which KD represents the transmissivity (Skaggs, 1976; El-Mowelhi and Van Schilfgaarde, 1982; Buckland et al., 1986). The present paper derives equations for the estimation of drainable pore space utilizing field measured data on hydraulic head and drain discharge both as a function of time. The tests are made utilizing data from two experimental fields in India.

THEORY

According to the Glover-Dumm equation (Dumm,1954), the drainage discharge from a drainage system can be calculated by the relationship

$$q = \frac{8.0 \, K \, d \, h_o}{S^2} \exp(-at) \qquad (1)$$

here, q is the drainage rate, m/day; K is the hydraulic conductivity, m/day; d is the equivalent depth in relation to D which is the actual depth to impermeable layer m, h_o is the initial hydraulic head, m; S is the drain spacing, m; and t is the time, since moment of recharge, days; 'a' is the reaction factor given by $\pi^2 \, K \, D/fS^2$, $days^{-1}$ and f is the drainable pore space in fraction. The relationship between initial head and head, h_t, at any future time is given by the relation

$$h_t = 1.17 \, h_o \exp(-at) \qquad (2)$$

Salient parameters used in Eq. (1) are described in Fig. 1.

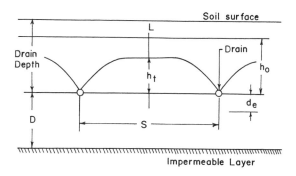

Fig. 1   Salient parameters used in Glover Dumm equation

Approach 1

The cumulative discharge, from the drain line can be calculated by integrating Eq. (1) with respect to time, t, between any time period say t_1 to t_2 such that

$$q(int) = \frac{8.0\ K\ d\ h_o}{\pi^2\ KD} (\exp(-a\ t_1) - \exp(-a\ t_2)) \qquad (3)$$

Here $q(int)$ represents the cumulative discharge per unit area drained by the drain line. In practice, when discharge is measured daily, $q(int)$ can be determined by the relation

$$q(int) = \sum_{i=1}^{n} q_i \qquad (4)$$

Incorporating Eq. (2) in Eq. (3) and simplifying, we get:

$$f = q(int)/0.638\ (h_1 - h_2) \qquad (5)$$

such that h_1 and h_2 are the hydraulic heads at time t_1 and t_2, respectively. It can also be shown that based on the cumulative discharge starting from time $t = o$ to any time t_1, the relation is given as

$$f = q(int)/(0.81\ h_o - 0.638\ h_1) \qquad (6)$$

Approach 2

This approach is based on the rate of change of the drainage rate. Such an approach has been commonly used in the context of inverse problems in solute transport; for example for estimating hydrodynamic dispersion coefficient (Kirkham and Powers, 1972). In order to utilize this technique, Eq. (1) is differentiated with respect to time, t, such that

$$\frac{dq}{dt} = \frac{8.0\ K\ d\ h_o\ (-a)}{s^2} \exp(-a\ t) \qquad (7)$$

At time t_1 Eq. (8) can simply be written as

$$a = (-)\ (dq/dt)/q_1 \qquad (8)$$

or

$$f = \frac{\pi^2\ K\ d}{s^2} (q_1)/(dq/dt) \qquad (9)$$

Equation (8) suggests that the reaction factor `a' can be obtained by plotting the discharge versus time data on an arithmatic graph paper such that $dq/dt$ and $q_1$ are read at an instant of time say `t_1'. The rate of change of q, dq/dt, can be obtained by drawing a tangent to the curve at that instant of time. The negative sign in Eq. (8) will be taken care of by the term dq/dt as it decreases with increasing time. Thus several values of drainable pore space could be evaluated from the same q-t curve and Eq. (9).

MATERIAL AND METHODS

A subsurface drainage experiment was installed to cover an area of 8 ha at Mundlana in Sonepat district (Haryana, India). The site represents waterlogged saline soils of the Indo-Gangetic plains. The soils at the experimental site are alluvial in nature. The texture of the soils is sandy loam to at least a depth of 1.0 m. Further down the texture is slightly heavier (Rao et al., 1986). The design parameters i.e. the hydraulic conductivity and the drainable pore space were estimated using auger hole method and soil texture-drainable pore space chart, respectively. For design purpose, average hydraulic conductivity for the profile was taken as 0.8 m/day. The drainable pore space was taken as 0.10. Drains at three drain spacings of 50, 67 and 84 m were laid out at a depth of 1.75m below the ground surface. The drain discharge-hydraulic head or drain discharge-time data for 10 day period were observed in all the three drain spacings.

A similar experiment has been laid out at Sampla in Rohtak district (Haryana, India). The soils of this site are also alluvial and the texture is sandy loam to at least 1.75m. Between 1.75 and 3.0m depth the texture is lighter (loamy sand). The hydraulic conductivity of the top layer is 1.0 m/day while the lower layer is 7.5 m/day. The drainable pore space is estimated to be 0.14. The hydraulic head versus time and drain discharge versus time data were collected for the 3 drain spacings of 25, 50 and 75m. According to Rao et al. (1976), a considerable fraction of drain discharge is received from outside the area from below the drains. Such fraction of discharge is maximum in 25m followed by 50m and 75m drain spacing.

Data from the site at Mundlana for the 3 drain spacings and from the site at Sampla for the 2 drain spacings of 50m and 75m form the test data for the purpose of this paper.

RESULTS AND DISCUSSIONS

The most commonly used equation for estimating drainable pore space is that of El-Mowelhi and Van Schilfgaarde (1982) which is given as

$$f = \frac{q(int)}{(h_1 - h_2)} \quad (10)$$

The differences in Eq. (5), Eq. (6) and Eq. (10) are apparent. While Eq. (10), assumes that water table dropped uniformly from h_1 to h_2, Eq. (5) and Eq. (6) takes into account the initial and final shape of the water table. The constants in Eq. (6) compared to Eq. (5) differ as a consequence of the fact that the shape of the water table at the start would be different than the shape of the water table at time t_1 or t_2. On this basis, it is suggested that the proposed equations are more flexible and are an improvement over the existing equation because shape of the water table is taken into account.

The proposed relation given by Eq. (9) require graphical evaluation of the rate of change of drainage rate. It could be calculated by drawing tangents at pre-decided times and measuring the slope at this time. No doubt, the drawing of tangents for estimating slope could be prone to judgement errors; yet, technically the procedure is sound. The main advantage of this technique would be in calculating drainable pore space as a function of the hydraulic head.

The estimated values of the drainable pore space utlizing various procedures for the 3 drain spacings at Mundlana are given in Table 1 and for Sampla in Table 2. Variation in the drainable pore space are noticed on account of drain spacing. Within a given drain spacing, estimated drainable pore space also varies with the procedure used. The deviation in the drainable pore space are also

Table 1   Comparative estimates of drainable pore space by existing and proposed procedures (Mundlana)

Equation number	Drainable pore space (dimensionless)		
	Drain spacing (m)		
	50	67	84
Proposed procedures			
Proposed Eq. (5)	0.22	0.17	0.17
Proposed Eq. (6)	0.12	0.08	0.07
Proposed Eq. (9)			
(a) 2 days	0.12	0.07	0.09
(b) 6 days	0.30	0.17	0.10
Existing procedures			
q-t plot on semilog paper[1]	0.18	0.13	0.10
Eq. (10)[2]	0.14	0.11	0.11

[1] Rao et a,. (1986); [2] El-Mowelhi and Van Schilfgaarde (1982).

noticed when Eq. (9) is used at 2 days and at 6 days for calculating drainable pore space. The differences are quite significant at Mundlana. From the analysed data reported in

Table 1 and Table 2, some important conclusions can be drawn.

i) For the Mundlana site one is inclined to believe that Eq. (5) is more closer to reality than Eq. (6). On the other hand for the Sampla site proposed Eq. (6) appears to be more closer to reality.

ii) While there is large variation in drainable pore space with spacing and with time in the case of the proposed Eq. (9) at Mundlana, there appears to be no such difference at Sampla.

iii) The average value of the drainable pore space for Sampla is 16 percent which is close to the upper limit of the drainable pore space for sandy soils (Dieleman and Trafford, 1976).

Table 2 Comparative estimates of drainable pore space by existing and proposed procedures (Sampla)

Equation number	Drainable pore space (dimension less)	
	Drain spacing (m)	
	50	75
Proposed procedures		
Proposed Eq. (5)	0.24	0.21
Proposed Eq. (6)	0.17	0.14
Proposed Eq. (9)		
(a)	0.18	0.12
(b)	0.17	0.17
Existing procedures		
q-t plot on semilog paper[1]	0.14	0.14
Eq.(10)	0.15	0.13

[1]Rao et al. (1980); [2]El-Mowelhi and Schilfgaarde (1982)

The estimated drainable pore space for the Mundlana site, was plotted on a arithmatic paper against depth to water table. For this purpose, the values of final hydraulic heads used in the calculations for drainable pore space were taken as the representative hydraulic heads for which estimated drainable pore space would be valid. There is a linear increase in the drainable pore space with increase in depth to water table (Fig. 2). To verify the findings for the Mundlana site, hydraulic heads were predicted with the proposed relation for drainable pore space as well as by assuming that drainable pore space is constant (0.1). For this purpose, Hooghoudt equation was integrated utilizing Bouwer and van Schilfgaarde (1963) approach. A closer match is obtained with the proposed relation than with the constant drainable pore space (Fig. 3).

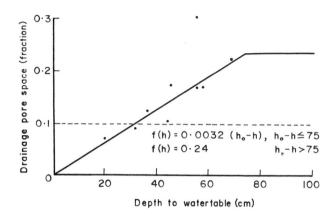

Figure 2   Relation between drainable porosity and depth to water table

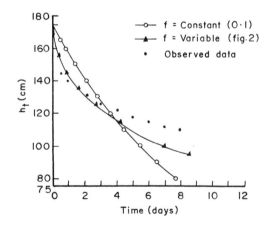

Figure 3   Observed and predicted water table decay curves

For the site at Mundlana, an exercise was also made to reestimate the design drain spacing with variable drainable pore space. The drainage criteria for this purpose required that the water table should be brought down to 30 cm below the ground surface within 2 days of its reaching the ground surface. The design drain spacing was found to be 82m compared to the design spacing of 50m when it was assumed that drainable pore space is constant. Thus, there would be a 50 percent increase in design lateral drain spacing, if variability in the drainable pore space is taken into account in design of subsurface drainage systems. As investments on drainage continue to bother developing countries, such improvements in drainage design would provide necessary impetus for implementing drainage schemes.

REFERENCES

Bishay, B.G., M.S. Abdel Dayem and A.I. El-Shabassy. 1978. The effect of different drainage and amelioration techniques on the chemical and physical properties of a salt affected and heavy textured soil in Egypt. Agric. Res. Rev., 56: 63-74.

Bouwer, H. and Van Schilfgaarde, J. 1963. Simplified method of predicting fall of water table in drained land. Trans. ASAE, 6:288-296.

Buckland, G.D., D.B. Harker and T.G. Sommerfeldt. 1986. Comparison of methods for determining saturated hydraulic conductivity and drainable porosity of two southern Alberta soils. Can. J. Soil Sci. 66: 249-259.

Dieleman, P.J. and B.D. Trafford. 1976. Drainage testing. FAO Irrig. and Drain. Paper No. 28, Rome.

Dumm, L.D. 1954. Drain spacing formulae. Agric. Engg. 35: 726-730.

El-Mowelhi, N.M. and J. van Schilfgaarde. 1982. Computation of soil hydrological constants from field drainage experiments in some soils of Egypt. Trans. Am. Soc. Agric. Engg. 25: 984-986.

Kirkham, D. and W.L. Powers 1972. Advanced soil physics. Wiley-Interscience, New York.

Rao, K.V.G.K., O.P. Singh, P.S. Kumbhare, S.K. Kamra, R.S. Pandey, and I.P. Abrol. 1986. Drainage investigations for salinity control in Haryana. Central Soil Salinity Research Institute, Karnal, Bull. No. 10. 95 pp.

Skaggs, R.W. 1976. Determination of hydraulic conductivity drainable porosity ratio from water table measurements. Trans. Am. Soc. Agric. Engg. 19: 73-80.

Taylor, G.S. 1960. Drainable porosity evaluation from out flow measurements and its use in drawdown equations. Soil Sci., 9: 338-343.

Zhang, W. 1986. New drainage formulaes considering delayed gravity response and evaporation from shallow water table. In Land Drainage: Proc. 2nd Int. Conference, Southampton University, U.K.(Eds. K.V.H. Smith and D.W. Rycroft): 35-47.

SUBSOIL WORKABILITY OF HEAVY SOILS WITH SHALLOW WATER TABLES

L. Müller, U. Schindler[*]

ABSTRACT

Studies of subdrainage and subsoiling have been conducted on alluvial clay soils influenced by groundwaters over more than ten years. Results are presented and discussed of the reachable loosening depth, the construction of mole channels and the recompaction of subdrainage backfills. Conclusions are drawn on the need of permeable backfill and the efficiency of additional subsoiling technique.

Natural backfills of drain trenches show better soil physical parameters than the adjacent soil only for 12 years on the average. The realizable loosening depth of tined tools can be influenced by the geometry of the tools but on a small scale. Its value is 0.50 m on the average. Mole channels are constructable below the loosening depth of the tine and above the water table. Favourable conditions for their durability are present in a depth of about 0.6 to 0.9 m below the soil surface. The soil consistency dependent on the water table is the decisive parameter to assess the subsoil workability.
KEYWORDS. Heavy soils, Water table, Subsoil workability.

INTRODUCTION

For carrying out land improvement measures, such as subdrainage, mole drainage or deep loosening the subsoil has to be suitable in status. Climatic and hydrologic conditions, as well as the soil type are important for this. Following the conception of the 'critical depth' (Spoor and Godwin, 1978) the geometry of the soil disturbing implements is also decisive. Below the critical depth no soil loosening, but deformation occurs. This decreases the water permeability (Boels, 1980, Tureckijj, 1985). It was to test which values of critical depth are to be expected in dependence on soil and tool parameters, and whether this depth is simply measurable in the field.

Because subsoiling measures including mole channels are usable for land drainage (Cannell et al. 1984, Sommerfeldt and Chang, 1987, Zajjdelman, 1991), the working effect of tined deep working tools and expanders for forming mole channels had to be studied more intensively. The soil parameters of old drain trenches backfills provide information on the durability of new ones, as well as the necessity of permeable backfills on the whole, and were to be studied therefore.

MATERIAL AND METHODS

Field experiments with rigid deep working tools and expanders for mole channel formation have been conducted on several mostly alluvial clay and loam soils in lowland areas of the Oder and

---

[*] Dr. agr. L. Müller, Group Leader, Scientist, and Dr. agr. U. Schindler, Scientist, Institute of Hydrology, Centre for Research on Agricultural Landscapes and Land Use (ZALF), Müncheberg, Germany.

Table 1. Site and Tool Parameters.

Site No.	Clay Content (%)		Dry Bulk Density Dd (g/cm³)		Consistency Index Ic		Plasticity Index Ip (% weight)	Penetrometer Tip Resistance DWo (MPa)
	a)	b)	a)	b)	a)	b)		
1	33.7	31.8	1.16	1.35	1.06	0.95	38.6	1.27
2	33.7	31.8	1.16	1.35	1.06	0.95	38.6	1.27
3	55.0	52.5	1.15	0.89	1.00	0.54	65.7	0.66
4	55.0	52.5	1.15	0.89	1.00	0.54	65.7	0.66
5	53.4	46.8	1.37	1.23	1.06	0.81	34.9	1.28
6	55.0	53.6	1.33	1.20	1.12	0.88	40.1	1.30
7	60.2	52.4	1.08	0.82	0.95	0.51	61.4	0.46
8	51.1	54.1	1.34	1.10	1.11	0.86	61.6	1.38
9	35.3	35.8	1.51	1.27	1.29	0.91	33.4	1.90
10	63.1	65.3	1.00	0.79	1.06	0.35	58.3	0.07
11	40.6	32.1	1.21	1.41	1.01	0.85	34.0	1.24
12	25.8	19.1	1.22	1.48	1.11	0.96	27.2	1.39
13	27.6	26.0	1.17	1.29	1.07	0.86	38.9	1.23
14	11.2	11.8	1.61	1.61	1.12	0.88	15.0	1.65
15	11.2	11.8	1.61	1.61	1.12	0.88	15.0	1.65

Continuing Table 1.

Site No.	Number of the Variant	Averaged Values of the Site				
		Working Depth Zz (cm)	Loosening Depth Zl (cm)	Tine Width bw (cm)	Tine Aspect Ratio Zz/bw	Tine Factor Kw
1	10	67	54	5.6	12.3	11.0
2	4	50	41	7.1	8.4	26.3
3	11	65	38	6.2	11.2	9.7
4	5	57	38	7.8	9.3	26.8
5	4	85	45	7.0	12.2	17.5
6	11	50	40	12.4	4.1	10.0
7	4	72	30	9.4	7.8	8.6
8	5	71	64	7.2	9.8	7.4
9	5	76	72	6.4	11.9	12.9
10	5	85	10	7.1	13.9	31.8
11	6	71	55	6.5	11.0	11.1
12	5	66	58	5.7	11.5	10.6
13	6	64	41	6.2	11.1	10.6
14	3	80	50	6.0	13.3	19.0
15	4	74	59	9.0	8.2	11.7

a) = above the loosening depth   b) = below the loosening depth
Kw = (Zz/bw) * tangens alpha
Soil types after FAO classification: No. 1 to 13: Eutric Fluvisols, No. 14, 15: Gleyic Luvisols

Unstrut rivers. These areas are located in the New German Bundesländer Brandenburg and Sachsen-Anhalt. Soil and tool parameters were widely varied (Table 1). All variants were dug and soil physical analyses were done. The loosening depth is diagnostable in the field by beginning the formation of a slot with remoulded walls (Fig. 1).

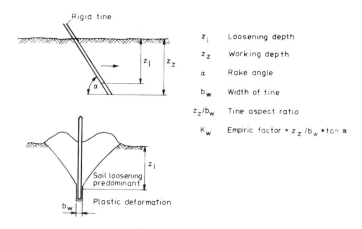

Figure 1. Soil Loosening Depth (Critical Depth).

In the Oderbruch area about 40 drain trenches of different age were analysed. The site conditions and soil physical analyses are described in Müller et al. (1989, 1992) and, partly, in related papers of this Conference (Müller and Schindler, 1992).

RESULTS

Realizable Soil Loosening Depth

The realizable soil loosening depth (critical depth) is to be understood as the horizontal division line between loosening and plastic deformation in the soil (Fig. 1). This depth of tined tools in heavy groundwater influenced soils is limited, even in summer, and can be increased by the geometry of the tools only on a small scale (Fig. 2).

On the sites under study the loosening depth ranges between 20 and 70 cm, and its value on clay soils is 45 cm on the average. In alluvial loam soils its value is about 50 cm and in the loess-loam soil under study about 55 cm. The realizable loosening depth does not increase significantly during the summer and autumn, because the moisture content in the subsoil decreases only slightly throughout the year, whereas there are marked differences between sections with differing drainage states at the same time. The soil loosening depth can be calculated from soil strength parameters, for example, the tip resistance Dwo of a penetrometer (Fig. 2). A further simple soil parameter to assess the loosening depth is the depth the soil becomes plastic (Müller and Mittelstedt, 1986). The influence of the tine expressed by the empiric factor k_W is also significant, but comes second to the influence of the soil (Fig. 2). Because the loosening depth in the soils under study averages 50 cm below the surface and has a distinct spatial variability in dependence on the drainage status, deep loosening measures do not produce satisfactory results. Deep working tines form pseudo-mole channels disintegrating within two years (Müller, 1985). Tines are more useful combined with expanders to install mole channels. If the soil has a slow plasticity (sites 14 and 15 from Table 1), a complete structural breakdown may occur, visible as soil flow. In such cases no deep loosening nor mole channel formation is possible.

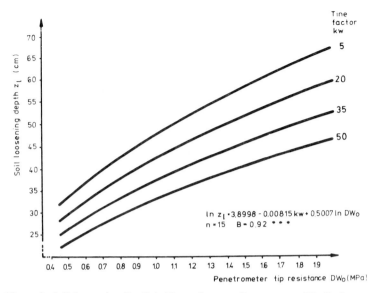

Figure 2. Soil Loosening Depth in Dependence on Penetrometer Tip Resistance, Dwo, and Tine Factor, Kw.

Trenchless drain-laying machines work in every case 10 to 30 cm below their loosening depth (Müller and Mittelstedt, 1986). The water flow towards drains is reduced in dependence on the soil plasticity and consistency. In soft-plastic soils nearly impermeable remoulded zones are formed (Boels, 1980, Mittelstedt, 1987). Their heigth needs to be considered in the drainage performance (Fry and Spoor, 1983). In the stiff-plastic consistency state cracks occur in the slot walls below the loosening depth, and the water flow towards drains is not hampered (Mittelstedt, 1987). If the soil has a slow plasticity (plasticity index lower than 15 % of weight), there is a distinct risc of a primary silting of pipe drains (Samani and Williardson, 1981, Balla, 1985).

It can be concluded that all field parts planned to use deep working implements are carefully to study. The soil plasticity and consistency evaluation based on the methods according to Casagrande and Atterberg is suited for this purpose.

Soil Suitability for Mole Channel Construction

In general, the construction of mole channels is possible, if the soil has a sufficient plasticity (plasticity index greater than 20 % of weight) and a suitable consistency (whole plastic range, consistency index lower than 1). The working depth of the expander must be at least 15 cm deeper than the loosening depth of the tine. In the stiff-plastic consistency state ($Ic = 0.75$ to 1.0) cracks occur in the smear and press zones providing a fast hydraulic function of the channels. The moling process causes a temporal loss of the soil strength of the channel wall. The values of soil strength, however, almost recover after about one day (Müller et al., 1989). Conditions favourable for formation and good durability of channels are present in a depth of about 0.6 to 0.9 m below the soil surface. From the viewpoint of durability this is also a suitable depth.

Testing various tine - expander combinations for subsoiling on working effect and traction requirements, a comparatively narrow tool with a small rake angle, plane edges and basally-fixed expander has revealed to be the most suitable under predominantly plastic but heterogeneous subsoil conditions.

Durability of Natural Drain Backfills

Natural backfills of drain trenches show soil physical parameters better than the adjacent soil during 12 years on the average. The values of water permeability and air volume of backfills younger than 12 years are higher than the undisturbed soil, and later they have roughly the same level. Backfills older than 12 years are distinctly more compact and dense than the adjacent soil (Fig. 3, Fig. 4).

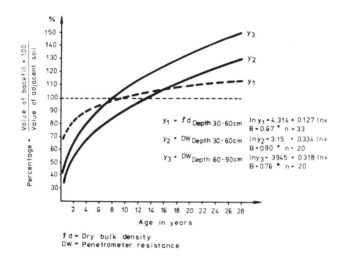

Figure 3. Backfill Recompaction, Dry Bulk Density and Penetrometer Resistance.

The recompaction process varies in dependence on drainage status. Under bad drainage conditions and arable land use there is a quick and intensive recompaction of backfills in the topsoil and the upper subsoil due to heavy machinery, as well as in the lower subsoil due to swelling.

Since wetness by stagnant water occurs in such cases, permeable backfill is useful. Under better drainage conditions (medium-term water table 0.9 to 1.4 m below the surface) recompacted backfills are no problem, because the water permeability of the whole site is better and wetness by perched or ponded water does not occur. The distinct differences in soil strengths of backfills older than 12 years and younger than 6 years to the adjacent soil may be used for finding drain lines in the field using simple vertical-penetrometers.

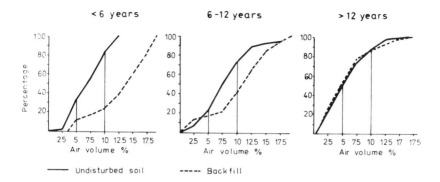

Figure 4. Decrease of Air Volume in Backfills.

REFERENCES

1. Balla, D. 1985. Untersuchungen zum Verschlammungschutz von Dränanlagen. Diss. Müncheberg. 76 p.

2. Boels, D. 1980. Effect of soil deformation in trenchless drain laying. Research Digest. ICW, Technical Bulletin 117, Institute for Land and Water Management, Wageningen, The Netherlands. 154-159

3. Cannell, R.Q., M.J. Goss, G.L. Harris, M.G. Jarvis, J.T. Douglas, K.R. Howse and S. Le Grici. 1984. A study of mole drainage with simplified cultivation for autumn-sown crops on a clay soil. J. agric. Sci. Camb. 102, 539-559

4. Fry, R.K. and G. Spoor. 1983. The influence of soil compaction and smear in the vicinity of a pipe on drain performance. J. Agric. Engng. Res. 28, 207-216

5. Mittelstedt, R. 1987. Arbeitseffekt und Zugkraftbedarf von Meliorationswerkzeugen in alluvialen Tonböden unter besonderer Berücksichtigung von Pressvorgängen. Diss. Müncheberg. 84 p.

6. Müller, L. 1985. Efficiency of measures for subsoil treatment of an alluvial clay site. Archives of Agronomy and Soil Science, Berlin, 29, 2, 99-106 (in German, Engl. summary)

7. Müller, L. and R. Mittelstedt. 1986. Studies in the critical depth for the use of tine-shaped tools in lowland soils with high clay contents. Archives of Agronomy and Soil Science, Berlin 30, 8, 485-492 (in German, Engl. summary)

8. Müller, L. and U. Schindler. 1992. Durability of Agro-ameliorative measures on mineral lowland soils with heterogeneous substrate and hydrologic conditions. Papers of the Sixth International Drainage Symposium. Nashville, Tennessee, December 13-15, 1992.

9. Müller, L., P. Tille and H. Heim. 1992. Investigations in the influence of the water regime on the performance of soils rich in clay of the Oderbruch region for agriculture. Journal of Rural Engineering and Development. Hamburg and Berlin, 33, 107-116 (in German, Engl. summary)

10. Müller, L., R. Mittelstedt, U. Schindler and S. Heim. 1989. Construction and durability of mole channels in heavy alluvial soils. Journal of Rural Engineering and Development, Hamburg and Berlin, 30, 394-403, (in German, Engl. summary)

11. Samani, Z.A. and L.S. Williardson. 1981. Soil hydraulic stability in a subsurface drainage system. Transactions of the ASAE, St. Joseph, Michigan, 666-669

12. Sommerfeldt, T.G. and C. Chang. 1987. Desalinization of an irrigated, mole drained, saline clay loam soil. Canadian Journal of Soil Science 67, 5, 263-270

13. Spoor, G. and R.S. Godwin. 1978. An experimental investigation into the deep loosening of soil by rigid tines. J. Agric. Engng. Res., London, New York 23, 243-258

14. Tureckijj, R.L. 1985. Optimizacija uglovykh parametrov nosevogo rabochego organa drenazhnojj mashiny. Mekhanizacija i ehlektrifikacija sel'skogo khozjajjstva, Moskva 3, 8-12

15. Zajjdelman, F.R. 1991. Ecologo-meliorative pedology of humid landscapes. Moscow. Agropromizdat, 320 p. (in Russian, Engl. summary)

TRAFFICABILITY AND CROP YIELDS ON GROUNDWATER INFLUENCED HEAVY SOILS

L. Müller[*]

ABSTRACT

The influence of the water table has been studied on trafficability and crop yields at alluvial clay soils over six to ten years under the climatic conditions in the Eastern part of Germany (annual rainfall 450 to 510 mm, mean temperature 8.3 °C). Both trafficability and crop yield are closely related to drainage status. Water table control is necessary and often possible, but mostly missing. Medium-term springtime water tables are required of 0.9 to 1.1 m below the surface for arable farming. Deeper water tables do not much improve the conditions of trafficability. Meanwhile most sites are overdrained due to missing water table control and the precipitation deficit during several years. The current water tables deeper than 1.4 m tend to regional decreases in yield and long-term soil damage. Crop yields depend on the water capacity of the 'reclaimed soil depth'. Using this conception optimizing the regional water table is possible under the typical heterogeneous soil and hydrologic conditions of lowland areas.
KEYWORDS. Heavy soils, Water table, Trafficability, Crop yields.

INTRODUCTION

To optimize land use knowledge on the main processes and relationships between the site conditions and management activities is required. On lowland soils water table control is an important part of an ecosystem management. Data on tolerable wetness or drought are needed in decision models for medium-term water table control.

Some work exists to simulate available time for field work and crop yield level (Nolte et al., 1982) and to estimate soil and plant parameters for drainage simulation models (Baumer and Rice, 1988, Carter et al., 1988, Drablos et al. 1988, McDaniel and Skaggs, 1988).

It was in our study to quantify the dependence of trafficability and subsoil workability on significant soil and hydrologic conditions.

MATERIAL AND METHODS

Site Conditions

The study on relationships between the water table and both trafficability and crop yield limitations has been conducted on alluvial clay soils under the climatic conditions in the Eastern part of Germany (annual rainfall 450 to 510 mm, mean temperature 8.3 °C). The experimental site of about 1,500 ha is located in the Oderbruch area. Clay soils are dominant with parameters after Table 1.

The clay content (particle size < 0.002 mm) in the soils under study ranges from 30 to 65 %.

---

[*] Dr. agr. L. Müller, Group Leader, Scientist, Institute of Hydrology, Centre for Research on Agricultural Landscapes and Land Use (ZALF), Müncheberg, Germany.

The water tables vary from 0.3 to 1.6 m. The thickness of the clay layer ranges from 0.4 to 3.0 m. Besides the clay soils, sandy soils occur with water tables deeper than 1.2 m.

Table 1. Soil Conditions of the Experimental Clay Site.

Parameter		Soil Group [1]		
		A Good Drainage Status	B Medium Drainage Status	C Poor Drainage Status
Clay Content (%)	a)	41.4	48.1	49.0
	b)	51.6	53.1	60.0
	c)	51.2	56.0	55.6
Organic Matter (%)	a)	2.8	3.9	6.1
	b)	2.0	2.4	4.8
	c)	2.5	4.7	6.8
Dry Bulk Density in March (t/m^3)	a)	1.29	1.24	1.05
	b)	1.26	1.07	0.96
	c)	1.15	0.78	0.78
Consistency Index in March	a)	0.93	0.88	0.65
	b)	0.83	0.84	0.63
	c)	0.94	0.48	0.42
Water Table (m Below the Surface)		0.95 - 1.3	0.55 - 0.95	< 0.55

[1] = Arithmetic mean of 4 - 6 profiles.
a) = Ap horizon. b) = Subsoil, 0.30 - 0.55 m. c) = Subsoil, 0.55 - 0.80 m.

Soil Physical and Hydrologic Studies

Soil substrate and structural conditions were estimated by several soil profiles including analyses of disturbed and undisturbed samples. The water table and the soil moisture were continuously measured once per week using water observation wells and piezometers resp. gravimetric moisture estimation.

Trafficability and Crop Yield Estimation

The trafficability was evaluated by a method according to Müller et al., 1990. This method is combined from an assessment of the soil consistency state at different depths and penetrometer measurements. Crop yields have been estimated at 28 plots over ten years. The zero-plots of the experimental field were used as described in Müller and Schindler (1992).

RESULTS

Drainage Status and Trafficability

Trafficability is related to the drainage status (Table 2). Medium-term springtime water tables of 0.9 to 1.1 m below the surface are suitable for arable farming. Deeper water tables do not much improve the conditions of trafficability.

Table 2. Tractable Days per Month at a Clay Site.
(6-Year Average, Arable Use)

Drainage Status	Month											
	1	2	3	4	5	6	7	8	9	10	11	12
A/B [1]	10	11	8	19	17	21	24	26	30	20	16	7
C [2]	8	6	3	10	10	5	23	22	30	20	13	7

[1] Springtime water table 0.7 - 1.4 m.
[2] Springtime water table 0.35 - 0.55 m.

Only at field sections with high springtime water tables (< 0.55 m below the surface, drainage status C) very distinct limitations occur in trafficability. Those sites are characterized by wetness caused by groundwater, additionally by perched and ponded water and non-gravitational water. The first day of trafficability is about two weeks later than at normal drained soils. Infiltration and profile drainage are hampered by a high swelling status and mostly missing biogene macropores.

Crop Yields in Dependence on Hydrologic and Substrate Conditions

Crop yields are also closely related to the drainage status (Fig. 1). Yields markedly depend on the depth of the root zone. The latter is limited by high springtime water tables in clay soils.

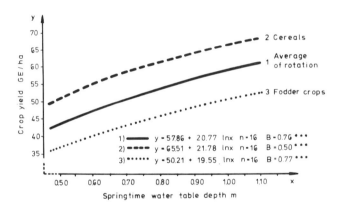

Figure 1. Drainage Status and Crop Yield.

The trend after Fig. 1 is not expected to continue with deeper water tables (Renger et. al., 1984). On clay or loam underlain by sand, the rooting depth and crop yields are limited by the depth of these cohesive substrates. If we go out from the fact that both shallow water tables and underlying sand substrate limit the rooting depth, the crop yield of the site as a whole depends on the water capacity of the 'reclaimed soil depth' (Müller and Tille, 1990) considering the capillary water supply. Figure 2 demonstrates the relationship between available soil water supply and crop yield. This complex parameter consists of the water storage capacity of the soil profile and can be used to assess the yield potential and amelioration need of the site.

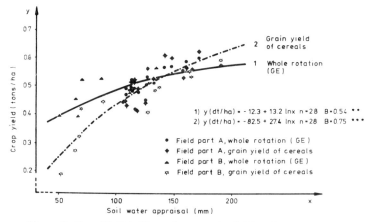

Figure 2. Crop Yield in Dependence on Available Soil Water Supply.

Meanwhile most sites are overdrained due to missing water table control and the precipitation deficit during several years (Table 3). Current water tables deeper than 1.4 m tend to regional yield declines and long-term soil damages.

Using the conception of the reclaimed soil depth, optimizing the regional water table is possible under the typical heterogeneous soil and hydrologic conditions of this and other lowland areas.

Table 3. Current Water Regime in the Whole Oderbruch Area.

Group	Drainage Situation	% of Area [1]
1	Very deeply drained, water table deeper than 2 m below the surface and deeper than 1 m below the clay cover	53
2	Medium deeply to deeply drained, water table 1 - 2 m below the surface	33
3	Shallow drained, watertable 0 - 1 m below the surface	8
4	Very shallow drained, water table 0 - 0.5 m below the surface, partly Artesian conditions	6

[1] Total considered area = 836 km^2.

DISCUSSION

Crop yield limitations from very shallow water tables can be explained by the lack of soil air volume limiting rooting. A site of the drainage status A (water table about 1.1 m) contains more than 6 % of air volume in the subsoil, whilst a site with a very shallow springtime water table of 0.3 m contains only 0 - 4 % air volume.

Such differences are diagnostable in the field by differences in the macro-morphologic soil structure. In alluvial clay soils, as a main feature compared with soils without shrinking and swelling processes, the changes in the soil water regime due to the medium and long-term structure development have to be taken into consideration. A drainage depth of 0.9 to 1.1 m below the surface improves the trafficability and increases yields attributable to a clearly manifested polyhedric soil structure. Such a structure has a better ecological quality than the coherent subsoil structure typical for a poor drainage status (Heim and Müller, 1988). Short-term water table regulations have less effect on the trafficability of clay soils, as the moisture loss in the upper subsoil is very small (Schindler, 1983). Drainage deeper than mentioned leads to too large and wide cracks allowing water and solutes quickly to migrate.

Current work deals with the riscs on agrarian production concerning the water balance and solute migration. The discussed relationships will be considered in a 'crop yield term' and a 'trafficability term' for the local validation of seepage rates. Besides this, relationships must be established to the parameters of the water management model DRAINMOD (Skaggs, 1980).

REFERENCES

1. Baumer, O. and J. Rice. 1988. Methods to predict soil input data for DRAINMOD. ASAE paper No. 88-2564.

2. Carter, C.E., R.L. Bengtson and J.S. Rogers. 1988. Drainage needs as indicated by high water tables. Transactions of the ASAE 31, 5, 1410-1415.

3. Drablos, C.J.W., K.D. Konyha, F.W. Simmons and M.C. Hirschi. 1988. Estimating soil parameters used in DRAINMOD for artificially drained soils in Illinois. ASAE paper No. 88-2617.

4. Heim, H. and L.Müller. 1988. Field studies on the structure of alluvial clay soils as a precondition for the determination of the drainage situation. Archives of Agronomy and Soil Science, Berlin, 32, 3, 141-151 (in German, Engl. summary).

5. McDaniel, V. and R.W. Skaggs. 1988. Corn root response to high water tables. ASAE paper No. 88-2609.

6. Müller, L. and P. Tille. 1990. Effects of substrate and water regime on the crop yield of an inhomogeneous alluvial soil in need of amelioration. Archives of Agronomy and Soil Science, Berlin, 34, 2, 103-112 (in German, Engl. summary).

7. Müller, L. and U. Schindler. 1992. Durability of agro-ameliorative measures on mineral lowland soils with heterogeneous substrate and hydrologic conditions. Papers of the Sixth International Drainage Symposium, Nashville, Tennessee, December 13-15, 1992.

8. Müller, L., P. Tille and H. Kretschmer. 1990. Trafficability and workability of alluvial clay soils in response to drainage status. Soil & Tillage Research, Amsterdam, 16, 273-287.

9. Nolte, B.H., N.R. Fausey and R.W. Skaggs. 1982. Time available for field work in Ohio. ASAE paper No. 82-2076.

10. Renger, M., O. Strebel, H. Sponagel and G. Wessolek. 1984. Einfluß von Grundwasserabsenkungen auf den Pflanzenertrag landwirtschaftlich genutzter Flächen. Wasser und Boden, 10, 816-824.

11. Schindler, U. 1983. Untersuchungen zum entwässerbaren Porenvolumen und zur nutzbaren Wasserkapazität grundwasserbeeinflußter Auenböden und Hinweise zur Regulierung des Wasserhaushaltes im Frühjahr. Arch. Acker- und Pflanzenbau u. Bodenkd., Berlin, 27, 6, 351-360.

12. Skaggs, R.W. 1980. DRAINMOD reference report. U.S. Department of Agriculture. North Carolina State University. Raileigh, N.C., 265 p.

EFFECT OF DRAIN ENVELOPES ON THE WATER ACCEPTANCE OF WRAPPED SUBSURFACE DRAINS

L.C.P.M. Stuyt[*]

ABSTRACT

Wrapped drain sections with surrounding soil were non-destructively sampled at 45 locations. All sections had been functioning in the field for approximately five years. The cores were examined with x-ray Computerised Tomography (CT) through 50 adjacent slices each, yielding three-dimensional (3D) mappings of the most permeable areas inside drain envelopes and surrounding soils. The results were used to study water flow patterns into drain pipes in a qualitative way, and to investigate the effect of envelopes on the development of such patterns. Water flow and mineral envelope clogging were found to be quite heterogeneous and were largely determined by soil structural features.

Keywords. Agricultural Drainage, Envelope Materials, Geotextiles, Drain Filters, Mineral Clogging, Computerised Tomography, Image Processing.

INTRODUCTION

In drainage engineering, installation techniques and machinery as well as pipe materials have been continuously improved during the past decades. The design of drain filters has not progressed to the same degree. This is attributed to the limited understanding of the complex and dynamic interactions occurring within soil/envelope systems, notably in weakly-cohesive soils which, in turn, is caused by the inability to monitor such systems in their natural state at an appropriate scale. Progress, made with analog and mathematical models in laboratories has been limited, simply due to a lack of field data against which to validate results (Stuyt, 1992a). In an attempt to obtain such field data, a CT scanner was used to generate sequences of CT-images of 45 soil cores, containing wrapped drain sections. After a service life of five years, these cores were retrieved from three experimental fields, located in areas with very fine sandy soils which have low structural stability. The sampling locations were chosen after internal, visual inspection of 9,600 m of lateral drains with a miniature TV-camera (Stuyt, 1992b). Each CT-sequence is a three-dimensional (3D), geometrically precise mapping of the interior density variations inside drain envelopes and the surrounding soils. These variations were located and quantified using 3D image analysis techniques.

DATA ACQUISITION TOOLS AND PROCEDURES

X-ray computerised tomography (CT) is a non-destructive and non-invasive imaging technique.

---

[*]L.C.P.M. Stuyt, Regional Water Management Engineer, The Winand Staring Centre for Integrated Land, Soil and Water Research (SC-DLO), P.O. Box 125, NL-6700 AC Wageningen, The Netherlands. Phone: +31.8370.74298, Fax: +31.8370.24812, e-mail: stuyt@sc.agro.nl

Transmission measurements of a narrow beam of x-rays, made at several different angles or projections around a given object, are used to resynthesize slices of interest within this object with the aid of a computer program. CT is able to resolve small differences in density and water content over distances of a few millimeters, and is therefore an appropriate tool for observation of distributions of density and water content in soils. A thorough discussion of CT is found in Herman (1980).

The use of CT in soil and water related sciences

In the beginning of the 1980s, CT technology has become available in the soil and water related sciences. CT-scans are computed as maps of linear attenuation coefficients for x-rays, providing quantitative information about features of an object like mineral densities and their distribution in the soil. Petrovic et al. (1982) established a linear relationship between mean soil bulk density and mean x-ray attenuation rate. Bergosh et al. (1985) found that all open and partially open macropores greater than 0.5 mm in width can be detected with CT. Macropores in soils have been characterized with CT by Anderson et al. (1988, 1990) and Grevers et al. (1989). Phogat and Aylmore (1989) and Phogat et al. (1991) have investigated the sensitivity, linearity, spatial resolution and suitability of CT as a method for assessing the structural status and the water content of a soil.

CT scanning of core samples

The set of CT-scans consisted of 45 sequences of 256 x 256 x 16 bit cross sections. Each sequence contained 50 scans. They were recorded with a Philips Tomoscan 350 scanner as 3 mm thick slices. All cross sections were taken at 3 mm pitch, effectively scanning the entire volume of a 150 mm wide central section of the cores lengthwise. The 3D images are digitized into rectangular parallelopipeds, usually referred to as volume elements or 'voxels'. They are described in terms of a Cartesian coordinate system. All image processing was done in this coordinate system: distances, volumes and surface areas were expressed as numbers of voxels. The dimensions of the image space are: width (x) 218 mm, height (y) 218 mm and depth (z) 150 mm. Voxel dimensions are x=y=0.85 mm by z=3 mm. The voxels were calibrated in Hounsfield units (H.U.) [-] in which x-ray attenuation rates in CT are expressed. Voxels containing air were mapped with -1000 H.U., the 'CT number' of water is 0 H.U., envelopes range from -500 to +500 H.U. and soils from 500 to approximately 2200 H.U..

RECOGNITION AND QUANTIFICATION OF PIPE AND ENVELOPE PARAMETERS

In the 3D image space, many features were automatically recognized, sampled and quantified, such as *dimensions of pipes* (horizontal and vertical diameter, eccentricity), *volumetric areas* of envelopes, *macroporosity statistics* in so-called regions of interest, centered around the mapped elliptical cross-section of the pipe wall: above it ('T'), below it ('B'), at its 'right' side ('R') and at its 'left' side ('L') (cf. Fig. 1), etc.. Image analysis procedures are discussed in Stuyt (1992a).

SOIL POROSITY AND -MACROPOROSITY

Within a well structured soil matrix, two major types of soil pores may be distinguished: *textural pores* inside soil aggregates, and *macropores* (voids, cracks) which separate these aggregates. **Macropores** have a strong effect on important soil properties like infiltration capacity, aeration, root development and saturated hydraulic conductivity. The hydraulic conductivity of a soil is associated with the macropore volume rather than with the total pore volume. In this study, the saturated hydraulic conductivity of the soil near drains is important. Hence, our interest was centered on the *macropore* volume and its spatial distribution around drains. Soil macroporosity was classified following Phogat & Aylmore (1989), as follows. The macroporosity of soil regions with minimum CT density, i.e. voids, is 100% (minimum x-ray attenuation rate, i.e. -1000 H.U.). The macroporosity of soil regions with maximum CT density, i.e. inside aggregates (having a textural porosity of e.g. 50%), is 0% (maximum x-ray attenuation rate, i.e. 2188 H.U.). Intermediate macroporosities were calculated from x-ray attenuation measurements. The regression line of x-ray attenuation rate and macroporosity is linear.

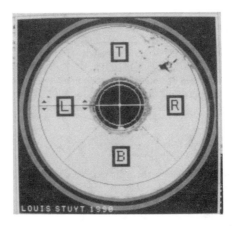

Figure 1.   Example of regions of interest in a CT scan image. The region is bordered by an elliptical curve and segmented into four equally large regions around the drain: 'T' = top, 'R' = right, 'B' = bottom and 'L' = left.

LIMITING MACROPOROSITY ('LMP') CONCEPT AND WATER ACCEPTANCE

3D macroporosity data may, in principle, be converted into estimates of saturated hydraulic conductivity, creating a database for numerical simulation of saturated water flow toward drain sections. The relationship between macroporosity and saturated hydraulic conductivity is however ill-defined. High macroporosity does not always coincide with high conductivity. Similarly, there is no direct correlation between hydraulic conductivity and the Hounsfield Unit (Hunt & Engler, 1987). Therefore, no attempt was made to simulate water flow into the sampled drain sections with a numerical model. Instead, a procedure was developed to visualise and quantify the spatial distribution of macroporosity in drain envelopes and surrounding soils. This procedure provides estimates of the water acceptance of wrapped drains in weakly-cohesive soils where the traditional concept of entrance resistance does not hold due to soil anisotropy and heterogeneity.

The basic idea of the LMP concept is the rather drastic assumption that water flow from any soil- or envelope unit towards the drain proceeds along the trajectory with the *highest overall hydraulic conductivity* between this unit and the drain. This means that bypass flow through other, less favourable channels is ignored. Along this trajectory, the flow is *governed* by the soil or envelope unit with the *lowest* (=limiting) hydraulic conductivity which acts as a throttle.

Hydraulic conductivities cannot be quantified from x-ray data. As an approximation however, *hydraulic conductivity* was considered to be *proportional* to *macroporosity* which, in turn, is *inversely proportional* to *x-ray attenuation*. Hence, water flow from any soil or envelope unit towards the drain is assumed to be governed by the unit with the lowest (=limiting) macroporosity ('LMP') along its trajectory. The LMP, associated with each soil- or envelope unit in the 3D image must be located downstream of this unit. The frequency distribution of macroporosity (and LMP) contains 29 classes, ranging from -1000 Hounsfield Units ('H.U.') (=100% macroporosity) to 2188 H.U. (0% macroporosity).

An automated search procedure was used to find all soil- and envelope units which are associated with a given LMP. At the end of this procedure, an LMP is assigned to each unit in the digitized 3D image. By approximation, the water acceptance of the upper and lower drain sections is estimated from the *weighed average* of all LMP's of all units above and under the drain. The weighing factors are the *numbers* of soil- and envelope units which are associated with these LMP's. The procedure is given by Stuyt (1992a).

RESULTS

Dimensions of pipes (horizontal and vertical diameter, eccentricity), volumetric areas of envelopes, macroporosity statistics in the regions of interest including average limiting macroporosities (LMP) etc. are given in Stuyt (1992a). Most drains are slightly compressed in the vertical direction. In 18 cases out of 45, the average macroporosity of the trench was lower than that of the subsoil. Two types of soil structural features were commonly found in the subsoil: horizontal layering and vertically oriented macropores (Fig. 2). Typical frequency distributions of macroporosity around drains are plotted in Fig. 3. The heterogeneity of mineral clogging of voluminous envelopes is illustrated in Fig. 4 in the form of transformed images depicting the envelopes as flat surfaces.

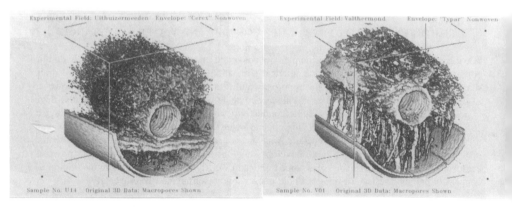

Figure 2.   Example of a layered subsoil (left) and of a subsoil with vertically oriented macropores, presumably developed as a result of plant roots (right). Parts of the plexiglass rims of both the sample container and the sample holder of the scanner were cut away by image processing techniques.

Regions in the 3D image space which were assigned particular LMP's have complicated geometric shapes. Such regions are important because they represent heterogeneous flow patterns towards drains and show the effect of the soil structure and the envelope material on such patterns (Fig. 5). In Stuyt (1992a), six such regions are printed stereographically with the anaglyph method.

DISCUSSION

Visual interpretation of regions which are associated with various LMP's (Fig. 5) shows that these regions may be geometrically complex and that they are largely determined by structural features

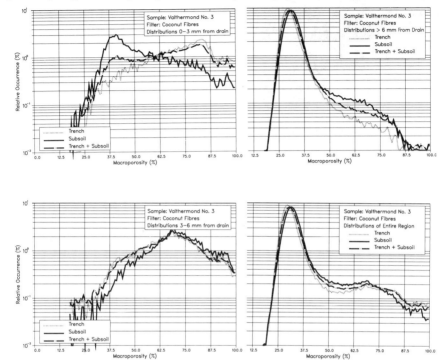

Figure 3. Frequency distributions of the macroporosity (MP) of the envelope and the soil around the drain in four regions of interest. Further away from the drain the trench backfill has a higher density than the subsoil. The inside area of the envelope under the drain appears to be clogged with soil particles; the outside area is 'clean'.

of the soil. Contrary to assumptions and results of studies made with mathematical models, the effect of an envelope on the water flow pattern towards a drain is limited as is its effect on radial and entrance resistance. Hence, differences in these flow resistances must be ascribed to structural features of the soil, i.e. macroporosity and its distribution near the drain. Study of all (90) water flow patterns into the drains supports the numerical data: there is no evidence that envelope specifications have a significant effect on the geometry of these patterns (Stuyt, 1992a)[1]. During this Symposium, various flow patterns will be presented as computed video animations (rotating objects).

Envelope specifications (composition, effective opening size, thickness, etc.) had a limited effect on the average LMP. Given the systematic difference between MP and LMP (10-15% on average) drainage resistance is more likely to be determined by soil macroporosity than by hydraulic

---

[1] In a concurrent field study, however, Stuyt (1992b) established that the so-called effective opening size ('O_{90}') of envelope pores has a significant effect on pipe sedimentation rates.

properties of the envelopes. *The soil around the drain may often be the major throttle to water flow into the drain.*

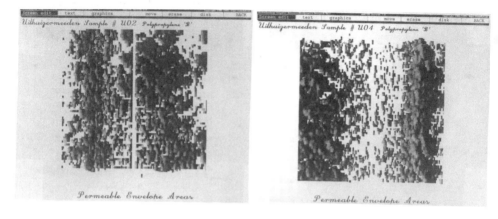

Figure 4. Examples of the heterogeneity of patterns of mineral clogging inside voluminous envelopes as depicted in transformed images in which these envelopes are displayed as flat surfaces. Envelope regions with average macroporosity higher than the median macroporosity are mapped as solid, shaded dark areas and are considered the most permeable. Other regions which are (partly) clogged are not depicted.

The successful use of the LMP as a qualitative indicator to analyse the water acceptance of drains in a heterogeneous medium demonstrates that a 'traditional' analysis in 2D cross-sections through such drains is inadequate and must be replaced by an analysis in 3D space. In heterogeneous media,

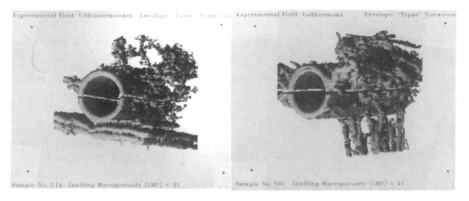

Figure 5. Image areas containing all voxels with Limiting Macroporosity LMP $\geq$ 37% (left). Subtle banding is evident under the drain. The trench contains some geometrically complex areas. Image areas containing all voxels with Limiting Macroporosity LMP $\geq$ 41% (right). Water access to this drain proceeds through a series of parallell vertically oriented macropores. Not all macropores are involved at this LMP, however, see Figure 2.

the mere use of macroporosity as such or the use of (bulk) density data to estimate hydraulic conductivities is likely to yield erroneous modelling results because it neglects the complex 3D geometry of density differences in the soil matrix and the resulting flow patterns.

CONCLUSIONS

Computerised tomography, combined with 3D image analysis is a powerful technique to investigate and quantify the physical interaction between drain filters and surrounding fine-sandy, weakly-cohesive soils. A visualisation study revealed that water flow patterns near drains are often heterogeneous. They depend on soil structure rather than on envelope type. Patterns of mineral clogging of envelopes were also found to be heterogeneous. Envelopes merely act as permeable constraints which support the soil near the drain. Good installation practice is therefore likely to be a decisive factor to secure a long service life of wrapped drains. Good envelopes will not cancel out adverse effects of poor installation practice. Installation under wet conditions must be avoided, if possible, in all situations.

REFERENCES

1. Anderson, S.H., C.J. Gantzer, J.M. Boone and R.J. Tully. 1988. Rapid nondestructive bulk density and soil-water content determination by computed tomography. Soil Sci. Soc. Am. J. 52:35-40.

2. Anderson, S.H., R.L. Peyton and C.J. Gantzer. 1990. Evaluation of constructed and natural soil mocropores using x-ray computed tomography. Geoderma 46:13-29.

3. Bergosh, J.L., T.R. Marks and A.F. Mitkus. 1985. New core analysis techniques for naturally fractured reservoirs. Soc. Petrol. Engin. SPE 13653. Bakersfield, CA USA March 27-29, 8 pp.

4. Grevers, M.C.J., E. De Jong and R.J.St. Arnaud. 1989. The characterization of soil macroporosity with CT scanning. Can. J. Soil Sci. 69:629-637.

5. Herman, G.T. 1980. Image Reconstruction from Projections, the Fundamentals of Computerized Tomography. Computer Science and Applied Mathematics, A series of Monographs and Textbooks. Academic Press, New York, USA

6. Hunt, P.K. and P. Engler. 1987. Computed Tomography as a Core Analysis Tool: applications and Artifact Reduction Techniques. Soc. Petrol. Engin. SPE 16952. Dallas, TX USA, Sept. 27-30, 8 pp.

7. Petrovic, A.M., J.E. Siebert and P.E. Rieke. 1982. Soil bulk density analysis in three dimensions by computed tomography scanning. Soil Sci. Soc. Amer. Journ. 46(3): 445-450.

8. Phogat, V.K. and L.A.G. Aylmore. 1989. Evaluation of Soil Structure by using Computer Assisted Tomography. Aust. J. Soil. Res. 27:313-323.

9. Phogat, V.K., L.A.G. Aylmore and R.D. Schuller. 1991. Simultaneous measurement of the spatial distribution of soil water content and bulk density. Soil Sci. Soc. Am. J. 55:908-915.

10. Stuyt, L.C.P.M. 1992a. The Water Acceptance of Wrapped Subsurface Drains. Ph.D.-Thesis, Agricultural University, Wageningen/The Winand Staring Centre for Integrated Land, Soil and Water Research (SC-DLO), P.O. Box 125, NL-6700 AC Wageningen, The Netherlands, 314 pp.

11. Stuyt, L.C.P.M. 1992b. Mineral clogging of wrapped subsurface drains, installed in unstable soils: a field study. Proc. 5th Int. Drain. Workshop, February 1992, Lahore, Pakistan.

NITRATE LEACHING IN DRAINED SEASONALLY WATERLOGGED SHALLOW SOILS

M.P. Arlot* D.Zimmer*

ABSTRACT

Nitrate content of subsurface drainage water has been monitored for several years in three French experiments. Soil type affects the seasonal occurence of nitrate losses. In loamy soils losses are mainly measured in winter, at the beginning of the drainage season; losses in spring are very low. In clayey soils winter losses are often less important, but spring losses can be very high, due to soil cracking. In all types of soils, instantaneous values of water nitrate concentration are highly variable especially during peak flows. But the mean concentration, defined as the ratio of cumulated losses versus cumulated discharges, is often constant during long periods, and sometimes during the whole drainage season. This is particularly true in fairly permeable loamy soils. In these soils the value of the mean contration is closely related to the mineral nitrogen content of the soil at the beginning of the drainage season. A simple nitrogen balance of the previous cropping season can be an accurate indicator of the mean nitrate concentration. Seasonal changes of the mean concentration are due to climatic and agricultural events. They are more frequent in clayey soils because of peculiar water flow patterns.

Key words: leaching, field experiment, nitrate, water flow patterns

INTRODUCTION

In France as in many other countries, nitrate leaching from agricultural land has become a major concern during the 1980's after surface and groundwater nitrate levels have risen in many regions. Subsurface drainage waters convey nitrates and other solutes which can affect the surface water quality. The impact of nitrate losses has been studied in many drainage field experiments mainly aiming at quantifying nitrates losses and at understanding the influence of factors like climate, soil type, drainage technique and agricultural practices on these losses (e.g. Baker et al., 1981; Belamie and Vollat, 1986; Bengston et al., 1988; Harris et al., 1984). One of the conclusions drawn from all these studies (Arlot, 1989) is that interactions between these factors lead to a great variety of situations resulting from differences of the amount of leachable mineral nitrogen and of the processes leading to its leaching.

This paper presents a summary of the results of three intensive experiments monitored in France in the past recent years to quantify nitrates losses in drainage waters and to understand the mechanisms and the periods of occurence of these losses with the help of hydraulic studies.

FIELD EXPERIMENTS AND METHODS

Three field experiments, Arrou, Courcival and La Jaillière, representative of french soils and conditions will be considered herein. They are seasonally waterlogged shallow soils: a perched water table forms in winter and early spring

---

* M.P. Arlot, Agricultural Engineer and D.Zimmer, Head Drainage Division, PHD Soil Physics, CEMAGREF, B.P. 121, 92185 ANTONY Cedex, France

in the upper one meter of the soil. Subsurface drainage technique involves 10 m spaced and 0.8-0,9 m deep pipes. The crop rotations are based on winter wheat and maize; the fertilization rates are close to each other. The experimental plot areas range between one and two hectares.

Arrou and La Jaillière experiments are located on loamy soils developed on a plateau loam and on a weathered schistic bedrock respectively. Both soils are alfisoils with an argillic horizon lying between 0.40m and 0.80m in depth. Their hydraulic conductivity decreases with depth, the impervious barrier depth being roughly equal to the drain depth. These soils represent more than 60% of the drained soils in France. Courcival experiment is located on a very clayey soil and is classified as a vertic haplaquept. It is representative of very clayey soils with high swelling-shrinking capacities. This soil represents less than 5% of the drained soils in France. Their impervious barrier is located at the bottom of the plough layer during most of the winter period.

Monitoring of field experiments is comparable to that of others (Cannell et al., 1984). Rainfall intensities, drainage flow rates and water table levels are hourly monitored from late automn until early spring. Drainage water is sampled at an eight hour time step for nitrate (and sometimes other chemicals) concentration determination. Other measurements and samples are collected during more limited periods : (i) hourly water pressures are measured by use of tensiometers, (ii) weekly soil water samples by porous cups and weekly to monthly soil samples are collected to determine their nitrate contents.

EXPERIMENTAL RESULTS

Water flow patterns

Knowledge of the water flow patterns is necessary to achieve a good comprehension of nitrate leaching mechanisms. Different water flow patterns are observed in drained shallow soils (Zimmer and Lesaffre, 1989). Regarding the three field experiments dealt with herein the situation is as follows.

(1) In Arrou and La Jaillière, the impervious barrier is located at the drain depth (0.80-0.90 m); the perched water table forms on this barrier; in this water table the flow is horizontal in agreement with Dupuit-Forchheimer assumption; during rainfall events, the water table is recharged by a vertical downward flow. In the following this flow pattern will be referred to as the *classical flow pattern*.

(2) In Courcival the location of the impervious barrier is variable; at the beginning of the winter enough cracks remain open to convey water to the drain pipe; this situation comes to an end when swelling closes up cracks located below the plough layer: a water table forms in the plough layer; water flows along the plough layer before percolating downards to the drain pipe through cracks in the vicinity of the drainage trench; in spring cracking starts again and preferential flow can induce important drainage events. Drainage of this heavy clay soil is as a shallow drainage process with very short time-responses.

Hydrologic functioning of shallow seasonally waterlogged soils

In Arrou and La Jaillière field experiments, more than 90 % of the drainage events and nitrate losses occur during a period called the *intense drainage season* and characterized by a high and constant drainage efficiency (Lesaffre and Morel, 1986). In France the intense drainage season generally starts in January and ends in early spring. In clayey soils like in Courcival drainage discharges and leaching can be important in spring too (Harris et al., 1984; Schwab et al., 1973); they can reach 40 to 60 % of total nitrate losses. However

Field	year	crop	drainage season (day/month)	IDS	D_{IDS} mm	% D_{DS}	N_{IDS} kg	% N_{DS}	C_w mg/l
ARROU	76-77	maize	21/12-->12/04	10/01-->07/04	118	82	34	81	128
	77-78	wheat	11/12-->19/05	23/01-->31/03	225	99	54	95	157
	78-79	barley	19/01-->05/05	19/01-->10/04	197	98	25	98	70
	79-80	maize	08/12-->10/04	02/12-->01/04	276	99	10	92	21
COURCIVAL	84-85	maize	25/10-->20/05	25/10-->07/02	-	-	-	-	40
	85-86	wheat	20/12-->30/04	06/01-->09/03	40	59	16	43	295
	86-87	maize	13/12-->30/06	13/12-->04/01	38	80	15	84	101
	87-88	wheat	15/10-->30/06	05/12-->17/01	34	71	2	27	40
LA JAILLIERE	88-89	wheat	21/02-->28/04	21/02-->13/03	88	85	13	93	71
	89-90	ryegrass	14/12-->26/02	22/12-->26/02	279	100	20	90	35
	90-91	wheat	26/12-->08/05	29/12-->21/01	108	99	52	87	161

Table 1. Major characteristics of the drainage seasons in the three field experiments.
IDS : intense drainage season (day/month)
D_{IDS} : total drainage discharge during IDS (mm)
% D_{DS} : ratio D_{IDS}/Total drainage discharge during drainage season
N_{IDS} : total nitrate losses in drainage water during IDS (kg N-NO$_3$/ha)
% N_{DS} : ratio N_{IDS}/total nitrate losses during IDS (mm)
C_w : mean annual nitrate concentration, ratio total N losses versus total discharge (mg/l)

cumulative drainage discharges and consequently nitrate losses are lower in Courcival (Table 1).

Sequences of nitrate concentrations

During the intense drainage season, nitrate concentrations are quite variable at an 8-hour time-step. Three main typical sequences are observed (Figure 1). A and B types can be observed in all three field experiments whereas C type is peculiar to Courcival.

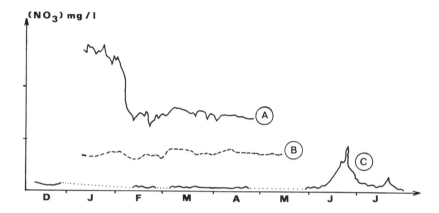

Figure 1. Typical sequences of nitrate concentrations in shallow soils during the intense drainage season. Water sampling time-step 8 hours. (A) and (B) : Field experiment of Arrou, 1976-1977 and 1979-1980 winters. (C) : Field experiment of Courcival, 1985-1986 winter.

Nitrate concentration during peak flows can evolve in three different ways (Figure 2) the frequency of occurence of which differs between fields.

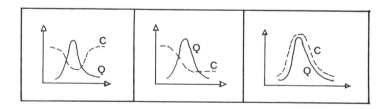

Figure 2. Patterns of nitrate concentration (C) evolution during drainage events (drain flow Q).

"A-pattern" is characterized by a dilution process during peak flow (Figure 2); it is mainly observed with the classical water flow pattern in Arrou and less often in La Jaillière. Analysis of soil water sampled by ceramic cups helps understanding this behavior (Figure 3). Dilution occurs when low concentrated rainfall water percolates through already leached soil zones (i.e. plough layer and whole profile at drain mid-spacing) before flowing through a high concentrated zone surrounding the drain (Figure 3b). This nitrate repartition results from a double process evolution: before the intense drainage season (Figure 3a), a nitrate front moves vertically from the plough layer as the wetting of the soil starts; at the beginning of the intense drainage season (Figure 3b) the water table forms, inducing an horizontal flow and a gradient of concentration between drain and drain mid-spacing.

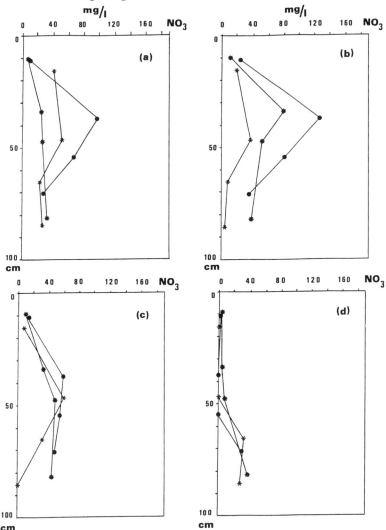

Figure 3. Profiles of nitrate concentrations in soil solution sampled by ceramic cups in La Jaillière field experiment. (a) 12/22/89, before intense drainage season starts; (b) 01/04/90 and (c) 01/15/90, intense drainage season; (d) 03/02/90, end of intense drainage season. ● trench location; ✳ 1 meter from trench; ✱ 5 meters from trench i.e. drain mid-spacing.

"B-pattern" shows a non-reversible impoverishment of the soil leachable nitrogen during the drainage event (Figure 2); the corresponding decrease of nitrate concentration is shown on figure 3d. A more moderate decrease with an homogeneization of the water table concentrations profiles between drain and drain mid-spacing is generally observed (Figure 3c).

Fertilizer dressing or mineralization can occur in spring at the end of the intense drainage season inducing high nitrate concentration in the plough layer while the nitrate concentrations are low in depth. This repartition combined with the second water flow pattern explain the "C-pattern" (Figure 2). This pattern is not observed with the classical water flow pattern, which is likely due to the fact that the watertable seldom reaches the high concentrated plough layer in that case. This suggests that leaching due to the watertable itself is more important than leaching due to the recharge of the water table.

Mean leaching nitrate concentration

Sequences of nitrate concentrations can be described by double mass curves of cumulative drainage discharges versus cumulative nitrate losses in drainage water. The slope of these curves is the *mean leaching concentration*.

Although nitrate concentrations are variable at the 8-hour time step, *the mean leaching concentration can remain constant* throughout the intense drainage season (Figure 4 and Figure 1, case B). In this case a mean *annual* leaching concentration can be defined. *This mean value is close to the instantaneous concentrations measured in drainage waters during peak flows*; as it remains constant it is determined as soon as the drainage season starts.

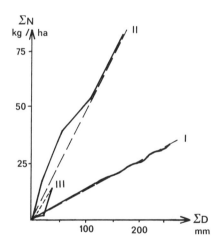

Figure 4. Typical double mass curves of nitrate losses (N) versus drainage discharges (D). Type I, Arrou 1979-1980; Type II, Arrou 1976-1977; Type III, Courcival.

The mean concentration can also be affected by sudden changes (Figure 4, cases II, III and Figure 1, cases A and C). An increase can for instance occur following a "C-pattern" drainage event; this is therefore more frequent in clayey soils. A decrease can be in relation with a decrease of soil nitrate content caused either by an intense leaching period occuring during a "B-pattern" drainage event, or by nitrogen removal by denitrification or reorganization.

Toward an indicator of the mean annual nitrate concentration

In search of a predicting indicator of the annual mean nitrate concentration in drainage water, a simplified mineral nitrogen balance of the previous cropping season -nitrogen brought by organic and mineral fertilizer minus crop consumption- was related to annual mean concentrations. For this purpose the mean concentration was calculated as the ratio of total drainage discharges versus total nitrate losses whatever the changes of trend of the double mass curve.

For all fields, good relations were found (Figure 5), except for two years following autumn droughts (Courcival, 1985-1986 and La Jaillière, 1990-1991). In these cases, the quantity of nitrate mineralized between the harvest and the beginning of the drainage season needs to be taken into account (Figure 5, encircled point).

The relation between nitrogen balance and nitrate mean concentration is of practical interest. First, its slope can be interpreted as a sensitivity of the soil toward nitrate leaching. For instance the higher slope of Courcival's soil is likely the result of the peculiar flow pattern which results in the leaching of a greater proportion of the nitrates stored after harvest or fertilization. Second, it helps determining the target nitrogen balance corresponding to a given nitrate concentration not to be exceeded.

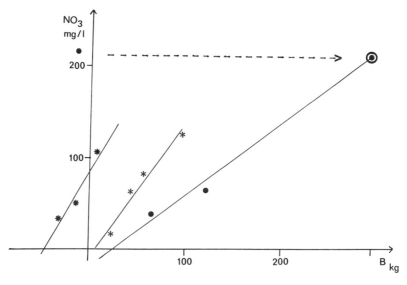

Figure 5. Relation between mean annual nitrate concentration (NO_3) and simple mineral nitrogen balance (B). ● La Jaillière; ✱ Arrou; ✻ Courcival.

CONCLUSIONS

Hydraulic and hydrologic studies carried out in three french field experiments in shallow soils helped distinguishing different nitrate leaching behaviours related to water flow patterns. In low permeable loamy soils (K_{sat} about 0.3 m/d) high losses are observed at a rather constant rate throughout the drainage season which occurs during winter and early spring months. In these soils a very simple mineral nitrogen balance of the previous cropping season is often enough to predict the mean leaching concentration of drained water during the drainage season. In heavy permeable soils (K_{sat} less than 0.01 m/d), losses alike drainage discharges are generally lower. But the shallow drainage process that occurs in these soils leads to a great sensitivity of nitrate leaching to climatic events or agricultural practices leading to ponctual nitrate enrichment of the soil.

These results must however be considered as tentative ones. However, they represent quite well the range of the situations encountered in France, and, more widely, the range of seasonaly waterlogged soils with a shallow impervious barrier. Drained field experiments also offer the opportunity of easy measurements of nitrate losses and water transfers under given agricultural and climatic conditions. Further studies ought to determine if the above conclusions can be extrapolated to non waterlogged soils, where predominant water and nitrate movement is vertical.

REFERENCES

1. Arlot M.P., 1989 - Caractérisation et limitation de l'impact du drainage agricole sur la qualité des eaux. Recensement des études, inventaire des méthodes et état des connaissances. Rapport CEMAGREF, Ministère de l'Environnement, 3 fascicules, 170p.

2. Baker J.L., Johnson H.P., 1981 - Nitrate-nitrogen in tile drainage as affected by fertilization. J. Environ. Qual., 10:519-522.

3. Belamie R., Vollat B., 1986 - Etude de la qualité des eaux de drainage, périmètre expérimental d'Arrou (Eure et Loir). Etudes du CEMAGREF, Hydraulique Agricole, 1:5-44.

4. Bengston R.L., Carter C.E., Morris H.F., Bartkiewicz S.A., 1988 - The influence of subsurface drainage practices on nitrogen and phosphorus losses in warm, humid climate. Trans. ASAE, 31(3):729-733.

5. Cannell R.Q., Goss M.J., Harris G.L., Jarvis M.G., Douglas J.T., Howse K.R., Le Grice S., 1984 - 1. Background experiment and site details, drainage systems, measurement of drain flow and summary of results, 1978-1980. J. Agric. Sci. Camb., 102:539-559.

6. Harris G.L., Goss M.J., Dowdell R.J., Howse K.R., Morgan P., 1984 - 2. Soil water regimes, water balances and nutrient loss in drain water, 1979-1980. J. Agric. Sci. Camb., 102:561-581.

7. Lesaffre B., Morel R., 1986 - Use of hydrographs to survey subsurface drainage networks ageing and hydraulic operating. Agric. Water Management seminar. Arnhem, The Netherlands. Balkema Ed., 175-189.

8. Schwab G.O., Mc Lean E.O., Waldron A.C., Whiter K., Michener D.W., 1973 - Quality of drainage water from a heavy textured soil. Trans. ASAE, 16(6):1104-1107.

9. Zimmer D., Lesaffre B., 1989 - Subsurface drainage flow patterns and soil types. ASAE-CSAE Summer meeting. Paper 892139, 17 p.

EFFECT OF DEEP SEEPAGE ON DRAINAGE FUNCTIONING AND DESIGN IN SHALLOW SOILS

D. Zimmer*

ABSTRACT

Deep seepage rates and mechanisms have been analysed in two field experiments located on seasonally waterlogged shallow soils in France. A deep seepage function has been introduced in drainage governing equations. These equations predict that a nil drainflow rate can be measured for mid-spacing water table elevations as high as 0.2 to 0.4 m above drain level. This was actually observed in field experiments. Based on these results deep seepage has been introduced in SIDRA model. Conclusions are as follows: (i) peak flow rates and, therefore drainage design rates are hardly affected by deep seepage and (ii) medium and low water table levels are more affected by deep seepage than high water table levels.

Key words: deep seepage, modeling, drainage design, aquifer recharge

INTRODUCTION

Subsurface drainage models often assume that a simple water balance "drainage discharge equals rainfall minus evapotranspiration" is fulfilled. This assumption is generally valid in case of permanent water table in absence of artesian flow and if surface runoff losses are negligible at field scale. However in case of perched water tables in shallow soils deep seepage may not be negligible. This is in particular the case in loamy low permeable soils (Zimmer et al, 1991) which are the most commonly drained soils in France. This results in a double concern.

1. As regards drainage design, what is the influence of deep seepage on watertable levels and drainflow rates; can field effective soil hydraulic properties (conductivity, drainable porosity) be measured in situ in this case?

2. In regions like South West of France, most of the irrigation water is pumped from shallow aquifers. Since the area had to face a severe drought during the past recent years, the recharge of these aquifers during winter season became a major concern. Questions arose about the relations between drainflow and deep seepage rates and about the consequences of drainage on aquifer recharge.

This research finds other applications in the case of landfill drainage, the objectives of which being to prevent water seepage through a geomembrane or inside an impervious material. The concern is in this case to be able to detect deep seepage as precisely and/or as quickly as possible.

In this paper answers will not be given to all these questions. The aim is to present preliminary results on deep seepage mechanisms and consequences on water table movements. Drainage will be tentatively simulated in presence of deep seepage the consequences of which will be discussed.

FIELD EXPERIMENTS AND METHODS

Results obtained in two field experiments, Arrou and Bouillac, are presented. Arrou is located in the Parisian Basin in Northern France, 200 km west of Paris

---

* D. Zimmer, Head Drainage Division, PhD Soil Physics, CEMAGREF, B.P. 121, 92185 ANTONY Cedex, France

city. Bouillac is located in the Aquitaine Basin in the South West of France, in the vicinity of Toulouse city. Both fields are located on loamy parent materials where leached soils (alfisoils) develop. Below the plow layer (0-0,25m) the subsoil includes an albic horizon (0,25-0,5 m) lying over an argillic horizon. The argillic horizon has a prismatic structure in Arrou; in Bouillac two types of structures, one prismatic and the other vertic, are found (Favrot et al, 1990). In depth the bedrock is made in Arrou either of flint clay or of the original loam, and in Bouillac of a mixture of loamy and gravel layers. This bedrock is low permeable and acts as a semi-permeable barrier. The mean ground slope is about 1 % in Arrou and 0,4 % in Bouillac.

The climate is oceanic, and the total annual rainfall depth amounts to 620 mm/yr in Arrou and 581 mm/yr in Bouillac. Waterlogging usually starts in december and lasts until late april. In summer the depth of the deep water table is at least 10 m in both fields. The perched watertable is drained with 10-20 m and 10-30 m spaced laterals in Arrou and Bouillac. The depth of the laterals ranges between 0,75 m and 0,85 m. Eight plots are monitored in Arrou, twelve plots are monitored in Bouillac.

Drainflow rates are measured in Arrou with V-notch weirs associated with an ultrasonic head-level sensor. In Bouillac they are measured with a tipping bucket recorder. Rainfall depths are measured with tipping bucket recorders in both sites. Water table heights are measured with tubewells and water pressures are measured with tensiometers. Data are generally hourly recorded. The field experiments of Arrou and Bouillac have been monitored respectively since 1973 and 1977.

RESULTS AND DISCUSSION

Deep seepage occurence and mechanisms

The hydrologic behavior of seasonally waterlogged soils under temperate climate is characterized by an "*intense drainage season*" as defined by Lesaffre and Morel (1986), during which the overall drainage efficiency is higher and remains nearly constant. The beginning and the end of this season are marked by abrupt changes of the drainage efficiency; at the beginning the change is related to the saturation of the profile; at its end, in spring, it results from the increase of evapotranspiration rate.

Evidence of deep seepage is supported by the spatial variability of the drainage efficiencies of the experimental plots of both fields during the intense drainage season. In Arrou two types of behaviors can be related to the depth of the flint clay material: the plots with a deep (2-4 m) flint clay material have drainage efficiencies ranging between 30 and 50%; those with a shallow (1-1.5 m) flint clay material have drainage efficiencies ranging between 60 and 90%. In Bouillac drainage efficiencies are related to the structure of the argillic horizon (Favrot et al, 1990). In case of dominant vertic structure they range between 30 and 50%; in case of dominant prismatic structure they range between 10 and 30%.

Water pressure monitoring indicates that drainage efficiency is related to the depth of a wetting front that slowly moves downwards during the winter season. This wetting front roughly correponds to the water table floor. The intense drainage season starts after the wetting front has reached the drain level; afterwards the front goes on moving down at different rates depending on the hydraulic conductivity of the deeper layers. In Arrou's plots with high drainage efficiency, the downward movement of the front slows down soon after the drainage season has started as the front enters the flint clay layer. The water flow pattern derived from measurements in situ (Figure 1) differs from the well-known flow pattern in case of deep impervious barriers: *no ascendant flow in the vicinity of the drain pipe is measured.*

Double mass curves of discharges of the two types of plots in Arrou (Figure 2) show *a constant and unique trend* throughout the intense drainage seasons. This trend helps reckoning the deep seepage rate at about 0.4 mm/d in Arrou. This order of magnitude is consistent with other results (Guiresse, 1989). However at the beginning of the intense drainage season the drainage efficiencies of both types of plots are similar. The explanation is that deep seepage occurs in both types of plots at this beginning (Lesaffre and Zimmer, 1988).

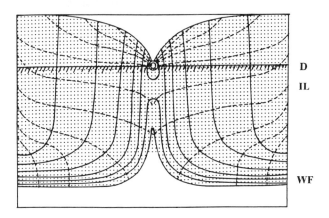

Figure 1. Water flow pattern in a shallow soil with perched water table as derived from tensiometric measurements. **WF** Wetting front; **D**. drain level; **IL** impeding layer; — equipotential lines; --- flow lines.

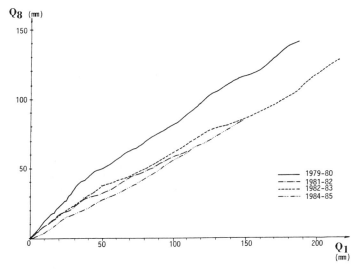

Figure 2. Cumulative discharges of plot Nr 8 (Q_8) - with deep seepage - versus cumulative discharges of plot Nr 1 (Q_1) - without deep seepage - in Arrou during intense drainage seasons 1979-1980 to 1983-1984.

Drainage modeling

When the pipes lay on shallow impervious barriers hydraulic modeling is based on Boussinesq's approach (Boussinesq, 1904) also described by Glover (in

Dumm, 1954), Guyon (1961) or Van Schilfgaarde (1965) and many other authors. The hydraulic system taken into consideration is presented in Figure 3. The basis of the approach is the integration of Boussinesq's equation without linearization assuming that the time and space variables are separable according to the equation :

$$h(x,t) = H(t).W(X) \tag{1}$$

where   h(x,t)   water table height at abscissa x
        H(t)   water table height at drain mid-spacing
        W(X)   nondimensional water table height at X=x/L,
                L being the mid-spacing

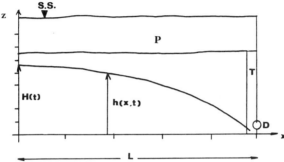

Figure 3. Schematic of subsurface drainage to laterals. S.S. soil surface; P plow layer; D drain pipe; T trench backfill.

This assumption will be hereafter referred to as *the constant water table shape assumption*. Its validity in case of deep seepage has been tested by use of tensiometers in Arrou: the ratio of hydraulic heads measured in the vicinity of the drain pipes versus hydraulic heads measured at drain mid-spacing remains constant over the intense drainage season (Figure 4). This had already been verified by use of tubewells by Guyon (1983) who measured a nearly elliptical water table shape in a plot with deep seepage.

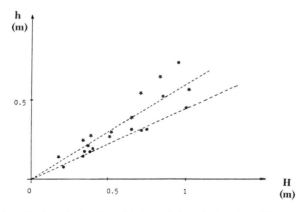

Figure 4. Hydraulic heads in the vicinity of the drain pipe (h) versus hydraulic heads at drain mid-spacing (H) during 87-88 winter in Arrou. ✶ plot 7: low deep seepage rate;   • plot 4: high deep seepage rate

Two other usual assumptions are to be discussed in case of deep seepage.

(1) Dupuit-Forchheimer assumption is to be valid at least midway between drains. This assumption is verified as indicated by the water flow pattern (Figure 2). The deep seepage rate (about 0.4 mm/d) is low as compared to the saturated hydraulic conductivity at the drain depth (0.05 - 0.1 m/d); thus the vertical hydraulic gradients are very low too.

(2) The downward flow at the bottom boundary of the hydraulic system (Figure 3) is generally neglected. In case of deep seepage, this flow has to be taken into consideration. Two assumptions can be made: (i) the downward flow is dependant on the water table height or (ii) it is independant. This second assumption will be considered herein since the deep seepage rate remains constant during most of the drainage season. A relation with the water table height is likely to exist at the beginning of the drainage season but will not be considered here.

The detailed procedure of solution is based on the following approach detailed in Guyon (1980), Lesaffre and Zimmer (1988) and Zimmer (1989).

- the deep seepage rate is introduced in the continuity equation alike the possible recharge rate of the water table;
- the equivalent hydraulic conductivity is introduced to take account of the vertical heterogeneity of the soil; on the basis of theoretical considerations not detailed herein, the hydraulic conductivity and drainable porosity variations with depth are further assumed to be satisfactorily described by power functions of the elevation, the power of the conductivity profile being twice the one of the drainable porosity;
- the continuity equation is integrated twice with respect to x, using the constant water table shape assumption; these two integrations yield two water table shape coefficients (B and C) resulting from the two integrations of W(X) with respect to X.

This integration yields the following equation:

$$\frac{dH}{dt} = \frac{1}{C\,f(H)} \left[R(t) - D_s - K_{he}(H) \frac{H(t)^2}{L^2} \right] \qquad (2)$$

where
D_s deep seepage rate (L.T^{-1})
K_{he} equivalent saturated horizontal hydraulic conductivity (L.T^{-1})
$R(t)$ water table recharge rate (L.T^{-1})
$f(H)$ drainable porosity

The relationship between drainflow rate, water table height midway between drains, recharge rate and deep seepage rate reads:

$$Q(H) = A\,K_{he}(H) \frac{H(t)^2}{L^2} + (1-A)[R(t) - D_s] \qquad (3)$$

where $A = B/C$ third water table shape coefficients, A is less than 1, 0.87 for an elliptical water table for instance

During recession stage (R(t)=0), as (1-A) is positive, Eq.(3) means that the drainflow rate becomes equal to zero for a water table height (H_f) equal to:

$$H_f = \sqrt{\frac{(\frac{1}{A} - 1)\,D_s\,L^2}{K_{he}(H_f)}} \qquad (4)$$

This has actually been observed in Arrou and in Bouillac in plots with deep seepage. In Arrou Guyon (1983) observed that the drainflow stops as the water table height is still 0.34 m above the drain pipe level. This value compares well with the value reckoned from Eq.(4) with the hydraulic conductivity of plots without deep seepage, 0.20 m. The discrepancy could partly be due to the unaccuracy of the hydraulic conductivity value as well as to the uncertainty of the water table height measured with the tubewells, these instruments having long time responses. In Bouillac the drainflow stops when the water table height is about 0.30 m above drain level in a plot with drainage efficiency of 30-50% (Zimmer et al., 1991).

The numerical solution of Eq. (2) - without deep seepage term - was at the origin of SIDRA model described by Lesaffre and Zimmer (1988). Deep seepage was introduced in SIDRA model in order to test its influence on drainage functioning. Results of Arrou's plot number 8 affected by deep seepage are presented. Numerical parameters calibrated during recession stage in plot number 1 (without deep seepage) were utilized; they are as follows. (1) equivalent hydraulic conductivity at the bottom of the plow layer: 0.5 m/d; hydraulic conductivity of the plow layer 2 m/d; (2) drainable porosity at the bottom of the plow layer and inside the plow layer: 3% and 5%; (3) exponent of the power function describing the decrease with depth of equivalent hydraulic conductivity and drainable porosity below the plow layer: 0.5 and 0.25; (4) deep seepage rate: 4.10^{-4} m/d. The model was run to simulate the whole intense drainage season of the 1982-1983 winter.

The fitness of the simulation is first checked (Figure 5) by use of double mass curves displaying cumulative simulated and measured drainage discharges versus cumulative net-recharges (P-ET). The water balance is well simulated except for two periods: (1) a 15 mm discrepancy follows a long lasting recession stage; this is due to unexplained bad simulation of the water table elevations during that period; (2) at the end of the drainage season the simulated discharges are lower than the measured ones which could be due to a lowering of the deep seepage rate.

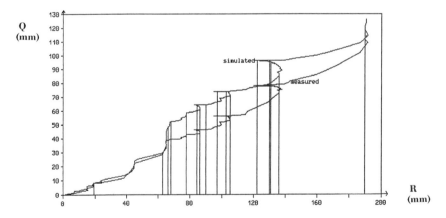

Figure 5. Cumulative drainage discharges (Q) versus cumulative net recharges (R) in Arrou's plot Nr.8 from 12.06.1982 until 04.11.1983.

Water table elevations during one representative drainage event of March 1983 (Figure 6) show that: (1) high water table levels and consequently high drainage discharges are hardly affected by deep seepage; (2) medium and low water table levels simulation is not adequate when deep seepage is not taken into consideration; (3) with deep seepage the simulation is much better although discrepancies are observed for low water table levels.

CONCLUSIONS

The quality of the simulation and of the model is hampered by at least two factors.

(1) <u>unaccuracy of soil hydraulic parameters</u>: parameters of plots with and without deep seepage may indeed differ; the problem is however the lack of a proper method for parameter calibration in case of deep seepage; one should also bear in mind that in case of dep seepage even tubewell measurements may not give accurate information on the real water table elevation in case of deep seepage;

(2) <u>unrelevant assumptions</u>: for very low water table levels the constant water table shape assumption is probably not valid since no more drainage discharges occur.

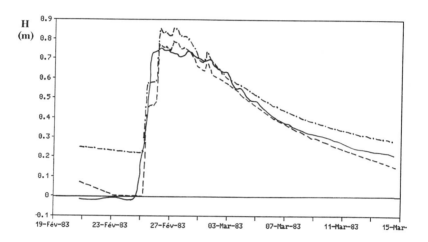

Figure 6. Hourly sequences of water table elevations in Arrou's plot Nr.8 from 02.19.1983 until 03.15.1983. —— measured values; - - - simulated values with deep seepage; —·—·— simulated values without deep seepage.

Nevertheless the following conclusions can be drawn.

High water table levels and drainflow rates are hardly affected by deep seepage. This result is valid at least for deep seepage rates that correspond to fairly low drainflow rates, about 20 times less than the drainage design rate in the cases studied herein. This means that *the drainage design rate should not be modified for such deep seepage rates*. As regards drain spacing calculation *a steady state approach will not result in any significant change* since deep seepage rate is in general low as compared to the drainage design rate. *In case of a transient state approach significant increases of the spacing can on the contrary be allowed* even for fairly low deep seepage rates.

The functional relation between water table heights and drainflow rates during tail recession is strongly modified: even for low deep seepage rates the drainflow is discontinued for water table levels of 0.10 to 0.30 m above drain level. This behavior could be relevant to check the existence of deep seepage in case of landfill drainage for instance.

Regarding drainage influence on aquifer recharge, the main result is that deep seepage rate does not seem to be related to water table levels since constant rates are generally observed throughout the drainage season. This also means that interferences between drainage and deep seepage rates are limited.

REFERENCES

1. Boussinesq J., 1904 - Recherches théoriques sur l'écoulement des nappes d'eau infiltrées dans le sol ; compléments. Jour. de mathématiques pures et applqiées, 10(1), 5-78 et 10(4), 363-394.
2. Dumm L.D., 1954 - Drain spacing formula. Agric. Eng., 35, 726-730.
3. Favrot J.C., Bouzigues R., Longueval C., Chossat J.C., Bourgeat F., 1990 - Structure du sol, drainage et environnement. Rôle de la structure du sol sur le fonctionnement hydraulique et la qualité des eaux d'un réseau de drainage. Actes du 14eme congrès de la CIID, Rio de Janeiro, Brésil, vol I-B, 367-379.
4. Guiresse A.M., 1989 - Drainage en sols de boulbènes. Relation entre les caractéristiques morphologiques et les propriétés hydrodynamiques des sols. Thèse de l'INP Toulouse. 177p + annexes.
5. Guyon G., 1961 - Quelques considérations sur la théorie du drainage et premiers résultats expériementaux. B.T.G.R., 52, 44p. + annexes.
6. Guyon G., 1980 - Transient state equations of water table recession in heterogeneous and anisotropic soils. Trans. ASAE, 23(3),653-656.
7. Guyon G., 1983 - Le périmètre expérimental de drainage d'Arrou. Aspects hydrauliques. Etudes du CEMAGREF, 5,45p.
8. Lesaffre B., Morel R., 1986 - Use of hydrographs to survey subsurface drainage networks ageing and hydraulic operating. Agric. Water Management seminar. Arnhem, The Netherlands. Balkema Ed., 175-189.
9. Lesaffre B., Zimmer. D., 1988 - Subsurface drainage peak flows in shallow soil. J. Irrig. Drain. Div., ASCE, 114(3), 387-406.
10. Van Schilfgaarde J., 1965 - Transient design of drainage systems. J. Irrig. Drain. Div., ASCE, 91(3),9-22.
11. Zimmer D., 1989 - Transferts hydriques en sols drainés par tuyaux enterrés. Compréhension des débits de pointe et essai de typologie des schémas d'écoulement. Thèse Univ. Paris VI. Etudes du CEMAGREF, Hydraulique Agricole, 5,321p.
12. Zimmer D., Bouzigues R., Chossat J.C., Favrot J.C., Guiresse A.M., 1991 - Importance et déterminisme des infiltrations profondes en luvisols-redoxisols drainés. Incidence sur les modalités de drainage. Science du Sol, 29(4),321-337.

WATER TABLE MANAGEMENT RAINSHELTER PROJECT RESEARCH

H.W. Belcher, T. L. Loudon, G.E. Merva[*]

ABSTRACT

This paper describes the capabilities of a unique water table control research facility under construction in central Michigan, USA. The facility includes rainfall control by lightweight movable buildings, a rainfall simulation system, and independent water table control for 32 small plots and 3 field scale plots. The underground pipe system allows the depth to the water table to be independently established from 0.3 m to 1.0 m for each plot. Each research plot has the capability to measure and sample overland flow, subdrain flow and water table elevation at 10 minute intervals. The small plots are instrumented to monitor the volumetric soil water content at 0.15 m increments to 0.60 m and soil temperature at 0.10 m, 0.30 m and 0.60 m depths. Root dynamics will be monitored in 16 plots by horizontal minirhizotron tubes.

KEYWORDS. Water Quality, Monitoring, Water Table Management, Water Table Control, Subirrigation, Rain Shelter

INTRODUCTION

The United States Department of Agriculture (1987), reports there are over 38 million acres of existing cropland that benefit from improved subsurface drainage in the United States. The improved drainage is achieved by installing underground pipe systems that remove excess water from the soil profile. Those drainage systems reduce the cost of producing crops. Sometimes they also provide a way for a part of the crop fertilizer to leave the field.

Subirrigation is a method of using the underground pipes to provide irrigation water when the fields are dry during summer months as well as provide drainage when the fields have too much water in the spring and fall. A subirrigation system is usually developed by modifying the existing subsurface drainage system.

Research conducted in Michigan and elsewhere shows subirrigation is a viable method of supplementing natural rainfall to reduce the cost of agricultural production. Recently Michigan researchers have initiated studies to evaluate the effect of subirrigation on water quality. Preliminary findings indicate water table management, by combination subirrigation and subsurface drainage systems, reduces runoff, erosion, and agricultural chemical movement to receiving surface waters (for example, loss of nitrate-nitrogen from the field has been reduced by 64% at one Michigan location and 58% at another compared to conventional subsurface drainage). However, it has become apparent that to develop the subirrigation system management guidelines that will optimize water quality benefits, greater control of research variables is necessary.

The United States Department of Agricultural (USDA) Cooperative State Research Service sponsored project, "Water Quality Impacts of Water Table Management" provides that control. Construction of a subirrigation/rainshelter research facility with capability to control the rainfall and water table for 32 small plots and three field scale plots was initiated in

---

[*] H.W. Belcher, Visiting Associate Professor, T.L. Loudon, Professor, and G.E. Merva, Professor, Dept. of Agricultural Engineering, Michigan State University, E. Lansing, Mi.

1991. The project will provide data that can be used to develop computer models that are used to formulate water table management system design and operation guidelines for a variety of crops, soils, and climatic areas. The project couples small plot research with field scale studies and demonstrations to validate the models and demonstrate the applicability of the models and guidelines to production scale agricultural units. The project supports research by a multi-disciplinary team of Michigan State University scientists from the Departments of Agricultural Engineering, Crop and Soil Sciences, and Agricultural Economics as well as scientists from Ohio, N. Carolina and Iowa. The research is being conducted in close cooperation with the USDA Soil Conservation Service, Cooperative Extension Service and Agricultural Research Service; the Michigan Departments of Agricultural and Natural Resources; and the Michigan Association of Conservation Districts, all of whom have participated in the planning for the project. For 1992, the team of scientists are conducting a comprehensive study of the homogeneity of the research area. The project work planned for subsequent years of the five year project will be to conduct research and demonstration directly related to the project objectives.

The research site was chosen because of soil type, level topography, and geographic location. The soil series at the site has been identified as Tappan (Fine-loamy, mixed, calcareous, mesic Typic Haplaquolls) by the USDA Soil Conservation Service. Tappan is one of the most extensive and agriculturally productive soils in Michigan. Tappan is also very suitable for water table management systems that include subirrigation.

RESEARCH OBJECTIVES

1. Evaluate the short and long term effect of subirrigation on:

    a. fate and transport of agricultural chemicals;
    b. soil properties and trafficability parameters;
    c. surface and subsurface water and sediment movement;
    d. biological dynamics of the root zone soil system;
    e. economics of production of field crops and vegetables;
    f. above and below ground plant development.

2. Through field research and simulation modeling develop nutrient, pesticide, tillage, residue and genotype management recommendations for agricultural producers and their advisors.

3. Share the research results with agricultural producers and their advisors, agricultural and environmental governmental agencies, and other subirrigation research oriented individuals and groups.

FACILITY CAPABILITIES

The 15 ha site was selected and construction began in August, 1991. Formal construction contracts were awarded for development of the small plot area, fabrication and installation of the rain shelter buildings and installation of a conventional subsurface drainage system within the 13 ha field scale area of the research site. The work required by those contracts is complete. Figure 1 provides a plain view of the site showing research facilities in place. Details of the small plot area is shown by the plan view in Fig. 2. A cross-sectional view of the small plot area is shown in Fig. 3.

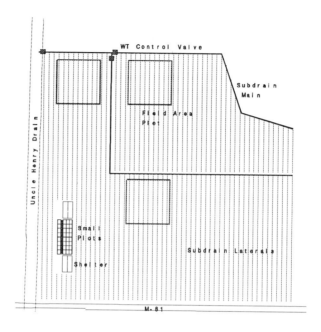

Figure 1. Plan view of research site showing research facilities installed in 1991.

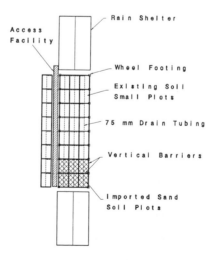

Figure 2. Plan view of small plot area at the rain shelter research site.

Rainfall Control

The knowledge gained from research designed to understand the processes and effects of water table depth and fluctuation is accelerated when rainfall is controlled. Rainfall control allows the researcher to gain knowledge for a dry year, normal year and wet year in a single growing season. Without rainfall control, the experiments must be repeated many years to achieve the same results.

The capability to control rainfall is provided by two light-weight movable buildings. The buildings are each 15.5 m wide by 25.6 m long with a clear height of 2.4 m. Together the buildings shelter 24 research plots. Between rainfall events, the buildings rest at the ends of the plot area. Rainfall causes hydraulically driven stationary wheels to rotate which drive the buildings toward each other. Cessation of rainfall causes the buildings to return to their resting location.

The hydraulic drive system is automatic with manual override. The amount of rainfall to trigger building movement and building travel distance can be modified as desired. In the future, a sprinkler system will be added to the building framework to provide irrigation when rainfall does not meet the research needs.

Small Plot Research Area

The 24 research plots with rainfall control and eight additional plots without rainfall control are each 4.5 m by 6.1 m in size. A bentonite/fabric curtain was installed around the perimeter of each plot. The curtains extend 0.3 m into the naturally occurring compacted till layer at 1.2 m depth.

A precast concrete access facility was installed underground along the length of 16 of the plots (eight of which are rain sheltered). The access facility is 51 m in length with a wall to wall width and floor to ceiling depth of 2 m. The access facility top will support the weight of a small farm tractor.

Each plot has a single 75 mm corrugated plastic drain pipe at 0..9 m depth and a surface water inlet, both of which outlet to the access facility. Each drain pipe outlet also serves as an irrigation water inlet. Instrumentation within the access facility provides capability to independently monitor the water table at four locations within each plot, monitor the underground pipe flow (drainage and irrigation) and overland flow from each plot and collect proportional overland flow and pipe composite flow samples. The water table control and water flow monitoring/sampling system is computer based and is a modification of the bubbler system described by Goebel (1986). The time between observations is user modifiable with a 10 minute minimum time interval. Instrumentation computers will be networked with capability for data access by a modem.

Field Scale Research Area

The field scale research area has, in place, a conventional subsurface drainage system. To provide independent water table control for three areas of the field; water table control structures, water table observation wells, and a water supply system will be installed and interfaced with a computer control system. The system will automatically monitor and control the water table independently within the three areas. Each area has a 0.5 ha subdivision for overland flow isolation. The water table control system includes capability to monitor underground pipe flow and overland flow from each area and sample those flows proportionally.

Figure 3. Cross-section of small research plots showing root observation access facility.

Climatological Monitoring Instrumentation

A commercial weather station with the capability to monitor daily rainfall, incoming solar radiation, wind speed and direction, relative humidity and daily minimum, maximum and mean air temperature has been installed at the site. It has the capability to interface with on site and off site computers by modem. In addition to the central weather station, up to four rain gauges with capability to monitor precipitation hourly are located at other locations within the research area.

On Site Computer Network and Data Base Management System

A central personal computer at the rain shelter site has been dedicated to data collection and storage. The system is networked with other computers and data loggers on site to allow automatic transfer of data to the central computer. A data base management system (DBMS) to support the scientific work at the Saginaw Bay rain shelter has been developed. The system will allow project researchers and other scientists efficient access to collected data.

The rain shelter site computer provides researchers a secure location to store the results of their work. The automatic updating of the central data base from automatic data loggers at the site will provide a backup to the storage function of those individual machines. Researchers doing chemical, plant tissue or other analysis off site will, through the use of high speed modems, be able to include their data in the central data base.

After data are stored, retrieval in a convenient and timely manner is critical. Researchers need to be able to monitor the progress of their experiments and the functioning of their equipment frequently. The system's capability to allow remote access provides the opportunity for monitoring performance of the on site data collection as well as retrieving data for analysis.

Water Sampling

Samples of rainfall, irrigation water, subdrain flow and overland flow are analyzed for nutrients and pesticides. Drainage samples are analyzed for total phosphorous, orthophosphate phosphorous, nitrate-nitrogen and ammonium-nitrogen. All surface runoff samples are analyzed for the same parameters plus total kjeldahl nitrogen and sediment.

Samples are analyzed for the herbicide used.

To quantify the effects of water table management on the mobilization of organic and inorganic colloids and on colloid-enhanced transport of nutrients and pesticides, subsamples of selected subdrain and runoff samples are collected. These samples are analyzed for clay mineralogy and carbon (C) content. The < 0.45 micron size fraction is frozen immediately on site and retained for ultrafiltration and analysis of C and inorganic materials. Soluble C is determined using a soluble C analyzer. Inorganic colloids are analyzed for mineralogy and elemental composition. Colloid size fractions are made available to other collaborators for analysis of colloid-bound nutrients and selected pesticides. This research provides a baseline for monitoring change in soil structure over the project time span in terms of change in movement of colloids and solubilized cementing agents.

Soil Water Monitoring

A key to the success of subirrigation is the proper management of the water table both for crop growth and water quality. Specific knowledge about the flow characteristics of the soil and water content changes over time are essential to the design of sustainable subirrigation management systems. It is also essential in the modification and verification of crop growth and water quality models needed to evaluate management alternatives. Volumetric water content will be measured in all 32 small plots using time domain reflectometry (TDR). TDR offers the potential for the automated, high quality, non-destructive measurement of soil water contents at time intervals not before possible (Baker and Allmaras, 1990; Topp et al., 1980). The TDR system has the potential to determine water flux rates (Topp et al., 1983). Therefore, these data will contribute to the quantification of the flow characteristics of these soils along with other proposed measurements.

TDR moisture probes consist of two stainless steel rods spaced 5 cm apart connected to a cable interfaced to a multiplexer port. Each plot will have a set of TDR probes installed at depths of 0-0.15, 0.30, 0.45 and 0.60 m for a total of 4 probes per plot. The small plots will also be instrumented with neutron probe access tubes (as will the field area plots) for determination of volumetric water content using a neutron probe. Neutron probe readings will allow for monitoring while the TDR system is being assembled and installed, will serve as a back-up system to the TDR and allow measurement of water contents at deeper depths.

Below Ground Plant Development

Nondestructive and continuous developments in plant root growth, branching, maturation, death and disappearance will be monitored at frequent intervals by the minirhizotron and microvideo camera methods (Smucker, 1990). Transparent minirhizotron tubes, 50 mm in diameter, have been installed at three depths in each of the 16 plots adjacent to both sides of the underground access facility. Eight plots receiving the minirhizotrons are located outside the confines of the shelter with the remainder adjacent to the shelter area. Each minirhizotron tube was installed horizontally through access ports in the precast concrete walls of the access facility, through the bentonite/fabric barrier and 2.3 m into the plots.

Two stepper motor-controlled video cameras have been custom designed for these experiments. Automated measurements of root intersections of the minirhizotron tubes are needed to save on labor costs during the numerous field observations and to video record continuous root changes during soil wetting and drying processes. These data will provide the specific root branching, dying and growth of roots by plants subjected to the multiple treatments available to the project. Video recorded images will be quantified by a computer image analyzer in the Root Image Processing Laboratory at Michigan State University (Smucker, 1990 and 1991). Specific root branching and turnover rates will be the primary measurements made during these studies. Additional morphological parameters can also be extracted by the image analysis computers.

Soil Temperature Instrumentation

Soil temperature is the predominant controlling factor of soil biological activities in the Saginaw Valley region of Michigan. This is partially true in the early Spring and late Autumn. Additionally, seasonal and diurnal thermal gradients greatly influence most of the chemical and biological transformations of nitrogen, carbon and many of the synthetic organic compounds which are potentially hazardous to the ground water. Soil thermal data is also essential for interpreting plant root responses to spatial and temporal soil water regimes. The rates of root growth, root turnover, plant residue decomposition and soil respiration are all directly controlled by thermal gradients in the upper 60 cm of the soil profile. Once soil temperature gradients are known, for the duration of the growing season and during the fallow periods before and following the cropping season, then the rates of ion flow into the ground water can be combined with soil water data and incorporated into computer algorithms which accurately predict ion movements and retention within the soil.

Duplicate thermocouples will be inserted into the soil at 10 and 30 cm below the soil surface. Each plot will have four thermocouples which are uniformly spaced within the total volume of soil. Each thermocouple will be connected to the central computer by a network of transmission lines through the access tubes at the corner of each plot. Data loggers will be used to collect the thermal data at hourly intervals. Accumulated data will be dumped into the central computer and held for processing or forwarded to the laboratories at Michigan State University.

ACKNOWLEDGEMENT

The following scientists are providing direct research support to the project:

From the Agricultural Engineering Department, Michigan State University:
Dr. H. W. Belcher, Dr. T. L. Loudon, Dr. G. E. Merva

From the Agricultural Economics Department, Michigan State University:
Dr. S. M. Swinton

From the Crop and Soil Sciences Department, Michigan State University:
Dr. S. J. Anderson, Dr. J. R. Crum, Dr. O. B. Hesterman, Dr. F. J. Pierce, Dr. K. A. Renner, Dr. J. T. Ritchie, Dr. A. J. M. Smucker, Dr. M. L. Vitosh

The following research scientists have participated in planning the project and have committed to cooperating in the research effort:

Dr. N. Fausey, USDA, Agricultural Research Service
Dr. R. Kanwar, Iowa State University, Agricultural Engineering Department
Dr. W. Skaggs, North Carolina State University, Agricultural Engineering Department

REFERENCES

Baker, J. M. and R. R. Allmaras. 1990. System for automating and multiplexing soil moisture measurements by time-domain reflectometry. Soil Sci. Soc. Am. J. 54:1-6.

Smucker, A. J. M. 1990. Quantification of root dynamics in agroecological systems. In: V. S Goeld and J. M. Norman (eds.) Instrumentation for Studying Vegetation for Remote Sensing in Optical and Thermal Infrared Regions. Remote Sensing Review 5: 237-248.

Smucker, A. J. M. 1991. Nitrogen and tillage modifications of corn root dynamics in a stratified loam soil. Agron. J. (submitted).

Topp, G. C., J. L. Davis, and J. H. Chinnick. 1983. Using TDR water content measurements for infiltration studies. In Advances in Infiltration. ASAE, ST. Joseph, MI. Publ. No. 11-83:231-240.

Topp, C. C., J. L. Davis and A. P. Annan. 1980. Electromagnetic determination of soil water content: Measurements in coaxial transmission lines. Water Resour. Res. 16:574-582.

USDA, 1987. Farm drainage in the United States: History, Status and Prospects. Miscellaneous Publication No. 1455. ESDA. Washington, D.C.

Water Conveyance Capacity Of Corrugated Plastic Drain Pipe with Internal Diameters Between 38 and 75 mm

R.S. Broughton[1] B.W. Fuller[1] R.B. Bonnell[1]
Member CSAE,ASAE   Member CSAE,ASAE

ABSTRACT

This paper presents results of a test performed to determine the water conveyance capacity of some 38mm, 50mm and 75mm diameter corrugated PE (polyethylene) drain pipes. The data points were plotted on graphs of discharge vs. hydraulic gradient. The data were then analyzed to determine the best fit to the curves $Q=KS_e^{1/2}$ and $Q=KS_e^b$. A Manning's roughness coefficient, n, was determined for each pipe.

Introduction

Smaller diameter (38mm-75mm) corrugated PE pipes are currently being used for intensive drainage of golf course fairways, athletic fields, and horticultural fields. One of the important design features for the drainage of these areas is the water conveyance capacity of the corrugated plastic pipe used. The capacities of some European manufactured 38mm-75mm corrugated PVC (polyvinyl chloride) pipes have been determined (Wesseling and Homma, 1967; Hermsmier and Willardson, 1970) as well as a few Canadian manufactured pipes of 80mm diameter or larger (Ami et al., 1978). Irwin (1982) reported the results from flow tests on pipes from 35 mm to 300mm I.D., with the small diameter pipes being of European manufacture.

---

[1] R.S. Broughton, Professor, B.W. Fuller, M.Sc. student and R.B. Bonnell, Research Associate, Agricultural Engineering Dept., McGill University, Montréal, Canada.

The research reported here was conducted to determine the functional water conveyance capacities of some newer small diameter (38mm, 50mm, 75mm) corrugated PE pipes. This included the determination of a Manning's roughness coefficient, n, for each diameter tested.

Materials and Methods

The tests were executed outdoors near the Agricultural Engineering/Soil Science Shop. The water was supplied by a fire hydrant. The water flowed past a riser into the corrugated pipe and finally into the discharge reservoir. A schematic of the test configuration is illustrated in Fig.1. The measurements were taken from full

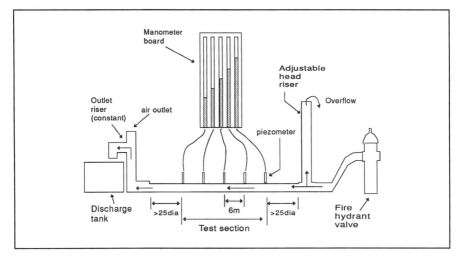

Figure 1: Experimental setup

flowing 30-35 metre lengths of non-perforated corrugated polyethylene pipes. The non-perforated pipes were assumed to have the same roughness factor and full water conveyance as perforated pipes of the same diameter and configuration. The test section was separated by a length of more than 25 diameters from any region of flow disturbance (see Fig.1). This was to avoid any effects of junction turbulence on the flow in the test section.

Five piezometer stations were installed at approximately 6 metre intervals along the test section of the pipe. Clear PVC pipes (6mm I.D.) were used to connect the piezometer taps to a manometer board. The piezometer stations were made with 4 openings spaced evenly around the pipe circumference on the crest of the corrugation. They were placed to the sides of the pipe to avoid the introduction of any stray air bubbles (see Fig.2). Final distances between the piezometric stations was measured after the piezometers were installed since the corrugation pitch usually did not allow this distance to be exactly 6 metres apart.

Pressure readings were observed from the manometer board. For low flows, the manometer board was tilted at 45° to provide greater accuracy.

The total length of the test section was approximately 24 metres. However, it was the middle 12 metres that was used for the evaluation of the hydraulic gradient. The other stations were used to check for any irregularities in the head loss along the tested pipe length.

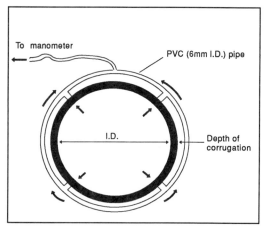

Figure 2: Cross-section of piezometer station

The discharge rate was controlled by adjusting the height of the riser. The riser established the upstream head. With time, the downstream head settled to a steady-state position, implying that the flow was fully established (constant). Any excess flow from the hydrant spilled over the top of the upstream riser. Thus, only the fire hydrant valve and the riser were used to regulate the flow.

To measure discharge, the discharge tank was calibrated for volume differences of 20 and 40 litres. Thus, the discharge could be determined by measuring how long it took to fill a certain volume of the tank.

Air was bled out of the piezometer lines with a high discharge before proceeding with the observations. The hydraulic gradients were adjusted from 0.02% to 6.0% and the discharges measured. The water temperature was noted during the tests so that the water viscosity could be determined from published data. The Reynolds numbers, *Re*, were calculated from the average water velocity, pipe inside diameter, and water viscosity.

Seven pipes from three different pipe manufacturers were tested. Pipe nominal diameters ranged from 38mm to 75mm. The measured dimensions of the different pipes are given below in Table 1 (also see Fig.3 and Fig.4).

Manufacturer A's 75mm pipe had a corrugation profile with an asymmetrical step profile (see dimension e in Fig.3). In order to determine whether there was any difference due to the flow direction with respect to the step orientation, this pipe was tested in both directions. The two directions are defined in Figure 4.

The range of Reynolds numbers over which the experiments occurred are noted in Table 2, along with the range of the approximate roughness factor, defined as h/(I.D.).

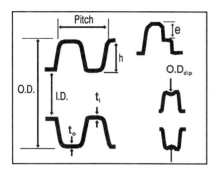
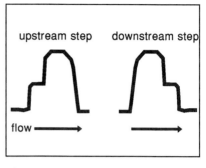

Figure 3: Pipe dimensions measured

Figure 4: Definition of step profile with respect to flow direction.

Table 1: Pipe dimensions

Pipe Manufacturer	Nominal Diameter	O.D. (mm)	O.D$_{dip}$ (mm)	I.D. (mm)	Pitch (cm/corr)	t$_o$ (mm)	t$_i$ (mm)	h (mm)	e (mm)
A	38mm	50.06	n/a	38.27	0.851	0.60	1.05	5.57	n/a
	50mm	60.24	n/a	51.84	0.850	0.54	0.88	3.66	n/a
	75mm	88.83	n/a	74.17	1.275	0.59	1.17	6.74	3.42
B	38mm	48.36	n/a	38.74	0.785	1.08	0.94	3.73	n/a
	50mm	58.83	n/a	50.06	1.041	0.64	0.73	3.79	n/a
	75mm	84.11	82.56	73.35	1.342	0.74	1.13	3.87	n/a
C	50mm	62.73	n/a	51.51	1.004	0.64	0.95	4.97	n/a

Table 2: Range of Reynolds numbers and roughness factors in experiments.

Pipe Nominal Diameter	Re$_{min}$	Re$_{max}$	Min. h/(I.D.)	Max. h/(I.D.)
38mm	2047	19980	0.096	0.146
50mm	3146	41119	0.071	0.096
75mm	3106	64179	0.053	0.091

Analysis of Data

The discharge of a steady flowing water pipe can be determined by Manning's equation:

$$V = \frac{R^{2/3} S_e^{1/2}}{n} \tag{1}$$

where: V = average velocity of the water in the pipe (m/s)
  R = hydraulic radius of the pipe (m)
  S_e = hydraulic gradient
  n = Manning's roughness coefficient (s/m$^{1/3}$)

Equation 1 can be altered so that it directly relates discharge and hydraulic gradient.

$$Q = AV = KS_e^{1/2} \qquad (2)$$

where: Q = the discharge from a full-flowing pipe (m^3/s)
  A = the flow area of the pipe (m^2)
    = $\pi(I.D.)^2/4$
  K = a constant for a given pipe (m^3/s)
    = $(AR^{2/3})/n$

In his studies of water conveyance in corrugated plastic pipes, Pelletier (1984) found that the discharge could be related to the hydraulic gradient by:

$$Q = KS_e^b \qquad (3)$$

where: b = a constant

Equation 3 provides more flexibility for analyzing the differences in water conveyance capacity which are a result of different pipe designs and diameters (Reynolds numbers). It can be that, for a pipe with constant inner diameter (*I.D.*), Eq.2 and Eq.3 are identical when $b=½$. Thus, the analysis can be continued by working with Eq.3.

If the logarithm of Eq.3 is taken, it becomes:

$$\log Q = b \log S_e + \log K \qquad (4)$$

which is similar in form to the equation of a straight line.

$$y = c_1 x + c_2 \qquad (5)$$

where: y = $\log Q$
  x = $\log S_e$
  c_1 = the slope of the line = b
  c_2 = the y-axis intercept = $\log K$

Two approaches were used for data analysis. First, the logarithms of the data points were taken and a linear regression was performed with a fixed slope ($b=½$).

The constant K was determined from the y-axis intercept of the best-fit line. Then a second linear regression was performed, with both the slope b and the constant K determined by the best-fit line.

Finally, it was desirable to know Manning's roughness coefficient, n, for use in Manning's equation (Eq.1). The coefficient was determined by rearranging the definition of K.

$$n = \frac{AR^{2/3}}{K} \qquad (6)$$

The calculated Manning's roughness coefficients and the constants from the two linear regressions are presented in Table 3. The characteristic curves for $Q=KS_e^{1/2}$ and $Q=KS_e^b$ can be seen in Fig.5 and Fig.6 respectively.

Table 3: Results of linear regression analysis of Discharge-Hydraulic gradient relationships

Pipe Manufacturer	Nominal Diameter	Manning's n	$Q = KS_e^{1/2}$	$Q = KS_e^b$	
			K	K	Slope, b
A	38mm	0.0203	0.00256	0.00282	0.522
	50mm	0.0142	0.00821	0.00764	0.521
	75mm (step upstream)[a]	0.0155	0.01952	0.01698	0.473
	75mm (step downstream)[a]	0.0159	0.01904	0.02644	0.568
B	38mm	0.0163	0.00327	0.00327	0.500
	50mm	0.0165	0.00646	0.00576	0.477
	75mm	0.0154	0.01912	0.01825	0.491
C	50mm	0.0153	0.00746	0.00993	0.554

[a] see Fig.4

It is noted that there is very little difference in the water conveyance capacity for the 75mm pipe with the step facing the upstream versus facing downstream. Therefore, flow direction relative to the step is not important.

The authors suggest that the Manning's roughness coefficients presented in Table 4 might be appropriate for design purposes when using small diameter corrugated PE pipes.

Table 4: Suggested Manning's Roughness Coefficients, n, to be used in design work.

Pipe Nominal Diameter	Manning's roughness coefficient, n
38mm	0.018 - 0.020[a]
50mm	0.016
75mm	0.016

[a] suggested design n value for deeper corrugations (Pipe Manufacturer A)

Conclusions

The experimental apparatus produced good results. The R-squared values for the best-fit lines, $Q=KS_e^b$, were all 0.99 or better. This suggests that Fig.6 could be used with confidence to give the full flow water conveyance capacities of these pipes for a wide range of hydraulic gradients.

References

Ami, S.R. and R.S. Broughton, A.M. Shady. 1978. Designing and instilling of subsurface drainage laterals less than 100mm in diameter. Can. Agric. Eng. 20:16-19.

Hermsmier L.F. and L.S. Willardson. 1970. Friction factors for corrugated plastic tubing. Journal of Irr. Drainage, Proc. ASCE, 96:265-271.

Irwin R. 1982. Hydraulic roughness of corrugated plastic tubing. Proceedings of 2nd International Drainage Workshop, Washington, D.C. U.S.A., December 5-11, 1982.

Pelletier, M.A. 1984. Roughness factors and water conveyance capacities of corrugated plastic tubing. M.Sc. Thesis, Macdonald Campus, McGill University, Montréal.

Wesseling J. and F. Homma. 1967. Hydraulic resistance of drain pipes. Bull. 50, Neth. J. Agric. Sci. 15.

Acknowledgements

The authors acknowledge the financial support of the Canadian National Sciences and Engineering Research Council and the companies which donated the pipes, namely Plastidrain Inc., Big 'O' Inc., and Soleno Inc. The authors are also thankful for the help of Stamatia Gazetas and Ray Cassidy in assembling apparatus and assisting with measurements.

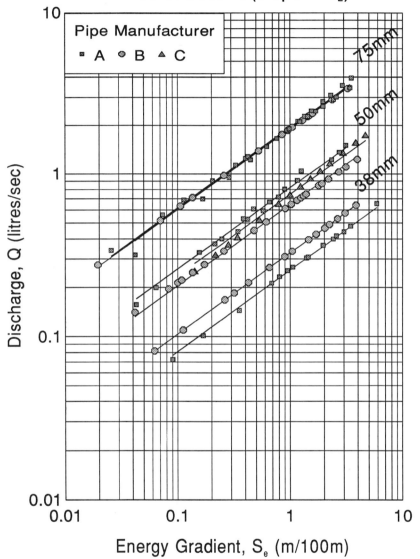

Figure 5: The full-flow water conveyance - hydraulic gradient relationship with fixed slope of ½ for pipes of nominal diameters: 38mm, 50mm, 75mm.

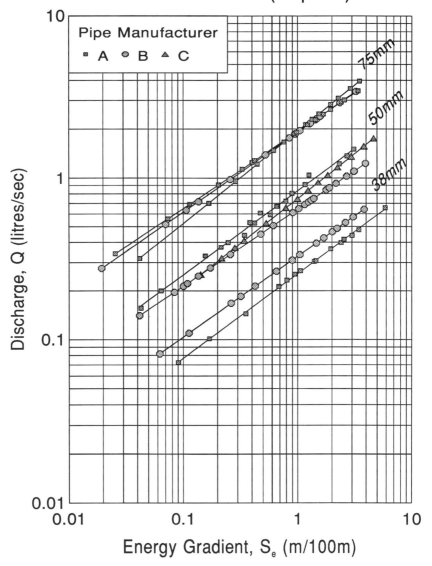

Figure 6: Full-flow water conveyance - hydraulic gradient relationship with slope=b (any constant) for pipes of nominal diameters: 38mm, 50mm, 75mm.

WEATHERING TESTS ON U-PVC AGRICULTURAL DRAINS : AN ARTIFICIAL VERSUS A NATURAL TEST

J. C. Benoist[*]   P. Beccaria[*]

ABSTRACT

An accelerated ageing test (AAT) was developed to check the resistance of U-PVC (unplasticized polyvinyl chloride) agricultural drains to weathering more quickly than using the natural weathering test (NWT) and more cheaply than other types of AAT. This test reaches satisfactory correlation with NWT. It was feasable with a low pressure mercury steam lamp sending UV rays from 275 to 365 nm with a power of 0.7 mW/cm^2, at 38 °C and 60 % moisture during 168 hours. If the products pass the test, they can be stored outside in EEC 9 months without protection.

Keywords : U-PVC agricultural drain pipe, natural weathering test, artificial ageing test.

INTRODUCTION

After the elaboration of the first reliable artificial test to guess the resistance to weathering of high density polyethylene tubing in storage (Desmond and Schwab, 1986), CEMAGREF developed an adapted method for U-PVC drain pipe. Collecting knowledge on the weathering of this material (Fougea, 1970; Lemaire et al., 1979 & 1981; Aumasson, 1983), it applied it to simulate a natural weathering test already standardized for agricultural drain (AFNOR, 1980). Due to the reduced growth of agricultural drain market, it also made an effort to find a simple and cheap apparatus. Finally, a satisfactory correlation was compulsory so that every manufacturor accepted that his products which have a particular formula and/or corrugation can be tested by the same method.

The purpose of this article is to present the choice of exposure parameters of NWT and AAT and the results of their correlation.

TESTING METHODS

The natural weathering test

The principle of the NWT is to expose products outside during the warmest and sunniest seasons. 100 samples of length 200 mm are laid on a frame in such a way that they keep their coil curvature (AFNOR, 1987). After exposure, they are submitted to an impact test (AFNOR, 1986) on the exposed generating line. The required breakage rate should not exceed 10 % (AFNOR, 1981). As control lot, 100 other sheltered samples are also impact tested. The exposure energy includes all sun radiation - IR, visible and UV - and is expressed in GJ/m^2.

---

[*] J. C. Benoist, standardization technician, and P. Beccaria standardization engineer, Drainage Division, CEMAGREF (Institute of Agricultural and Environmental Engineering Research), Antony, France.

Basic assumption, apparatus and calibration of the AAT

A critical period of the NWT has a certain variation of the combined natural weathering factors. The basic assumption to define the U 51-159 AAT was that the sum of their effects can be reproduced by a constant artificial combination.

The apparatus were built and regulated (AFNOR, 1990) to keep :
- the exposure to UV rays between 275 and 365 nm with a power of 0.7 mW/cm^2 ± 7 % using a low pressure mercury steam lamp ;
- the temperature at 38°C ± 1°C ;
- the relative humidity at 60 % ± 10 %.

Proportional to the time, the AAT exposure level is then expressed in hours. Compared to the Weatherometer xenon-arc for instance (ASTM G26-77), its light spectrum is less faithful and there is no water cooling. Unlike the Xenotest, AAT has no variation of the factor combination and so is more than 20 times cheaper.

The AAT was then carried out with the following durations : 50, 100, 150, 200 and 250 hours. It is between 150 and 200 hours that 10 % of the exposed samples break when tested. The AAT duration ranges then between 6 and 9 days which is much shorter than NWT.

CORRELATION BETWEEN THE NWT AND THE AAT

First testing period

During the first testing period, NWT of 1.75 and 3.14 GJ/m^2 were respectively compared with AAT of 150 and 200 hours. The 1.75-GJ/m^2 NWT corresponds to an exposure of 6 months from 1st July to 31th December 1987. The experiment carried on until 1st June 1988 so that the received energy reached 3.14 GJ/m^2. This figure lies between 2.9 GJ/m^2 and 3.4 GJ/m^2 which respectively correspond to 6 and 9 months of the warmest and sunniest days of the French climate. Seven products having nominal diameters (DN) of 50 mm or 65 mm were tested.

High correlation coefficients were calculated (Diag. 1 & 2 hereafter) : **0.87** for the pair NWT(1.75 GJ/m^2)/AAT(150 hours) and **0.80** for the pair NWT(3.14 GJ/m^2)/AAT(200 hours). These coefficients are significant for a risk threshold of 5 % but not for one of 1 % (Snedecor & Cochran, 1971, table VA, quoting R. A. Fisher). The correlation coefficient of the first pair is the best and misses being significant at 1 % by only 0.014. The standard deviations of the regression line parameters are also better for the first pair. Finally, the average breakage rates are nearer for the first pair (11.29 % & 10.86 %) than for the second (20.43 % & 26.14 %)(Tables 1 and 2 hereafter).

Second testing period

The objective of the second testing period was to find exposures for the NWT and the AAT which could be more convenient than those of the first campaign and that yielded a good correlation between them. Firstly, the NWT ought to be carried out from 15th February to 15th November because the received energy between these two dates is the nearest to 3.4 GJ/m^2. In 1988 during this period, the received energy was 3.43 GJ/m^2. Secondly, an AAT shorter than 200 hours is preferable so that the breakage rate could be nearer to that of the NWT(3.4 GJ/m^2). A period of exactly one week, i. e. 168 hours, was tested as it is convenient for the

Table 1 - Results of the NWT(1.75 GJ/m²) and the AAT(150 hours)

Product reference	1	2	3	4	5	6	7	Average
Nominal diameter	50	65	50	50	50	50	50	
% breakage :								
Control lot	8.5	4.5	7.5	4.5	0	3.5	6	4.9
x1=NWT(1.75 GJ/m²)	18	2	19	18	1	10	11	11.3
y1=AAT(150 hours)	17	1	26	15	3	11	3	10.9

Diagram 1 : Relation between NWT and AAT breakage rates

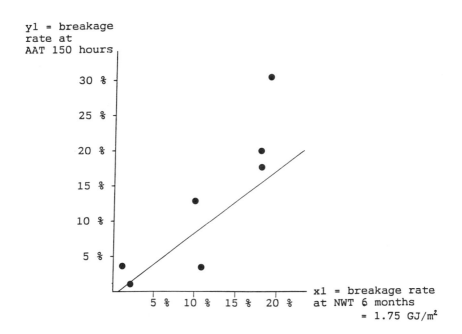

Correlation coefficient :
r = **0.8653** (Threshold 5 % = 0.754, threshold 1 % = 0.874)

Regression line :
Constant = - 0.983 (σ = 5.038)
Coefficient = 1.049 (σ = 0.272)

Table 2 - **Results of NWT(3.14 GJ/m²) and AAT(200 hours)**

Product reference	1	2	3	4	5	6	7	Average
Nominal diameter	50	65	50	50	50	50	50	
% breakage :								
Control lot	8.5	4.5	7.5	4.5	0	3.5	6	4.9
x1=NWT(3.14GJ/m²)	40	2	43	27	4	16	11	20.4
y1=AAT(200 hours)	23	12	63	38	2	27	18	26.1

Diagram 2 : **Relation between NWT and AAT breakage rates**

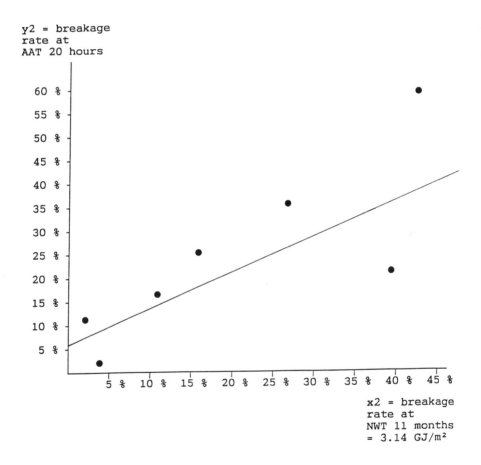

Correlation coefficient :
r = **0.7992** (Threshold 5 % = 0.754, threshold 1 % = 0.874)

Regression line :
Constant = 6.643 (σ = 13.053)
Coefficient = 0.955 (σ = 0.321)

Table 3 - Results of NWT(3.34 GJ/m²) and AAT(168 hours)

Product reference	8	9	10	11	12	13	14	15	16
Nominal diameter	50	50	50	50	50	50	65	65	65
% rate :									
Control lot	4	9	1	2	0	8	2	0	0
x3 = NWT(3.34 GJ/m²)	19	19	4	1	0	13	3	0	1
y3 = AAT(168 hours)	12	18	5	11	1	28	9	3	0

Product reference	17	Average
Nominal diameter	65	
% breakage		
Check lot	7	3.3
x3 = NWT(3.34 GJ/m²)	25	8.5
y3 = AAT(168 hours)	28	11.5

Diagram 3 : Relation between NWT and AAT breakage rates

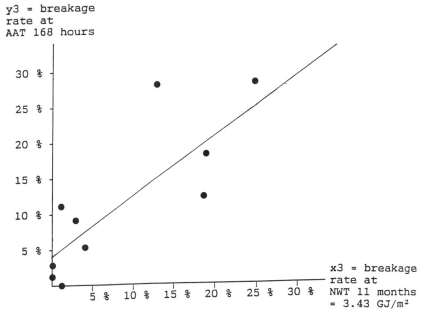

Correlation coefficient :
r = **0.809** (Threshold 5 % = 0.632, threshold 1 % = 0.765)

Regression line :
Constant = 4.098 (σ = 6.406)
Coefficient = 0.871 (σ = 0.224)

management of the testing laboratories.

The manufacturers were informed of the results of the first period so they could improve the resistance to weathering in order to obtain a breakage rate of less than 10 %.

Using 10 products of DN 50 and DN 65, the second period gave the expected results. The correlation coefficient is **0.81** (Diag. 3) which is rather lower than the best previous period. However due to a larger number of products, it is significant with a risk threshold of 1 %. The standard deviations of the regression line parameters are nearly the same as those of the first period (NWT 1.75 GJ/m^2/AAT 150 hours). AAT and NWT breakage rates - respectively 8.5 and 11.5 % - are close to each other and to the critical rate (Table 3).

For the second period, the overall weathering resistance is higher. For an energy level equivalent to that of the second part of the first testing period, the average breakage rate halves. The results of the control lot are also better (Tables 2 and 3, average breakage rates).

Furthermore, the four products whose breakage rate is higher than 10 % for the NWT (products n° 8, 9, 13 and 17) have a breakage rate higher than 10 % for the AAT as well. Three of them (products n° 9, 13 and 17) already have a check lot whose breakage rate is higher than 10 %. These three products generally exit the overall average rate of just over 10 % (Table 3).

Dispersion of the results

A dispersion around the correlation lines is observed mainly due to the impact test inaccuracy. It is not possible to choose another test to determine the natural weathering or the accelerated ageing resistance as it is the decrease in impact resistance which is the main quality problem due to weathering.

The rather sophisticated shape of a drain causes the variability of impact test results as well as the irregularity of its production quality. 100 vertical impacts with a precise definition of failure raises the problem of the confidence interval of binomial distribution (Snedecor & Cochran, 1971, p 6).

Apart from the mechanical inaccuracy and the random dispersion, the colour could also partially explain some cases. Indeed, the products 3 (Diag. 1) and 4 (Diag. 2) which are grey are situated above the regression line. Because of a higher surface temperature during the NWT, there is a distorsion of the correlation. On the contrary the products 1 (Diag. 2) and 8 (Diag. 3) which are yellow have better results with the AAT and then are under the regression line. Other recipe elements make possible that grey, yellow and even white pipes have well correlated results. Anyway, the standards are supposed to select products according to the results and not according to the means of their production.

This problem of accordance between the NWT and the AAT was actually raised for three products. Product 2 in 1st testing period-part 2 and product 11 passes the NWT (respectively 2 & 1 % breakage) and narrowly fails the AAT (respectively 12 & 11 % breakage). On the contrary, product 7 in 1st testing period-part 1 is largely passes the AAT (3 % breakage) but also just fails the NWT (11 % breakage). In French certification practice, one or two additional impact tests would be done in the case of narrow failure to palliate the distorsion between the NWT and the AAT. This is already done for single impact test NF U 51-154. This implies that 3 x 100 samples would need to be compulsory exposed for NWT and that an AAT could sometimes last three weeks.

Possible confirmations and improvements

Additional testing periods could be done. Increasing the number of tested products is also possible using those from factories outside France. In this way, the significance of the correlation could be confirmed.

Regarding the apparatus and method, the impact test should particularly be changed. The staircase method (ISO, 1991) could solve the both mechanical and random problems. Firstly, this method makes impacts only on the weakest points of the drain pipe i. e. seam and perforation lines. Secondly, realized by a single series of 36 impacts, this method requires then one third of the impacts of NF U 51-154 and avoids additional series. Charpy's method (AFNOR, 1976) measures the kinetic energy absorbed in breakage using a striker mounted on a pendulum. So, each impact records a value reducing greatly the random dispersion.

Secondarly, it is possible to exclude 275 to 290-nm rays to reproduce more faithfully the sun radiation (Lemaire et al., 1981). For this purpose, a suitable filter or another lamp can be chosen. This should improve the correlation with the NWT. Improvement to the AAT regulation system could also be made.

Nevertheless, all these improvements are more intricate and, except for the staircase method, more expansive. The interested parties would have to choose a good compromise between the price, the accuracy and the convenience of the specification mode.

New opportunities offered by the AAT

It is difficult to manage the NWT between several countries even on same land mass. The climates are indeed too different to apply the same exposure duration and/or the same specification. This issue of environment causes problem for trade between countries. Taking into account more precisely the factors, AAT could be a means to research other ways so as to define another minimal requirement which could be valid for a wider area. Relevant investigations on UV rays, temperature and moisture in several countries and adequate discussions on reasonable storage conditions could be done in this perspective.

Moreover because of its short duration, AAT can be used to check the quality of extrusion in parallel with melt flow rate (ISO, 1981, 1991) which checks the material fitness for extrusion.

CONCLUSION

As standardized in document U 51-159, the AAT is cheap and rapid. These three first testing periods show that it appears to be a satisfactory test. Despite the technical accuracy and random dispersions of the impact test, the correlation is good. The AAT maybe needs confirmation and improvement but makes it possible to improve the quality checks on PVC-U drain pipes.

REFERENCES

1 AFNOR. 1976. *NF T 51-035, Matières plastiques - Détermination de la résistance au choc - Méthode Charpy*. Paris. 12 p.

2 AFNOR. 1981. *NF U 51-101, Drainage agricole - Spécification*. Paris. 5 p.

3 AFNOR. 1st edition. 1980. *NF U 51-158, Drainage agricole - Essai de vieillissement naturel*. Paris. 7 p. 2nd edition. 1987.

4 AFNOR. 1986. *NF U 51-154, Agricultural drainage - Impact resistance*, (translation). Paris. 5 p.

5 AFNOR. 1990. *U 51-159, Drainage agricole - Essai de vieillissement accéléré*. Paris. 12 p (experimental standard).

6 Aumasson M.. 1983. Le vieillissement solaire, la lumière et ses effets. *Revue Générale des Caoutchoucs et plastiques*. (631). Paris. pp 91-94.

7 ASTM. 1977. *ASTM G26-77, Practice for operating light exposure apparatus "Xenon-arc type" with and without water for exposure of non-metallic material*. Philadelphia. 9 p. (new version : ASTM G26-90)

8 Desmond E. D. & Schwab G. O..1986. Ultraviolet degradation of corrugated plastic tubing. *Transactions of the ASAE*. Vol. 29(2). St Joseph - MI (USA). pp 467-472.

9 Fougea D.. 1970. Essai de caractérisation du vieillissement naturel des polyclorures de vinyle. *Cahier du CSTB*. Vol 925 (107). Paris. 12 p.

10 Gardette J. L. & Lemaire J.. 1987. Acquis récents dans l'étude du vieillissement du PVC. *Revue Générale des Caoutchoucs et Plastiques*. (652). Paris. pp 133-137.

11 ISO CD11173. 1991. *Thermoplastic pipes - Determination of resistance to external blows - Staircase method*. Geneva. 9 p.

12 ISO 4440. 1981. *Polyethylene pipes and fittings - Determination of melt flow rate*. Geneva. 2 p.

13 ISO DIS 4440. 1991. *Polyethylene pipes and fittings - Determination of melt flow rate - Part 1 : Test method - Part 2 : Test parameters*. Geneva.

14 Lemaire J., Arnoud R., Gardette J. L., Gindhal J. M., Fanton E., 1979. Vieillissement des polymères : science ou empirisme ? *Revue générale des caoutchoucs et plastiques*. (631). Paris. pp 147-152.

15 Lemaire J., Arnoud R. & Gardette J. L.. 1981. Vieillissement des polymères, principe d'études du photovieillissement. *Revue Générale des Caoutchoucs et Plastiques*. (613). Paris. pp 87-92.

16 Snedecor G. W. & Cochran W. G.. 1971. *Méthodes statistiques* (translation from English by H. Boelle & E. Camhji). Association de Coordination Technique Agricole. Paris. 649 p.

BOUNDARY MODELING OF STEADY SATURATED FLOW

TO DRAIN TUBES[1]

Sinite C. Yu and Kenneth D. Konyha[2]

ABSTRACT

A boundary model solution to the Laplace equation has been developed to determine flow nets in drainage and subsurface irrigation systems. The model predicts water table position, hydraulic head loss at the drain, as well as flux and potential along the boundary. The program can analyze flow problems involving layered soils, trench effects, subsurface irrigation, and drainage, considering steady infiltration or evaporation. Mass balance errors are less than 0.13%. Three applications of the model are made: 1) comparison to Hooghoudt's, Kirkham's, and Hammad's solutions of steady drainage, 2) determination of steady flow in a combined subsurface irrigation and drainage field, and 3) comparison of steady drainage through homogeneous soils versus steady drainage through layered soils to demonstrate the effect of soil heterogeneity on chemical transport.

INTRODUCTION

Numerical methods are frequently used to solve ground water flow problems. In the area of shallow ground water flow and drainage, solutions are usually based on finite difference or finite element approximation methods. We present a numerical approximation model which is based on boundary model techniques. The boundary method (BM) described is a solution to the Laplace equation for flow in multi-domain soils. This model can determine the location of the water surface profile, drainage flux, and head loss near drains.

BM modelling has been used in ground water flow problems for many years (Brebbia, 1978), but it has not been frequently applied to drain flow problems. The advantages of BM modelling are 1) fewer node points are needed, as compared to finite difference and finite element methods, 2) applicability to a wide variety of flow problems, 3) very good mass balances, and 4) few assumptions. In this paper, we present a very brief explanation of the BM, compare BM predictions to analytical solutions for the case of steady infiltration to drains, and demonstrate the model's utility by applying it to two drainage problems.

---

[1]This research was sponsored by the Illinois Water Resources Center, Urbana, Illinois.

[2]Sinite C. Yu, Graduate Student and Kenneth D. Konyha, Assistant Professor, Dept. of Agricultural Engineering, University of Illinois, Urbana-Champaign.

MODEL DESCRIPTION

Boundary method solutions of the Laplace equation have been used for many years, and detailed discussions of the BM can be found in several texts (Brebbia, 1978; Liggett et al., 1983; Gipson, 1987). The model used in this paper was developed by Gibson (1987) and has been modified by the author for unconfined flow problems (Yu, 1992). A detailed description of the model can be obtained from the authors.

The boundary method solution of the Laplace equation requires knowledge of either the flux or the piezometric head along the boundary of the flow domain. It also requires the location of the boundary and the hydraulic conductivity within the flow domain. Flow in layered soils and in trenches are solved by defining separate homogeneous sub-domains coupled by conditions at the interface. The location of the free water surface is determined by an iteration process--assuming a location for the boundary, solving the system of BM equations, then adjusting the boundary based on the magnitude and sign of the resulting error. Flow to drain tubes are not modelled explicitly. Instead, we use the effective radius concept (Skaggs, 1978), modelling the actual drain tube as open tube of much smaller radius.

The BM program is written in Fortran-77. The size of the program depends on the number of boundary nodes, but flow problems involving up to 240 boundary nodes can fit within the 640K memory limitations of the DOS environment. All solutions presented here were solved using a 33 MHz '386 computer with math coprocessor. Solution times depend on the problem, but are generally solved within 1 to 6 minutes.

STEADY INFILTRATION TO DRAINS

The BM can be applied to the familiar problem of steady-state infiltration through an isotropic, homogeneous soil to a drain (see Fig. 1). [For homogeneous soils, the conductivity of the layers are identical]. The boundary conditions for this case are 1) zero flux along the impermeable layer (O-A) and the vertical boundaries of the domain (O-F and A-E) except 2) at the drain (B-C) which has radius r_e and a specified hydraulic head. The boundary conditions for the free water surface (E-F) are 3a) a known flux equal to the component of infiltration normal to the water surface and 3b) a total hydraulic head equal to the elevation of the water surface. The position of the free water surface is not known at any point, even above the drain.

Steady infiltration was solved for the following conditions; infiltration (R) was 0.0092 m/d, the half-drain spacing (L) was 7.5 m, saturated hydraulic conductivity (K) was 1.08 m/d, the depth of the soil below the drain (D2) was 7.5 m, 4.8 m, 3.75 m, 2.4 m, 1.2 m, 0.75 m, 0.45 m, or 0.24 m, and the effective drain radius (r_e) was 0.0005 m, 0.0015 m, 0.005 m, or 0.015 m. The BM used 72 boundary nodes; 25 nodes on the water table, 37 nodes along the zero flux boundaries and

10 nodes along the circumference of the drain. Solutions to the 32 cases were found. These solutions are compared to the analytical solutions of Hooghoudt (1940), Kirkham (1958), and Hammad (1963). Hooghoudt's equation is the common method of determining the mid-point water table elevation (Dm-D2); Kirkham's method provides an analytical solution to the flow net; Hammad's solution provides a solution to the water table elevation above the drain (Dc-D2).

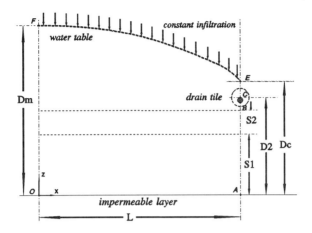

Figure 1. Flow domain and boundary conditions for steady infiltration through soils to subsurface drain tubes

Mid-point Water Table Elevation

Fig. 2 compares predictions of the heights of the mid-point water table mound (Dm-D2) for all three analytical methods and the BM for those cases where the effective radius was 0.005 m. The height of the mound predicted by BM ranged from 0.21 m to 0.55 m. For all cases, mound heights predicted by the BM agreed very closely with those predicted by Hooghoudt's equation. The difference between Hooghoudt and BM mound heights ranged from -0.02 m to +0.03 m. Kirkham's predicted mound heights were always greater than those of Hooghoudt's, and Hammad's predicted mound heights were always smaller. These results agree with the evaluation of these equations made by Lovell et al. (1984).

Water Table Elevation Above the Drain

The head loss near the drain can be substantial enough to result in a water table elevation at the drain that is several centimeters above the drain. The BM model and Hammad's method can find this height. For the cases examined, this mounding above the drain ranged from 0.038 m for a shallow soil (D2=0.24m) and a large drain opening (r_e=0.005m) to 0.128 m for a deep soil (D2=7.5m) and a small drain opening (r_e=0.0005m). The BM results are shown in Fig. 3 along with predictions from Hammad's equation. The BM results show a slightly lower mound height above the drains for shallow soils than for deep soils, a relationship not predicted by Hammad's equation. The BM predictions bracketed Hammad's predictions at all drain diameters and were in good agreement with Hammad's for all cases.

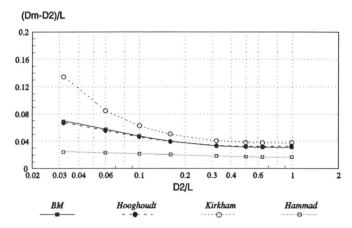

Figure 2. Comparison between BM prediction and analytic equations for the water table depth at midpoint between drains (re=0.005 m)

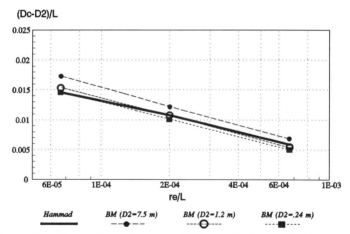

Figure 3. Comparison between BM prediction and Hammad's solution for the water table depth at the drain

The above examples demonstrate that BM predictions agree well with analytical solutions of steady infiltration to drains. The mass balance errors for these solutions ranged from 0.02% to 0.13%.

APPLICATIONS

The true utility of the BM model is shown by its ability to solve flow problems that have no analytical solutions. We present two examples.

Combined Drainage and Subsurface Irrigation

The first application is an analysis of water movement from a subsurface irrigation line to a drainage line. This application is an evaluation of wastewater effluent through soil. The wastewater is applied to an agricultural field via a system of subsurface irrigation lines. Treatment occurs as the effluent moves laterally through the soil to an independent system of drains. Fig. 4 shows this system.

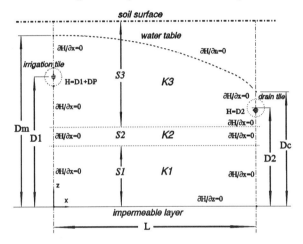

Figure 4. Combined subirrigation and drainage in a three-layered soil

The soil at the site was modelled as a three-layer soil; the top layer (S3) was 1.5 m thick and had a saturated hydraulic conductivity (K3) of 1.5 m/d, the second layer (S2) was 0.3 m thick and had a conductivity of 7.5 m/d, and the deepest layer (S1) was 0.8 m thick and had a conductivity of 1.0 m/d. Below the third layer was an impermeable layer. The subsurface irrigation line (D1) was 1.3 m above the impermeable layer and water in the subsurface irrigation tube had a pressure (H=D1+DP) of 1.7 m, relative to the impermeable layer. The drain line (D2) was 1.0 m above the impermeable layer and pressure inside the drain was atmospheric (H=D2). The spacing (L) between the subsurface irrigation line and the drain was 7.8 m. The boundary conditions for this case are shown in Fig. 4. This problem was modelled using 78 nodes; 8 nodes at the drain line, 9 nodes at the subsurface irrigation line, 15 nodes along the free water surface, and 30 nodes along the zero flux boundaries. There were also 16 nodes along the interfaces of the soil layers.

The flow net for the BM solution of this problem is shown in Fig. 5. The flow through the soil was predicted to be 0.193 m²/d, equivalent to an infiltration rate of 24.7 mm/d. As seen from the stream lines, fifty-five percent of the flow was restricted to the highly conductive second layer and only 22 percent came in contact with the deepest layer. Because most of the flow was in the 0.3 m thick second layer, the apparent velocity in this layer was high (0.35 m/d) compared to the other layers (0.09 m/d). The residence time of water was, therefore, quite short and the treatment of the wastewater would be less than in a homogeneous soil with equivalent hydraulic properties. The entrance head loss as the water leaves the subsurface irrigation line was 46 mm and the exit head loss as the water enters the drain line was 68 mm.

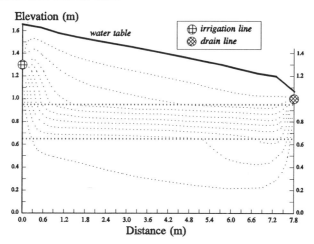

Figure 5. Water table profile and contours of stream functions for combined subirrigation and drainage in a three-layered soil

<u>Comparing Steady Infiltration to Drains for Homogeneous versus Layered Soils</u>

The second application is an analysis of water and chemical movement for steady infiltration to a drain (Fig. 1). Two cases are compared: (case 1) flow in homogeneous soil and (case 2) flow in layered soil. The boundary conditions are identical to the steady infiltration example discussed earlier.

The data for this case were L = 11.5m, D2 = 0.95m, r_e = 0.005 m, infiltration (R) = 0.01 m/d, S1 = 0.6 m, and S2 = 0.3 m. For the homogeneous soil, K1 = K2 = K3 = 0.5968 m/d. For the layered soil, case K1 = K3 = 0.3862 m/d and K2 = 1.545 m/d. (Note: The conductivity for the homogeneous case is the equivalent lateral conductivity for the layered case. Using Hooghoudt's equation, the watertable at the midpoint should be the same for both cases). A total of 70 nodes were specified along the boundary and interfaces.

The resulting flow net for the homogeneous soil case is shown on Fig. 6 and the flow net for the layered case is shown on Fig. 7. The mid-point water table elevations were similar, 1.96 m for case 1 versus 2.00 m for case 2. The water table profiles were similar, but the water table was 0.258 m above the drain for the homogeneous flow case and 0.404 m above the drain for the layered case. The flow nets were quite different, with the highly conductive middle layer of case 2 carrying much more of the total flow than the equivalent flow region in the homogeneous case.

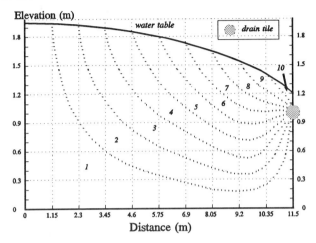

Figure 6. Flow profile and contours of stream functions for steady infiltration to drains in a homogeneous soil (L=11.5 m, D2 = 0.95 m, K = 2.487 cm/hr)

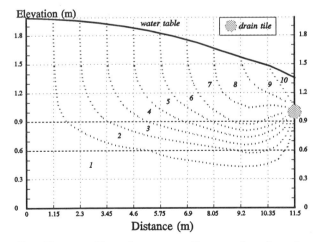

Figure 7. Flow profile and contours of stream functions for steady infiltration to drains in a three-layered soil (L=11.5 m, D2= 0.95 m, S1=0.6 m, and S2=0.3 m)

Are these two flow nets significantly different? To answer this question, we can look at chemical transport for the two cases. As a rough estimate of chemical transport, we assume one-dimensional steady chemical transport occurs within each stream tube (i.e. lateral dispersion is neglected). The flow regime was divided into ten parts, one part for each stream tube. (The ten stream tubes are numbered on Fig. 6 and Fig. 7). A one-dimensional chemical transport model, developed by Beljin (1990), was used to predict the relative concentration of a non-adsorptive, stable chemical delivered to the drain by each stream tube. The model required the average apparent velocity, which was determined from the flow net. The model also required porosity, a pulse duration, and dispersivity. Using the average width of the stream tube and the known flux to determine velocity, assuming a dispersivity of 0.1 m and a porosity of 0.45, the concentration of the chemical in the drain line was determined as a function of time. Fig. 8 shows relative concentration versus time for the first 250 days of flow for the two cases. The differences between these two cases were due solely to differences in the average widths of the stream tubes and the resulting differences in velocity in each stream tube.

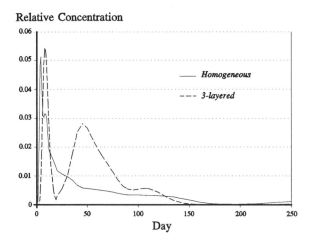

Figure 8. Chemical outflow for steady infiltration to drains in a homogeneous and a three-layered soil

The first peak in chemical outflow occured as the chemical was transported through the stream tube nearest the drain; i.e. stream tube 10. For the homogeneous flow case, chemicals from stream tubes entered the drain tile at increasingly later times and lower concentrations because both the length and width of the stream tubes steadily increased as you move away from the drain. By day 200, 80% of the total chemical applied had reached the drain. For the layered flow case, the length of the stream tube increased as you move away from the drain, but average width decreased and velocity, therefore, increased in much of the flow regime. The stream tube width decreased because the high conductivity layer carried flow from these tubes. The result was faster transport times for these drain tubes which resulted in a second peak at day 50 (Fig 8). By day 200, 90% of the chemical applied to the layered soil had reached the drain.

Although the one-dimensional transport model is only a rough approximation of actual transport, this example demonstrates the potential significance of layered soils on the chemical transport process. The ability of BM modelling to quickly generate flow nets for complex flow situations makes the method a valuable tool for drainage engineers.

REFERENCES

1. Brebbia, C.A., 1978. The Boundary Element Method for Engineers. Pentech Press, London. 46-48 pp.

2. Beljin, M.S., 1990. A Program Package of Analytical Models for Solute Transport in Groundwater. Holcomb Research Institute. Butler University, Indianapolis, Indiana.

3. Gipson, G.S., 1987. Boundary Element Fundamentals. Computational Mechanics, Southampton, UK. 53-55 pp.

4. Hammad, H.Y., 1963. Depth and spacing of tile drain system. Transactions of ASCE, 128 (III):585-629.

5. Hooghoudt, S.B., 1940. Contributions to the knowledge of some physical problems of the soil. 7, General discussion of the problem od drainage and infiltration by means of parallel drains, trenches, ditches and canals. Versl. Landbouwkd. Onderz., 46:515-707.

6. Kirkham, D., 1958. Seepage of steady rainfall through soil into drains. Transactions, American Geophysical Union, 39(5):892-908.

7. Liggett, J.A., Liu, P.L-F., 1983. The Boundary Integral Equation Methods for Porous Media Flow. George Allen & Unwin Ltd., London. 17-18 pp.

8. Lovell, C.J., Youngs, E.G., 1984. A comparison of steady-state land-drainage equations. Agricultural Water Management, 9(1):1-21.

9. Skaggs, R. W., 1978. Effect of drain tube openings on water table drawdown. Journal of Irrigation and Drainage Engineering, ASCE, 104(IR1):13-21.

10. Yu, S.C., 1992. Boundary modeling of saturated flow during steady application of municipal effluent in a subsurface irrigation and drainage field. University of Illinois at Urbana-Champaign, Ph.D. thesis.

RESEARCH NEEDS FOR WATER QUALITY/ SALINITY/WETLANDS MANAGEMENT

Dale A. Bucks[1]
Member ASAE

ABSTRACT

President George Bush of the United States (U.S.) launched the President's Initiative on Enhancing Water Quality, in early 1989, to protect ground and surface waters from contamination by agricultural chemicals. The authorized program is based on principles that express concern about reducing the risks of contaminating ground and surface water resources and supporting the development of programs to provide accurate information and technical assistance.

The goal of the United States Department of Agriculture (USDA) for water quality is to prevent agricultural chemicals from contaminating the Nation's waters. In 1990, USDA designed and implemented programs to update farming practices and alter management systems that should result in a more environmentally sensitive and efficient farming community. As a part of the President's Water Quality Initiative, research programs are being conducted in conjunction with actions of cooperating state research and educational institutions, other Federal agencies, state and local governments, and farmers themselves. However, there is widespread opinion that more needs to be done in a short period of time to protect the food supply and environment from degradation by agricultural chemicals. Science must improve its ability to transfer water quality, salinity, and wetlands research to large landscapes if farmers and managers are to meet environmental quality standards and production goals.

INTRODUCTION

In many parts of the world, there is public concern that agricultural and forestry activities are contributing to the contamination of ground and surface waters. Pesticide and other chemical contamination of water is a national problem that

---

[1] Dale A. Bucks is National Program Leader for water quality and water management for the U.S. Department of Agriculture, Agricultural Research Service, Beltsville, MD 20705.

needs timely and rational solutions. The extensive use of agricultural chemicals, including pesticides and fertilizers, has been brought about by economic factors and farmers' efforts to obtain a fair return on their investment in crop production systems.

There is evidence that agricultural chemicals do indeed leach or run off and contaminate ground and surface waters. These include nitrate, phosphorus, certain pesticides, salts and toxic trace elements. Although occurrences of serious degradation may have reached levels of major concern in only isolated situations, it is recognized that:

> Once contaminated, cleanup of ground water is difficult, expensive and long term.
>
> Detection of potential problems is arduous, costly and possible only at discrete points in a three-dimensional system.
>
> Prediction of the migration of a problem is rigorous, high-priced and not reliable.
>
> Several years may elapse before a contamination problem reaches a ground water aquifer or a surface water stream, lake or estuary.

An integration of technologies for efficient crop production and reduced pollution is required. Integrated methodology is urgently needed to manage soil, water, ground cover, pesticide applications, fertilizer, livestock waste applications, etc. in such a way that pesticides and nutrients are contained in their functional "action zones" of the soil profile (Fouss and Willis, 1990). It should be possible to manage and control root-zone soil-water content such that plant nutrient use efficiency is enhanced, thereby decreasing fertilizer or livestock waste requirements and reducing nutrient pollution potential. There also is an opportunity to reduce pollution by overall pesticide management practices, particularly for those pesticides that degrade as a function of soil moisture and aeration conditions. The reduced need for applying agricultural chemicals could increase operating profits to farmers. However, there are policy matters and competition among supply and demand production considerations that may not make the enhancement of water quality easy.

POLICY SETTING IN THE 1990's

World demands for food will increase at least 20 percent because population is expected to grow from about 5 to 6 billion by the year 2000. Continued economic growth will increase food demand even more, particularly in rapidly developing countries. As it has in the past, increases in food supply will have to come from improved technology and high yields per land area. This will increase the pressures on our natural resources.

World concerns for food supply and environmental quality will require that future technology in agriculture be sustainable and protective of the environment. Efficient uses of agricultural chemicals will remain a necessary part of that future technology. We will need to learn how to use them so they will not contaminate ground or surface waters to an unacceptable level in relation to human and environmental quality.

U.S. agriculture is an important contributor to the world's food supply and a major component of the U.S. economy. Nearly one-fifth of the world's agricultural exports are shipped from the U.S. We have a heritage of productive agricultural lands and favorable climate which provides us with a comparative advantage over many other countries in producing basic food and feed grains and many other important crops.

Agricultural chemicals, especially nitrate and pesticides or their breakdown (degradation) products, are being detected with increasing frequency in ground and surface waters throughout the U.S. During the last 18 months, reports have been issued by the U.S. Environmental Protection Agency (USEPA), U.S. Geological

Survey (USGS), and Monsanto Chemical Company. These reports do not indicate that alarming percentages of our public and private water supplies are contaminated with agricultural chemicals in excess of the drinking water standard or Maximum Contamination Level (MCL). National standards or advisory levels have not been established for aquatic ecosystems and habitats.

The USEPA's well water survey was released in November 1990 (U.S. Environmental Protection Agency, 1990). The survey reflected concern that pesticides and excess nitrate can pose a health problem to people and the plants and creatures that live in or near water. Chemical analysis for the presence of 126 pesticides and their degradates in drinking water wells revealed detectable residues in 10.4 percent of the community wells and 4.2 percent of the rural domestic wells. Pesticide contamination at levels of concern to human health was found to be less than 1.0 percent in the rural domestic wells. Over one-half of the community wells and rural domestic wells contained nitrate above the detectable level of 0.15 milligrams per liter (mg/L). Nitrate, as nitrogen, exceeded the MCL of 10 mg/L in about 1.2 percent of the community wells and 2.4 percent of the rural domestic wells.

The USGS has reported water quality from 149 sites in 122 river basins in the midwestern U.S. They were collected during a harvest phase (October and November 1989); a preplanting phase (March and April 1990); and a postplanting phase (May and June 1990). Thurman et al. (1991) indicated that "large concentrations of herbicides were flushed from croplands and were transported through the surface water system as pulses in response to late spring and early summer rainfall." About 50 percent of the sampling sites had concentrations of dissolved atrazine in excess of the MCL of 3 micrograms per liter (ug/L). Such elevated levels of atrazine were associated with rainfall and subsequent runoff.

During the first runoff following atrazine application, 29 percent of the samples had atrazine concentrations greater than 12 ug/L. Later monitoring data from the Mississippi River and some of these same tributaries showed that atrazine concentrations above the MCL were sustained for 4-6 weeks from about mid-May to mid- or late June in the lower Platte River in Nebraska, the lower White River in Indiana, the lower Illinois River in Illinois, and the lower Missouri River in Missouri.

CURRENT U.S. WATER QUALITY POLICY

USDA is developing programs to deal with the evidence that agricultural nutrients and pesticides are contaminating water supplies in vulnerable situations. Although abuses and accidents in handling some chemicals have occurred, agricultural nonpoint source problems cannot be blamed solely on careless management of resources and farm inputs. They have also occurred as a result of normal agricultural production practices. Unfortunately, we have an imperfect understanding of several key physical, chemical and biological processes involved in water pollution, especially ground water contamination.

In the U.S., the majority of the rural water quality data has been obtained from farm wells. However, we are unsure how much contamination is caused by practices near wellheads, such as chemical storage, chemical mixing or cleaning of chemical application equipment. We suspect that septic systems, livestock operations, and livestock barnyards and manure storage facilities on the farmstead contribute to excessive nitrate levels in some

wells because coliform bacteria are found in many high-nitrate wells. Further, we have not fully mastered the art of optimizing the use of plant nutrients and crop production chemicals required for an efficient and economically viable agriculture.

USDA wants to base water quality policy on sound science. Yet, there are too many questions for which we lack answers: What are the maximum tolerable levels of contaminants in water? What are the real risks to both human and ecological health from chronic exposure to low-level concentrations of nitrate or pesticides and their breakdown products? What are the risks of cumulative exposure to several pesticides? What is the ultimate fate of chemicals applied to the land? What physical, chemical and biological factors influence that fate? How can these factors be managed to achieve the most desirable fate? What can farmers realistically do to minimize or avoid contaminating water without jeopardizing their economic viability? Until we answer these and other basic questions, we cannot be sure we are targeting our resources in the most efficient way. We cannot wait for all the answers before developing definitive solutions. From a policy standpoint, we must work with the tools that are available to us now and continue to refine and improve those tools as we progress toward our goal.

USDA's voluntary approach, which is supported by the present Administration, is to inform the farmers what we do know about preventing contamination and to help them apply the best available technology. As we acquire new knowledge, we will move new technologies into the field through a program of education, technical and financial assistance.

A complementary approach is to rethink the goals of our farm policy. The U.S. Congress began to make major policy shifts with the 1985 Farm Bill. Current policy is aimed at motivating farmers to plant crops to meet market demands rather than support prices. These changes could help to mitigate some environmental concerns over a period of time. The 1990 Farm Bill also introduces a provision called the Water Quality Incentives Program, a concept of land management for the environment as an alternative to land retirement for the environment. The 1990 Farm Bill is taking a step toward inducing the farmer to become not only a producer but an environmental manager.

Regulation of agricultural chemicals is seen as an alternative to the voluntary approach and by others as a complementary approach. USEPA plays an important role, but final responsibility for water quality rests with each individual state. There are already regulations affecting agriculture to some extent in many states. U.S. Administration policy recognizes that where regulations exist, scientific and technical support must be provided to the farmers and state regulatory agencies to help them identify best available technology.

U.S. PRESIDENT'S WATER QUALITY INITIATIVE

In February 1989, President George Bush made a commitment to protect ground and surface waters from contamination by agricultural chemicals. The President stated, "The protection of the environment and the conservation and wise management of our natural resources must have a high priority on our national agenda. But given sound research, innovative technology, hard work, sufficient public and private funds, and--most important of all--the necessary political will, we can achieve and maintain the environment that protects the public health and enhances the quality of life for us all."

The President's Initiative on Enhancing Water Quality is based on these principles quoted directly from his State of the Union address (U.S. Administration, 1989):

> The President is committed to protecting the Nation's ground water resources from contamination by fertilizers and pesticides without jeopardizing the economic vitality of U.S. agriculture.
>
> Water quality programs must accommodate both the immediate need to halt contamination and the future need to alter fundamental production practices.
>
> Ultimately farmers must be responsible for changing production practices to avoid contaminating ground and surface waters. Federal and state resources can provide valuable information and technical assistance to producers so

that environmentally sensitive techniques can be implemented at minimum cost.

As the lead agency for the President's Water Quality Initiative, USDA has established close working relationships with the USEPA, USGS, National Oceanic and Atmospheric Administration (NOAA), U.S. Army Corps of Engineers, Tennessee Valley Authority (TVA), U.S. Fish and Wildlife Service (FWS) and, most importantly, with private industry and the individual states themselves to assure success of the Initiative. The USDA Working Group on Water Quality coordinates the Initiative. The Agricultural Research Service (ARS) and Cooperative State Research Service are the major participants in research; the Agricultural Stabilization and Conservation Service (ASCS), Extension Service (ES), and the Soil Conservation Service (SCS) lead the work on education, technical and financial assistance; the Economic Research Service (ERS) and the National Agricultural Statistics Service provide resource evaluation and assessment.

CURRENT USDA WATER QUALITY RESEARCH PROGRAM

A USDA water quality research plan (U.S. Department of Agriculture, 1989) was developed jointly by ARS and CSRS with major contributions from other agencies of the department and State agricultural experiment stations (SAES). An effective partnership has been accomplished through collaborative, administrative and research efforts with USGS, USEPA, and state groups including the SAES. The research plan addresses two types of research: Priority Components and Selected Geographic Systems--Management Systems Evaluation Areas (MSEAs).

The research strategy of the President's Water Quality Initiative addresses five primary problem areas as follows:

Assessment, Sampling, and Testing Methods. Develop improved, inexpensive methods of risk assessment for site-specific potential problem areas, sampling, measuring, and evaluating water quality.

Fate and Transport. Identify and increase understanding of factors and processes that control fate and transport of agricultural chemicals.

Management and Remediation Practices or Systems. Develop new and modified agricultural production management practices and systems including remediation techniques that substantially

reduce the movement of potentially hazardous chemicals into ground and surface waters and are cost effective for farmers, ranchers, and foresters to adopt.

Regional Application and Transferability of Results. Develop and adapt procedures, models, and decision aids to apply and transfer water quality research results to other locations or at larger scales by researchers and user agencies.

Social, Economic, and Policy Considerations. Evaluate the economic, social and political impacts of alternative agricultural production practices and systems, policies, and institutional strategies to control ground and surface water quality.

ARS has funded 62 projects at 26 locations in fiscal years (FY) 1990, 1991, and 1992. For the three years, CSRS has awarded 167 competitively selected projects, of which 142 were funded for two or three years. Many of the ARS and CSRS projects involve researchers from other agencies and scientific institutions. ARS and SAES also have other research that addresses potential water contamination.

The current research program is placing greater emphasis on socioeconomic research. The lack of understanding of the impact of ground and surface water quality issues on economics and state, county, and local government policy was apparent. The research program also is placing more emphasis on nitrate in ground water and herbicides in surface water. All aspects of nitrogen management are under review and improved soil tests for nitrogen are being evaluated for efficacy in a national study. An expanded effort on pesticide contamination of surface waters is planned based on recent findings. New or modified pesticide management practices are needed for both ground and surface water protection.

The multiagency, long-term MSEA program is a model for agency cooperation. Agencies actively participating in the program include, ARS, CSRS, SAES, USGS, USEPA, SCS, ES, Cooperative Extension System (CES), and various state and local agencies. MSEA projects are farm- and field-size test sites that are used to evaluate the environmental and economic performance of corn and soybean production systems developed for the purpose of reducing the risk of agricultural chemical contamination. The 10 field sites are operated by five coordinated research teams in Iowa, Minnesota, Missouri, Nebraska, and Ohio. The Iowa MSEA has three sites and the Minnesota MSEA has sites in Minnesota, North Dakota, South Dakota and Wisconsin.

The MSEAs have installed state-of-the-art field equipment to monitor soil and water parameters, characterize the weather and determine the effects of various crop management systems on water quality. Modifications of prevailing cropping systems were developed for each MSEA site to study the impacts of these systems on ground and surface water contamination. These cropping systems are specifically suited to the soil, geology, climate, irrigation, nitrogen, and pesticide needs, and are a unique feature at each site. Improved pesticide and nitrogen applications are being stressed, although social and economic considerations will ultimately be the dominant factors in the adoption of modified cropping systems by farmers. Quality assurance and control procedures have been initiated to ensure a standard for data collection. Soil and water tests are providing valuable data concerning the fate and transport of agricultural chemicals in the soil profile.

FUTURE USDA WATER QUALITY RESEARCH PRIORITIES

USDA has maintained a major research effort in water quality assessment and protection for more than two decades. During this time, ARS and SAES scientists have made a number of significant contributions to the total national effort on water quality. There is general agreement that this national effort has materially improved the quality of the Nation's water resources. Nonetheless, widespread beliefs still exist within the U.S. community that more needs to be done to protect both the food supply and environment from degradation by agricultural activities. As long as these real or perceived threats to human health and the environment are not adequately addressed through research, education, technical and financial assistance programs, public concern about the state of the environment will persist.

The July 1991 issue of the journal "Agricultural Engineering" includes an article by Jack King entitled "A Matter of Public Confidence" (King, 1991). The article focuses on the results of a survey conducted by the American Farm Bureau in 1990 on the public's attitudes and perceptions regarding food safety, agricultural chemicals and their onfarm use. The Farm Bureau's survey showed that more consumers (89 percent) are concerned about pesticide residues in food than any other food quality issue. This fear of chemicals, even though difficult to justify on a scientific basis, spills over into the community's attitude toward environmental degradation by chemicals.

Water Quality and Conservation

At the present time, USDA research on water quality is directed toward reducing the potential for water contamination by agricultural chemicals. A smaller, but still substantial, effort is being made to control water quality degradation by other contaminants such as sediments, salts and toxic elements. The current research agenda is primarily farm oriented and focuses on the development and promotion of farming practices and systems that improve both water quality and the efficiency of chemical use. The farming practices that are being evaluated include: adjustments to the timing, placement and rates of application of chemicals; new pesticide formulations; changes in crop rotations, tillage, and irrigation and drainage water management; increasing crop rooting depths to improve yields and efficiency of nutrient use; and the expanded use of nonchemical alternatives for controlling plant pests.

Because of the high cost of watershed scale research, limited studies have been conducted on assessing and controlling the offsite effects of agricultural production systems. One major component of the USDA research program on water quality that is being conducted on a watershed scale is the MSEA program which has already been discussed. However, it will be some time before the experimental data from these sites will permit objective conclusions to be reached on offsite effects. Even then, it will add only peripherally to our knowledge of the fate and transport of contaminants in streams, rivers and larger water bodies. ARS and CSRS also are collaborating, through scientist involvement and limited financial support, with ASCS, ES, and SCS in Demonstration Projects and Hydrologic Unit Areas which are major efforts within the education, technical and financial assistance program under the President's Water Quality Initiative. ARS and CSRS will continue to explore opportunities to strengthen their support for the SCS and ES programs and, depending on the success of future funding proposals, will expand their water quality activities to other parts of the country.

Is the present USDA research program adequate in scope and vision to serve the needs of the farm community, other USDA agencies, and the Nation? This question continues to be asked. Some insight into how others view the progress that has been made and the deficiencies of current water quality programs can be gained from the June 1991 Interim Report by the Member Organizations of Water Quality 2000 entitled "Challenges for the Future." The findings in the report are implicitly supportive of the research agenda. Many of the knowledge and information-related impediments to water quality protection are being addressed by USDA research. However, a few are not. For example, the report indicated that "Sediment from agriculture and other nonpoint sources accounted for 42 percent of the impaired river miles, nutrients for 26 percent, while pesticides accounted for 10 percent," and that "Economic studies place the cost of sediment damages in the billions of dollars per year." ARS and SAES have maintained a major research effort on landscape erosion control and have successfully demonstrated the benefits of reduced tillage in controlling sediment releases to rivers, lakes and estuaries. However, other major sources of sediments have received less than adequate support. For example, the financial and human resources available in ARS for research on gully and channel erosion and control has decreased substantially even though in some areas, such as the Yazoo River Basin in Mississippi, this is the dominant source of stream sediments.

Another central issue that has not been fully addressed is the scale at which the USDA water quality and natural resource programs ought to be conducted. Because implementation of much of the new science and technology that is developed by research directly benefits the individual farmer, field-scale research will continue to be a dominant component of the ARS program. Nevertheless, the projected loss of rural representation in the Congress, and the continued strength of the environmental movement should provide enough incentive for us to give more attention to the broader natural resource concerns of our urban community. What is the role of USDA research in protecting aquatic ecosystems and habitats? How great is the risk that too narrow a perspective on the strategies to be used in solving water quality problems will prove suboptimal, and perhaps lead to ineffective or unnecessary protection measures? This is particularly the case where a large part of the watershed is in agriculture and the urban areas are increasing.

Finally, and perhaps as crucial as any water quality issue, is the future role and function of ARS and CSRS in water conservation research and management. Can USDA afford to be complacent about the water quantity challenges of the next few decades? What effect will our changes in farming practices and systems have on water yields, peak stream flows, and low flows? Are we satisfied with what has been done to mitigate the effects of droughts and floods? For the present, our water resources agenda is dominated by community concern for the quality of the environment. As the more serious of these problems are solved, and the perceived problems are addressed through effective assessment and educational programs, changes in societal priorities are inevitable.

Salts and Toxic Elements

Irrigation has contributed greatly to the production of food and fiber and to a richer and more diversified diet. Because irrigated plants consume water and nutrients and leave behind in the soil those salts contained in the water, irrigation always

degrades water quality. Without some drainage, soil salinization would soon reduce crop production. Therefore, society generally must sacrifice some environmental or economic value for irrigation to be sustainable. The discovery in the 1980's of severe adverse impacts on fish and birds from selenium carried in irrigation drainage water to a wildlife refuge in California has added a new dimension to environmental concerns. Studies since then have shown similar problems with trace elements at several other locations across the western U.S. and shows that these potential toxins, of geological origin, are dissolved from the soil complex after irrigation is introduced.

These problems have been identified before and some research and demonstration is underway to find solutions or to minimize the problems. An elaborate list of technologies has been developed to improve irrigation efficiency and management. For example, highly sophisticated and carefully controlled microirrigation systems can often increase crop yields while minimizing offsite impacts. Water reuse technologies take advantage of the fact that some crops in some stages of growth can make effective use of saline water often considered of little agricultural value. Similarly, many institutional changes have been suggested to improve the delivery and use of water resources through market mechanisms and in other ways and to enhance societal returns.

The technology is there, or waiting to be discovered. The real issue is what should be the future of irrigation in the U.S., with a focus on constraints posed by all the water users. USDA research has the capability to develop new technologies to identify sites that are high in salts and trace elements, to develop better management practices and systems that reduce the impact of irrigation on water quality, to use predictive models to determine the impact of irrigation on water quality, and to determine the suitability for using marginal water quality for irrigation.

Another water quality aspect is the increasing use of irrigation for nonagricultural purposes. Irrigation has become critical for greenhouses, golf courses, highway medians, parks and other urban purposes, and the use of reclaimed water has proven to be an effective substitute for higher quality water in these cases, as well as for certain agricultural crops. Research, education and technical assistance programs on irrigation water management for farmers and urban dwellers are generally inadequate.

Wetlands and Riparian Areas

The USDA research program on wetlands and riparian areas, though productive and instructive, has not attracted a high level of internal support. In the absence of a clear national policy in the past, there has been limited interest in supporting research on these ecosystems. More recently, as a result of farm and environmental legislation and executive policies on wetlands, there has been a growing awareness of the value of these ecosystems and the need for providing them greater protection. The important role that riparian areas have in removing plant nutrients from surface water and shallow ground water has been recognized for some time and ARS has made several significant contributions to our present knowledge of their effectiveness in maintaining surface water quality. Even so, major knowledge gaps remain. CSRS competitive research grants also are supporting research projects on Wetland and Riparian areas as a means of enhancing water quality. For example, the capacity of riparian areas to reduce the potential contamination of surface waters by pesticides is unknown.

Despite strong expressions of support at the national level in the last two years, it is unlikely that USDA will be in a position to support a major research effort aimed at protecting existing wetlands from farming operations. USDA research will, of course, continue to be responsive to requests from the SCS, ES, FWS, and other action agencies for assistance in developing guidelines for managing wetlands, restoring degraded wetlands and creating artificial wetlands. Wetland research offers many challenges and opportunities to the research community. For example, in summarizing current limitations on research and development, the Water Quality 2000 report concludes that "It remains unclear whether the creation of restoration of wetlands is technically or scientifically feasible."

In the future, the research community would be in a stronger position to establish direction for increased work on the management, restoration, and protection of wetland ecosystems. Meanwhile, scientists need to determine how best to design and operate wetlands and riparian areas for water quality protection and base their recommendations in terms of overall societal benefits. The following quotation from the Water Quality 2000 report might help: "Our tendency, as a society, is to underestimate the cost of pollution in currently less populated areas, such as wilderness and aquifer recharge areas, because there are fewer immediately measurable impacts on human health and because we tend to undervalue the impacts on biological communities and their habitats."

SUMMARY

The President's Initiative on Enhancing Water Quality is making significant progress toward the protection of ground and surface waters from contamination by agricultural chemicals. An integration of technologies for efficient crop production and reduced pollution is required. As new knowledge and technologies are developed, they will need to be moved into the field through education, technical and financial assistance programs.

The current USDA water quality research program is placing emphasis on the development of new and modified agricultural production management practices and systems, the development of procedures models and decision aids to apply and transfer water quality research results, and the evaluation of economic, social, and political impacts of alternative agricultural production practices and systems, policies, and institutional strategies to control ground and surface water quality.

Future USDA water quality research priorities which need increased emphasis in the opinion of the author are as follows:

> Assess and control the offsite effects of agricultural production management systems at a watershed scale.

> Develop improved, inexpensive methods and practices for reducing soil erosion to decrease sediments and agricultural chemicals in surface waters.

> Expand water conservation and management research to mitigate effects of droughts and floods and address competition among water users.

> Develop improved irrigation management systems to minimize the effects of salts and trace elements in ground and surface water quality.

> Develop knowledge and technologies for the management, restoration, and protection of wetland and riparian areas.

> Integrate water quality, salinity, and wetlands research into agriculture and natural resource systems.

> Expand technology transfer of water quality, salinity, and wetlands research to larger landscapes if farmers and managers meet environmental quality standards and agricultural production goals.

The clientele that benefits from USDA water quality research includes farmers, scientists, other Federal and state agencies, policy decisionmakers, the agricultural industry, special interest groups--and most importantly--the general public. Greater attention needs to be placed on the economic, social, and environmental aspects of water quality and natural resources research in terms of societal benefits and requirements.

REFERENCES

Fouss, J. L. and G. H. Willis. 1990. Research need on integrated system for water and pest management to protect groundwater quality. Amer. Soc. Civil

Engineering, Proceedings of 1990 National Irrigation and Drainage Division Speciality Conference, pp. 288-296.

King, J. 191. A matter of public confidence. Agr. Engr. 72(1): 16-18.

Thurman, E. M., D. A. Goolsby, M. T. Meyer,, and D. W. Koplin. 1991. Herbicides in surface waters of the midwestern United States: The effects of spring flush. J. Environ. Sci. and Tech. 25(10): 1794-1796.

U.S. Department of Agriculture. 1989. USDA research plan for water quality. Agricultural Research Service and Cooperative State Research Service in cooperation with State agricultural experiment stations, USDA, Washington, DC.

U.S. Administration. 1989. Building a better America. Report to U.S. Congress on Enhancing Water Quality, p. 92-93.

U.S. Environmental Protection Agency. 1990. National survey of pesticides in drinking water wells. EPA 570/990015. U.S. Government Printing Office. Washington, DC.

DRAINMOD-N: A NITROGEN MODEL FOR ARTIFICIALLY DRAINED SOILS

M. A. Breve[*]	R. W. Skaggs[*]	H. Kandil[*]	J. E. Parsons[*]	J. W. Gilliam[*]
Student Member	Fellow ASAE		Member ASAE	

ABSTRACT

Computer simulation models promise to be an effective means of evaluating pollutant transport in drained soils. *DRAINMOD-N* is a pseudo two-dimensional, functional model which was developed to predict nitrogen transport, uptake, and transformations in artificially drained soils. The model uses *DRAINMOD* to calculate vertical soil water fluxes in the profile. An explicit solution to the advective equation is used to predict nitrogen transport. Functional relationships are used to represent net mineralization, denitrification, plant uptake, and fertilizer dissolution. The reliability of *DRAINMOD-N* was tested by comparing predictions with those of a more complex numerical model. Results were in good agreement. The model was applied to evaluate the effect of soil type, and water management and fertilizer practices on nitrogen transport. Simulated results showed controlled drainage resulted in lower nitrate-nitrogen (NO_3-N) losses through subsurface drains, and in higher denitrification and lower mineralization rates, as compared to free drainage. Controlled drainage also decreased NO_3-N concentrations in the soil solution, but the effect was more evident in a clay loam than in a sandy loam. Split fertilizer application, as compared to a full application, predicted lower nitrate-nitrogen concentrations only in the upper layer, and had no effect on cumulative denitrification, mineralization, and drainage losses.

INTRODUCTION

The movement of agricultural chemicals in soils is a major concern worldwide because of potential contamination of surface and ground water. This concern has recently been accentuated for artificially drained soils. Although some of the USA's most productive agriculture is on artificially drained soils, agricultural drainage is increasingly perceived as a major contributor to detrimental impacts on water quality. However, the water quality effects of improved drainage can be positive or negative depending on many factors such as conditions prior to artificial drainage, design and management of drainage systems, cultural practice, soil type, climate, etc. (Gilliam, 1987; Skaggs and Breve, 1992).

Computer simulation models can be used to evaluate the effects of certain agricultural practices on pollutant transport and fate. Many models are available to simulate the movement of pesticides and nutrients in agricultural soils. The majority of these models are deterministic and have either functionally- or numerically-based algorithms. Very few models, however, adequately

---

[*]Ph.D. Candidate, William Neal Reynolds Professor and Distinguished University Professor, Ph.D Candidate, and Assistant Professor, Biological and Agricultural Engineering Department, and Professor, Soil Science Department, North Carolina State University, Raleigh, NC.

incorporate the effect of drainage systems on the fate and movement of agricultural pollutants in shallow water table soils. Among the functional models that consider artificial drainage are those by Kanwar et al. (1983), Johnsson et al. (1987), Parsons et al. (1989), Chung et al. (1991), Karvonen and Peltomaa (1992), and Tremwel and Campbell (1992). Their application has been limited because they are either too simplistic, somewhat site-specific, confined to the root zone, or have not been widely tested. Numerical models that handle drains have been developed by Harmsen et al. (1991), Munster et al. (1991), and Clemente and Prasher (1992), among others. These models are useful for research purposes, but their practical application is limited due to high input requirements and extensive user training necessary for their operation. The objective of this paper is to describe and demonstrate *DRAINMOD-N*, a pseudo two-dimensional, functional simulation model for predicting nitrogen transport, uptake, and transformations in shallow water table soils with artificial drainage.

MODEL DESCRIPTION

The transport of a reactive solute in a one-dimensional, unsaturated, homogeneous medium can be described by the advective-dispersive-reactive *(ADR)* equation:

$$\frac{\partial(\theta C)}{\partial t} = \frac{\partial}{\partial z}(\theta D \frac{\partial C}{\partial z}) - \frac{\partial(qC)}{\partial z} + \Gamma \tag{1}$$

where C is the solute concentration [M L^{-3}], D is the coefficient of hydrodynamic dispersion [L^2 T^{-1}], θ is the volumetric water content [L^3 L^{-3}], q is the vertical water flux [L T^{-1}], Γ is a source/sink term used to represent additional processes (plant uptake, transformations, etc.), z is the coordinate direction along the flow path [L], and t is the time [T]. If a dispersion-free field is assumed, Eq. (1) is reduced to:

$$\frac{\partial(\theta C)}{\partial t} = - \frac{\partial(qC)}{\partial z} + \Gamma \tag{2}$$

Assuming z is positive in the downward direction and water flows downward in the soil profile, Eq. (2) can be solved as follows:

$$C_{i_{new}} = \frac{C_{i_{old}} \theta_{i_{old}}}{\theta_{i_{new}}} + \frac{(q_{I_{new}} C_{i-1_{old}} - q_{I+1_{new}} C_{i_{old}}) \Delta t}{\theta_{i_{new}} \Delta z} + \frac{\Gamma \Delta t}{\theta_{i_{new}}} \tag{3}$$

where C_{old} and C_{new} are the previous time step and resulting solute concentrations [M L^{-3}], respectively, i corresponds to the layer where the concentration is being estimated, I corresponds to the interface between layers i and $i-1$, and Δz and Δt are space and time discretizations, respectively. An additional term is added for the saturated zone to represent lateral mass flow. Eq. (3) then becomes:

$$C_{i_{new}} = \frac{C_{i_{old}} \theta_{i_{old}}}{\theta_{i_{new}}} + \frac{(q_{I_{new}} C_{i-1_{old}} - q_{I+1_{new}} C_{i_{old}} - q_{l_{new}} C_{i_{old}}) \Delta t}{\theta_{i_{new}} \Delta z} + \frac{\Gamma \Delta t}{\theta_{i_{new}}} \tag{4}$$

where q_l, the difference between the vertical fluxes entering and leaving the corresponding layer, is the lateral flux going to the drain which is also used to compute solute losses at the drain.

For upward flow the solution is similar to Eq. (4), except that $q_{I,new}C_{i-1,old}$ becomes $q_{I,new}C_{i,old}$, $q_{I+1,new}C_{i,old}$ becomes $q_{I+1,new}C_{i+1,old}$, and the q_l term vanishes, except when water is flowing from the drains as it happens in some cases for controlled drainage. In that case, the current model version assumes the water flowing into the domain has a zero solute concentration.

As the name implies, *DRAINMOD-N* uses a modified version of *DRAINMOD* (Skaggs, 1978) to calculate fluxes and water contents necessary to solve Eq. (4). Based on the drained-to-equilibrium assumption, Skaggs et al. (1991) developed an algorithm that computes average groundwater fluxes at any distance between the soil surface and the water table depth by breaking the profile into increments, calculating the change in volume of water for each increment from a volume drained-water table depth relationship, and applying conservation of mass principles in the unsaturated zone. For the saturated zone, vertical fluxes are linearly decreased from Hoodghout's drainage flux at the depth of the water table to zero at the impermeable layer depth. In addition, a profile of water contents is also generated from this version of *DRAINMOD* using soil-water characteristic data based on the assumption that hydrostatic conditions are prevalent in the profile at the end of the day. This approach for computing fluxes and water contents proved to be reliable for shallow water table soils as indicated by comparisons with numerical solutions to the Richards equation for saturated and unsaturated flow (Skaggs et al., 1991; Kandil et al., 1992).

Since *DRAINMOD* fluxes may be computed at midpoint between the drains or as the average vertical flux in the zone between drains depending on the drainage algorithm used, the predicted solute concentrations correspond to the same location. An average concentration at the drain is approximated by dividing the total lateral mass transport in the saturated zone ($\Sigma\ q_l\ C$) by the estimated drainage rate.

Because ammonium-nitrogen losses are generally low in poorly drained soils, only nitrate-nitrogen is considered in this version of the model. *DRAINMOD-N* uses functional relationships to quantify other processes besides NO_3-N transport, as follows:

$$\Gamma = \Gamma_{fer} + \Gamma_{mnl} - \Gamma_{upt} - \Gamma_{den} \tag{5}$$

where Γ_{fer} stands for fertilizer dissolution, Γ_{mnl} for net mineralization, Γ_{upt} for plant uptake, and Γ_{den} for denitrification. Fertilizer dissolution is quantified by a zero-order function:

$$\Gamma_{fer} = K_{fer} \frac{A_{fer}}{D_{fer}} \tag{6}$$

where K_{fer} is the dissolution rate [T^{-1}], D_{fer} is the depth at which the fertilizer was incorporated [L], and A_{fer} is the amount of fertilizer present in D_{fer} [$M\ L^{-2}$].

Net mineralization is also represented by a zero-order term:

$$\Gamma_{mnl} = K_{mnl}\,\rho\,O_n \tag{7}$$

where K_{mnl} is the net mineralization rate [T^{-1}], ρ is the soil bulk density [$M\ L^{-3}$], and O_n is the concentration of organic nitrogen present in the i layer [$M\ M^{-1}$]. O_n is estimated using the following expression by Davidson et al. (1978):

$$O_n = O_{n_{max}}[\exp(-0.025z)] \qquad (8)$$

where On_{max} is the maximum organic nitrogen concentration in the top layer.

Plant uptake is estimated using a relationship similar to that employed by Shaffer et al. (1991):

$$\Gamma_{upt} = \frac{Ry\ \%N\ \Delta ft}{Rz} \qquad (9)$$

where Ry is a relative yield value [$M\ L^{-2}$] obtained from *DRAINMOD*, $\%N$ is the percentage of nitrogen present in the plant/crop, Rz is root length [L], and Δft is a fractional N-uptake demand given by an N-uptake versus growing season curve presented by Shaffer et al. (1991).

Denitrification is approximated by a first-order equation, as follows:

$$\Gamma_{den} = K_{den}\ \theta_{i_{old}}\ C_{i_{old}} \qquad (10)$$

where K_{den} is the denitrification rate [T^{-1}].

Soil moisture factors are also used to account for the effect of aerobic or anaerobic conditions on the different reaction rate coefficients. The functional relationships presented by Johnsson et al. (1987) for denitrification and mineralization are adopted in *DRAINMOD-N*. These factors increase and decrease the denitrification and mineralization rates, respectively, when soil water saturation is approached, and vice versa.

Rainfall deposition of N is also considered in *DRAINMOD-N*. Water infiltrating into the soil is assumed to have a NO_3-N concentration which is used in the mass balance of the first layer.

A global mass balance is performed at the end of the simulation: total nitrate-nitrogen amounts present in the soil solution at the beginning and end of the simulation, and cumulative amounts for rainfall deposition, fertilizer dissolution, plant uptake, net mineralization, denitrification, and drainage losses are computed to yield a simulation mass balance error.

MODEL VALIDATION

The purely advective component of *DRAINMOD-N* was tested by comparison with numerical solutions for long-term solute transport (Kandil et al., 1992). Both models use DRAINMOD to compute soil water fluxes.

The validating simulations consisted of assigning a uniform initial concentration of 1 mg l^{-1} to a clay loam soil with subsurface drains and applying a continuous infiltration rate of 0.3 cm d^{-1} for 30 days. The infiltrating water was assumed to have a concentration of 2 mg l^{-1}. The drainage system parameters and soil properties used in the validation are given in Table 1.

Table 1. Inputs to *DRAINMOD-N*.

		Sandy Loam	Clay Loam
1.	Soil Properties:		
	Sat. Hyd. Cond. (m d^{-1})	0.960	0.024
	θ_{sat} (cm^3 cm^{-3})	0.459	0.453
	θ_{wp} (cm^3 cm^{-3})	0.044	0.210
	Bulk Density (g cm^{-3})	1.6	1.1
	Organic Matter (% wt)	2.0	10.0
	Max. Org-N (µg g^{-1})	1000	3000
2.	Drainage System Parameters:		
	Drain Depth (m)	1.0	
	Drain Spacing (m)	20.0	
	Depth to Imperm. Layer (m)	1.6	
	Effective Drain Radius (cm)	1.5	
3.	Controlled Drainage Parameters:		
	Weir Depth Set to 50 cm between May 20 and Aug 15		
4.	Corn Parameters:		
	Desired Planting Date	April 15	
	Length of Growing Season (d)	130	
	Max. Effective Root Depth (cm)	30	
	Nitrogen Content in Plant (%)	1.55	
	Fertilizer Rate (kg ha^{-1})	150 (Full Appl.)	100+50 (Split Appl.)
	Date Fertilizer Application	April 15	April 15 and May 22
	Depth Fertilizer Incorporated (cm)	10	
5.	Nitrogen Transport and Transformations Parameters:		
	K_{min} (d^{-1})	0.00002	
	K_{den} (d^{-1})	0.03	
	K_{fer} (d^{-1})	0.15	
	NO$_3$-N of Rain (mg l^{-1})	0.80	

Considering the lesser computational effort required by *DRAINMOD-N*, Fig. 1 shows a reasonable agreement between concentrations of a non-reactive solute simulated by *DRAINMOD-N* and by a numerical model (Kandil et al., 1992). Field experiments are underway to test the model under conditions of conventional drainage and controlled drainage.

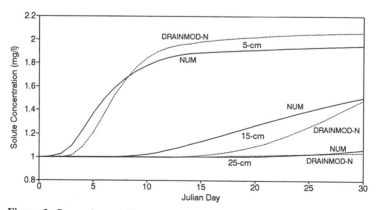

Figure 1. Comparison of Simulated Solute Concentrations Using *DRAINMOD-N* and a Numerical Model for a Period of Continuous Infiltration in a Clay Loam.

MODEL APPLICATION

DRAINMOD-N was used to simulate the effects of drainage and fertilizer management on the movement and fate of nitrogen in two soils. Simulations were conducted for corn production during a relatively wet year (1959) in Wilson, N.C. Inputs to *DRAINMOD-N* are summarized in Table 1. The reaction rates and corn N-content were determined from the literature (Johnsson et al., 1987; Harmsen et al., 1991; Meisinger and Randall, 1991). An initial NO_3-N concentration in solution, linearly varying from 20 mg l^{-1} at the surface layer to 0 mg l^{-1} at the depth of the impermeable layer (160 cm), was assumed at the start of the simulation.

Eight hypothetical scenarios were simulated with *DRAINMOD-N* to evaluate the effects of several factors influencing nitrogen transport. These scenarios consisted of interactions of the following treatments: two soil types (a sandy loam and a clay loam), two water table management practices (conventional and controlled drainage), and two fertilizer applications (full and split). Detailed information for all treatments is presented in Table 1.

Daily NO_3-N concentrations, calculated for 5-cm increments and averaged within four layers (0-40, 40-80, 80-120, and 120-160 cm), and rainfall amounts are presented in Fig. 2-7.

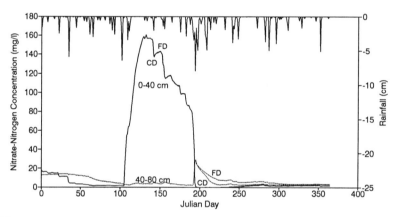

Figure 2. Nitrate in Top Layers for Sandy Loam with Full Fertilizer Application.

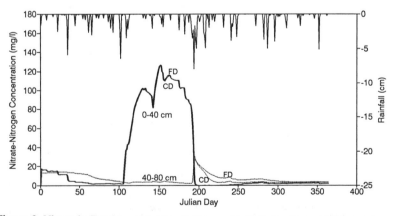

Figure 3. Nitrate in Top Layers for Sandy Loam with Split Fertilizer Application.

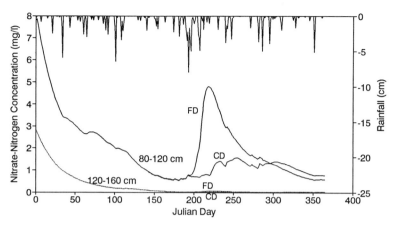

Figure 4. Nitrate in Bottom Layers for Sandy Loam with Full Fertilizer Application.

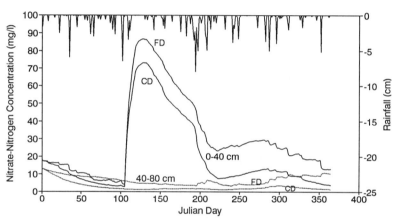

Figure 5. Nitrate in Top Layers for Clay Loam with Full Fertilizer Application.

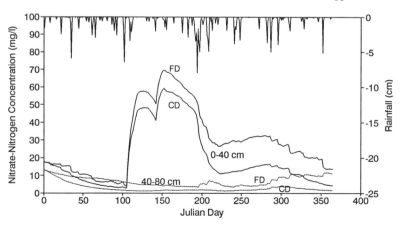

Figure 6. Nitrate in Top Layers for Clay Loam with Split Fertilizer Application.

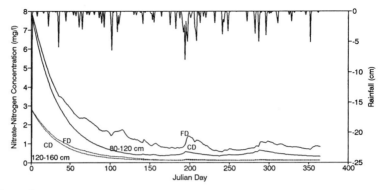

Figure 7. Nitrate in Bottom Layers for Clay Loam with Full Fertilizer Application.

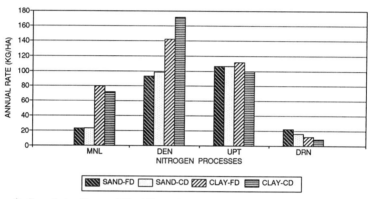

Figure 8. Cumulative Rates of Net Mineralization (MNL), Denitrification (DEN), Plant Uptake (UPT), and Drainage Losses (DRN) for Two Soil Types and Two Water Managements Practices.

Figures 2-7 indicate controlled drainage resulted in lower NO_3-N concentrations in the soil solution as compared to free drainage, but the effect was more evident in the clay loam (Fig. 5-7). This indicates denitrification, as induced by shallower water tables caused by controlled drainage, played a significant role in lowering those concentrations. This is consistent with results of previous research (Skaggs and Gilliam, 1981; Gilliam and Skaggs, 1986).

Split fertilizer application, as compared to a full application, predicted lower NO_3-N concentrations only in the upper layer (Fig. 2,3,5 and 6). Nitrate-nitrogen concentrations at depths of 80-160 cm in both soils were nearly identical for full and split fertilizer application, thus results for only the full application are shown in Fig. 4 and 7. The impact of splitting the fertilizer application was more pronounced for the sandy loam which was probably due to its greater subsurface drainage intensity. Splitting the application of fertilizer reduced the peak NO_3-N concentrations for the sandy loam from 167 to 140 mg l^{-1} in the top 40 cm (Fig. 2 and 3).

Figure 8 displays cumulative rates of net mineralization, denitrification, plant uptake, and subsurface drainage losses for both soil types and water management practices. Total mass balance errors ranged from 6.8 to 18.7 % and averaged 15.4 % for the one-year period of simulation. No significant differences in cumulative rates were found due to fertilizer application. Results shown in Fig. 8 are associated with the full fertilizer application. Net mineralization and denitrification rates were consistently lower in the sandy loam. Drainage losses were higher for the sandy loam. Relative yields were different between treatments (100 and 84% for the sandy and clay loam soils with free drainage, and 100 and 78% with controlled drainage, respectively) which caused a small variation in plant uptake rates. Controlled drainage resulted in lower mineralization and higher denitrification rates, especially for the clay loam. Controlled drainage also reduced drainage losses, but more significantly for the sandy loam. Higher denitrification rates and lower drainage losses associated with controlled drainage is consistent with results reported in the literature (Skaggs and Gilliam, 1981; Gilliam and Skaggs, 1986; Gilliam, 1987; Skaggs and Breve, 1992).

DRAINMOD-N shows promise of being a useful tool for evaluating the water quality effect of the design and management of drainage systems, and fertility and cultural practices. However, the current version of the model has several limitations: it does not consider ammonium transport and transformations, dispersive transport, and temperature-dependent reaction rate coefficients. These processes are important in some situations. Results from current field studies will be used to test the reliability and further modify the model.

SUMMARY AND CONCLUSIONS

A simulation model, *DRAINMOD-N*, was developed to predict nitrogen transport, uptake, and transformations in shallow water table soils with improved drainage. The model uses *DRAINMOD* to calculate vertical soil water fluxes in the profile. An explicit solution to the advective equation is used to predict nitrogen transport. Functional relationships are used to represent net mineralization, denitrification, plant uptake, and fertilizer dissolution. *DRAINMOD-N* was tested by comparison with a more complex numerical model. Results were in good agreement. The model was applied to evaluate the effect of soil type, and water management and fertilizer practices on nitrogen transport. Simulated results showed controlled drainage resulted in lower nitrate-nitrogen (NO_3-N) losses through subsurface drains, and in higher denitrification and lower mineralization rates, as compared to free drainage. Controlled drainage also decreased NO_3-N concentrations in the soil solution, but the impact was more pronounced in a clay loam than in a sandy loam. Split fertilizer application, as compared to a single application, predicted lower nitrate-nitrogen concentrations only in the upper layers, and had no effect on cumulative denitrification, mineralization, and drainage losses. *DRAINMOD-N* shows promise of being a useful tool for evaluating the water quality effect of design and management of drainage systems, and fertility and cultural practices. Results from current field studies will be used to test the reliability and further modify the model.

REFERENCES

1. Chung, S.O., A.D. Ward, N.R. Fausey, and T.L. Logan. 1991. Evaluation of the pesticide component of the ADAPT water table management model. ASAE Paper 91-2632, Am. Soc. Agric. Eng., St. Joseph, MI.

2. Clemente, R.S. and S.O. Prasher. 1992. *PESTFADE*, a model for simulating pesticide fate and transport in soils. A paper presented at the Beltsville Symposium XVII on agricultural water quality priorities, a team approach to conserving natural resources. May 4-8, 1992, Beltsville, MD.

3. Davidson, J.M., D.A. Graetz, P. Suresh, C. Rao, and H.M. Selim. 1978. Simulation of nitrogen movement, transformations, and uptake in plant root zone. U.S. Environ. Prot. Agency, Environ. Res. Lab., Athens, GA. EPA-600/3-78-029.

4. Gilliam, J.W. 1987. Drainage water quality and the environment. Keynote address. Proc. Fifth Nat. Drain. Symp., Am. Soc. Agric. Eng., ASAE Publ. 7-87. pp. 19-28.

5. Gilliam, J.W. and R.W. Skaggs. 1986. Controlled agricultural drainage to maintain water quality. J. Irrig. Drain. Div., 112, 254-263.

6. Harmsen, E.W., J.W. Gilliam, and R.W. Skaggs. 1991. Variably saturated 2-dimensional nitrogen transport. ASAE paper 91-2630, Am. Soc. Agric. Eng., St. Joseph, MI.

7. Johnsson, H., L. Bergstrom, and P.E. Jansson. 1987. Simulated nitrogen dynamics and losses in alayered agricultural soil. Agric. Ecosys. Environ., 18, 333-356.

8. Kandil, H., C.T. Miller, and R.W. Skaggs. 1992. Modeling long-term solute transport in the unsaturated zone. Water Resour. Res. (In press).

9. Kanwar, R.S., H.P. Jonhson, and J.L. Baker. 1983. Comparison of simulated and measured nitrate losses in tile effluent. Trans. ASAE, 26, 1451-1457.

10. Karvonen, T. and R. Peltomaa. 1992. Testing of a simulation model aimed at predicting the environmental impact of subirrigation and controlled drainage. In: Proc. Int. Conf. Subirrig. Controlled Drain., Michigan State Univ., East Lansing, MI, Aug. 12-14. (In press).

11. Meisinger, J.J., and G.W. Randall. 1991. Estimating nitrogen budgets for soil-crop systems. In: R.F. Follet, D.R. Keeney, and R.M. Cruse (eds.): Managing nitrogen for groundwater quality and farm profitability. Soil Sci. Soc. Amer., Inc., Madison, WI. pp. 85-124.

12. Munster, C.L., R.W. Skaggs, J.E. Parsons, R.O. Evans, and J.W. Gilliam. 1991. Modelling aldicarb transport under drainage, controlled drainage, and subirrigation. ASAE Paper 91-2631, Amer. Soc. Agric. Eng., St. Joseph, MI.

13. Parsons, J.E., R.W. Skaggs, and J.W. Gilliam. 1989. Pesticide fate with *DRAINMOD/CREAMS*. In: D.B. Beasley, W.G. Knisel, and A.P. Rice (eds.): Proc. *CREAMS/GLEAMS* Symp., Publ. 4, Agric. Eng. Dept., Univ. Ga., Coastal Plain Exp. Sta., Tifton, GA. pp. 123-135.

14. Shaffer, M.J., A.D. Halvorson, and F.J. Pierce. 1991. Nitrate leaching and economic analysis package (*NLEAP*): model description and application. In: R.F. Follet, D.R. Keeney, and R.M. Cruse (eds.): Managing nitrogen for groundwater quality and farm profitability. Soil Sci. Soc. Amer., Inc., Madison, WI. pp. 285-322.

15. Skaggs, R.W. 1978. A water management model for shallow water table soils. Rept. 134, North Carolina Water Resour. Res. Inst., North Carolina State Univ., Raleigh, NC.

16. Skaggs, R.W. and M.A. Breve. 1992. Environmental impacts of water table control. Keynote address. In: Proc. Int. Conf. Subirrig. Controlled Drain., Michigan State Univ., East Lansing, MI, Aug. 12-14. (In press)

17. Skaggs, R.W. and J.W. Gilliam. 1981. Effect of drainage system design and operation on nitrate transport. Trans. ASAE, 24, 929-934.

18. Skaggs, R.W., T. Karvonen, and H. Kandil. 1991. Predicting soil water fluxes in drained lands. ASAE paper 91-2090, Am. Soc. Agric. Eng., St. Joseph, MI.

19. Tremwel, T.K. and K.L. Campbell. 1992. *FHANTM*, a modified *DRAINMOD*: sensitivity and verification results. ASAE Paper 92-2045, Am. Soc. Agric. Eng., St. Joseph, MI.

AN EVALUATION OF THE ADAPT WATER TABLE MANAGEMENT MODEL

S.O. Chung	A.D. Ward	N.R. Fausey	W.G. Knisel	T.J. Logan
Member	Member	Associate		
ASAE	ASAE	Member		
		ASAE		

ABSTRACT

A subsurface water quality model, ADAPT (Agricultural Drainage and Pesticide Transport), has been developed by modifying GLEAMS and extending its use by adding drainage and subirrigation algorithms from DRAINMOD. Major processes of plant uptake, transformation and transport of nutrients and pesticides are included in the model. The hydrologic and pesticide components were evaluated with measured data from three water table management field facilities in northern Ohio. With little or no calibration, the ADAPT model provided reasonable estimates of pesticide concentration in the perched water and pesticide losses due to surface runoff, subsurface drainage, plant uptake, and decomposition.

Key Words: ADAPT, water table management, model, water quality, subsurface drainage, subirrigation, nutrients, pesticides, erosion

INTRODUCTION

Total U.S. agricultural production is more than 262% of what it was in 1930, while agricultural chemical usage has increased twenty-fold. Annually about 19 billion kilograms of fertilizer and 450 million kilograms of pesticide active ingredient are used in the United States. The increased use of agrichemicals has had an impact on the environment. In 1988, 46 pesticides were found in ground water resources of 26 states (Office of Technology Assessment, 1990). Water table management practices to help maintain the agricultural productivity and profitability without causing any degradation of water quality, may be required throughout the United States. In 1985, 44 million hectares of agricultural land benefited from drainage improvements.

The authors are: Sang-Ok Chung, Assistant Professor, Agricultural Engineering Department, Kyungpook National University, Taegu, KOREA; Andrew D. Ward, Associate Professor, Agricultural Engineering Department, The Ohio State University; Norman R. Fausey, Research Leader, USDA-ARS Soil Drainage Research Unit; Walter G. Knisel, Senior Researcher, Agricultural Engineering Department, The University of Georgia, and Terry J. Logan, Professor, Agronomy Department, The Ohio State University.

The CREAMS model (Knisel, 1980) was developed to evaluate non-point source pollution from field size areas. The migration of agricultural chemicals into the ground water system necessitated the modification of CREAMS into the GLEAMS model (Leonard et al., 1987). To incorporate water table management practices the ADAPT (Agricultural Drainage and Pesticide Transport) model was developed by extending GLEAMS to provide a comprehensive model to simulate the quantity and quality of flows associated with water table management systems (Ward et al. 1988, Chung et al. 1991 and 1992). This paper provides an overview of the model and evaluation studies on its hydrology and pesticide components.

MODEL DESCRIPTION

The ADAPT model is a daily simulation model and has three components: hydrology, erosion, and chemical transport. The hydrology component of the ADAPT model includes snowmelt, surface runoff, macropore flow, evapotranspiration, infiltration, subsurface drainage, subirrigation, and deep seepage. Detailed descriptions of these can be found in Chung et al. (1991 and 1992). The ADAPT model is PC based and written in FORTRAN language with modular programming techniques.

Hydrologic Component

The soil system modelled is from the soil surface down to an impermeable or restrictive layer. The top 10 mm is taken as the first layer, the rest of the effective rooting depth is equally divided into 6 layers, and the profile from the bottom of the root zone to the impeding layer is divided into two layers - making a total of 9 computational soil layers.

The snowmelt component is based on the theory presented by Anderson and Crawford (1964) and Viessman, Jr. et al. (1989). Snowmelt occurs by radiation, rainfall, conduction, convection, and condensation. Surface runoff is assumed to occur only if there is sufficient rainfall to fill the depression storage on the soil surface. Two options are provided for determining surface runoff depths based on SCS curve numbers and antecedent soil moisture. The first option is taken from the GLEAMS model and the second option is the original SCS method as modified by Schmidt and Schulze (1987). An approach is included in the model to account for surface sealing following intense rainfall on dry soils. In the ADAPT model, soil surface cracks due to dry soil condition are considered in the form of a runoff adjustment factor following Pathak et al. (1989).

Potential ET can be calculated by either the Ritchie or Dorenbos-Pruitt method. The latter is an added option in ADAPT which is not found in GLEAMS. After determining the PET, evaporation and transpiration are computed separately as a function of the leaf area index (Knisel, 1980). Plant transpiration is determined for each soil layer with respect to root zone depth based on the approach in GLEAMS. The volume of water available for infiltration is rainfall and surface ponding minus runoff and macropore flow. A modified Green-Ampt equation (Mein and Larson, 1971) is used to calculate infiltration time. The assumption is made that the wetting front advances to the next layer when soil moisture content in a layer is at field capacity. When the wetting front reaches the water table, any additional infiltration will raise the water table height as pore spaces are filled from field capacity to saturation. If the total volume of available water does not infiltrate within 24 hours, the remainder carries over to the following day as surface ponding which is subject to evaporation and infiltration.

Subsurface drainage and subirrigation algorithms are based on DRAINMOD (Skaggs, 1980). When the water table is at the soil surface, Kirkham's equation is used. Hooghoudt's steady state equation is used when the water table is below the soil surface. Both ADAPT and DRAINMOD use Darcy's equation to determine deep seepage through the impermeable layer but the modeling approaches are not identical.

Chemical Transport

Nutrient and pesticide transport in the system is generally the same as in the GLEAMS model. Each day pesticide partitioning and degradation are calculated. The model estimates the concentration and the mass of pesticide transported by runoff, sediment, subsurface drainage, and deep seepage. It also calculates the mass of pesticide uptake by plants and loss by decomposition. Decomposition includes biodegradation and hydrolysis of a pesticide. The decomposition of the pesticide in the soil and on the plant leaf is assumed to follow first order kinetics. It is assumed that the pesticide concentrations of solid and solution phases in the soil profile are under equilibrium condition during the simulation period.

The processes of nutrients transformation in the ADAPT model were adopted from the GLEAMS new nutrient model (Knisel et al., 1992). Two elements of nutrients, nitrogen and phosphorus, are included in the model. Common processes for both elements are mineralization from crop residue, soil organic matter, and animal waste, immobilization to crop residue, plant uptake, nutrient partitioning between soil and solution phases, and nutrient transport by various mechanisms. On the other hand, the nitrogen has unique processes such as nitrogen fixation by legumes, denitrification, nitrogen in rainfall, ammonia volatization from animal waste, ammonification, and nitrification.

Nitrogen mineralization is considered as a first order ammonification process and a zero order nitrification process. Denitrification, the change of soil nitrate to nitrogen gases, occurs when soil water content exceeds field capacity. This is a first-order process with a rate constant as a function of organic carbon, water content and temperature. The processes of nitrogen transport include nitrogen in rainfall and fertilizer, nitrogen in runoff and sediment, those in plant uptake, evaporation, subsurface drainage, and deep seepage flows. The processes of phosphorus transformations include mineralization and immobilization, while processes of phosphorus transport include phosphorus in runoff and sediment, plant uptake, evaporation, subsurface drainage, and deep seepage flows. Nutrient and pesticide concentrations discharging from a subsurface drain are very difficult to determine. Currently, nutrient and pesticide concentration in subsurface drainage is assumed to be the solution concentration in the water table layer.

The vertical movement of pesticides and nutrients includes macropore flow, infiltration, and deep seepage. In macropore flow, water and chemical moves down to the water table within 24 hours. During infiltration, nutrients and pesticide are modeled as moving downward in sequence from one soil layer to the next. In each soil layer, nutrients and pesticides are added by infiltrating water from above, it equilibrates between the solid and solution phases, and then the solution moves downward to the next layer. This process is repeated until no further downward movement of water exists. Evaporating water transports pesticides in solution upward in the soil profile. The plant uptakes nutrients and pesticides in solution by transpiration in each layer depending on the root distribution in the soil. ADAPT can be used to model pesticide injection in the subirrigation line. During subirrigation, it is assumed that pesticide moves upward along with the subirrigated water from the bottom layer to the water table layer.

MODEL INPUTS AND OUTPUTS

To use the model, input data for weather, soil, crop, subirrigation/drainage system, and pesticide and nutrient parameters are required. Weather data include daily rainfall, air temperature, radiation, and wind speed. Soil data are soil texture, thickness of horizons, organic matter content, soil water characteristics, and hydraulic conductivity. Crop data such as effective rooting depth and leaf area index as a function of growth stage are required. Subirrigation/drainage system input parameters include drain depth, spacing, diameter, outlet weir height, and depth to an impermeable layer. Surface storage depth and SCS curve number at antecedent moisture condition two are also necessary as inputs.

Pesticide parameters include pesticide application date, amount, method, pesticide water solubility, foliar and soil half lives and the adsorption constant. Nutrient parameters include fertilizer and manure application date, amount, method; a nutrient partitioning coefficient which is calculated internally based on the clay content; crop data such as name, leguminous, potentila yield, dry matter ratio, C/N ratio, N/P ratio, etc.; and nitrogen concentration in rainfall. Initial conditions are defined by parameters such as crop residue on the soil surface, base saturation, pH, calcium carbonate content, various nitrogen and phosphorous concentrations in the soil horisons.

Output data are monthly sums of surface runoff, subsurface drainage, combined surface runoff and subsurface drainage volumes, monthly rainfall, evapotranspiration, deep seepage, and subirrigation volumes. Pesticide output includes concentration and mass in surface runoff, sediment, subsurface drainage, and deep seepage. It also includes pesticide concentrations in the soil layers and masses decomposed and uptaken by the plant. Nutrient output includes concentration and mass in surface runoff, sediment, subsurface drainage, and deep seepage. It also includes nutrient concentrations in the soil layers and masses uptaken by the plant.

MODEL EVALUATION STUDIES

Hydrologic Component

Complete details of this study are provided by Chung et al. (1992). The hydrologic component was evaluated with subsurface drainage field data from Castalia, Ohio. Studies conducted by Schwab et al. (1975) and Skaggs et al. (1981) provided most of the input data for this study. The experimental site was located at the North Central Branch, Ohio Agricultural Research and Development Center near Sandusky. The site was nearly flat (0.2% slope) and the predominant soil type was a Toledo silty clay. Field plots were planted mostly in corn with conventional tillage. Exceptions occurred in 1965 and 1966, respectively, when soybeans and oats were planted. Field installations consisted of plots with surface drainage only, subsurface tile drainage only, and a combination of surface and subsurface drainage, each with four replications.

Simulations were for a continuous 10-year period. There were two major storms during the 10 year period: 191 mm on July 13, 1966 and 297 mm on July 5, 1969. Both storms exceeded a 100 year return period rainfall and the field monitoring systems were inundated resulting in a loss of flow records. Figure 1 shows comparisons of the seasonal sums of observed and predicted runoff, subsurface drainage, and combined discharges. In general, the results suggest that runoff is overpredicted and drainage underpredicted. However, considering no calibrated parameters were used in the model evaluation, the predicted combined values are in good agreements with the observations. A statistical analysis of observed versus predicted monthly runoff, drainage, and combined flows gave r-squared values of 0.87, 0.68, and 0.90 respectively. A sensitivity analysis was conducted with several of the input values and generally, observed values of runoff, drainage, and combined flows fell within the predicted ranges by using reasonable estimates for the tested model inputs.

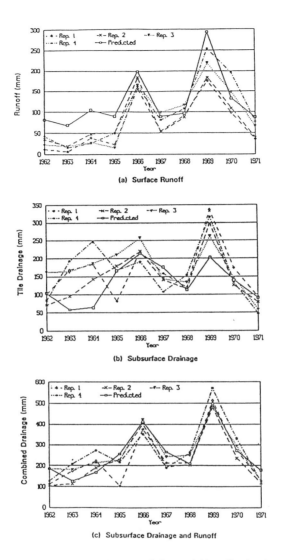

Figure 1. Comparisons of seasonal sums of observed (4 replications) and predicted values.

Pesticide Component

Complete details of this study are provided in Chung et al. (1991). The following two sets of data were used for the pesticide evaluation study: four years (1987-1990) of subirrigation data from Wooster, Ohio; and two years (1987-1988) tillage system and runoff-subsurface drainage data from Hoytville, Ohio. The experimental site at Wooster was located at the Ohio Agricultural Research and Development Center at Wooster, Ohio, on a poorly drained Ravenna silt loam soil. The site was nearly flat (less than 0.1% slope). Field plots were under conventional tillage and were planted in continuous soybeans until 1989, and starting in 1990, a soybean-corn rotation with each crop in half the plot. Only the soybean plots were included in this study. The other site was located at the North West Branch, Ohio Agricultural Research and Development Center, Ohio, on a Hoytville silty clay soil, a poorly-drained soil formed in late Wisconsin high-lime glacial till which is high in organic matter, has a neutral pH, and high fertility levels. Details of the site and soil characteristics are given by Logan (1987). Figure 2 shows a comparison of predicted and observed metolachlor concentration in the perched water at Wooster, Zone 2. Model predicted values are in acceptable agreement with the observed data considering order of magnitude variations are not uncommon at these low concentrations.

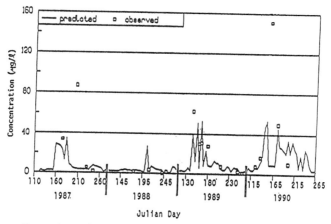

Figure 2. Comparison of observed and predicted metolachlor concentration in the perched water at Wooster, Zone 2.

Figure 3 shows observed and predicted monthly pesticide loads in the surface runoff at the Hoytville site. The metolachlor loads were higher than 1 g/ha only in one month during 1987. The monthly atrazine loads in surface runoff were less than 0.2 g/ha. Figure 3(a) represents a good fit case while 3(b) represents a poor fit case. Both Figure 3(a) and 3(b) show some discrepancies between observed and predicted values. The discrepancies could be due to a number of factors including inaccurate input parameter values, sampling error, analytical error, and model capability itself. Most of the discrepancies can be accounted for by making small adjustments in the hydrology and pesticide input parameters such 44as curve number for antecedent moisture condition two, hydraulic conductivity, pesticide soil half life, and adsorption constant. However, the model gives reasonable predictions and most predicted values are in acceptable agreements with the observed data considering the low loads and concentration level.

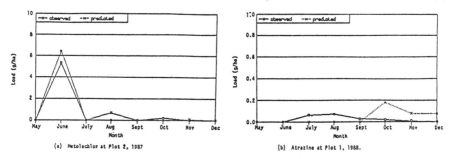

Figure 3. Comparison of observed and predicted monthly pesticide loads in surface runoff at Hoytville.

Table 1 shows a comparison of predicted and observed pesticide loss by surface runoff and subsurface drainage at Hoytville. The predicted flow values were within the range of variations in four zones. Two of the four observed subsurface flows in 1988 were zeros. The predicted pesticide losses were within the range of variations in four zones except metolachlor which showed a large value. Pesticide loss by subsurface drainage was less than 0.1% for all the pesticides applied indicating little pesticide transport to the depth of the subsurface drain system. Without any calibration, this analysis with ADAPT model provided reasonable estimates of pesticide concentration in the perched water and pesticide losses due to surface runoff, subsurface drainage, plant uptake, and decomposition.

Table 1. Comparison of predicted and observed pesticide loss at Hoytville, Plots 1 to 4.

Year	Parameter	Amount Applied	SURFACE RUNOFF			SUBSURFACE FLOW		
			Predicted	Observed		Predicted	Observed	
				Range	Mean		Range	Mean
1987	flow depth (mm)	NA*	9.80	5.40 - 29.1	14.6	9.80	2.90 -16.7	7.7
	Metolachlor (g/ha)	2250	7.21	0.84 - 5.48	1.92	0.20	0.19 - 0.61	0.37
	Metribuzin (g/ha)	430	0.80	0.25 - 1.29	0.80	30.07	0.07 - 0.25	0.13
1988	flow depth (mm)	NA*	19.5	5.2 - 58.8	24.5	0.20	0.0 - 13.9	6.3
	Atrizine (g/ha)	2250	0.33	0.91 - 14.79	5.84	0.00	0.0 - 0.04	0.01

*NA - Not applicable

FUTURE RESEARCH

An evaluation of the nutrient component of the model will be completed by early in 1993. Research is being conducted to include algorithms which account for macropore flow due to worm holes and organic matter. Research has also been initiated to link the model to the soybean crop development model SOYGRO (Jones et al.,1987). It is also planned to link the model with the corn crop development model CERES-MAIZE (Jones and Kiniry, 1986). When the linkage of ADAPT to the two crop models has been completed it will be used in a study which will evaluate the usefulness of satellite thematic mapper data to provide calibration and model input parameter information.

REFERENCES

Anderson, E. A. and N. H. Crawford. 1964. The synthesis of continuous snowmelt runoff hydrographs on a digital computer. Technical Report No. 36, Dept. Civil Engineering. Stanford University.

ACRE. 1991. Testing well water for contamination. ACRE Fact Sheet No. 19. Alliance for a Clean Rural Environment, P. O. Box 413708, Kansas City, MO 64179.

Chung, S. O., A. D. Ward, N. R. Fausey, and T. J. Logan. 1991. Evaluation of the pesticide component of the ADAPT water table management model. Paper No. 91-2632, ASAE Winter Meeting, Chicago, Illinois.

Chung, S. O., A. D. Ward, and C. W. Schalk. 1992. Evaluation of the hydrologic component of the ADAPT water table management model. TRANSACTIONS OF ASAE, 35(2):571-579.

Jones, C. A., and Kiniry, J. R. 1986. CERES-Maize: A simulation model of maize growth and development. Texas A&M University Press, College Station, Texas, 194 p.

Jones, J. W., K. J. Boote, S. S. Jagtap, G. Hoogenboom, and G. G. Wilkerson. 1987. SOYGRO v5.4, User's Guide. Florida Agricultural Experiment Station Journal No. 8304.

Knisel, W. G. 1980. (Ed.) CREAMS: A field scale model for chemicals, runoff, and erosion from agricultural management systems. U.S.D.A. Conservation Research Report No. 26.

Knisel, W. G., R. A. Leonard, and F. M. Davis. 1992. The GLEAMS model: Plant nutrient component. Part I. Model documentation. USDA-ARS Report (In preparation).

Leonard, R. A., W. G. Knisel, and F. M. Davis. 1990. The GLEAMS model - A tool for evaluating agrichemical ground-water loading as affected by chemistry, soils, climate and management. TRANSFERRING MODELS TO USERS, November issue, pp. 187-197. American Water Resources Association.

Logan, T. J. 1987. Tile drainage water quality: a long-term study in NW Ohio. Proc. Third Int. Workshop on Land Drainage. The Ohio State University, Columbus, C53-64.

Mein, R. G. and C. L. Larson. 1971. Modeling infiltration component of the rainfall-runoff process. Bull. 43, Water Resources Research Center, Univ. of Minnesota, Minneapolis.

Office of Technology Assessment. 1990. Beneath the bottom line: Agricultural approaches to reduce agrichemical contamination of groundwater. Summary. U. S. Government Printing Office. Washington, D. C.

Pathak, P., K. B. Laryea, and R. Sudi. 1989. A runoff model for small watersheds in the semi-arid tropics. TRANSACTIONS OF ASAE 32(5):1619-1624.

Schmidt, E. J. and R. E. Schulze. 1987. SCS-based design runoff-user manual. Agricultural Catchments Research Unit Rep. No. 25, Department of Agricultural Engineering, University of Natal, South Africa.

Schwab, G. O., N. R. Fausey, and C. R. Weaver. 1975. Tile and surface drainage of clay soils. II. Hydrologic performance with field crops (1962-72). III. Corn oats and soybean yields (1962-72). Res. Bull. 1081, Ohio Agricultural Research and Development Center, Wooster, Ohio.

Skaggs, R. W. 1980. A water management model for artificially drained soils. North Carolina Agricultural Research Service, Technical Bulletin No. 267, N. C. State University Raleigh.

Skaggs, R. W., N. R. Fausey, and B. H. Nolte. 1981. Water management model evaluation for north central Ohio. TRANSACTIONS OF ASAE 24(4):922-928.

Soil Conservation Service, USDA. 1972. National engineering handbook: Section 4, Hydrology. Washington, D. C.

Viessman, W., Jr., G. L. Lewis, and J. W. Knapp. 1989. Introduction to hydrology. 3rd Ed. Harper & Row, New York.

Ward, A. D., C. A. Alexander, N. R. Fausey, and J. D. Dorsey. 1988. The ADAPT agricultural drainage and pesticide transport model. Proceedings of Modeling Agricultural, Forest and Rangeland Hydrology. ASAE, December 12-13, 1988, Chicago, IL. pp. 129-141.

ANALYTICAL METHOD OF DETERMINING WETLAND HYDROLOGY

Salvador Palalay   and   Frank Geter*

ABSTRACT

The water management simulation model, DRAINMOD, was used to quantify the presence of wetland hydrology of a strip of ground between two parallel drains. It was concluded that on a specific type of Fallsington soil and subjected to the 1958-1973 rainfall and temperature at the Wilmington Airport, Delaware, there would be wetland hydrology if 120 cm (4 ft) deep drains were spaced no closer than 9750 cm (320 ft).

The result of the simulation was also used to identify easily measurable climatic variables that were significant in affecting the highest 7-day average water table. Regression analyses were done on several variables and some predictive equations were derived. The analyses showed that the 14-day rainfall (7-day AFTER and 7-day BEFORE the rise of the highest 7-day average water table) had the greatest impact. Rainfall of a prior period up to three months also had some effect and the coefficient of determination was slightly improved by including the prior 3-month rainfall as a second variable. **Keywords:** wetland, wetland hydrology.

INTRODUCTION

According to the Federal Manual for Identifying and Delineating Jurisdictional Wetlands (FMIDJW) (1989), three criteria must be met in order for a site to be classified as wetland. The soil must be hydric, there must be a prevalence of hydrophytic vegetation, and it must have wetland hydrology. Hydric soil properties are nearly permanent and vegetation will be present as long as there has been no clearing, cutting, or burning. On the other hand, wetland hydrology, indicated by flooding, inundation, or soil saturation, changes constantly throughout the year. During a field visit, wetland hydrology indicators may not be present. And even if they are, one would not know how long they will be there and how often.

Fortunately, in most situations the degree of flooding or saturation does not need to be quantified because field indicators are overwhelming. For example, if we find that the vegetation are all obligate wetland plants or there is standing water most of the time, then there would be no question that the site has wetland hydrology. However, if the vegetation has been removed or the landscape has been disturbed and changed the hydrology so that there is no standing water at the time of the

---

*Salvador Palalay, Planning/Hydraulic Engineer; and Frank Geter, Environmental Engineer, Soil Conservation Service, Northeast National Technical Center, Chester, PA

field visit, there may be some doubt that wetland hydrology existed or still exists. In this situation, an analytical method of determining the hydrology may be needed.

The manual states that wetland hydrology exists when there is inundation or saturation for a duration of 7 days or longer during the growing season at a probability of at least 50 percent in any year. And there is saturation to the surface when the water table is within a specified depth depending on the soil's drainage class and hydraulic conductivity.

One direct way of determining wetland hydrology is to monitor the groundwater table for some period of years and if the water table is within the specified depth for seven days for at least 50 years out of 100 years of monitoring, then the area has wetland hydrology. This procedure would be used most of the time if not for the costs and the need of clients for an immediate answer. Groundwater monitoring is expensive and if used exclusively, it must be done for many years to provide an answer. However, when used with a model, it may provide an answer in a relatively short time.

THE MODEL

The water management simulation model DRAINMOD (Skaggs, 1980) computes the water balance of a soil profile above a groundwater divide located mid-way between two parallel drains. These drains are usually man-made but they may be natural drains in few instances. The model computes and keeps track of the water table and considers rainfall, infiltration, surface and subsurface flows, storage and deep seepage. Inputs to the model include climatological data, soil parameters, crop parameters, and drain parameters. A brief description of the major input data follows:

Climatological Data: The accuracy of the prediction is dependent on the complete description of the rainfall. A basic time increment of one hour is used. Ideally, the data should represent as many years as possible to include wet and dry years.

Soil Parameters: The Green and Ampt equation is used to compute infiltration from rainfall. Parameters for this equation are the vertical saturated hydraulic conductivity, effective suction at the wetting front and initial moisture deficit from saturation. These parameters are physically based and can be obtained from measurable properties of the soil.

Soil Moisture Characteristic: This property is a measure of how tightly water is held in the soil matrix in the unsaturated state. It impacts on the value of how much water is released from storage when the water table is lowered and how much moisture can move upward for plant use from the water table. This property is second in importance only to hydraulic conductivity in modeling soil water movement.

Depression Storage: Surface runoff is dependent on the average depth of depression storage that must be satisfied before surface runoff begins. Depression storage is broken down into a micro component representing storage in small depressions due to surface structure and cover, and a macro component which is due to larger surface depressions which may be altered by land forming and grading.

Hydraulic Conductivity: Water flow into drain pipes or ditches depends on the saturated horizontal hydraulic conductivity (HHC) of the soil, drain spacing and depth, depth to a restrictive layer, and water elevation. The flow is estimated using Hooghoudt's steady state equation, as used by Bouwer and van Schilfgaarde (1963). By far, the saturated HHC is the most important basic soil property in modeling soil water flow. In layered soils, the HHC values may be different for each layer, and for some fields, the soil may be different at various parts of the field.

Evapotranspiration: Evapotranspiration is estimated using the empirical method developed by Thornthwaite (1957) which is based on minimum and maximum daily temperatures. First, the daily potential evapotranspiration (PET) is calculated and distributed on an hourly basis. The PET represents the maximum amount of water that will leave the soil system by evapotranspiration when there is a sufficient supply of soil water. If soil water is not limiting, ET is set to PET; otherwise, ET is set to the smaller amount that can be supplied from soil moisture.

DESCRIPTION OF PROCEDURE

Given a site specific climatological data, soil parameters, and drain parameters, the model was run with a trial drain spacing to evaluate whether the water table position at the midpoint satisfies the wetland hydrology criterion. If not, a different spacing was tried and the evaluation repeated until the criterion is satisfied. Drain depth of 4 feet was evaluated.

Rainfall used was a 16-year data of hourly rainfall from 1958 to 1973 measured at the Wilmington Airport, Delaware. (Presently, this length of record is the only available data set for this area. SCS is updating this data set to include about a 30-year period). The growing season for wetland hydrology determination was taken as March 1st thru October 31st.

The soil evaluated was Fallsington, a hydric soil that occurs in the coastal plains of New Jersey, Delaware, Maryland, and Virginia. Based on the soil's drainage class and hydraulic conductivity, saturation to the surface would occur when the water table is within 45 cm (1.5 feet) from the surface. Wetland hydrology would, therefore, exist when the water table was within 45 cm (1.5 feet) from the surface for more than 7 days during the growing season for at least 8 years out of the 16-year period.

A spacing of 8840 cm (290 ft) was used for the first simulation. For each year of the simulation, the daily output from March 1st thru October 31st were examined to find out the number of days that the water table was within 45 cm (1.5 ft) from the surface. Table 1 shows an example of a daily output for April 1961. The rainfall of 5.74 cm on the 13th caused the water table to rise as indicated under the column DTWT (distance to water table) from 64.25 cm to 10.13 cm, all distances measured from the ground surface. The water table was within 45 cm through the 19th, for a period of 7 days.

Table 1. Example of a DRAINMOD'S Daily Output for April 1961

DAY	RAIN	INFIL	ET	DRAIN	TVOL	DDZ	DTWT	STOR	RUNOFF	WLOSS
1	.05	.05	.06	.34	5.09	.00	61.08	.00	.00	.34
2	.00	.00	.04	.32	5.45	.00	65.40	.00	.00	.32
3	.00	.00	.07	.29	5.81	.00	69.77	.00	.00	.29
4	.00	.00	.05	.27	6.13	.00	73.18	.00	.00	.27
5	.00	.00	.10	.25	6.48	.00	76.78	.00	.00	.25
6	.00	.00	.11	.23	6.81	.00	80.23	.00	.00	.23
7	.00	.00	.09	.21	7.11	.00	82.82	.00	.00	.21
8	.00	.00	.07	.20	7.37	.00	85.11	.00	.00	.20
9	.18	.18	.07	.18	7.45	.00	85.79	.00	.00	.18
10	2.39	2.39	.06	.28	5.41	.00	64.92	.00	.00	.28
11	.00	.00	.10	.29	5.80	.00	69.65	.00	.00	.29
12	.79	.79	.06	.28	5.35	.00	64.25	.00	.00	.28
13	5.74	5.74	.01	.58	.21	.00	10.13	.00	.00	.58
14	.00	.00	.13	.61	.95	.00	20.72	.00	.00	.61
15	.18	.18	.21	.55	1.53	.00	26.15	.00	.00	.55
16	.71	.71	.09	.55	1.46	.00	25.50	.00	.00	.55
17	.00	.00	.13	.53	2.13	.00	31.55	.00	.00	.53
18	.00	.00	.10	.49	2.71	.00	36.67	.00	.00	.49
19	.00	.00	.11	.46	3.28	.00	41.85	.00	.00	.46
20	.00	.00	.12	.43	3.84	.00	47.36	.00	.00	.43
21	.00	.00	.15	.40	4.39	.00	53.17	.00	.00	.40
22	.51	.51	.33	.36	4.56	.00	55.16	.00	.00	.36
23	.00	.00	.33	.35	5.24	.43	62.29	.00	.00	.35
24	.00	.00	.41	.30	5.96	1.46	69.31	.00	.00	.30
25	.66	.66	.42	.27	5.98	.00	71.63	.00	.00	.27
26	.58	.58	.31	.29	5.99	.00	71.73	.00	.00	.29
27	.00	.00	.16	.26	6.40	.06	75.94	.00	.00	.26
28	.48	.48	.14	.24	6.30	.00	74.95	.00	.00	.24
29	.13	.13	.16	.24	6.58	.00	77.88	.00	.00	.24
30	.00	.00	.15	.22	6.95	.17	81.33	.00	.00	.22

At a spacing of 8840 cm (290 ft), there were only 3 out of 16-years that the water table was within 45 cm (1.5 ft) from the surface for at least 7 days during the growing season, thus, wetland hydrology was not satisfied.

There was a need to increase the number of qualifying wet years in order for the site to have wetland hydrology. A wider spacing would reduce the flow of water to the drains thereby increasing the wetness of the area midpoint of the drains. We increased the drain spacing, re-ran the model, analyzed the output, and kept on adjusting the spacing until wetland hydrology was satisfied. After several trials, we determined that a spacing of about 9750 cm (320 ft) would result in 8 out of 16 of qualifying wet years thus, satisfying the wetland hydrology criterion, Table 2.

Table 2. Summary of Simulation at Drain Spacing of 9750 cm (320 ft), 120 cm (4 ft) drain depth, Fallsington soil.

Begin Date of Highest 7-day Average WT	Days That WT Within 45 cm From Surface	Begin Date of Highest 7-day Average WT	Days that WT Within 45 cm From Surface
Mar 20, 1958	16	Mar 17, 1968	9
Mar 12, 1959	0	Sep 3, 1969	3
Mar 1, 1960	0	Apr 14, 1970	12
Apr 13, 1961	10	Mar 1, 1971	11
Mar 12, 1962	6	Mar 1, 1972	2
Mar 6, 1963	2	Apr 4, 1973	11
Apr 20, 1964	2		
Mar 29, 1965	0		
Oct 19, 1966	7		
Mar 7, 1967	7		

WETLAND HYDROLOGY INDICATORS

Another purpose of the evaluation was to identify some readily measurable variables that affected depth to water table. Knowing these variables would help understand and interpret wetland hydrology indicators.

If we examine a daily output, for example, for April 1961 shown in Table 1, it showed that rainfall had the greatest influence on the water table, particularly, the rainfall that fell immediately before and during the rise of the water table. The daily evapotranspiration during that month was very small.

The variables analyzed are listed in Table 3. The highest 7-day average water table was the dependent variable and the rest were the independent variables.

Table 3.   Variable Names and Description

Variable Name	Description
WT_{7-day}	Highest 7-day average water table any time between March 1 thru April 30.
P_{7-day}	Precipitation that fell 7 days AFTER the start of the rainfall that caused the rise of the highest 7-day average water table
P_{14-day}	Precipitation that fell 7 days BEFORE and 7 days AFTER the rise of the highest 7-day average water table
P_{mo}	Precipitation for the month during occurrence of the rainfall that caused the rise of the highest 7-day average water table
P_{-1mo}	Precipitation for the month BEFORE the rainfall that caused the rise of the highest 7-day average water table
P_{-2mos}	Precipitation for the 2-months BEFORE the rainfall that caused the rise of the highest 7-day average water table
P_{-3mos}	Precipitation for the 3-months BEFORE the rainfall that caused the rise of the highest 7-day average water table
P_{-6mos}	Precipitation for the 6-months BEFORE the rainfall that caused the rise of the highest 7-day average water table
P_{annual}	Precipitation for the year during the occurrence of the highest 7-day average water table

Although there were two years during the 16-year period that the WT_{7-day} occurred during the fall, for this analysis I excluded those that occurred during that period and only included those that occurred during the spring months. The water table during the fall was significantly affected by evapotranspiration; therefore, the water table during these two periods were affected by different variables. Perhaps, including the temperature as another variable would improve the coefficient of determination, but including too many variables would defeat the purpose of finding a simple aid to improve the interpretation of other wetland hydrology indicators. Linear bivariate and multiple regression analyses were made and the results are shown in Table 4A and Table 4B.

Table 4A. Linear Bivariate Regression Statistics
Spacing = 9750 cm (320 ft),   Depth = 120 cm (4 ft)

Dependent variable: Highest 7-day average water table
                      Mean = 43.52 cm, Std deviation = 20.35 cm

Ind Variable	Regression Coefficients		Coefficient Of Determination	Standard Error Of Estimate
	c_0 (cm)	c_1		(cm)
$P_{7\text{-day}}$	69.24	-6.45	.49	15.04
$P_{14\text{-day}}$	88.39	-6.21	.70	11.61
P_{mo}	68.67	-2.62	.37	16.74
P_{-1mo}	65.04	-2.51	.23	18.52
P_{-2mos}	77.66	-2.26	.30	17.69
P_{-3mos}	95.48	-2.28	.39	16.39
P_{annual}	83.97	-0.40	.15	19.43

Table 4B. Linear Multiple Regression Statistics
Spacing = 9750 cm (320 ft), Depth = 120 cm (4 ft)

Dependent variable: Highest 7-day average water table
                      Mean = 43.52 cm, Std deviation =20.35 cm

Ind Variable	Regression Coefficients			Coefficient Of Determination	Standard Error Of Estimate
	c_0 (cm)	c_1	c_2		(cm)
$P_{7\text{-day}}; P_{-1mo}$	93.2	-6.7	-2.7	.75	10.92
$P_{7\text{-day}}; P_{-2mos}$	100.4	-6.2	-2.1	.75	10.96
$P_{7\text{-day}}; P_{-3mos}$	107.3	-5.5	-1.8	.74	11.23
$P_{7\text{-day}}; P_{annual}$	112.6	-6.7	-4.2	.66	12.80

DISCUSSION OF RESULTS

Results of the analyses showed that of the eight independent variables listed in Table 3, the 14-day precipitation had the greatest influence on the $WT_{7\text{-day}}$. For a 120 cm deep drain, the predictive equation would explain about 70 percent of the variation of the water table (coefficient of determination (COD) equals 70). The precipitation that occurred BEFORE the 14-day period showed a lower effect on the $WT_{7\text{-day}}$. Furthermore, the longer the prior period used, the lower the COD. Of particular interest was that the annual rainfall had a very low correlation with the $WT_{7\text{-day}}$.

Including the precipitation of previous months as a second independent variable slightly increased the accuracy of the predictive equation. The longer the period included, but only up to 3 months, the higher the coefficient of determination. After 3 months, the COD went down indicating that precipitation before the winter months had very little influence on the $WT_{7\text{-day}}$ of the following spring.

POTENTIAL USES OF THE MODEL

Guide for Field Office Use

The model was used to determine the minimum spacing between two parallel drains so that the strip of land midway between the drains would have wetland hydrology. These analyses were made for drain depths of 120 cm. Other depths could be analyzed the same way and using the results, a graph of spacings against depths could be developed for a specific soil.

Using this graph as a guide, a technician would be able to assess that where two parallel drains of a given depth are much closer than the spacing indicated by the guide, the land in between would probably not have wetland hydrology. And if the drains are much farther apart than the spacing indicated by the guide, then the area midway may have wetland hydrology. Those cases that fall near the line would require closer evaluation using other wetland indicators or using the model for more precise determination.

Other major wetland soils could be analyzed and guides developed for these soils. Developing these guides would reduce the need to use the program in field offices. The program would be run only for the border cases where more precise determination is needed.

Predictive Equations

The result of the simulation was also used to identify easily measurable climatic variables that were significant in affecting the WT_{7-day} and to derive the following predictive equation:

Drain Spacing at 9750 cm (320 ft) and Depth at 120 cm (4 ft)

WT_{7-day} (cm) = 88.39 $-6.21*P_{14-day}$ for .70 COD
or
= 93.2 $- 6.7*P_{7-day} -2.7*P_{-1mo}$ for .75 COD

For example, when a field visit is made on a potential wetland site, this equation would provide a guide where the water table should be. And if the observed water table is much different from the computed depth using the equation, the technician should evaluate why this is so and look for causes such as physical changes that may have affected the hydrology.

From this study, it was shown that the 14-day rainfall (7-day prior and 7-day after the rise of the water table) has the greatest impact on the WT_{7-day} during the spring months. This means that if there had not been any rainfall during the last 7 or so days, one should not expect to find a high water table in a wetland site in this soil.

Predictive equations could be developed for other high water table soils and used by technicians in the field to help interpret signatures in aerial photographs and other wetland hydrology indicators.

CONCLUSION

We believe that the value of this study was not so much for the specific spacing or predictive equations obtained but in showing how the model could be used to develop guides to determine wetland hydrology and to identify significant variables affecting wetland hydrology. Knowing these variables can help improve interpretation of other wetland hydrology indicators.

REFERENCES

1. Bouwer, H. and L. van Schilfgaarde. 1963. Simplified method of predicting the fall of water table in drained land. Transactions of the ASAE 6(4):288-291, 296.

2. Federal Interagency Committee for Wetland Delineation. 1989. Federal Manual for Identifying and Delineating Jurisdictional Wetlands. U.S. Army Corps of Engineers, U.S. Environmental Protection Agency, U.S. Fish and Wildlife Service, and U.S.D.A. Soil Conservation Service, Washington, D.C. Cooperative Technical Publication. 76 pp. plus appendices.

3. Thornthwaite, C. W. and J. R. Mather. 1957. Instructions and tables for computing potential evapotranspiration and the water balance. *In Climatology.* Drexel Inst. of Tech., Vol. 10(3): 185-311

4. Skaggs, R.W. 1980. DRAINMOD Reference Report. Methods for Design and Evaluation of Drainage Water Management Systems for Soils With High Water Tables. U.S.D.A. Soil Conservation Service, Washington, D.C. 1980. 169 pp. plus appendices.

CONCENTRATION OF AGRICULTURAL CHEMICALS IN DRAINAGE WATER

Safwat Abdel-Dayem[*]   Mohamed Abdel-Ghani[*]

ABSTRACT

Agricultural drainage systems will be completed in about 5.5 million acres by the year 2000 in Egypt. Subsurface field drainage systems of a composite layout discharge into a network of open main drains. Part of the drainage water is currently used in irrigation to meet the demands which exceed the fresh water supply. This will be expanded in the future for utilizing as mush as 7.0 billion cubic meter of drainage water for irrigation by the year 2000.

A monitoring program was carried out in an area with pilot drainage system in the Nile Delta to determine the concentration of agricultural chemicals in the drainage water. The temporal and spatial variations of these concentrations were observed and analyzed with respect to the agricultural practices and drainage intensity in the area.

Nitrates were found as the dominant fertilizer in the drainage water. Highest concentration was found in the discharge of deep lateral drains or drains closely spaced. The concentration of nitrates decreased while water was flowing through the drainage system due to mixing with waters with less concentration. Nitrates of relatively high concentrations were found in a shallow drinking groundwater well. No organophosphate pesticide residues was found in the water, the soil or the plants. However, residues of organochlorine pesticides which were baned about 12 years ago, were still present in the area.

INTRODUCTION

Subsurface drainage systems are implemented on a large scale in Egypt to sustain the irrigated agriculture and maximize crop yield. A total of 3.6 million acres is already provided with subsurface drains and another 2.0 million acres will be subsurface drained by year 2000. The covered field drainage systems discharge into the main open drains which transport the drainage effluent to the Nile River in Upper Egypt or to the North lakes and Mediterranean Sea from the Delta area. The total drainage water effluent was estimated in the early 1980's at 2 billion cubic meter from Upper Egypt and 14.0 billion cubic meter from the Nile Delta.

Irrigation water is supplied mainly by the River Nile and supplemented in few areas by groundwater. Egypt share from the Nile is fixed at 55.5 billion cubic meter and rainfall is negligible. The available limited water resources called for the re-use of the agricultural drainage water in irrigation to meet the future agricultural expansion and development needs. Plans were made to re-use up to 7 billion cubic meter of the agricultural drainage water by the year 2000. However, there are some concerns about the quality of this water and its possible adverse effect on the soil, crop, livestock and human life.

---

[*] SAFWAT ABDEL-DAYEM, Director and MOHAMED ABDEL-GHANI, Senior Researcher, Drainage Reserach Institute, Water Research Center, Delta Barrage (El-Khanater), P.O.Box 13621/5, Egypt.

The drainage water of the main system is subject to continuous monitoring to evaluate its quantity and quality (Amer and de Ridder, 1990). Focus has been made over the past years on the salinity and mineral chemicals in the drainage water. During the last few years, there was increasing attention towards the other contaminants which may exist in this water and there possible impact on the environment. This paper will present the results of a field investigation carried out at a drainage pilot scheme in the Nile Delta to determine the concentration of agricultural chemicals in the drainage water. It is a step in a program which will continue in the near future for clarifying the different issues related to the drainage water quality both at the farm and main system levels.

AGRICULTURAL CHEMICALS IN DRAINAGE WATER

Irrigation leads to water quality degradation as water flows over land surface and passes through the crop root zone. Highly productive lands usually receive large applications of fertilizers, pesticides and organic and inorganic amendments. Part of these materials can be leached by the irrigation water into the drainage water. Therefore, they are regarded as potential source of environmental pollution. In addition to salts, substances such as nitrates, phosphorus, potassium and pesticides are the most common pollutants leached into the agricultural drainage water.

Nitrogen losses leached in drainage water depend on climate and cultivated plants (Brink, 1990). It depends also on the rate of application and type of fertilizers, the sequence of plant absorption, amount of water applied for irrigation and soil physical characteristics. Nitrate has the general reputation of constituting almost all nitrogen in leached waters as it is normally quite mobile in soils and groundwater (FAO, 1979). Thus, improved surface drainage or controlling the discharge from subsurface drains may be recommended for reducing the leaching of nitrate (Skaggs and Gilliam, 1981).

Phosphorous and potassium are normally retained in soils. Thus, they are not considered water quality problems (Hoffman, 1990). However, phosphorous moves readily through sands and peats in the absence of clays, carbonates and oxides of iron or aluminium. Good subsurface drainage reduces the loss of phosphorous compared to poor subsurface drainage (Skaggs, 1979). Thus, it is usually recommended to provide better subsurface drainage in areas where the receiving water surfaces contain critical phosphorous levels.

There was a widespread opinion that pesticides do not usually leach through soils to shallow and deep groundwater. Contamination of groundwater wells and streams by pesticides were attributed to incorrect handling of application equipment and pesticides residues and of spray drift (Brink, 1990). As more groundwater is being sampled and analytical techniques become more advanced, more contamination by pesticides are reported (Bouwer, 1986). Field experiments showed that subsurface drains at a depth of 1.0 m reduced the loss of herbicides (atrzine and metolachlor) by slightly more than 50 percent compared with surface drained fields. However, significant quantities of the pesticides were still in the subsurface drains after number of years (Hoffman, 1990).

MATERIALS AND METHODS

System Layout and Agricultural Practices

The field drainage system in Egypt consists of subsurface drains installed according to a gridiron layout. Equally spaced field drains (laterals) of perforated plastic tubing are placed at an average depth of 1.30 m, spacing 30-60 m and length 200m. They discharge at right angles into covered collector drain of increasing diameter and length varying between 500 and 2500 m. The land is usually divided into small ownerships and thus one lateral drain usually serves number of individual farms.

Crops are cultivated in a mixed pattern following a two or three-years crop rotation. The main crops are wheat and berseem (Egyptian clover) in winter and maize, rice and cotton in summer. The crop intensity usually exceeds 200 percent. The winter season starts in October-November and the summer season starts in May-June except for cotton which starts on March-April. The study was carried out at a drainage pilot scheme in the south-east of the Nile Delta at Mashtul (Sharkia Governorate). It has lateral drains of different depths and spacings for testing and investigation purposes. The monitored fields had crops, drain depth and spacing as in Table 1. The area served by lateral drains in a field is usually cultivated with one crop type at a time. However, the individual farms within such area were irrigated and managed separately. Surface irrigation is applied for crops cultivated in basins or on furrows.

Table 1: Crops and Drainage Conditions

Crop	Wheat	Berseem	Cotton1	Cotton2	Maize	Rice
Spacing (m)	30	15	30	30	15	30
Drain Depth (m)	1.2	1.2	1.2	1.5	1.2	1.2

The common fertilizers used in Egypt are urea, calcium nitrate, potassium sulphate, supper phosphate and ammonium sulphate. Dates and doses of fertilizers application are given in Table 2. The fertilizer doze is sometimes divided into two applications separated by about one month especially in the case of calcium nitrate and ammonium sulphateee. This happens also in the case of urea and potassium sulphateee for wheat. Fertilizers and pesticides are supplied to farmers through the village co-operative at a subsidized cost.

Table 2. Fertilizer Doses (kg/acre) and Dates of Application

Fertilizer Type	Crop				
	Wheat	Berseem	Maize	Cotton	Rice
Supper Phosphate	100 (Nov)	50 (Oct.)	----	50 (March)	100 (June)
Potassium Sulphate	100 (Dec-Jan)	----	50 (May)	------	50 (June)
Urea	150 (Dec-Jan)	----	----	50 (Aug)	----
Calcium Nitrate	-----	----	150 (June-Jul)	100 (Jul-Aug)	----
Ammonium Sulphate	----	---	----	----	150 (May-Jul)

Organophosphorus and carbamate pesticides were applied in the area under investigation according to the common Egyptian practices. No pesticides are used for berseem. The types and dosages of pesticides used for other crops is shown in Table 3. Herbicides are applied before sowing in the case of wheat. Cotton is the crop receiving the maximum amount of pesticides.

Pesticides monitored during this study included also organochlorines previously used in the Egyptian pest control regimes. Although such group was prohibited from use since 1980. They exhibit chronic toxicity in people and livestock (Steenvoorden, 1976). Its long presistance and tendency to be adsorbed on soil particles neccessitate its adequate detection in soils, water and plant.

Table 3. Pesticides Used for Different Crops

Crop	Pesticide	Dosage	Application date	Group
Wheat	Bifenox	600 ml/acre	Dec.	Herbicide
Cotton	Pendimethalin	950 ml/acre	April	Herbicide
	Flumeturon	1.25 ml/acre	April	Herbicide
	Trifluralin	1.25 ml/acre	April	Herbicide
	Monocrotophos	1.0 ml/acre	April	Insecticide
	Triazophos	1.25 ml/acre	May	Insecticide
	Methamidophos	1.25 ml/acre	May	Insecticide
Rice	Simetryne	400 ml/acre	July	Herbicide
Maize	Methomyl	300 ml/acre	July	Insecticide

Sampling and Monitoring

Three observation wells were installed in each farm at the midway between the lateral drains to a depth of 2.0 m below ground level. Water samples were collected every two weeks from the groundwater observation wells, the outlets of the laterals and collector drain and from the main open drain. The water samples were kept into ice boxes and were transferred to the laboratory in the same day for analysis. Water samples from a drinking well in a nearby farm house were also collected for analysis. The Nitrate and potassium concentration in the water samples were determined. Analysis of pesticides residues were also carried out within two hours of sampling to avoid hydrolysis of the pesticides.

Soil and plant samples were also collected for pesticides residues analysis. The soil samples were collected at the beginning of the wheat and berseem seasons and at the mid of the cotton growing season. In each crop area, three representative samples were taken along a diagonal line across the lateral drains. Soil samples were taken at two consecutive depths from the surface namely; 0-30 cm and 30-60 cm. Five plant samples were collected from the berseem (a fodder crop) and wheat (a grain crop) at the time of harvest. The dates of irrigation, fertilizers and pesticides applications were recorded.

RESULTS AND DISCUSSION

Fertilizers

The nitrates concentration in groundwater and drainage water (lateral discharges) throughout the different crop growing seasons is shown in Figure 1. The time scale (horizontal axis) refers to the time of collecting the water samples after the sowing date of each crop. The concentration of nitrates fluctuates during the season with remarkable increase after each application of fertilizers. Pollution of the shallow groundwater with nitrates (NO_3) during both the winter and summer is very similar to the pollution of the drainage water at the farm level (lateral drains). However, the concentration of nitrates in the drainage water seems to be very much influenced by the drainage intensity. The concentration in drainage water of the cotton field with deeper drains (cotton 2) or the maize field with closer drain spacing reached higher peaks than those of the other crops.

The nitrates concentration in the groundwater and drainage water during winter (Fig. 1, a and b) is very small and seldom exceeded 25 ppm. This is due to the fact that berseem is not fertilized with nitrates. The overall intensity of berseem is usually higher than the wheat intensity over the total cultivated land in the Delta. Thus, the concentration of the nitrate in the collector and open main drains was small. The nitrate concentration in the drainage water of the rice field (Fig. 1, d) is relatively less than the rest of the summer crops. Continous flooded crops produces lower concentrations than intermittently irrigated crops probably due to dilution and denitrification (Bouwer, 1987).

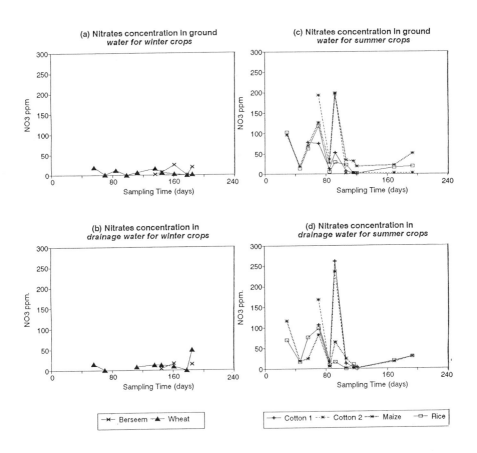

Figure 1. Nitrates Concentration in Shallow Groundwater and Field Drainage Water

The concentration of nitrates in the drainage water is reduced as the drainage water flows from the field drains into the collectors and then to the main drain. The peak nitrates concentration during summer at the outlet of the closed collector and the open main drain were 152 and 89 ppm, respectively. In the collector system, the field drainage water from different field crop gets mixed together. The open main drain usually receives fresh irrigation water losses and surface runoff which cause further dilution of the nitrates concentration. However, this concentration is sufficient to encourage and enhance the aquatic weeds growth in the Egyptian drains. It is worthy to mention that the water in the main open drainage system is the water to be re-used for irrigation according to the current practices.

Nitrates concentration in drinking water from a groundwater well (a pipe penetrating the clay cap to the sandy aquifer at about 10.0 m deep) is shown in Figure 2. The nitrates concentration is much less than in the shallow groundwater due to dentrification. The maximum concentration in winter was 35.5 ppm. It occurred in February following the application of Urea for wheat. In summer, a higher concentration of 52.5 ppm occurred in July which associated the application of Ammonium Sulphateee for rice and Calcium Nitrate for cotton and maize. These concentrations are critical and calls for better water management to control the groundwater pollution with nitrates. The current Egyptian water quality standards require that the concentration of nitrates discharged into surface fresh water bodies or groundwater aquifers not to exceed 30 ppm (Mab, 1983). The drinking well in this farm is too shallow and it might be more safe to adhere to the common practices of drilling deep wells for drinking purposes.

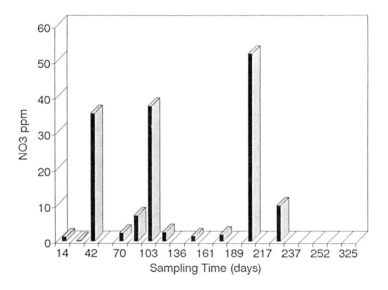

Figure 2. Nitrates Concentration in Groundwater Drinking Well

Potassium concentration in drainage water was low and the peak values in the different crop fields did not exceed 4.0 ppm. Potassium is susceptable to adsorption and ion-exchange on the soil. However, relatively higher concentrations were observed in the drainage water from the wheat field where 100 kg/acre of potassium sulphateee was applied. A peak concentration of 25 ppm was observed during this study.

Pesticides

The main contaminants found in water samples were organochlorine pesticides residues. HCH followed by DDT dominated the contaminants while endrine, heptachlor, heptachlor oxide and dieldrin were detected occasionally with low amounts. Lateral drainage water collected from the different sites showed fluctuations in residual amounts of organochlorine. Open drains showed less contamination with total organochlorines. The presence of DDT mainly in soil could be attributed to the higher solubility of HCH compared to the non soluble DDT having a half life of 22 years in clay soils.

Plant samples from berseem and wheat were contaminated with negligible amounts of HCH and DDT residues due to translocation from soil to plant. Table 4 gives the maximum concentration of organochlorine pesticide residues detected during this monitoring program. The maximum value of each residue in this table is not necessarily ocurring at the same site and time as the total organochlorine residues shown in the last line in the table.

Table 4. Organochlorine Pesticide Residues in Water, Soil and Plant (ppb)

Pesticide	Water			Soil	Plant	
	LD^a	CD^b	MD^c		Berseem	Wheat
Total HCH	5.57	1.68	0.92	46.00	20.30	21.09
Aldrin	0.12	0.00	0.12	0.00	0.00	0.00
Dieldrin	0.07	0.00	0.00	0.00	0.00	0.00
Heptachlor	0.17	0.00	0.07	2.00	0.00	0.00
Hepta. Epox.	0.03	0.05	0.07	277.00	1.90	0.00
Endrin	0.26	0.00	0.00	34.00	0.00	0.00
Total DDT	1.28	0.61	0.93	415.00	72.80	5.10
Total O. Chl's	5.57	2.01	1.16	415.00	86.50	26.19

(a) LD = Lateral drain water
(b) SD = Collector drain water
(c) MD = Main drain water

Most of the peak values of the total organochlorine residues in the different waters occurred during the winter time and particularly at the wheat farm. except that of the collector drainage. The corresponding collector drainage water total residues was, however, 1.58 (ppb) only. In the soils the maximum organochlorine residues occurred in the cotton field. They could be due to some old contaminated sites exhibiting long persistance. This should not exclude the possibility of illegal use of such groups of pesticides.

On the other hand, analysis of water samples did not reveal any organophosphate, carbamate pesticides and herbicides in the area under investigations. Monocrotophos residues were detected in only one soil sample from a cotton field. Wheat and berseem plants from the observed fields did not contain any pesticide residues of this group.

CONCLUSION

Subsurface drainage water is contaminated with nitrates due to the fertilizers used for crop production. Maximum concentrations occur during the summer where cotton, maize and rice are cultivated. The increase in drainage intensity results into higher nitrate losses. The concentration of nitrates decreases as the drainage water flows into the collector drains and eventually in the main drainage system. Pollution of deep groundwater with nitrate still occurs but at far less concentrations. There is a need for further investigations on the extent of drainage and groundwater pollution by agricultural chemicals. This will help for better and environmentally sound water management.

REFERENCES

1. Amer, M. H., and N. A. de Ridder (ed). 1989. Land Drainage in Egypt. Drainage Research Institute, DRI, Cairo, Egypt.

2. Bouwer, H. 1987. Effect of Irrigated Agriculture on Groundwater. Journal of Irrigation and Drainage Engineering, Vol. 113, No. 1, ASCE: 4-15.

3. Brink, N. 1990. Losses of Substances from Tile Drained Soils. Fourth International Drainage Workshop, Cairo, Egypt.

4. FAO, 1979. Groundwater Pollution. FAO Irrigation and Drainage paper No. 31, Rome.

5. Hoffman, G. F. 1990. Environmental Impacts of Subsurface Drainage. Fourth International Drainage Workshop, Cairo, Egypt.

6. Kassas, M. A., M. Abdel Hamid and M. I. M. Ibrahim (ed). 1983. Periodical Bulletin of the Egyptian National Committee for Man and the Biosphere. No. 3 and 4.

7. Skaggs, R. W. 1987. Principles of Drainage. In:Farm Drainage in the United States: History, Status and Prospects. USDA Economic Research Services, Miscellaneous Publication No. 1455, P. 62-78.

8. Skaggs, R. W. and J. W. Gilliam. 1981. Effects of Drainage Design and Operation on Nitrate Transport. Transaction of ASAE 24: 929-934, 940.

9. Steenvoorden, J. H. A. M. 1976. Nitrogen, phosphate and biocides in groundwater as infleunced by soil factors and agriculture. Institute for Land and Water Management Research, Technical Bulletin No. 97, Wagenigen, the Netherlands.

A FIELD MONITORING SYSTEM TO EVALUATE THE IMPACT OF AGRICULTURAL PRODUCTION PRACTICES ON SURFACE AND GROUND WATER QUALITY

R.S. Kanwar, D.G. Baker, G.F. Czapar, K.W. Ross, D. Shannon, M. Honeyman[1]
Member ASAE

INTRODUCTION

Among the inputs sustaining our agriculture in Iowa are vast acreages of dark relatively flat croplands, much of which needs artificial drainage. Artificial drainage is necessary to permit farming of some of Iowa's most productive soils. Drainage is needed to provide trafficable conditions for planting and harvesting and to ensure a suitable environment for plant growth during the growing season. At the same time excessive drainage is undesirable as it reduces soil water available to growing plants and leaches fertilizer and pesticides, carrying them to receiving streams and/or to deeper groundwater systems where they act as pollutants. Currently about 40 percent of Iowa's agricultural land under corn and soybeans is artificially drained. Kanwar et al. (1983, 1984) found in a study conducted in Iowa that inadequate drainage systems were responsible for average crop yield reductions equal to about one-third of the maximum yield potential on heavy soils. Since herbicides are used on approximately 98 percent of corn and soybean acres in Iowa there is a potential for ground water contamination from these artificially drained soils (Kanwar et al., 1990).

Recent studies in Iowa and in the north-central United States have shown the evidence of the presence of nitrate and pesticides in groundwater as a result of agricultural activities (Hallberg, 1986, 1989; Hallberg et al., 1987). Although ground water contamination has been documented in Iowa and other states, the mechanics of contamination is often unknown. Hallberg (1986) has suggested that infiltration recharge may be the primary delivery mechanism of agricultural related contaminants to ground water. Researchers are investigating the possibilities of developing Best Management Practices to protect water sources from chemical pollution while sustaining crop production at an optimum level. Agricultural management practices, such as tillage, crop rotations, chemical management and water table management practices are considered to reduce the negative effects of agricultural chemical use on Iowa's water resources. Several large scale field experiments are underway in the U.S. and other countries to study the effects of agricultural production systems on the quality of ground and surface waters (Belcher and Merva; Bengtson et al., 1991; Fausey et al., 1991; Kanwar et al., 1991; Skaggs et al., 1991; Soultani et al., 1991; Thomas et al., 1991; Vereecken et al., 1991). One of the major objectives of these studies is the develop a comprehensive data base for developing and testing models to predict the loss of agricultural chemicals to surface and ground water sources. This paper describes the layout and instrumentation of a large field experiment being conducted in Iowa to study the effects of several tillage and crop production systems on water quality. This experiment uses state of the art technology to understand the movement of water and chemicals to ground water.

---

[1]R.S. KANWAR, Professor; D.G. BAKER, former Research Associate, Dept. of Agricultural and Biosystems Engineering; G.F. CZAPAR, former Extension Associate, Dept. of Agronomy; K.W. ROSS, Superintendent; D. SHANNON, Manager; M. HONEYMAN, Coordinator and Assistant Professor, Outlying Research Farms, Iowa State University, Ames, Iowa 50010

MATERIALS AND METHODS

The experimental site for this study is located at Iowa State University's Northeast Research Center, Nashua, Iowa. This study site is on a predominantly Kenyon silty-clay loam soil with 3 to 4% organic matter. These soils have seasonally high water-tables and benefit from subsurface drainage. Pre-Illinoisan glacial till units of 60 m overlie a carbonate aquifer used for water supply. However, in some areas, bedrock is near the surface.

The site was selected because 36, 0.4 ha experimental plots had fully documented tillage and cropping records for the past twelve years. The subsurface drainage system has been in place at this site for nine of these twelve years. The tile lines were installed about 1.2 m deep at 28.5 m spacing eleven years ago, therefore, there should be little chance for contamination of lower soil levels because of soil disturbance.

Soils at the experimental site are from the Clyde-Kenyon-Floyd Association formed on upland by native prairie vegetation and are characterized by lengthy slopes, slightly rounded ridge tops and well defined drainage ways. Surface drainage is well developed, but this drainage is slow to medium due to moderate permeability and high available water capacities of the soils. Each soil has seasonally high water tables and benefits from subsurface drainage. Floyd, Kenyon, and Readlyn soils are capability subclass IIw, I, and I, respectively. Subsurface drainage is related to soil type (Floyd soils are somewhat poorly drained, Kenyon soils are well drained, and Readlyn soils are somewhat poorly drained).

Long-term conservation tillage studies were initiated at this site in the fall of 1976 to compare moldboard plow, chisel plow-disk, ridge till, and no-till systems. Crop rotations and tillage subplots were replicated three times in a split plot experimental design. Crop rotations are continuous corn and corn-soybean. Therefore, 12 plots are under continuous corn, 12 plots are under corn-soybean rotation, and 12 plots are under soybean-corn rotation. Herbicide, insecticide, and fertility treatments have been the same since 1976. Continuous corn received 200 Kg-N/ha and corn-soybean rotation received 168 Kg-N/ha. No nitrogen was applied in the year soybeans were planted in soybean-corn rotation. The continuous corn treatment has received Lasso + Atrazine at 2.2 Kg/ha and Counter for rootworm control. Corn in the corn-soybean rotation has received Lasso + Bladex at 2.2 Kg + 2.8 Kg/ha and no insecticide. Soybean plots received Lasso + Metribuzin at 2.2 Kg + 0.45 Kg/ha.

During the first two years after the beginning of the long term tillage experiment in 1976, it became evident that variability in soil drainage was going to create some problems in trying to accurately measure treatment effect. A decision was made to install subsurface drains at the end of the 1979 growing season to alleviate this problem. Since the tillage experiment was in progress, it was important to minimize soil disturbance during the tile installation. It was decided to run a tile line lengthwise through the center of each plot, using a trenchless tile plow. Another tile was run lengthwise along the border between plots, using a trencher. This system resulted in a 28.5 m spacing between tile lines. All plots were tiled, regardless of their position on the topography or apparent need for drainage and tile lines were installed at a depth of 1.2 m. Figure 1 shows the layout of 36, 0.4 ha plots and location of subsurface drains.

Thus, each 0.4 ha plot has one tile line passing through the middle of the plot and there is a tile line at each of the two borders. The tile lines at the borders should pick up any cross contamination from the surrounding plots. There are a maximum of ten 0.4 ha plots in a row and plot rows are separated by an uncultivated area of 0.9 m width. Therefore, the tile line installed in the middle of the ploy (which is intercepted for monitoring) drains about 0.2 ha area and is free from cross contamination from all four sides.

Subsurface drain lines passing through the middle of the 36 plots were intercepted and rerouted to ten collection sites in the fall of 1988 (Figure 1). Each 10.2 cm perforated plastic drainage tube was intercepted at the lower border of each plot and transported to one of the ten collection sites using 5.1 cm SDR 26 polyvinyl chloride (PVC) drain pipe. Connection of the perforated drain tube and the slip-joint PVC drain pipe was coupled using a 10.2 cm by

5.1 cm rubber coupling with stainless steel clamps and a 7.6 cm Schedule 40 PVC coupling inserted into the plastic drain tube to prevent its collapse.

Collection and meter sumps were installed at each of the ten collection sites. Collection sumps of 60 cm diameter corrugated black plastic culvert were erected close to other existing non-intercepted drain lines located along the sides of plot boundaries. These sumps were connected into the existing drain lines using 10.2 cm non-perforated drain tubes and 10.2 cm black plastic tees (Figure 2). Meter sumps were located near the collection sumps.

Two to six meter sumps, depending on topography, were installed around each of the ten collection sumps (Figure 1). Constructed of 38 cm PVC airduct tubing, each meter sump houses a pump and meter assembly as shown in Figure 3. Each meter sump was connected to one of the 36 plot drains and by an overflow to its collection sump using 5.1 cm slip-joint PVC drain pipe. Because the meter sumps may sometimes be located below the water table, rubber pipe boots with stainless steel clamps, as shown in Figure 3, were installed in the lower 5.1 cm pipes to ensure good seals against contamination of ground water at the collection site. Another 3.8 cm Schedule 40 PVC pipe was also connected from each sump to the collection sump for the pump discharge (Figures 2 and 3) and was sealed using a silicon sealer.

The electrical distribution system was installed to provide power to sump pumps and other units. Water meters were installed to monitor the volume of tile effluent in each meter sump. Each meter has the capability of recording effluent volumes mechanically as well as sending digital pulse signal to multi-channel dataloggers. Each datalogger has the capability to record tile flows from 11 to 13 plots.

To monitor tile flow on a continuous basis, each tile sump has a 110 volt effluent pump, water flow meter, and an orifice tube to collect water samples for water quality analysis. For water quality sampling, an orifice tube was designed to deliver about 0.2% of the tile water into a sampling bottle each time effluent is pumped from the sump. This way, we will not miss any drainage water from sampling and an accurate count on the loss of chemicals with drainage water can be made. Four H-flumes with automatic samplers were also installed to monitor the quality of surface runoff coming from four different tillage plots. In addition, soil samples are being taken from each plot (before planting, in July, in August, and after harvesting) using zero contamination tubes at depths of 15, 30, 45, 60, 90, 120, 150, and 180 cm for chemical analysis. Plant samples are also taken to determine N-uptake. For NO_3-N sampling, the frequency of sampling averaged 3 times a week when tile lines were flowing. For pesticide sampling, tile water samples were taken on weekly basis and after every major rainfall within 60 days of pesticide application. During the remaining part of the year, sampling frequency did not exceed once a week when tile lines were flowing. Any unusual rain storm is sampled for both, NO_3-N and pesticide analyses. Drainage water samples are analyzed for alachlor, atrazine, cyanazine, metribuzin, and nitrate.

Two piezometers (3.8 cm diameter plastic pipe with a 22.9 cm long screen at bottom) were installed 15 m apart in May 1988 at depths of 1.9 and 2.4 m. Piezometers have also been installed at various depths between 3.0 m and the bedrock aquifer to monitor water quality and to examine geochemical processes that effect recharge water to the bedrock aquifer. Piezometers will be used to measure hydraulic head gradients and to obtain water samples for nitrate, pesticide (metabolite) and major element analysis. Concern has been raised that some of the degradation products (metabolites) formed from pesticides are more persistent and more mobile than the parent compounds. Therefore, we will analyze water samples from all piezometers for pesticide metabolites using HPLC or GC technology. These data will help to determine the fate of chemicals in the larger graduation flow system and will help to determine the quantity and quality of ground water recharge to the bedrock aquifer.

Pesticide (metabolites), and major element and NO_3-N analyses are being conducted at the National Soil Tilth Laboratory at Iowa State University with duplicate samples of some treatments analyzed at the Iowa Hygienics Laboratory in Iowa City. Results of monitoring should allow us to determine if the products in question can leach in detectable quantities to lower depths of the root zone and to the bedrock aquifer.

RESULTS AND DISCUSSION

The years 1988 and 1989 were extremely dry, and rainfall, was well below normal, therefore, tile lines did not flow. But 1990 and 1991 were unusually wet years with total rainfall amounts of more than 104 and 97 cm, respectively (annual normal precipitation at the experimental site is around 75 cm). This rainfall caused all tile lines to flow through most of the growing season of 1990. Figure 4 gives the daily measured values of tile flow as a function of the tillage system from April 1 to November 30 for 1990 for continuous corn. This figure shows that peak tile flows occurred immediately after larger rain storms. The rapid increase in tile flows after heavy rain shows the preferential movement of water through macropores. Larger tile flows were observed under continuous-corn practice than under soybean-corn rotation for similar rain storms and tillage systems which indicated that preferential flow paths may be different under different crop rotations. Also, this figure shows that the no-tillage system seems to give highest peak flows for most of the storms during the growing season. Larger peak flows under the no-tillage system could result when macropores (worm or root holes and natural fractures) are not destroyed or disturbed because of lack of tillage; the macropores function as large channels to carry water to deeper soil depths in a matter of minutes. The type of instrumentation installed at this site gives us the capability to monitor preferential flow during and after rain storms.

Figure 5 gives the average NO_3-N concentrations in tile water as a function of tillage for 1990 for continuous-corn. This figure shows that under moldboard plow, NO_3-N concentrations in water were much higher in comparison with the other three tillage systems for both crop rotation practices (continuous-corn and corn-soybean rotation). Significantly higher NO_3-N concentrations in tile water of moldboard-plowed plots indicate that upon plowing and disking, macropore structure is eliminated and rain water has to pass through the soil profile according to the concepts of piston flow (rather than preferentially) allowing more time for water to carry soil NO_3-N down to tile lines. Figure 5 also shows that NO_3-N concentrations in tile water of no-tillage plots were lowest, which indicates that soil NO_3-N was bypassed by the water moving through macropores. Figure 6 gives annual tile flows and annual NO_3-N losses through tile lines as a function of tillage system and crop rotations.

Table 1 gives the monthly average NO_3-N losses through subsurface drainage water as a function of tillage and crop rotation. The seasonal total NO_3-N losses in drainage water ranged from 30.3 to 107.2 kg/ha for 1990. NO_3-N losses were much greater under continuous corn in comparison with corn-soybean or soybean-corn rotation. Similar losses for 1991 ranged from 29.1 to 75.4 kg/ha. Also, in 1990, the highest NO_3-N losses of 107.2 kg/ha with drainage water were observed from no-tillage plots under continuous corn, which is equal to approximately 50% of the applied nitrogen fertilizer for that year. In 1991, the higher NO_3-N losses of 75.4 kg/ha with drainage water were observed from chisel plow plots. This demonstrates the compounding effects of tillage, crop rotation, and higher N application rates on NO_3-N losses to subsurface drainage water. Although NO_3-N concentrations were greater under conventional tillage than under a no-tillage system, total NO_3-N losses through tile drainage water were much greater under no-tillage and chisel plow systems compared with conventional tillage because of a greater volume of water moving through the soil.

Table 2 gives losses of atrazine and alachlor with drainage water as a function of tillage and crop rotation for 1990. This table gives the average monthly as well as yearly total losses of herbicides with subsurface drainage water. These results show that atrazine losses were greatest in comparison with the three other herbicides. Also, no-till and ridge-till had greater losses under continuous corn, because of the preferential movement of these herbicides through macropores. Ridge-till and no-till systems also had larger losses for cyanazine, metribuzin, and alachlor under the soybean-corn rotation practice.

SUMMARY

Rapid increase in tile flows after heavy rains were observed under all tillage systems (moldboard plow, chisel plow, ridge till, and no-till). Larger tile flow peaks immediately after heavier rains indicate the preferential movement of rain water to the tile lines. The appearance of atrazine, alachlor, cyanazine, and metribuzin in the tile water in larger concentrations immediately after rainfall suggests the preferential movement of these herbicides to ground water. Also, results of this study (summarizing water quality data from 1990) indicate that leaching potential of chemicals to shallow ground water depends on many factors. These factors include not only the agricultural production systems (tillage systems and crop rotations) but also the chemical properties of herbicides (solubility, persistence, degradation, adsorption, uptake, mineralization, denitrification, volatilization, etc.). Several years of data will be needed to conclude the overall effects of tillage and cropping systems on nitrate and pesticide leaching. This experiment (with state of the art instrumentation) will make it possible to understand the chemical transport processes in the soil profile.

REFERENCES

1. Belcher, H.W. and G.E. Merva. 1991. Water table management at Michigan State University. ASAE Paper No. 91-2025. St. Joseph, ME: ASAE.

2. Bengtson, R.I., C.E. Carter, and J.L. Fouss. 1991. Water management research in Louisiana. ASAE Paper NO. 91-2021. St. Joseph, MI: ASAE.

3. Fausey, N.R., A.D. Ward, and L.R. Brown. 1991. Water table management and water quality research in Ohio. ASAE Paper No. 91-2024. St. Joseph, MI: ASAE.

4. Kalita, P.K. and R.S. Kanwar. 1990. Pesticide mobility as affected by water table management practices. ASAE Paper No. 90-2089, St. Joseph, MI ASAE.

5. Kalita, P.K. and R.S. Kanwar. 1992a. Shallow water table effects on photosynthesis and yield of corn. TRANSACTIONS of the ASAE (in press).

6. Kalita, P.K. and R.S. Kanwar. 1992b. Effect of water table management on the transport of nitrate-nitrogen to shallow groundwater. TRANSACTIONS of the ASAE (in press).

7. Kanwar, R.S. 1990. Water table management and groundwater quality research at Iowa State University. ASAE Paper No. 90-2065. St. Joseph, MI: ASAE.

8. Kanwar, R.S., H.P. Johnson, and T.E. Fenton. 1984. Determination of crop production loss due to inadequate drainage in a large watershed. Water Resources Bulletin 20(4):589-597.

9. Kanwar, R.S., H.P. Johnson, D.Schult, T.E. Fenton, and R.D. Hickman. 1983. Drainage needs and returns in north central Iowa. TRANSACTIONS of the ASAE 26(2):457-464.

10. Kanwar, R.S. and J.L. Baker. 1991. Long term effects of tillage and reduced chemical application on the quality of subsurface drainage and shallow groundwater. In Proc. Environ. Sound Agric. ed. A.B. Bottcher, K.L. Champbell, and W.D. Graham, 121-129.

11. Skaggs, R.W., R.O. Evans, J.W. Gilliam, J.E. Parsons, and E.J. McCarthy. 1991. Water management research in North Carolina. ASAE Paper NO. 91-2023. St. Joseph, MI: ASAE.

12. Soultani, M., C.S. Tan, J.D. Gaynor, C.F. Drury and T.W. Welacky. 1991. The design of a field experiment to evaluate the effects of an integrated management system on pesticide and nitrogen loss. ASAE Paper No. 91-2584. St. Joseph, MI:ASAE.

13. Thomas, D.L., M.C. Smith, G. Vellidis, C.D. Perry, and B.W. Maw. 1991. Status of water table management research in Georgia. ASAE Paper No. 91-2022. St. Joseph, MI: ASAE.

14. Vereecken, H., M. Vanclooster, and M. Swerts. 1991. Development of a nitrogen leaching model for regional studies. Agricultural Water Management (in press).

Table 1: Average monthly nitrate-nitrogen loading as a function of tillage and crop rotation for 1990

Month	Rainfall (cm)	Nitrate-nitrogen loading, kg/ha			
		Chisel	MB Plow	Ridge Till	No-till
CORN-SOYBEANS ROTATION:					
April	10.2				
May	10.9	11.71	4.16	6.33	8.96
June	17.5	11.65	10.47	8.53	11.56
July	34.0	16.99	11.47	8.78	9.78
August	19.5	7.28	6.62	4.07	5.35
September	5.3	1.67	1.82	1.22	0.62
October	5.4	2.42	2.46	1.23	0.20
November	2.0	0.63	0.95	0.17	0.01
Total	**104.8**	**52.35**	**37.95**	**30.33**	**36.48**
SOYBEANS-CORN ROTATION:					
April	10.2				
May	10.9	5.76	7.59	5.77	5.01
June	17.5	14.61	9.19	10.35	8.07
July	34.0	17.26	13.15	9.34	11.44
August	19.5	9.29	7.31	5.70	5.00
September	5.3	2.55	2.13	1.43	1.21
October	5.4	1.33	1.68	1.10	0.76
November	2.0	0.04	0.01	0.01	0.17
Total	**104.8**	**50.84**	**41.06**	**33.70**	**31.66**
CONTINUOUS CORN:					
April	10.2				
May	10.9	18.99	3.95	14.32	23.98
June	17.5	24.87	10.57	23.70	29.02
July	34.0	28.48	19.60	23.10	27.48
August	19.5	19.22	16.94	17.31	18.74
September	5.3	5.52	4.38	3.91	5.42
October	5.4	2.62	2.49	1.02	2.27
November	2.0	0.25	0.12	0.05	0.28
Total	**104.8**	**99.95**	**58.05**	**83.41**	**107.19**

Table 2: Average monthly pesticide loading as a function of tillage and crop rotation for 1990

Month	Pesticide loading, g/ha			
	Chisel	MB Plow	Ridge Till	No-till
CONTINUOUS CORN: ATRAZINE				
May	1.728 ab	0.083 a	1.467 ab	3.790 b
June	1.734 ab	0.287 a	3.534 ab	4.957 b
July	2.250 ab	0.820 a	2.747 ab	4.337 b
August	1.636 ab	0.751 b	3.080 a	3.490 a
September	0.338	0.170	0.497	0.634
October	0.1183 a	0.0551 ab	0.0217 b	0.2571 c
November	0.0061	0.0010	0.0008	0.0162
Total	7.810	2.167	11.348	17.481
CONTINUOUS CORN: ALACHLOR				
May	0.264	0.019	0.149	0.239
June	0.020	0.032	0.082	0.073
July	0.0045	0.0027	0.0043	0.0058
August	0.0053 a	0.0034 b	0.0058 a	0.0071 a
September	0.0013	0.0008	0.0013	0.0016
October	0.0005 ab	0.0004 ab	0.0001 b	0.0008 a
November	<0.0001	<0.0001	<0.0001	<0.0001
Total	0.296	0.059	0.301	0.328

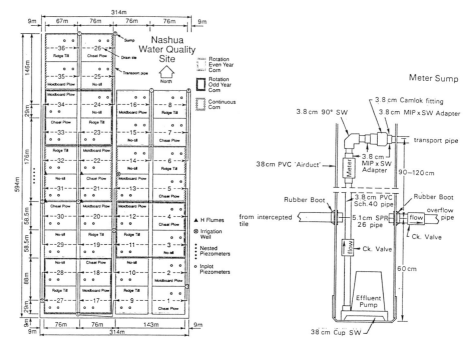

Figure 1. Layout of field plots with a water quality monitoring system

Figure 3. Subsurface drainage flow metering system with the sump pump

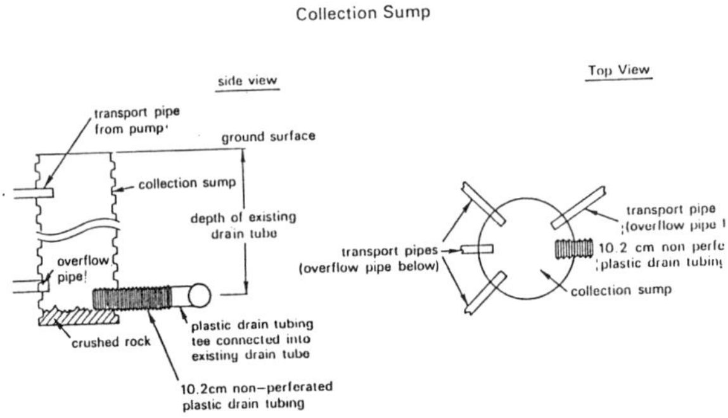

Figure 2. Side and top view of a subsurface collection sump

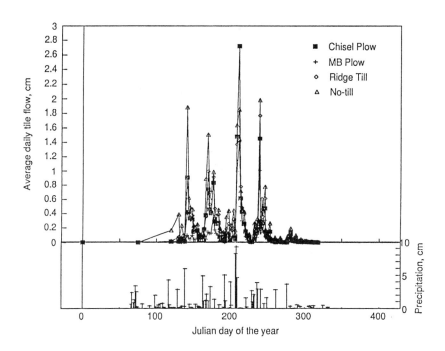

Figure 4. Measured tile flows as a function of tillage under continuous corn for 1990

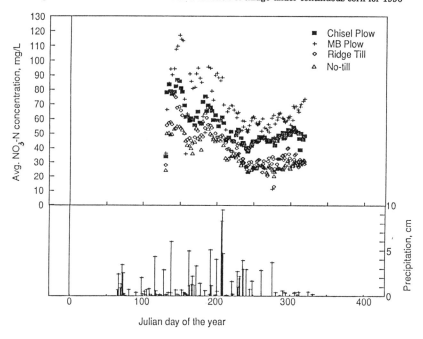

Figure 5. NO_3-N concentrations in tile water under continuous corn for 1990

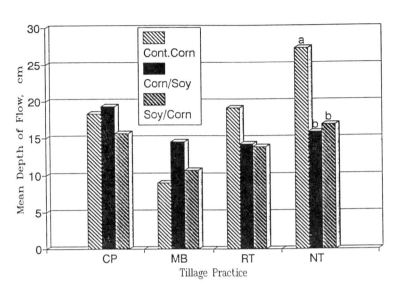

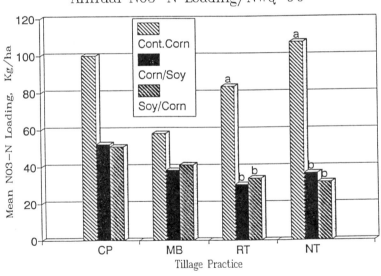

Figure 6. Annual tile flows and NO$_3$-N losses as a function of tillage and crop rotation

DRAINAGE EFFECTS ON WATER QUALITY IN IRRIGATED LANDS

J. E. Ayars[*] R. A. Schoneman[*]
Member ASAE Member ASAE

ABSTRACT

Disposal of drainage water from arid irrigated land is a major problem because of the presence of salt and other contaminants. In order to meet existing and proposed water quality standards in water bodies which receive drainage water, there is a need to know the relationship between the drainage flow and load of salt and each contaminant of interest in the drainage water. The objective of this study was to determine the load-flow relationships for individual drainage sumps in an irrigation districts on the west side of the San Joaquin Valley, Ca. and for the district as a whole.

Load-flow relationships were determined for drainage points in the Panoche Water and Drainage District by measurement error regression using the measured monthly salt, selenium (Se), and boron (B) loads and the discharge (Q). A linear relationship with one parameter (slope only) for each contaminant was found to adequately describe the load-flow rate data at each outlet for total salt, selenium, and boron load. Multivariate analysis of the slopes (rates) of the resulting load-flow relationships for each drainage point in the Panoche Water District determined that the relationships could be grouped into three or four major groups. The resulting relationships could be used in conjunction with irrigation efficiency studies to develop district wide management strategies to reduce potentially toxic loading of trace elements in the district drainage effluent.

Keywords: Irrigation, irrigation efficiency, selenium, boron, drainage disposal

Irrigated agriculture in the arid and semi-arid areas of the world relies on subsurface drainage to survive. In places where the natural drainage is inadequate, artificial drainage has been provided using open ditches, clay tiles and more recently corrugated tubing. Beside preventing waterlogging, drainage is needed to permit leaching of salts from the crop root zone which are transported in the irrigation water and which are naturally occurring in the soil.

The primary water quality issue for subsurface drainage water in the 1950's was salinity and boron. In subsequent years, the emphasis has shifted to nitrates, then pesticides and now to selenium and other toxic elements (Tanji, 1990). With the recent Presidential Initiative on Water Quality, the focus is being directed back to the presence of nitrates in the ground water.

---

[*] Agricultural Engineers, Water Management Research Laboratory, Fresno, Ca.

The drainage disposal problem facing managers in arid irrigated areas is typified by the situation in the San Joaquin Valley (SJV) where the drainage collector system serving 17000 ha in the Westlands Water District was plugged to prevent selenium contaminated drainage water from reaching Kesterson Reservoir (Moore, 1989). Kesterson was to be a regulating reservoir for drainage water flowing from irrigated areas in the San Joaquin Valley through the San Luis Drain to the San Francisco Bay.

For both environmental and economic reasons the drain was not finished beyond Kesterson which was then developed as a wildlife refuge. When toxicosis found in migratory waterfowl nesting at Kesterson (Ohlendorf et al., 1986) was determined to be a result of the accumulation of Se in the plants growing in the reservoir, the operation of the drainage system was stopped and the drains were plugged.

Subsurface drainage effluent from other irrigation districts on the west side of the SJV is often combined with runoff water from surface irrigation and discharged into the San Joaquin River. When water quality standards for trace elements are established for this river, it will become very difficult for these districts to use river discharge as a means of drain water disposal. Unless a means of disposal can be developed which does not violate quality standards, there will be a gradual salinization of the soil and loss of productivity within these districts.

Before any management system can be developed, the total mass of salts and contaminant elements being transported in the drainage water has to be determined as a function of the drain discharge, this is often called the load-flow relationship. Once this relationship is determined for the area as a whole and the individual components of the area, management alternatives to achieve a water quality goal related to the concentrations of salt and other elements can be developed. When the load-flow model is coupled with a water management model describing the interaction of drain flow and irrigation, it will be possible to develop a water management plan for the entire irrigation district.

The objective of this study was to determine the load-flow relationships for individual drainage sumps in irrigation districts located on the west side of the San Joaquin Valley, Ca. and for the district as a whole. This paper will report on the results of that study.

MATERIALS and METHODS

The sites selected for this study were the Panoche Water and Drainage District (PWDD) and the Broadview Water District (BWD), 16000 ha and 4000 ha irrigation districts, respectively, located on the west side of the San Joaquin Valley of California. The principal crops grown in these districts are cotton, processing tomatoes, alfalfa, dry beans, and melons. Cropping pattern data for each district were based on grower surveys and were available each year of the study.

Subsurface drains have been installed on approximately 7100 ha of the PWDD district and 2600 ha of the BWD. In both districts the drain lateral installation depth varies from 1.7 to 2.1 m with spacing up to 150 m between laterals. The length of tile installed in a field varies from roughly 24 m ha^{-1} to 207 m ha^{-1}. Effluent from each drainage sump is delivered to a main drain for disposal in the San Joaquin River along with collected surface waters. There are a total of 52 drainage points (dp) in the PWDD and 25 drainage sumps in the BWD.

In the PWDD drainage flow was monitored at each drainage point (dp). Discharge (Q, mm month^{-1}) was determined by either: measuring the flow with a water meter installed in the drainage point outlet pipe; estimating the flow using the power records from a pumped outlet in conjunction with measured pump performance; or by estimating the flow using a weir at a gravity outlet. These data were collected monthly and summarized quarterly (Summers, 1989). Sufficient data for both drainage flow and mass of salts in the drainage water were available for statistical analysis from only 45 drainage points. A maximum of 52 months of data were available for analysis at some sites.

In the PWDD the drainage water was sampled and analyzed for electrical conductivity (EC), boron (B), total dissolved solids (TDS), and selenium (Se) at the time that the flow and power meters and weirs were read each month. In the BWD the EC was read weekly and samples collected for analysis of EC, B, AS and SE every two weeks. Chemical analyses were done in a commercial testing laboratory using standard techniques (U.S.E.P.A., 1979). Total monthly salt, Se and B loads were estimated by integrating the flow and concentration values for Se, B, and TDS values at each dp with time.

For each dp, the total salt load (T), Se load (Se), and B load (B), all in (kg ha^{-1} month^{-1}) versus drainage flow (Q, mm month^{-1}) regressions were calculated as follows. A robust slope from measurement error line without an intercept was estimated with Kerrich's method (1966) modified by taking Gentle's least first power estimator (Gentle, 1977) for total salt load rate (S1, Eq. [1]), Se load rate (S2, Eq. [2]), and B load rate (S3, Eq. [3]).

$$T = S1 \times Q \quad (1)$$
$$SE = S2 \times Q \quad (2)$$
$$B = S3 \times Q \quad (3)$$

Factor and cluster analyses (see e.g. Hair et al., 1987) were used to find a suitable categorization for all 45 of the load-flow slope (rate) estimates into a few related groups; the variable sets included S1-S3, median Q, soil permeability rank (6 possible levels) and soil Se rank (6 possible levels) for each site. For each defined group, the simplest possible measurement error regression (see e.g. Fuller, 1987) was developed.

Because of the similarities between the two districts in soil types, cropping pattern, irrigation system and drainage system management, results from the PWDD will be used to discuss the findings of this study.

RESULTS

Cotton, alfalfa and tomatoes are the principal crops grown in the districts therefore any strategies used to manage drainage effluent will primarily involve these three crops. Alfalfa, melons and tomato are moderately sensitive to salinity while cotton is a salt tolerant crop (Maas and Hoffman, 1977). More leaching and consequently more deep percolation are required for salt sensitive crop production than is required for cotton for a given water quality.

An example of a discharge hydrograph from an individual dp is given in Figure 1. The peak drainage flow occurred during the irrigation season and decreased to a base flow rate during the winter months. The increase in flow during the winter was the result of deep percolation from pre-plant irrigation given to both replenish the soil water and to provide leaching. The monthly Se loading rate for the dp was plotted in Fig. 1 with the drainage flow. The Se load follows the drainage response very closely.

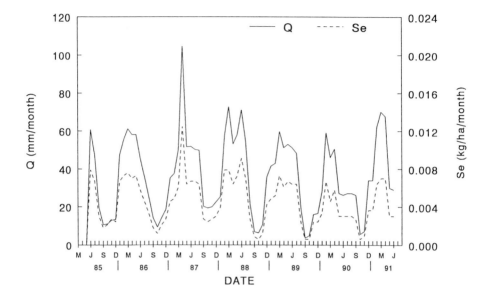

Figure 1. Discharge hydrograph and selenium loading rate from drainage point 18M in the Panoche Water and Drainage District.

When the total annual selenium discharge for the 1988 water year for the fields in the PWDD district was plotted in kg ha^{-1}, the emerging pattern reflected the influence of the total soil selenium more than that of the total drain flow (data not shown). The largest amounts of selenium transported occurred in areas which contained the highest values of soil selenium and not necessarily the largest flows.

Based on the correlation coefficient, r, (data not shown) almost all the load flow slopes were highly significant ($P<0.01$). A typical total load-flow relationship (S2) is shown in Figure 2. Factor analysis indicated that two variables, the flow rate slope and flow can adequately characterize the data. All total load -flow rate estimates, S1-S3, along with the median monthly flow, and suggested clustering into three groups** are summarized in Table 1 and shown in Figure 3.

A subsequent cluster analysis on the main sequence data indicated that this set can be subdivided into two subgroups (Table 1). The subgroup 1 data are low loading rates with high flows and the subgroup 2 data are the reverse, high loading rates

---

**The Figure 3 graph loosely resembles the pattern in the famous Hertzsprung-Russell diagram of astrophysics, hence the nomenclature of giant, main and dwarf sequence was adopted (e.g. Smith and Jacobs, 1973).

and low flows. For example, the average values for Q are 56.6 and 10.7 mm month^{-1} for subgroups 1 and 2 respectively, while the average values for S1 are 1.63 and 3.88 (kg mm^{-1} ha^{-1}) for subgroups 1 and 2, respectively. The subgroup 1 drainage points are primarily in areas containing soils with high permeability and lower values of Se. The subgroup 2 drainage points are in areas containing soils with low permeability and high values of Se.

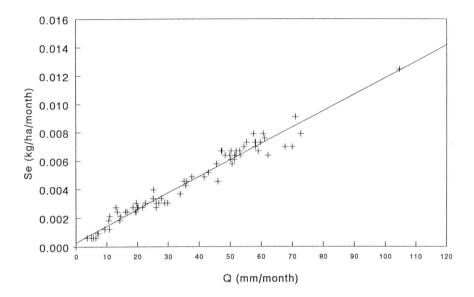

Figure 2. Load-flow regression for selenium load from drainage point 18M in the Panoche Water and Drainage District.

In Figure 3, the slope of the regression line used to characterize the grouping of the individual sumps is decreasing from the top group (Giant) to the bottom group (Dwarf). It is possible that this trend is related to the total time a subsurface drain has been installed and operating. This suggests that more leaching has taken place in the group labelled dwarf than in that labelled giants. It is also possible that the dwarf sites had less salt initially and the difference is not a result of the length of time the soil has been leached.

The categorization of the load-flow relationships suggests some options for managing total chemical loads in the drainage effluent from the district. Each group can be managed by using the group's regression relationship to limit loads by constraining drainage flows at each of the drainage points in the sequence. Alternatively, the main sequence can be grouped, the giants managed individually since there are so few, and the dwarves could possibly be left alone.

Since the load response in the drain water can be described mathematically as a function of the flow, the key to managing the chemical load in the drainage effluent will be to manage the deep percolation loss from irrigation which is the principal source of the drain water. The data in Figure 1 show the response of the drain flow to irrigation. Since surface irrigation methods (furrows and borders) are the principal irrigation methods used in the district, the immediate management objective would be to improve irrigation efficiency, i.e. reduce deep percolation losses, by better management of the existing systems.

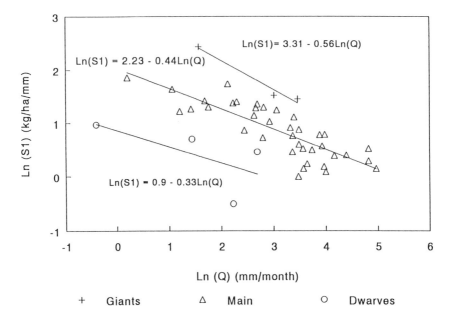

Figure 3. Categories and corresponding measurement error lines for the salt load-flow rate data in Table 1.

SUMMARY

The presence of toxic elements in drainage effluent has created problems related to the disposal of drainage effluent. It is no longer possible to dispose of drainage effluent into a body of water without considering the environmental consequences of that action.

The load-flow characteristics of the drainage effluent from the 16000 ha Panoche Water and Drainage District on the west side of the San Joaquin Valley were studied and the results reported in this paper. For each of the 45 discharge points, data on monthly drain flow (Q) and total mass of salts , Se and B were used to determine each respective load-flow rate (S1-S3) with a robust measurement error regression. Factor analysis suggested that S1 and median Q could adequately characterize groups for this data. Cluster analysis suggested that three or four categories can adequately characterize the load-flow relationships for these 45 sites.

The resulting groups could be used in conjunction with irrigation efficiency studies to develop district-wide management strategies to reduce potentially toxic loading of salts and trace elements in the district effluent.

Acknowledgements

This study was funded in part through a contract with the California Department of Water Resources. The authors would like to thank the staffs of the Panoche Water District and the Broadview Water District for their assistance in the data collection and help throughout the entire study. The statistical analysis was done by Mr. David Meek with the assistance of Dr. Bruce Mackey.

REFERENCES

Fuller, W. A. 1987. Measurement Error Models, Wiley, New York, New York, Chapter 1, pp. 1-99.

Gentle, J.E. 1977. Least absolute value estimator: an introduction. Commun. Statistics-Simula. Computa., B6(4):313-328.

Hair, J. F., Anderson, R. E., and Tatham, R. L. 1987. Multivariate Data Analysis, Macmillan, New York, New York, Chapters 6 and 7, pp. 233-348.

Kerrich, J. E., 1966. Fitting the line Y=αX when errors in observation are present in both variables. American Statistician 20, p.24.

Maas, E.V. and Hoffman, G.H. 1977. Crop salt tolerance-current assessment. J. Irrig. and Drainage Div.,ASCE 103(IR2):115-134.

Moore, S.B. 1989. Selenium in agricultural drainage: essential nutrient or toxic threat. J. Irrig. and Drainage Div., ASCE 115(1):21-28.

Ohlendorf, H.M., Hothem, R.L., Bunck, C.M., Aldrich, T.W. and Aldrich, J.F. 1986. Relationships between selenium concentrations and avian reproductions. Proceedings, 51st Annual Wildlife and Natural Resources Conference, Reno, Nevada.

Smith, E.P. and K.C. Jacobs. 1973. In Chapter 12, Spectral classification and the Hertzsprung-Russell diagram. p. 291-313. Introductory Astronomy and Astrophysics. W.B. Saunders, Philadelphia, Pa.

Summers Engineering, Inc. 1989. Drainage point discharge estimates, Panoche Drainage District, Summers Engineering, Inc, Consulting Engineers, Hanford, California.

Tanji, K.K. 1990. Nature and extent of agricultural salinity. In Salinity Assessment and Management, ASCE Manual on Engineering Practice, No. 71, pg 1-17.

Tidball, R.R. 1988. Unpublished selenium data. U. S. Geological Survey, Denver, Colorado.

U.S.E.P.A. 1979. Methods of Chemical Analysis of Water and Wastes, EPA-600/4-79-020, CERI, USEPA, Cincinnati, Ohio.

Table 1. Classification of drainage point load-low rates in the Panoche Water and Drainage District.

DP	GROUP		Flow mm/month	Load Rates (kg/ha/mm)		
				Total Load	SE	B
			Q	S1	S2	S3
4	Giant		20.379	4.5714	0.000413	0.009082
5	Giant		31.999	4.2727	0.000346	0.007192
35	Giant		4.798	11.4345	0.028665	0.025808
6	Main	1	48.727	2.1882	0.000179	0.005608
7	Main	1	29.146	2.1516	0.000124	0.006156
8	Main	1	52.911	1.2006	0.000093	0.003900
9	Main	1	38.226	1.2678	0.000114	0.002155
10	Main	1	53.682	2.2003	0.000589	0.003260
12	Main	1	50.992	1.7676	0.000178	0.003462
13	Main	1	64.492	1.4780	0.000117	0.004206
14	Main	1	42.072	1.6457	0.000157	0.003913
15	Main	1	123.524	1.6853	0.000287	0.003980
18	Main	1	28.910	1.5847	0.000131	0.004876
19	Main	1	80.956	1.4933	0.000098	0.004206
20	Main	1	32.725	1.8285	0.000321	0.004146
22	Main	1	35.636	1.1581	0.000184	0.002859
23	Main	1	143.166	1.1581	0.000120	0.003474
30	Main	1	32.342	1.6457	0.000388	0.004145
38	Main	1	32.760	1.5451	0.000204	0.004206
39	Main	1	27.817	2.4990	0.000541	0.005364
55	Main	1	35.333	1.6883	0.000146	0.003417
56	Main	1	54.708	1.0894	0.000054	0.003901
57	Main	1	123.425	1.3395	0.000116	0.003393
1	Main	2	14.345	3.6144	0.000271	0.008347
2	Main	2	14.772	3.8887	0.000252	0.009569
3	Main	2	16.706	3.6571	0.000392	0.009721
16	Main	2	16.383	2.0724	0.000276	0.006095
25	Main	2	8.405	5.6743	0.004494	0.012617
27	Main	2	5.400	4.1264	0.001801	0.006827
31	Main	2	9.305	3.9619	0.005490	0.008838
32	Main	2	2.888	5.1443	0.012994	0.010751
34	Main	2	1.220	6.4022	0.011975	0.014010
40	Main	2	9.966	4.0533	0.006505	0.006583
41	Main	2	4.139	3.5474	0.003775	0.005547
42	Main	2	29.818	3.0476	0.003263	0.004387
44	Main	2	13.854	3.1390	0.000985	0.006741
45	Main	2	3.331	3.3887	0.002166	0.004472
49	Main	2	21.440	3.4735	0.001411	0.007710
50	Main	2	5.792	3.6754	0.003679	0.005486
51	Main	2	11.534	2.3771	0.000447	0.005973
52	Main	2	18.705	2.8038	0.002101	0.003474
11	Dwarf		14.726	1.5847	0.000104	0.004083
33	Dwarf		4.212	2.0175	0.000802	0.003078
43	Dwarf		0.667	2.6392	0.003641	0.003777
48	Dwarf		9.280	0.6037	0.000243	0.001008

MONITORING SYSTEM FOR WATER TABLE MANAGEMENT

WATER QUALITY RESEARCH

H.W. Belcher, G.E. Merva, A.C. Fogiel[*]

ABSTRACT

This paper describes a flow monitoring / water sampling system that is presently being used at six field research locations. A personal computer based water quality monitoring system has been developed that has the capability of measuring rainfall, underground pipe flow, overland flow and water table elevation at 10 minute intervals with automatic proportional sampling of rain water, pipe flow and overland flow. The system is assembled from off-the-shelf components, costs less than $5,000 and is battery powered.

KEYWORDS. Water Quality, Monitoring, Water Table Management, Water Table Control, Subirrigation

INTRODUCTION

Water table management involves using underground drainage pipes to maintain a water table at a depth conducive to crop production. To evaluate the effect of water table management on water quality it is necessary to monitor the quality and quantity of: 1) discharge water from the underground pipe system, 2) overland flow from the water table managed field area, 3) irrigation water supplied to the underground system, and 4) precipitation. Water quality parameters of interest include nutrient, pesticide and sediment concentrations along with pH and temperature data. The water quality parameters are evaluated by standard laboratory analyses.

A field based monitoring system for water quality needs to be dependable, accurate, easy to service, relatively inexpensive and be functional without alternating current electrical service. Data storage needs to be in digital form and be easily transferred to office based personal computer systems. Important but not essential attributes include capability to provide off-site warning of system malfunction and capability to download data from remote locations. The system described has evolved since 1987 and is continuing to be improved. In varying degrees, the systems in place provide all of the above attributes.

SYSTEM DESCRIPTION

The current version of the water table management monitoring system is shown schematically in Fig. 1. The system is capable of automatically monitoring rainfall, overland flow (runoff), underground pipe flow and water table elevation at four locations, all at 10 minute intervals and with automatic proportional sampling of pipe and overland flow, and rainfall.

The functioning of the system depends on the accurate measurement of the height of a water column at each sensor. This is accomplished by providing a continuous flow of compressed

---

[*] H.W. Belcher, Visiting Associate Professor, G.E. Merva, Professor, and A.C. Fogiel, Graduate Assistant, Dept. of Agricultural Engineering, Michigan State University, E. Lansing, Mi.

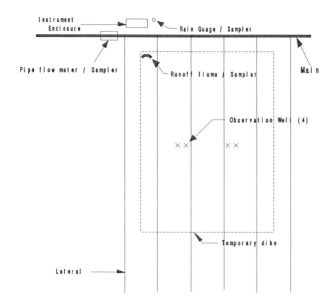

Figure 1. Schematic of water table management water quality monitoring system.

air through small diameter (3 mm) tubing to the bottom of the water column associated with each sensor where the air bubbles up through the water column. The air pressure in the tubing provides the height of the water column above the known elevation of the air tube outlet. The air pressure is measured by pressure transducers located in the instrumentation enclosure. The bubbler system is described in detail by Goebel (1986). Goebel determined that the height of water columns located up to 600 m from the pressure could be reliably measured.

The sensors that make up the system are further described as follows.

Pipe Flow Meter

The flow meter used for pipe flow (Fig. 2) ia an in-line orifice meter described by Protaswiewicz, et al., (1987). The equation used to model flow through the in-line orifice under full pipe flow is taken from Stearns (1951) and has the form:

$$Q = 2.086 * (d_2)^2 * K * (p*H)^{1/2} \qquad (1)$$

where: Q = Flow rate, l/min
 d_2 = Diameter of orifice, cm
 K = Orifice discharge coefficient (dimensionless)
 p = Density of fluid, g/cm^3
 H = Head loss across the meter (cm)

The orifice meter measures flow rates which are used to obtain proportional flow based tile water samples. The head loss across the orifice plate is determined by monitoring the height of the water column on each side of the orifice. The software calculates and accumulates the flow using equation (1). The system triggers the air activated sampling pump, located at the flow meter, at predetermined flow intervals.

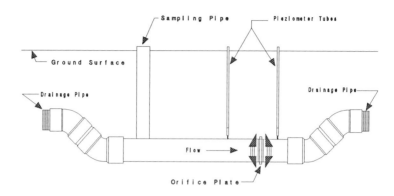

Figure 2. Schematic of underground pipe flow meter.

The water samples are composited into a four liter glass bottle placed below ground for frequent retrieval and transport to a laboratory for chemical analysis.

Overland Flow Meter

An H-flume, installed at the outlet of the diked field area, is used to monitor overland flow. The H-flume was laboratory calibrated to obtain the rating curve (Pruden and Fogiel, 1990). In 1990, six flumes were fabricated and calibrated. The non-linear regressions among the six flumes were almost the same and all yielded R^2's greater than or equal to 0.99. The equation used to calculate flow rate through those flumes has the form:

$$y = (0.009*(x^{2.036}) + 0.8)*0.003785 \qquad (2)$$

where: x = Depth of Flow, mm
                y = Flow Rate, m^3/min

The H-flume measured flow rates are used to obtain flow based proportional overland flow water samples. The bubbler system monitors the depth of flow through the flume by measuring the depth of water in the flume stilling well. The software calculates and accumulates the flow using equation (2). The system triggers the air activated sampling pump, located at the flume, at predetermined flow intervals.

Observation Wells

To monitor the performance of a water table management system, it is essential that the elevation of the water table be closely monitored. The water table is defined as the upper surface of groundwater or that level in the ground where the water is at atmospheric pressure (Soil Sci. Soc. Am. 1978). Four observation wells are installed at the approximate location shown by Figure 1. Two wells are within 1 m of an underground pipe lateral with the remaining two wells midway between laterals.

The observation wells are fabricated from 1.5 m lengths of 19 mm diameter galvanized steel electrical conduit with holes drilled throughout the length. The wells are wrapped with a thin spun fiberglass material to prevent soil movement into the well and are installed using a 100 mm diameter bucket auger and backfilled with sand. The wells are fabricated so that the top 0.40 m can be removed to allow field operations. After field operations, a magnetic locator is used to find the well and reinstall the top portion. Bubbler tubing is inserted into the observation well just below the top portion and runs to the bottom of the well.

Rain Gage

The rain gage consists of a 150 mm diameter sharp edged funnel draining to a vertical 5 mm diameter polyvinyl chloride (PVC) pipe, 0.65 m long, housed in a 150 mm diameter x 0.85 m long PVC pipe. A bubbler tube, mounted within the 5 mm pipe from the base of the funnel to the bottom of the pipe, monitors the rise in water column as rainfall accumulates.

SYSTEM OPERATION

Operation of the water table management monitoring system requires one time installation of the data acquisition instrumentation, the flow meter, the observation wells, the rain gage and the 3 mm plastic bubbler tubing. Temporary dikes are installed each year following spring field operations to isolate surface runoff. The H-flume, observation well tops and sampling units are removed and replaced for each field operation (tillage, seeding, cultivating and harvesting).

Annually, the elevation of the top of the observation wells, the pipe flow meter piezometer and the H-flume stilling well are surveyed. Periodically, at about two week intervals, the depth to the water in the observation wells and pipe flow meter piezometer is manually measured. Approximately monthly, the H-flume stilling wells and rain gage is calibrated by filling with a known volume of water. The date and time of these operations are recorded and the resulting data are used to develop or modify the conversion equations from pressure transducer digital output to water table elevation, rainfall and flow rates.

The two 12v deep cycle marine type batteries used to power the data acquisition and air compressor are replaced with fully charged batteries weekly. The two additional 12v deep

Table 1. System parts and approximate cost.

ITEM	SUPPLIER	UNIT COST	NUMBER NEEDED	COST
Toshiba Computer Model T1000		$650	1	$650
Starbuck Data Logger Model 8232	Starbuck Data Co. 9 Smith Street Wellesley, MA 02181 (617) 237-7695	$350	1	$350
Pressure Tranducer - 1 psi Honeywell Model 142PC01D	All-Phase Electric Co. P.O. Box 13128 Lansing, MI 48912 (517) 482-4449	$70	1	$70
Pressure Tranducer - 2 psi Honeywell Model 142PC02D	All-Phase Electric Co. P.O. Box 13128 Lansing, MI 48912	$70	7	$490
4 gal Air Tank Speedaire Model 4Z492	Grainger 5617 Enterprise Dr. Lansing, Mi 48911 (517) 394-2010	$35	1	$35
12 V. Air Compressor Speedaire Model 2Z880	Grainger 5617 Enterprise Dr. Lansing, Mi 48911	$65	1	$65
1/4" Air Regulator 2Z767	Grainger 5617 Enterprise Dr. Lansing, Mi 48911	$8	1	$8
12v Diaphragm Pump (2P366)	Grainger 5617 Enterprise Dr. Lansing, Mi 48911	$45	2	$90
Pump Pressure Switch (4A089)	Grainger 5617 Enterprise Dr. Lansing, Mi 48911	$17	1	$17
Air Pressure Switch (P61AA-6C)	The Tilford Co. 1629 E. Kalamazoo Lansing, Mi 48912 (517) 487-3900	$18	2	$36
PVC Orifice Meter		$200	1	$200
Observation Wells		$25	4	$100
H-flume		$225	1	$225
Water Sampler Enclosure		$75	2	$150
Instrumentation Enclosure		$150	1	$150
12v Marine Battery		$70	4	$280
TOTAL PARTS				$2,916
LABOR				$800
GRAND TOTAL				$3,716

cycle marine batteries used to power the pipe flow and overland flow samplers are replaced with fully charged batteries monthly.

Output from the data acquisition system consist of sensor identification, date, time and pressure transducer digital representation of pressure. Pressure readings that vary from the previous reading are automatically stored on cassette tape or floppy diskette. The tapes or diskettes are replaced weekly. In the office, data on cassettes is transferred to a personal computer for further transformation and analysis.

ACCURACY

The accuracy of the overall system is difficult to quantify. A conservative approach is to combine potential individual component errors in a way that the combined error is the maximum possible.

Sources of known error include pressure transducer error and analog to digital conversion error. The pressure transducers have a range from 0 to 700 mm with a manufactured specified error range of $\pm 0.15\%$ of full scale. The analog to digital converter has a resolution of 8 bits with a specified absolute accuracy of ± 3 least significant bits (LSB) maximum (± 1.5 LSB typical) with a 0 to 5 volt range. Assuming temperature does not effect accuracy and ignoring calibration errors, the error range for the combined components is ± 8.2 mm of water (700*1.0015*3/256) maximum and ± 2.7 mm typical. To increase accuracy, a pressure transducer with less operating range and/or less operational error can be used. Also, higher resolution analog to digital converters are readily available. For a 12 bit resolution analog to digital converter with a maximum ± 1 LSB error range, the error range for the combined components would be ± 0.2 mm of water (700*1.0015*1/4096).

For in-field systems, the accuracy of the results is likely to be more effected by calibration and environmental factors than by component limitations. This concern is addressed by making frequent water column measurements manually throughout the study period. These data are then correlated to numeric representations of pressure transducer voltage output to obtain the regression equations used for subsequent analysis. These data are also used to evaluate change in system performance with time.

RESULTS AND DISCUSSION

The system has been in use at a research site since April, 1990 (Fogiel, et al. 1992). Table 2 provides an overview of the functioning of the system during the 1990 and 1991 growing seasons. The pressure transducers used have a linear analog signal response to changes in pressure caused by variations in depth. The coefficients of determination (r^2) using the manually measured water column heights are presented in Table 2. In Table 2, under number of observations, the first number is the number of sensors in the field. The second number is the number of times each sensor water column was measured. The weighted mean r^2 was calculated by summing the sensor type r^2's and dividing by the number of that type sensor. The technique for the manual measurement of the water columns is described in the systems operation section.

The high r^2 values obtained for all of the sensors suggest that the data that resulted are at least as accurate as what would have been obtained by manual measurement of the sensor water columns.

Table 2. Summary of correlation coefficients for bubbler sensors monitored during the 1990 and 1991 growing seasons.

Sensor	1990 Growing Season			1991 Growing Season		
	Number of Sensors (Observations)	Minimum r^2	Average r^2	Number of Sensors (Observations)	Minimum r^2	Average r^2
pipeflow	2 (6)	0.955	0.969	2 (6)	0.994	0.995
overland flow	5 (6)	0.876	0.964	6 (6)	0.995	0.999
observation well	8 (4)	0.860	0.942	12 (6)	0.817	0.936
rain gage	2 (6)	0.912	0.952	2 (4)	0.997	0.998

Figures 3 and 4 provide examples of water table data converted to water table elevation and plotted against time. Figure 3 (from the conventional subsurface drainage treatment) shows the effect of rainfall on water table elevation. Figure 4 (from the subirrigation treatment) shows the increase in water table elevation resulting from beginning subirrigation in early July.

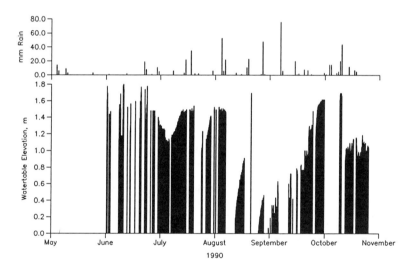

Figure 3. Plot of rainfall and water table for a conventional subsurface drainage treatment (1990 growing season, Unionville, 2.3 m from tile, taken from Fogiel, 1992)

ACKNOWLEDGEMENTS

The authors thank the Michigan State University Agricultural Experiment Station and the USDA Soil Conservation Service for funding to bring the water table management monitoring system to its present level of sophistication.

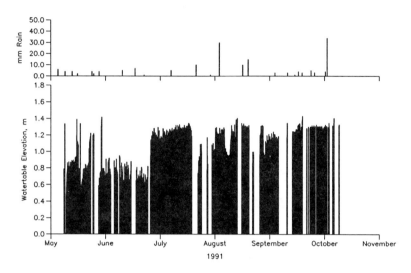

Figure 4. Plot of rainfall and water table for a subirrigated treatment (1991 growing season, Unionville, 2.3 m from tile, from Fogiel, 1992).

REFERENCES

Fogiel, A., H. Belcher and J. Crum. 1992. Water quality impacts of water table management systems - Unionville site annual report. Depart. of Agr. Engr., Michigan State University, E. Lansing.

Fogiel, A. C. 1992. Water quality and biomass impacts of water table management. M.S. Thesis, Michigan State Univ., E. Lansing.

Goebel, K.M. 1986. A bubbler system for water table monitoring. M.S. Thesis, Michigan State Univ., E. Lansing.

Protasiewicz, L.J., G.E. Merva and B.P. Darling. 1987. A flow meter to measure drain pipe discharge. ASAE Paper No. 87-2608. Presented at the ASAE International Winter Meeting held at Chicago, Illinois, Dec. 15-18, 1987.

Pruden, T.M. and A.C. Fogiel. 1990. Flume calibration and flow measurement using the bubbler system. ASAE Paper No. 90-2641. Presented at the ASAE International Winter Meeting held at Chicago, Illinois, Dec. 18-21, 1990.

Soil Science Society of America. 1978. Glossary of soil science terms. Madison, Wisconsin.

Stearns, R.F. 1951. Flow Measurement with Orifice Meters. Van Nostrand Company, NY, pp. 350.

EFFICACY OF ACID RECLAIMANTS IN COMBINATION WITH NONCONVENTIONAL FERTILIZERS FOR SALINITY CONTROL.

M.H.K. Niazi, N. Hussain, S.M. Mehdi, M. Rashid, G.D. Khan.

Soil Salinity and sodicity are the major limiting factors of crop production in the irrigated arid/semi arid regions of the world including Pakistan. The problem has become very acute and poses a serious threat to agriculture. To address the problem a field trial was conducted in a saline-sodic soil for assessing the impact of H_2SO_4 (100, 200 Kg ha^{-1}) and HCl (325, 650 kg ha^{-1}) in combination with ammonium sulphate and ammonium chloride as source of nitrogen (125 kg N ha^{-1}) on the yield of rice (IR 6). The results indicated that both the reclaimants in combination with N sources and SSP (50 kg ha^{-1}) increased the paddy yield significantly over control. Post harvest analysis of the soil revealed that ECe, ESP and pH values of the soil were lowered through sprinkling H_2SO_4 and HCl in the flooded water during rice growth. Acid-Grain Ratio and Value-Cost Ratio indicated a clear superiority of H_2SO_4 over HCl.

Key Words: Salimty, Acid Reclaimants, Fertilizer, Rice.

INTRODUCTION:

Salinity and sodicity are among the major identified factors restricting land utilization and causing low crop production in Pakistan. But the dilemma is that the situation will remain in the days to come as long as the present climatic conditions (low precipitation coupled with high temperature) prevail. Therefore planners, researchers and farmers will be ever-busy in the fight against salinity/sodicity. Enough information has been generated and packages of technology do exist for the reclamation of almost all classes of salt affected soils. However, the developed recommendations could not find their way into practice due to limited financial resources of the farmers as well as the federal and provincial governments. Hence the problem has been aggravated and more and more hectares of land are drifting out of cultivation. This fact demands that a simple low-cost technology be evolved to manage the salted soils in which inputs are directly used on crop production instead of prior reclamation and then growth of crops. Such a peculiar type of management is called the "Saline Agriculture".

Ameliorants are necessarily required to convert the salty soil environment into a favourable medium for crop production. A variety of chemical reclaimants (Gypsum, Calcium Chloride, Iron Sulphate, Aluminum Sulphate, Sulpher Sulphuric Acid, Hydrocloric Acid, Pyrites etc.) have been investigated and each proved beneficial under certain set of conditions (Mathers, 1970, Pooma and Bhumbla, 1974, Prather, 1978).

Soil Salinity Research Institute, Pindi Bhattian - PAKISTAN

Acids or acid forming amendments perform better on calcareous soils (Milap et al., 1977, Prather 1978) through bodily dissalution of $CaCO_3$, Ca replacing sodium from soil complex and producing an aggregated Ca saturated soil with good air and water premeability. Soils of Pakistan are generally calcareous having 5 to 12% lime in Punjab and Sind and 20% or more in NWFP and Baluchistan provinces. Thus acids will be preferred using soil Ca potential for reclamation. Rice plant is well adopted to salinity/sodicity and possesses reclaiming effect by increasing percolation rate, biological activity of roots and reduction in soil ESP (Chabara and Abrol, 1977, Abrol and Bhumbla, 1978). Therefore, this crop is preferably placed in selected rotations under "Saline Agriculture". Hence, acids when applied in standing water used for rice production increase the process of improvement on calcareous soil and enhance the yields (Rashid and Majid 1979).

Chemical fertilizers are basic inputs in modern agriculture and needed in higher quantities on saline/sodic soils (Rashid and Bhatti, 1985). Efficiencies of fertilizers are low in Pakistan and lowered further under saline conditions (Ahmad, 1985). Acids can increase this efficiency as a results of decreased pH, more availability of P and micronutrients and improving the clay complex of these soils. Present studies were carried out to investigate the probabilities of ascending fertilizer efficiency, achieving reclamation and increasing amen-ability of calcareous saline sodic soils for rice production through use of acids. Thus, a simple low-cost practicable technology was sifted in field studies.

MATERIAL AND METHODS.

A field study was conducted on a saline sodic soil situated in the typic rice growing area of Pakistan. This site was analysed for various soil characteristics before transplantation of rice. Plant bed was prepared as desired for low land rice and IR-6 variety was tested under flooded conditions. Thirty days old seedlings were transplanted for this purpose at a distance of 20 x 20 cm. Nitrogen and Phosphorus were applied at the rate of 100 kg ha^{-1} and 50 kg ha^{-1} respectively seven days after transplanting. Source of basal phosphorus dose was single super phosphate. Nitrogen was added as ammonium chloride (NH_4Cl) and ammonium sulphate ($(NH_4)_2SO_4$). Sulphuric acid (H_2SO_4) and hydrochloric acid (HCl) were also sprinked in the standing water as reclaimants. Concentration of HCl was 30% and that of H_2SO_4 was 98%. A supplemental dose of nitrogen was also applied as 25 kg ha^{-1} to all treatments 30 days after transplanting of seedlings. Following were the details of the treatments:-

Sr. No.	Treatments		
	N	P_2O_5	Acid
	Kg ha^{-1}		
1.	125 (NH_4Cl)	50 (SSP)	-
2.	-do-	-do-	325 (HCl)
3.	-do-	-do-	650 (HCl)
4.	-do-	-do-	100 (H_2SO_4)
5.	-do-	-do-	200 (H_2SO_4)
6.	125 ($(NH_4)_2SO_4$)	-do-	0
7.	-do-	-do-	325 (HCl)
8.	-do-	-do-	650 (HCl)
9.	-do-	-do-	100 (H_2SO_4)
10.	-do-	-do-	200 (H_2SO_4)

The design of the experiment was randomised complete block comprising of three replications. The plots were kept under continuous flooded conditions, 5-6 cm water layer always standing on the surface. The low land rice crop was harvested at maturity and yield data was recorded. Soil samples were obtained after the harvest and analysed in detail. Analysis of variance and economic analysis of the yield data was also carried out by computing increases over control, acid-grain ratio and value-cost ratio. Formulae used were as under:-

$$\text{Increase over control (\%)} = \frac{\text{(Yield of Acid Treatment-Yield without acid)}}{\text{Yield without acid}}$$

$$\text{Acid-Grain Ratio} = \frac{\text{Yield of Acid treatment-Yield without acid}}{\text{Quantity of acid added}}$$

$$\text{Value-Cost Ratio} = \frac{\text{Value of increased yield}}{\text{Cost of acid used}}$$

RESULTS AND DISCUSSION:

Soil Properties.

The experimental soil having ECe of 5.5 dS m^{-1} and ESP of 35.7 was greatly improved as a result of various treatments (table-1). A noticeable decrease was observed in ECe, pH, ESP, Na$^+$, HCO$_3^-$, Cl$^-$ and SO$_4^{-2}$ while an increase was recorded in soluble Ca^{2+} + Mg^{2+}. All the treatments of the experiment proved useful causing full or partial reclamation, as adjudged by the criteria layedout for classification of salt affected soils. Application of $(NH_4)_2SO_4$ + SSP + H_2SO_4 (200 kg ha^{-1}) to rice was the best observation of the experiment bringing ECe from 5.5 to as low as 2.5 dS m and ESP from 35.7 to 9.5. Same was the case with other parameters. Sulphuric acid was superior to HCl in combination with $(NH_4)_2SO_4$ as well as NH_4Cl. However, HCl also favoured reclamation and affected positively. Higher doses of both the acids were more efficient in this regard. Growing rice with the addition of fertilizers (ammonium sulphate or chloride + SSP) was also a good practice to improve the properties of a saline sodic soil but remained the most inferior among the treatments of these studies. Ammonium sulphate proved superior to NH_4Cl. Similar results were also reported by Akram et al (1990).

When H_2SO_4 or HCl acid is applied to calcareous soils, the following reactions may take place (Agri. Hand Book No. 60, 1954).

$$CaCO_3 + H_2SO_4 \longrightarrow CaSO_4 + CO_2 + H_2O$$

$$2Nax + CaSO_4 \longrightarrow Cax_2 + NaSO_4$$

$$CaCO_3 + HCl \longrightarrow CaCl_2 + CO_2 + H_2O$$

$$2Nax + CaCl_2 \longrightarrow 2NaCl + Cax_2$$

TABLE-1. SOIL PROPERTIES BEFORE AND AFTER HARVEST OF RICE.

Treatment Acids (Kg ha^{-1})	$Ca^{2+} + Mg^{2+}$	Na^+	CO_3^{2-} meL^{-1}	HCO_3	Cl^-	SO_4^{2-}	ECe dS m^{-1}	pH	ESP
POST HARVEST ANALYSIS.									
Control (NH_4Cl + SSP)	4.5	32.5	Nil	4.5	6.0	26.5	4.5	8.3	27.5
325 (HCl) + -do-	7.0	19.5	---	3.5	7.5	15.5	3.5	8.2	18.5
650 (HCl) + -do-	7.5	18.5	---	3.0	6.5	14.5	3.2	8.1	15.5
100 (H_2SO_4) + -do-	7.0	12.5	---	3.0	5.0	11.5	3.1	8.2	11.5
200 (H_2SO_4) + -do-	8.5	12.0	---	2.5	5.5	12.5	3.1	8.1	10.5
Control ($(NH_4)_2SO_4$ + SSP)	4.0	31.0	---	4.5	9.0	21.5	4.5	8.3	25.0
325 (HCl) + -do-	7.0	18.0	---	3.5	9.0	12.5	3.6	8.2	19.5
650 (HCl) + -do-	8.0	15.0	---	3.1	8.5	11.5	3.2	8.1	15.0
100 (H_2SO_4) + -do-	7.5	13.0	---	3.0	6.0	11.5	3.0	8.2	12.0
200 (H_2SO_4) + -do-	8.0	12.5	---	2.5	5.0	13.0	2.5	8.1	9.5
ANALYSIS BEFORE FLOODING THE FIELD.									
	3.0	51.0	Nil	5.0	12.5	35.0	5.5	8.3	35.7

TABLE-2 RICE YIELD AND ECONOMIC ANALYSIS.

Treatment Acids (kg ha^{-1})		Grain Yield (t ha^{-1})	Increase over control (%)	Acid:Grain ratio	Value Cost ratio
Control	(NH_4Cl + SSP)	2.12 f	---	---	---
325 (HCl) +	-do-	2.98 ef	40.6	1:1.6	1:3.8
650 (HCl) +	-do-	3.89 de	83.4	1:2.8	1:4.1
100 (H_2SO_4) +	-do-	4.51 hcd	112.9	1:23.9	1:13.3
200 (H_2SO_4) +	-do-	4.95 abc	133.6	1:14.1	1:7.9
Control	($(NH_4)_2SO_4$ + SSP)	2.27 f	---	---	---
325 (HCl) +	-do-	2.81 f	23.9	1:1.7	1:2.4
650 (HCl) +	-do-	4.18 cd	83.9	1:3.1	1:2.4
100 (H_2SO_4) +	-do-	5.22 ab	129.9	1:29.5	1:16.4
200 (H_2SO_4) +	-do-	5.50 a	141.9	1:16.1	1:8.9

Cd_1 0.915

Cd_2 1.254

a. NH_4Cl or NH_4SO_4 @ N 125 Kg ha^{-1}

b. Single Phosphate @ P_2O_5 50 Kg ha^{-1}

Thus the reactions of acids convert different salts into soluble forms which are leached downs into lower profile after subsequent flooding. The root action (physical and biological) of rice plant, decay and production of organic acids further put the conditions into favourable ones and soil amelioration took place. Similar were the claims of Mathers (1970), Chhabra and Abrol (1977), Abrol and Bhumbla (1978), Prather et al. (1978) and Rashid and Majid (1979).

RICE YIELDS:

The sifted treatments of this trial caused significant differences in rice yeilds (table-2). Sulphuric acid at the higher dose (200 kg ha^{-1}) combined with $(NH_4)_2SO_4$ was the best strategy to produce 5.5. tona ha^{-1} rice grains with an increase of 141.9% over control. This acid also signified the yields at the lower level (100 kg ha^{-1}) in combination with $(NH_4)_2SO_4$ as well as NH_4Cl. Observed deviations among both the doses were not appreciable statistically. Hydrochloric acid was regarded as inferior to H_2SO_4 in the light of observation of these studies (Fig-1).

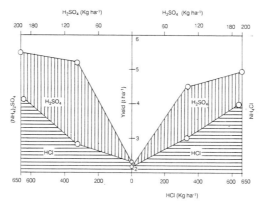

Fig-1: Comparison of acids and Nitrogen sources for rice production on a saline sodic soil.

Its lower dose (325kg ha^{-1}) remained non-significant to control as well as to the upper level while the latter was significant to control when additions were practised with NH_4Cl. However, its combination with $(NH_4)_2SO_4$ at the higher dose caused the yield increase measurable in statistical terms. Respective yields of 2.12 and 2.27 t ha^{-1} were recorded when NH_4Cl and $(NH_4)_2SO_4$ were applied alone while minimum increase of

23.9% was noticed in HCl + $(NH_4)_2SO_4$ treated plots. Enhanced yields due to acid applications were also noted by Milap (1977) and Rashid and Majid (1979).

Acid supplementation to fertilizers improved the soil characteristics, as it has been noticed in the earlier part of this discussion. Therefore, the soil medium was converted to the extent that it suited well for rice growth. The applied acids coupled with organic acids produced as a result of biochemical processes (more quicker under reduced conditions) decreased the pH of calcareous alkaline saline sodic soil and possibly increased the availability of some macro and micro nutrients. Better soil complex and favourable conditions of soil solution were depicted into elevated yields.

ECONOMIC ANALYSIS:

The computed parameters of Acid-Grain Ratio and Value-Cost Ratio indicated that acid application alongwith fertilizer were very useful economically (Table-2). By adding one kg of HCl over NH_4Cl, 2.6 and 2.8 kg of rice grains were obtained in the lower and higher levels respectively. Similar ratios for $(NH_4)_2SO_4$ were calculated as 1.7 and 3.1. Respective ratios for H_2SO_4 were very high with both the fertilizers and at both the levels. Maximum acid-grain ratio for this acid as well as in the results of the trial was 1:29.5 at the lower level combined with $(NH_4)_2SO_4$.

Value-Cost ratio revealed that by spending one additional rupee on HCl 3.8 and 4.1 rupees were earned in case of NH_4Cl respectively at the lower and higher dose of the acid. Similar values for H_2SO_4 were 13.3 and 7.9 depicting a clear superiority over HCl. Likewise was the case of $(NH_4)_2SO_4$ where these ratios were still more higher for H_2SO_4. Maximum value-cost ratio of 16.4 translated the fact that applying 100 kg of H_2SO_4 in addition to $(NH_4)_2SO_4$ + SSP probably would be the most beneficial strategy in rice production on a calcareous saline sodic soil.

CONCLUSIONS:

Primary and secondary data of the field trial revealed that soil attributes of a saline sodic soil were greatly improved as a result of HCl and H_2SO_4 supplementation over $(NH_4)_2SO_4$ + SSP and NH_4Cl + SSP and subsequent growth of rice crop under flooded conditions. Rice yields were also elevated to a mentionable extent due to these treatments. Higher levels of acids were generally more efficient than the lower one. However, lower level of H_2SO_4 and higher level of HCl proved more economical dictating wide Acid-Grain Ratio and Value-Cost Ratio. H_2SO_4 may be claimed superior to HCl in all the observed parameters.

REFERENCES.

1. Ahmad, N. 1985. Fertilizer efficiency and crop yields in Pakistan. Fert. and Agri. No. 89.

2. Akram, M., G. Hussain, M. Ashraf and Ch., Ehsan-ul-Haq 1990. Effect of gypsum and sulphuric acid on soil properties and dry matter yield of wheat grown in highly saline sodic soil. Proc. 2nd National Congress of Soil Sci. Faisalabad, Dec, 20-22, 1988. PP 80 - 84.

3. Abrol, I.P. and D.R. Bhumbla 1978. Crop response to different gypsum applications in highly sodic soil and the tolerance of several crops to exchangeable sodium under field conditions. Soil Sci. 127: 79 -85.

4. Chhabra, R. and I.P.Abrol 1977. Reclaiming effect of rice grown in sodic soils: Soil Sci. 124: 49 - 55.

5. Mathers, A.C. 1970. Effect of Ferrous sulfate and sulphuric acid on grain sorghum yields. Agron. J. 62: 555-556.

6. Milapchand, I.P. Abrol and D.R. Bhumbla 1977. A comparison of the effect of eight amendments on soil properties and crop growth in a highly sodic soil. Ind. J.Agric. Sci. 47(7): 348 - 354

7. Pooma, S.R. and D.R. Bhumbla 1974. Effect of H_2SO_4, HCl and $Al_2(SO_4)_3$ on the yield, Chemical composition and Ca uptake from applied $CaCO_3$ by Dhaincha in a saline-alkali soil. Plant and soil 40: 557 - 564.

8. Prather, R.J., J.O. Goertzen, J.D. Rhoades and H.Frenkel 1978. Efficient amendment use in sodic soil reclamation. Soil Sci. Soc. Amer J.42: 782 - 786.

9. Rashid, M. and A. Majid 1979. Use of sulfuric acid for economical rice production. Paper presented in rice research and production seminar at Islamabad on 18-22 March. 1979.

10. Rashid, M. and H.M. Bhatti 1985. Use of fertilizer in saline sodic soils. J.Agric Res. 23(2): 678 - 92.

11. U.S. Salinity Lab. Staff 1954. Diagnosis and Improvement of Saline and Alkali Soils. U.S. D.A. Hand Book No. 60. Washington D.C.

SOLUTE CONCENTRATION PREDICTION IN AGRICULTURAL DRAINAGE LINES UNDER A STRUCTURED SOIL

Gil Shalit, Tammo S. Steenhuis, Jan Boll, Larry D. Geohring, Hans A.M. Hakvoort, and Harold van Es[1]

ABSTRACT

A drainage experiment was conducted in soil with preferential pathways to study solute movement to tile lines. Three tillage practices were examined: no till, plowed, and incorporated (plowing, applying the salt, and incorporating with disk tillage). A nonadsorbed salt was surface-applied to 16 plots with subsurface drainage lines, which were subsequently irrigated for up to 22 hours. Insights into the mechanisms involved in preferential flow were achieved through examining the different outflow patterns resulting from these practices. A simple mixing layer model was employed in order to predict solute concentrations in the drainage lines. Initial moisture content, as well as the tillage systems, were found to influence solute concentration in the drainage lines and amounts lost as a result of irrigation. Up to 32% of the salt applied to a plot with an initially high moisture content was lost to drainage.

LITERATURE REVIEW

The first overview of solute loss through the soil was carried out more than 100 years ago by Lawes, Gilbert, and Warrington (1882). In today's terminology, their theory means that water collected in drains can be separated into two constituents: preferential and matrix flows. Preferential flow is precipitation that passes with little modification through soil channels consisting of cracks near the surface, and of deeper channels that are formed by either roots or earthworms. Matrix flow is water discharged from the pores of the saturated soil.

Preferential and matrix drainage water may have different solute contents. The chemical composition of preferential flow reflects the concentration of the water near the surface, while the matrix drainage flow represents the concentration of water around the drain. Thus, when a salt is evenly distributed throughout the soil, the salt content of matrix drainage is higher than that of preferential drainage which is characterized by the rainfall composition. The opposite is true (i.e., preferential flow has a higher solute content than the matrix flow) when a fertilizer or tracer has been surface-applied shortly before it rains. Consequently, the drainage water solute concentration depends on the ratio of preferential and matrix drainage water.

We are interested in predicting the concentration of preferentially moving solute from the surface to groundwater shortly after the chemicals were applied. To that end, we make the assumption that water enters macropores when it is at, or close to, atmospheric pressure. In other words, when there is free water at the soil surface. A similar assumption was made by Jarvis et al. (1991). A simple approach in which the free water at the surface and the percolating water are admixed in a layer of depth h_{mix} below the surface was suggested by Steenhuis and Walter (1980). The concentration of a nonadsorbed tracer left in the mixing layer at time t, $C_T(t)$, while runoff is occurring is given by:

---

[1] Research Assistant, Associate Professor, Research Assistant and Senior Extension Associate, Department of Agricultural and Biological Engineering; Visiting Scholar and Assistant Professor, Department of Soil, Crop and Atmospheric Sciences, Cornell University, Ithaca, NY 14850.

$$C_T(t) = C_0 \exp\left[-\frac{R(t-t_{sa})}{h_{mix} \cdot mcs}\right] \qquad (1)$$

t_{sa} is the time needed for saturation of mixing layer:

$$t_{sa} = \frac{h_{mix}(mcs-mci)}{R} \qquad (2)$$

and the initial concentration, C_0, is given by:

$$C_0 = \frac{M_0}{V \, mcs} \qquad (3)$$

where:
- h_{mix} = depth of mixing layer, m
- R = irrigation rate, m/hr
- mcs = volumetric moisture content at saturation
- mci = initial volumetric water content
- M_0 = original amount of tracer per unit area, kg/plot
- V = the zone volume (plot area multiplied by h_{mix})

The assumptions made above, mean that the concentration of solute in the water flowing through the preferential pathways to the groundwater and drainage lines will be the same as in the mixing layer.

EXPERIMENTAL METHODS

The general experimental procedure used in this study is similar to that used in a previous study, in the summer of 1990 (van Es et al., 1991). Both studies were conducted on the same tile-drained plots at Cornell University's Experimental Research Farm in Willsboro, northern New York State. The procedure of the earlier study consisted of the application of a 1 m band of a non-adsorbed tracer (bromide) on two sides of plots planted with alfalfa. In the current experiment, another non-adsorbed tracer (chloride) was surface-applied over the whole plot with three different tillage treatments and followed immediately by irrigation. Tile outflow and tracer concentration were determined throughout the duration of outflow.

Sixteen, 18x18 m hydrologically independent, drained plots, bounded on all sides by impermeable 8 mm PVC plastic liner to a depth of 1.8 m, were used. Corrugated drainage lines were installed in 1985 at a depth of 0.9 m and horizonal spacing of 6 m. The soil is a deep somewhat-poorly drained Kingsbury sandy clay loam with a slowly permeable layer at 1.2 m depth. The plots were plowed and harrowed and an alfalfa/timothy mixture was seeded in August 1988, after which no soil disturbance occurred until the present experiments.

Four of the 16 plots (plots 3, 6, 12 and 13) were irrigated throughout the summers as part of a different study. The water table in those plots was kept at the tile line depth of 90 cm. This resulted in a higher and more uniform initial moisture content in those plots. In the plots that were not summer irrigated, the water table was well below the tile lines.

The chloride tracer was applied to the total plot area at a rate of 30 kg chloride per plot, using $CaCl_2 \cdot H_2O$ flakes with 3 different tillage treatments (see Table 1):

- No Till (NT): Applying the tracer on the undisturbed soil surface.

- Plowed (PL): Plowing the plots, disking them, followed by the tracer application.
- Plowed and Incorporated (IN): Plowing the plots, applying the tracer and then disking, thus incorporating the tracer into the upper layer of the soil body. The plowing and disking were to a depth of approximately 0.15 m.

In addition, a fourth, unreplicated treatment was the check treatment (CH) in which no tracer was applied and the soil was undisturbed.

The first 8 plots were irrigated for 22 hours (run 1, see Tab. 1) and the other 8 plots were irrigated subsequently for 14 hours (run 2). Actual irrigation rates were determined with rain gauges placed near the center of each plot and sampled at 1 hr intervals after drainage outflow commenced. These measurements indicate actual irrigation rates of 3.6 to 8.9 mm/hr.

Drain outflow was sampled at changing intervals during the irrigation period. Outflow rates were determined by timing the filling of a graduated beaker. Outflow samples were analyzed for chloride and bromide (bromide was applied in previous years). Bromide concentrations in the outflow samples were negligible. Soil samples were analyzed for moisture content.

RESULTS AND DISCUSSION

Outflow rates and chloride concentration in the outflow are presented in Fig. 1 for one plot from each treatment. These plots had the lowest irrigation rate in each treatment. Different scales are used to emphasize the breakthrough patterns. Drainage outflow began 4.3-19 hours after the commencement of irrigation due to differences in initial moisture content and irrigation rate, as well as the different tillage treatments as will be discussed later. The highest chloride concentration observed was 1015 mg/l in plot 5 soon after outflow had begun. Most concentrations were in the 200-800 mg/l range. The

	Plot	Run	Irrigation Rate [mm/hr]	Water [mm] Applied	Drained	Summer Irrigated
IN	5	2	4.4	110	19.1	
	7	2	7.4	134	4.9	
	4	1	8.0	190	4.9	
PL	8	2	5.0	110	9.9	
	15	1	5.0	105	1.8	
	10	1	5.5	111	1.3	
	3	2	4.2	85	16.5	yes
	13	1	5.0	117	64.9	yes
	6	2	5.7	120	24.5	yes
NT	2	2	3.6	71	0.6	
	11	1	4.8	107	5.1	
	16	2	5.0	122	6.3	
	9	2	5.9	120	4.6	
	1	1	6.5	176	33.3	
	12	1	5.1	111	28.3	yes
CH	14	1	8.9	203	30.8	

Table 1: The study plots, arranged by treatment and irrigation rates. Applied water includes precipitation.

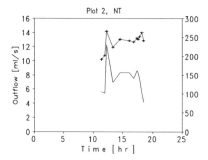
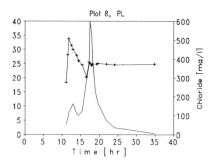

depth of precipitation needed to start outflow, which we shall call the delay, is presented in the last column of Tab. 2. When plotted against the irrigation rate in Fig. 2, inspection of the data revealed two distinct different behaviors: summer irrigated plots (+) and plots not receiving any additional water during the summer (□). A positive linear relationship emerged between irrigation rate and delay for the two groupings although, for the summer irrigated plots, the linear regression had a low R^2 value.

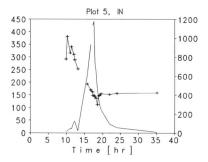

Figure 1: Cl Concentration (+) and outflow rates (full line) in drainage outflow for three plots.

Regression equations of the following forms were obtained:

$D = 14.1 * R$ nonsummer irrigated
$\qquad\qquad\qquad\qquad (R^2 = 0.75)$
$D = 7.0 * R$ summer irrigated
$\qquad\qquad\qquad\qquad (R^2 = 0.22)$

Where:
$\qquad D$ = delay [mm]

Using the regression equations, the average time delay was 14.1 hrs for the nonsummer irrigated plots and 7 hrs for the summer irrigated plots. Thus the time and not the volume of water applied determines when the outflow starts for each grouping. This is unexpected as one would assume that outflow would have started after a certain volume of water raised the water table level up to the tile line level. Although more research is needed, it is not unlikely that the time delay is the time required for the soil to swell and close up its many cracks. At

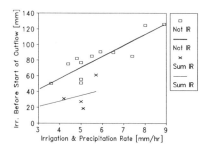

Figure 2: Depth of water needed to start outflow (delay), versus irrigation rate. IR denotes the plots which were summer irrigated. R^2 is 0.22 and 0.75 for the summer and non summer irrigated plots, respectively.

depths below 1 m, many fine cracks were observed, which probably close up when the soil becomes wet. This is unlike the upper 1 m layer, in which there were wormholes and cracks filled with roots and other material that had fallen in from the surface, preventing the closing of the cracks. This mechanism can further explain how a small volume (as little as 5 cm in the plots which were not summer irrigated) can bring the water table up from a depth of at least 3 m, to the tile line level. The following hypothesis can, therefore, be formulated as follows:

> The length of time which elapses before outflow can start is dependent only on previous wetting history and is independent of the rate at which water is applied.

Chloride Concentration In The Outflow

In the previous study on the same plots (van Es et al., 1991), a positive correlation was found between outflow rate and the concentration of the surface applied tracer in the drainage outflow. This is in accordance with the description presented in the literature review, as the higher flow rates are associated with the increased preferential flow that carries the high solute concentration downwards. Compared to the current experiments, the amounts of water applied in that experiment were much lower (3.5 cm) and the initial moisture contents were higher so that outflow occurred much sooner. Correlations between chloride concentration and outflow rates for the current study are presented in Tab. 2. Only four plots demonstrated a positive correlation (associated with a surface-applied salt). Most plots had a negative or non-conclusive correlation between concentration and outflow. This is most evident in the IN treatment where all three plots had a negative correlation with a relatively high R^2. Three of the four plots with a positive correlation were NT plots (2, 9 and 11). The fourth plot was 13, a summer irrigated PL plot. The positive correlation in those four plots is most evident at the beginning of the outflow period. A few more plots demonstrated a positive correlation at the start of the outflow (see plots 8 and 5 in Fig. 1).

If PL is divided into two treatments, one that received summer irrigation (PL-IR) and one that did not (Tab. 1), the following result emerges. The irrigation rates presented in Tab. 1 suggest that a positive correlation occurred in the plots with the lowest irrigation rate of each treatment: plots 8 in PL, 2 11 and 16 in NT, 5 in

	A	B	R^2	Delay [mm]
NT				
2	5.407	198.3	0.265	50.2
11	1.549	230.2	0.172	82.2
16	-0.265	248.9	0.034	55.6
9	0.460	381.9	0.183	90.7
1	-0.777	406.1	0.339	89.7
12 (IR)	-0.608	351.0	0.340	18.9
PL				
8	-0.650	396.1	0.19	77.0
15	-4.972	625.0	0.248	50.9
10	-3.748	244.8	0.127	85.3
3 (IR)	-0.079	616.7	0.002	31.2
13 (IR)	0.319	456.5	0.323	27.4
6 (IR)	-0.008	538.7	2.e-5	61.1
IN				
5	-0.842	665.9	0.307	75.3
7	-1.230	431.6	0.323	84.9
4	-0.809	330.1	0.201	124.6

Table 2: Correlations between Chloride concentration in outflow, and outflow rates (Y=AX+B). Delay is the depth of irrigation water needed before drainage outflow is started. IR denotes summer irrigation.

IN and 12 which was the only summer irrigated NT. Thus, the relationship between flow rate and solute concentration is not as simple as postulated in the previous study (van Es et al., 1991). Therefore, we will now explain the observed phenomena with the aid of the model described earlier (Equations 1 and 2).

Predicting The Concentration In The Tile Lines

Dye infiltration studies on a nearby field with very similar soil (Steenhuis et al., 1989) are useful in understanding the flow path of water and solutes under different tillage practices. In plowed plots, we can expect mixing of new and old water in the plowed layer, where a capillary fringe exists above the plow pan. This is contrary to the no till plots where water and solute flow over the surface to the macropore openings with little mixing with old water.

The chloride hydrographs have three clearly discernable parts. The first, a rapid rise, is noticeable mainly in the plots which start with a positive correlation between outflow rate and solute concentration, as discussed above. The second part is an exponential decrease in concentration, and the third is a constant or linearly decreasing concentration. These three parts could easily be recognized in the previous experiment (van Es et al., 1991). The exponential part occurs when the flow is high and is, thus, related to the preferential flow, whereas the linear part takes place when the flow is low and matrix flow predominates. To explain this further, an analogy with the flood hydrograph is useful here (Fig. 3): Matrix flow is analogous to the base flow and preferential flow to surface runoff (in accordance with Lewas et al., 1882 and Everts and Kanwar, 1990). Comparing the flood hydrograph in Fig. 3 with the hydrographs in Fig. 1, we note that the concentrations decrease exponentially until the outflow hydrographs reach point b, which is the point where surface runoff (or preferential flow in our case, Fig. 1) stops. We hypothesize, therefore, that the exponential part of the chloride hydrograph is water flowing from the surface to the tile lines, and the linearly decreasing part is matrix flow from the area around the tile lines. The distinctly different characteristics of those three parts are in agreement with the

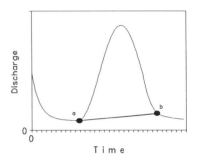

Figure 3: Schematic stormflow hydrograph. Base flow increases linearly between a and b, and the flood flow increases rapidly and decreases exponentially.

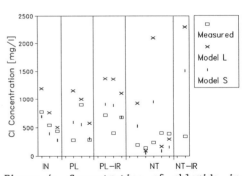

Figure 4: Concentration of chloride in outflow at the onset of outflow. h_{mix} is 15 cm for IN and PL, and 4 cm for NT. 'Model L' and 'Model S' use initial moisture content from neutron probe measurements or saturation.

observations of Lawes et al. We shall further show that the exponential decrease is also in agreement with the model predicting the concentration (Equations 1 and

2).

To do that, we examine if these equations are valid in the period before the tile outflow commences. We will try to predict, with Equations 1 and 2, the initial salt concentration in the tile outflow. To estimate h_{mix} we use observations made in a previous study of preferential flow through large core samples taken from the Willsboro Experimental Research Farm on a similar soil (Andreini and Steenhuis, 1990). The cores were subjected to 2 cm/day irrigation for 42 days after which they were dissected by cutting successive horizontal layers from the core's bottom surfaces. In cores extracted from no till plots, a layer 1 cm deep under the soil surface was found to be close to saturation. In cores extracted from conventionally tilled plots, this layer was 15 cm deep. In both cases, the soil under these layers was not uniformly moist and showed preferential flow patterns. These depths of 1 cm for NT and 15 cm for PL and IN were taken as values for h_{mix}. The saturated moisture content, mcs, was 0.45 cm³/cm³ for NT and 0.49 cm³/cm³ for PL and IN. Because of uncertainty about the measured moisture content, two sets of values for the initial moisture content were taken. The lower value set is the neutron probe measurements taken 17 days prior to the experiment (determined individually for each plot), and the second set is the upper bound for moisture content - saturation. The initial concentration (C_0 in Eq. 1) is the amount of chloride initially applied to each plot - 30 kg/plot divided by the total initial amount of water in the mixing layer for the whole plot. The time (t in Eq. 1) is the time outflow started for each plot (t=0 when irrigation starts).

Results of the model predictions (Equations 1 and 2) for concentrations of chloride in the drainage outflow at the onset of outflow with the two sets of initial moisture contents are compared with the measured values in Fig. 4. For the plowed and incorporated (IN) treatment, the measured results fall within the two model predictions. This is not surprising, as the assumption of "complete mixing" was best met in this practice. For the plowed (PL) treatment, the measured values were generally lower (especially for the summer irrigated plots), than the predicted values. Mixing was not complete and some preferential flow occurred in the plowed layer between clods, as observed in dye experiments in a nearby field, reducing salt concentration in its soil water faster then predicted by Equations 1 and 2. The value of 1 cm for h_{mix} in the NT treatment gave a poor correlation with the measured values. This is to be expected, as the no till treatment was sustained for a much longer period in the Andreini and Steenhuis (1990) study. By evaluating the model with different values, a depth of 4 cm was found to give the best correlation (Fig. 4).

Predictions and measured values of tile outflow concentration for one of the IN plots are shown in Fig. 5. The measured concentration is enveloped by the two model predictions (lower and upper bound of initial moisture content) until the inclination point of the outflow hydrograph (point b in Fig. 3). This is once again an indication that, up to that time, most of the drainage outflow originates in the upper soil layers. The results from these tracer experiments are encouraging

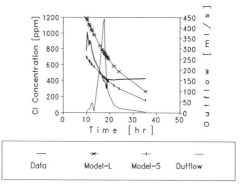

Figure 5: Chloride concentration in drainage outflow of plot 5 and those predicted with the mixing model. 'Model L' indicates the use of measured initial moisture conditions. 'Model S' uses saturation.

401

and may lead to predictive equations.

SIGNIFICANCE AND CONCLUSIONS

Surface-applied chloride was readily available for loss. If it were nitrate, it would have caused high levels of contamination while not being available for plant consumption. In the plowed plots, an initially higher moisture content was associated with higher loss of chloride to tile lines. This finding might have significance for the development of Best Management Practices for the reduction of loss of chemicals to drainage lines.

A simple model was used to predict the solute concentration in the tile lines under plowed plots where the salt was incorporated into the upper layer prior to irrigation. As the model performs well only while preferential flow is the main source of flow, it can be used to distinguish between matrix and preferential flows.

Tillage systems, and initial moisture content affected solute flow patterns. These factors can be controlled by land users and should be taken into account for reduction of solute loss to the drainage system.

REFERENCES

1. Andreini, M.S. and T.S. Steenhuis. 1988. Preferential Flow under Conservation and Conventional Tillage. ASAE Paper No. 88-2633. American Society of Agricultural Engineers, St. Joseph, MI. 22 pp.

2. Everts, C.J. and R.S. Kanwar. 1988. Quantifying Preferential Flow to a Tile Line with Tracers. Paper No. 88-2635. American Society of Agricultural Engineers, St. Joseph, MI. 15 pp.

3. Jarvis, N.J., L. Bergström and J. Stenström. 1991. A Model to Predict Pesticide Transport in Macroporous Field Soils. Conference Proceedings, National Symposium on Preferential Flow. Chicago, IL. Dec. 16-17, 1991.

4. Steenhuis, T.S., W. Staubitz, M. Andreini, R. Paulsen, T. Richard, J. Surface, N. Pickering, J. Hagerman and L.D. Gehoring. 1989. Preferential Pesticide Movement under Two Tillage Practices. American Society of Civil Engineers Journal of Irrigation and Drainage 116(1):50-66.

5. Steenhuis, T.S. and R.E. Muck. 1988. Preferred Movement of Nonadsorbed Chemicals on Wet, Shallow, Sloping Soils. J. Env. Qual. 17:376-384.

6. Steenhuis, T.S. and M.F. Walter. 1980. Closed Form Solution for Pesticide Loss in Runoff Water. Transactions of the ASAE 23:615-612, 628.

7. van Es, H.M., T.S. Steenhuis, L.D. Geohring, J. Vermeulen and J. Boll. 1991. Movement of Surface Applied and Soil Embodied Chemicals to Drainage Lines in a Well Structured Soil. Proceedings of the National Symposium on Preferential Flow. Chicago, IL. Dec. 16-17, 1991. pp. 59-67.

EVALUATION OF AN ULTRASONIC RANGING SYSTEM FOR

MONITORING WATER TABLE LEVELS

J.D. Eigel S.A. Marquie B.A. Vorst
Assoc. Member ASAE

INTRODUCTION

The measurement of water table levels is an important aspect of drainage research. In addition, field determination of lateral saturated conductivities of water bearing strata often requires the measurement of water table levels in either single or multiple wells over a period of time. Water table level measurements are also important in the determination of the direction of groundwater flow. This problem has increased significance when the goal is to predict the migration of a contaminant plume in the vicinity of water supply wells. Water table measurements are also useful in field tests of analytical and detailed numerical models of soil water movement.

Hall (1978) describes many common methods of liquid level measurement. Measurement techniques can be classified as contact or non-contact detection devices. Contact methods include capacitance and conductive probes, float-actuated potentiometers and chart recorders, and pressure transducers that are submerged in the fluid. Non-contact methods include radiation, sonic, optic, and pressure transducers coupled with a gas bubbling system (Dedrick and Clemmens, (1984).

A float and chart recorder is perhaps the most common contact method of water table level measurement. Some researchers, such as Tromble and Enfield (1971), Van der Weerd (1977), and Munster, et al.(1991), have integrated a potentiometer with a chart recorder so that a voltage proportional to the stage of water in the well is generated. Goebel and Merva (1985) give a comprehensive review of water table measurement techniques.

Tinham (1989) gives an overview of recent developments in level measurement and control that concentrates on developments in ultrasonic systems. Tinham reviews two commercial units based on time-of-flight measurements. These units feature auto-referencing, and compensation for temperature, vapor, product build up, and obstructions. The range of the instruments are on the order of 0.2 to 15 m. One unit can be modified to extend the range to 60 m. These units can be multiplexed and provide digital outputs. One unit supports packet radio transmission of data that allows sensors to be located several hundred meters from the recording system. He also discusses optical, acoustical, radio wave actuated, and pump motor load sensors for level control.

Purdue Agricultural Experiment Station Journal Paper No. 13519
The authors are J.D. Eigel, Assistant Professor, and S.A. Marquie, Research Associate, B.A. Vorst, Undergraduate Technical Assistant, Agricultural Engineering Dept., Purdue University, West Lafayette, IN.
The authors are grateful for the assistance throughout this project of Mr. Matt Carroll, Research Engineer, Robot Vision Laboratory, Electrical Engineering Department , Purdue University.

Wang, et al. (1991) developed a simple, inexpensive, non-contact device for monitoring liquid surface levels. Their problem was to determine the level of a water surface relative to electrodes in experiments on electrochemical behavior in spread monolayers at the air-water interface. These measurements require a non-contact method capable of accuracy in the 10 - 100 μm range. Their device was capable of a precision of 10 μm. The device used two 45 kHz transducers manufactured by Polaroid®* Corporation. They used the phase shift in the reflected wave to detect the level of the water surface.

Ramana and Radhakrishna (1990) sensed microlevel changes in groundwater levels. They were interested in measuring groundwater level fluctuations in response to surface temperature and barometric pressure fluctuations as well as the influence of tidal variations. They implied that micromeasurements of groundwater levels may be useful as a tool for predicting earthquakes. They used a 1 MHz ultrasonic transducer submerged in an observation well. The transducer faces upward so that ultrasonic pulses are reflected at the air-water interface in the well. Travel time measurements were made using a 100 MHz oscilloscope. The sensitivity of their instrument ranged from 0.001 to 0.009 cm. The depth of the transducer below the water surface was 2.3 to 8.5 cm. They used the system to measure groundwater level fluctuations in response to temperature and barometric pressure variation over a three-day period at a site in Hyderabad, India.

Erhst (1991) used an ultrasonic transducer to measure the level of liquid in a tube. The transceiver is mounted in the bottom of the tube and is used to measure distance from the transducer to the liquid/air interface. A commercial unit from France is capable of measuring water levels in wells with resolution of 1mm (CR2M, 1986). Two units are offered. One is based on a 1 MHz transducer. It has a range of 45 to 1500 mm with 1 mm resolution. The second unit is based on a 500 kHz transducer. It has a range of 90 to 2500 mm and has a resolution on the order of 2 to 3 mm. Both of these units are submerged in the well tube. Measurements are based on the propagation of ultrasonic waves through water above the transducer.

This paper presents an evaluation of an ultrasonic ranging system for use as a low cost electronic water table level monitoring system. Goals for the project include the development of a water table monitoring system with sensors that are portable, easily maintained and serviced, and have low DC power requirements. The tests reported here are an initial evaluation of the system for application as a stand-alone sensor network that has the capability of remote sensing via packet radio transmission to a computer controlled transceiver.

THE ULTRASONIC RANGING SYSTEM

The ultrasonic ranging system consists of a Polaroid® 6500 Ranging Board (Part No.615077), Polaroid® Part No. 607281 Environmental Grade Transducer , a National Instruments® PC-TIO-10 timing card , and a Samsung® DeskMaster 386S/16 personal computer. A schematic of the ranging system is shown in Figure 1. A 74S00 NAND gate is used to condition the signal so meaningful information can be extracted using the PC-TIO-10 timing card. The ultrasonic sonar circuit and transducer are modified versions of the Polaroid® ranging system incorporated in some standard automatic focus mechanisms in cameras. The ranging board can be operated in single echo mode when the detection of a single target is required or can be operated in multiple-echo mode when the detection of more than one target is desired. Single echo mode was used for measurements in the experiments reported here.

*The use of tradenames does not imply endorsement of a product by the authors or Purdue University.

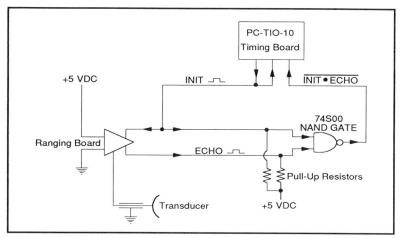

Figure 1. Schematic of Ultrasonic Ranging System.

The ranging system is triggered by an initiate (INIT) pulse generated by an external source, the ranging circuit produces a series of 16 pulses at a frequency of 49.4 kHz. The frequency of the transducer corresponds to a resolution of approximately 0.7 cm for an ultrasonic pulse travelling through air. This signal is amplified with a step-up transformer to 300 V peak-to-peak. with a 150 V DC bias to excite the transducer (Maslin, 1983). The INIT signal must be high for a minimum of 100 ms and then go low for at least 100 ms. The terms "high" and "low" refer to high logic level and low logic level signals, respectively. The initiate signal was generated using the output from the PC-TIO-10 timing board, using its 1 kHz clock. For each period, the INIT signal is high for 100 ms and low for 200 ms.

The processed echo output (ECHO) is also a logic level signal. The length of the high pulse ranges from 1.6 ms to 62.2 ms, corresponding to target distances of 15 and 1067 cm, respectively. Internal blanking inhibits the input of the RECEIVE input of the ranging control IC to eliminate ringing of the transducer as being detected as a return signal. The time between the leading edge of the INIT signal going high and the leading edge of the ECHO signal going high corresponds to the travel time of the ultrasonic pulse. The travel time can calculated by the difference between the time at which the INIT signal is generated and the time at which the ECHO signal is received, as shown in Figure 2. The internal blanking is effective for 2.38 ms after the INIT signal is received. This makes the effective range of the system 41 to 1067 cm.

The PC-TIO-10 timing card can be configured as an event counter, a frequency counter, and can be configured to make pulsewidth measurements. None of these modes of operation are acceptable for making direct measurements of the arrival time of a series of high level signals of the INIT and ECHO signals generated by the ranging board.

The logical operation $\overline{\text{INIT} \bullet \text{ECHO}}$ (INIT NAND ECHO) can be used to generate a logic level signal from which the travel time of the ultrasonic pulse can be derived. The truth table for $\overline{\text{INIT} \bullet \text{ECHO}}$ is given in Figure 2. As shown in Figure 2, the travel time, T_t, can be expressed as:

$$T_t = L_I - L_N$$

Where L_I is the width of high pulse in the INIT signal and L_N is the width of the logical low in the

INIT•ECHO signal. It represents the time period between the time at which the ECHO signal is received and the time at which the INIT signal goes low. In this application, L_I is 100 ms.

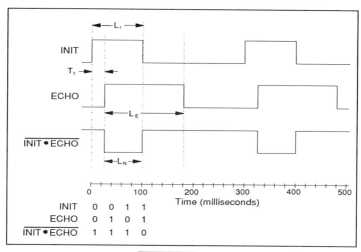

Figure 2. INIT, ECHO, and INIT•ECHO signals.

The INIT and ECHO signals are fed into a 74S00 NAND gate as shown in Figure 1. 2.7 kΩ pull-up resistors are used to bring the C-MOS level ECHO signal within the range of the TTL level of the 74S00 gate. The INIT signal is pulled up to the same level. Switching time of the gate is 3 ns, so errors resulting from switching of the gate are negligible in this application. The 74S00's output was then fed into the input of the Lab Windows timing board. The ranging board emits a pulse-generated signal that produces a back-wave on the V_{cc} line powering the system. Improved supply-line regulation was obtained by increasing power supply filtering and adding "by-pass" capacitors to the input of the NAND gate. Utilizing separate voltage supplies for the ranging board and the gate also produced satisfactory results. Care also needs to be taken to insure that no open ended inputs are present on the ranging board or 74S00 gate.

EXPERIMENTS

The goal of this project was to evaluate the performance of the ultrasonic ranging system in monitoring well tubes in the laboratory. Two well tubes were installed in the test fixture shown in the schematic of Figure 3. As many as four well tubes can be mounted vertically in the test fixture. Water is fed to each well tube through a manifold. Each well tube port on the manifold is fitted with a ball valve and a pipe union. The well tubes and ports are constructed from 1.9 cm (3/4 in nominal) steel pipe. A PVC plug was machined to fit the bottom of each well tube, then bored and threaded to accept a 1.9 cm (3/4 in) pipe nipple. Well tubes are attached to the manifold using the nipple and pipe union.

Only one tube is tested at a time. During a test, only the ball valve for the tube being tested is opened. The water level in the test tube is raised and lowered by opening and closing the inlet and outlet valves of the manifold. Water depths in the tube are measured using the clear plastic standpipe and scale. The scale can be adjusted vertically so that the zero mark of the scale coincides with the bottom of the ultrasonic transducer. The scale is marked in 1 mm increments and the position of the zero mark was determined by differential leveling.

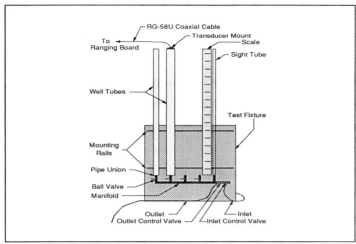

Figure 3. Schematic of laboratory test fixture.

Initially, the performance of the transducer was to be evaluated in 5.1 (2 in nominal) and 10.2 cm (4 in) diameter (inside) well tubes. However, initial tests on the 5.1 cm diameter tubes were not successful when the walls of the tube were wet. Water beads forming on the inside of the tube reflected false echoes back to the transducer. This problem was not experienced with the 10.2 cm diameter tube. The problem of false echoes from water droplets in the 5.1 cm diameter tubes may be alleviated by using a waveguide to concentrate the ultrasonic burst in the center of the tube. Further testing will be required to evaluate the use of the ultrasonic transducer in 5.1 cm tubes after prototype waveguides are developed. Use of alternative construction materials for small diameter tubes should also be evaluated.

Tests were run in vented 10.2 cm diameter well tubes and in open air. The results of the tube tests are compared with those from tests of the system in open air. The length of the tubes is 3 m. Three replicates of vented tube and open air tests were conducted. The order in which the tests were completed was chosen randomly. The target for the open air tests was a concrete block wall.

Each trial consisted of a calibration followed by a test in which soundings were used in conjunction with calibration data to predict the depth or distance to the target. Calibrations were made at 11 water depths in the tube: 40, 65, 90, 115, 140, 165, 190, 215, 240, 265, and 290 cm. In open air trials, the range of the test was extended to 800 cm. The sensor was calibrated at 40, 65, 90, 115, 140, 165, 190, 215, 240, 265, 290, 400, 500, 600, 700, and 800 cm. In each trial, the calibration readings were taken in random order. After the calibration, test measurements were made at levels chosen by random between 40 cm and 290 cm in the well tubes and between 40 cm and 800 cm in open air tests. Readings were taken at 20 levels in the well tubes and at 30 distances from the target in the open air tests. Pulsewidth measurements were made at each test and calibration level until 10 consecutive consistent readings of the $\overline{\text{INIT}\bullet\text{ECHO}}$ signal were made during well tube tests. Readings were taken until 10 consecutive readings, each within 10 μs of the first, were made in open air tests.

RESULTS AND DISCUSSION

The calculation of distances based on the measurement of the travel time of ultrasonic sound waves depends upon an "a priori" knowledge of the velocity of sound in air. The velocity of sound in air should remain relatively constant, but does vary with temperature and humidity. A temperature variation form -30 to 60° C results in a 14.6% difference in the speed of sound while it only varies 0.44% over the range of 0 to 100% relative humidity (Tinsey and Rohrbach, 1991) Air temperatures in the well tube varied from 22.5 to 26° C between the experiments. Other than variations due to temperature, the speed of sound should remain constant throughout the experiments.

Results of the calibrations for a trial of tube tests and a trial from an open air test are shown in Figure 4. The apparent sound velocity presented in Figure 4 is the sound velocity computed during the calibration readings. The apparent velocity, V_a, is given by:

$$V_a = 2D/T_t$$

In tube tests, D is the depth to the water surface. In open air tests, D is the distance between the transducer and target. T_t is the elapsed time between emission of the pulse from the transducer and the time at which the reflected echo is received. In all cases, the apparent velocity was low near the target. The change in apparent velocity is small at distances greater than roughly 140 cm. This phenomena was observed in all open air tests and in all well tube tests, so interference of sound waves in the well tubes does not appear to contribute to the error. Interference of echoes was not apparent in observations of the reflected echo from well tube tests when compared to echoes from open air tests.

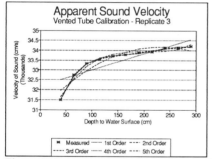 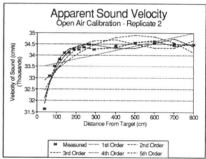

Figure 4. Apparent sound velocity for vented well tube and open air experiments.

The curves in Figure 4 are plots of a series of regression equations fitted to the calibration data from the trial. The equations predict the apparent sound velocity in terms of the travel time of the ultrasonic burst. Five regression equations were generated for each trial. Each regression equation can be expressed as:

$$V_a = a_0 + \sum_{i=1}^{n} a_i T_t^i$$

Here, a_0 is the intercept, and a_i is the coefficient on each of the i-th order terms. The 1st, 2nd, 3rd, 4th, and 5th order curves in Figure 4 represent regression equations for that trial with i = 1, 2, 3, 4, and 5, respectively.

The distance from the target in open air experiments and the depth of water in well tubes were predicted for each trial using the average sound velocity over the range of the calibration and each of the five regression equations to account for variations in the apparent velocity of sound. Table 1 presents the maximum absolute errors in measurements using the ultrasonic transducer compared to the measured distance to the target in open air tests and depth of water in well tubes. Absolute errors in Table 1 are given by:

$$E_a = D_u - D_s$$

D_u is the distance or depth determined from ultrasonic measurements and D_s is the distance measured by scale or the depth measured in the standpipe. The values given in Table 1 are the maximum values for each trial for a given calibration equation. The maximum relative errors in the measurements are given in Table 2. The "Average" columns refer to errors when the average velocity over the range of calibration measurements is used to predict depth or distance in a trial.

Table 1. Maximum absolute errors in open air and well tube tests*.

Trial	Average	1st Order	2nd Order	3rd Order	4th Order	5th Order
			Open Air Tests			
OA1	12.68	11.33	7.44	6.58	12.05	10.14
OA2	12.03	11.30	4.20	4.02	2.42	5.01
OA3	13.22	12.36	8.68	4.86	5.19	4.70
			Well Tube Tests			
TV1	5.53	1.48	1.80	1.78	1.16	1.65
TV2	4.85	1.79	1.22	1.35	0.95	0.62
TV3	2.84	2.06	1.42	0.69	0.81	0.58

*All values given in cm.

The relative error, E_r, is given by the following relationship:

$$E_r = (E_a/D_s) \cdot 100\%$$

Table 2 presents the maximum relative error for each trial for a given calibration equation. In these trials, all of the relative errors are greater than that specified by the manufacturer (1%). Only in the tube tests with 4th and 5th order calibration equations used to determine V_a are relative errors less than those specified by the manufacturer. None of the absolute errors are within the resolution of the transducer (0.7 cm).

Table 2. Maximum relative errors in open air and well tube tests*.

Trial	Average	1st Order	2nd Order	3rd Order	4th Order	5th Order
			Open Air Tests			
OA1	1.69	1.50	1.62	1.79	1.72	1.34
OA2	4.20	2.74	1.40	1.21	1.29	0.67
OA3	1.74	1.67	1.69	1.44	1.01	0.62
			Well Tube Tests			
TV1	2.58	2.43	1.34	1.14	0.71	0.61
TV2	4.99	2.36	0.91	0.61	0.53	0.58
TV3	3.29	1.85	1.28	0.95	0.62	0.52

*All values expressed as percent of measured distance or depth.

Preliminary tests in which the gain of the detector circuit is varied indicate that relative errors may be reduced to less than 1% using an single average value for the velocity of sound. These tests are incomplete at the time of this printing and can not be presented here. The decrease in apparent sound velocities near the target appears to be the result of the ECHO signal not being triggered by the same return echo pulse as targets at greater distances. Sixteen pulses are emitted during each sounding. If the ECHO signal is being generated at anytime other than the return of the first pulse, a timing error will result. Adjustments to the detector gain may prove to be sufficient to alleviate this problem. Work should also be carried out to determine whether or not wave guides will be effective in alleviating false echoes in smaller diameter well tubes. The use of large well tubes can be a detriment to field use of the system.

REFERENCES

CR2M. 1986. SAB 600 Series, Presentation of the Apparatus. CR2M. 72-74 rue Bernard Iskè, 92350 Le Plessis Robinson, France.

Goebel, K.M. and G.E. Merva. 1985. Bubbler system for water table monitoring. Paper No. 85-2563. Presented at the 1985 Winter Meeting of the American Society of Agricultural Engineers. December 17-20 at Chicago, Illinois.

Holbo, H.R., R.D. Harr and J.D. Hyde. 1975. A multiple-well water level measuring and recording system. Journal of Hydrology. 27:199-206.

Lovell, A.D., J.W. Ellis, R.R. Bruce and A.W. Thomas. 1978. Remote sensing of water levels in small diameter wells. Agricultural Engineering. 59(10): 44-45.

MacVicar, T.K., and M.F. Walter. 1984. An electronic transducer for continuous water level monitoring. Transactions of the ASAE. 27(1): 105-109.

Maslin, G.D. 1983. A simple ultrasonic ranging system. Presented at the 102nd Convention of the Audio Engineering Society at Cincinnati, Ohio. May 12, 1983.

Munster, C.L., R.W. Skaggs, J.E. Parsons, R.O. Evans, and J.W. Gilliam. 1991. Modeling Aldibcarb Transport under drainage, controlled drainage, and subirrigation. ASAE Paper No.91-3024. Presented at the 1991 International Winter Meeting of the American Society of Agricultural Engineers. December 17-20, 1991 at Chicago, Illinois.

Tinham, B. 1989. Novel sensing can give greater depth. Control & Instrumentation 21(8):37, 39.

Tinsey R.G. and R.P. Rohrbach. 1991. Non-contact proximity sensing using ultrasonic waves. ASAE Paper No.91-3024. Presented at the 1991 International Summer Meeting of the American Society of Agricultural Engineers. June 23-26 at Albuquerque, New Mexico.

Tromble, J.M., and C.G. Enfield. 1971. Adapting analog water stage recorders to digital data acquisition systems. Agricultural Engineering. 52(2):80-81

Wang, Y., C. Mingotaud and L.K. Patterson. 1991. Noncontact monitoring of liquid surface level with a precision of 10 micrometers: A simple ultrasound device. Rev. Sci. Instrum. 62(6): 1640-1641.

Ramana, Y.V. and I. Radhakrishna. 1990. An ultrasonic pulse-echo technique for sensing microlevel changes in groundwater. Acoustics Letters. 14(5): 86-91.

Van der Weerd, B. 1977. A registration unit for drain outflow, groundwater depth, and precipitation. Journal of Hydrology. 34: 383-388.

COMPARISON OF THE PERFORMANCE OF THICK AND THIN ENVELOPE MATERIALS [1]

Hany El-Sadany Salem[2] and Lyman S. Willardson[3], F.ASCE

ABSTRACT

Thick and thin synthetic envelope materials were tested against three coarse textured soil samples. Combinations of the envelope materials and soil samples were evaluated in a one dimensional flow permeameter where it was possible to control the hydraulic gradient by control of the flow rate. When a certain hydraulic gradient was reached, the thick envelope materials began to clog. The thin envelope materials started to clog shortly after the start of the test. The expected exit gradients under different field conditions were calculated. There is no risk of thick envelope materials clogging under field conditions for the soils tested. Some risk of thin envelope material clogging exists under the simulated field conditions.

Keywords: drain envelopes, drainage, geotextiles, subsurface drains

INTRODUCTION

One of the most important decisions to be made during the planning stage of a subsurface drainage project is whether drain envelopes are needed and, if needed, what material should be used with the existing drain pipes and soil types. Many natural origin materials (organic and inorganic) have been used as drain envelope materials (Willardson, 1974). Various problems have been found associated with the use of natural inorganic materials, mainly high price of transportation and scarcity. Inorganic materials such as fine gravel and coarse sand, despite being considered the ideal envelope material, are also heavy, bulky, and difficult to handle during construction. Organic materials are biodegradable and consequently, may be short-lived under field conditions.

Due to the problems mentioned above, synthetic fiber envelopes, made of polypropylene, polyethylene, polyester, and nylon, (called geotextiles) have been used as drain envelopes. Dierickx et al (1990) stated that synthetic fiber drain envelope materials are expected to prevent the movement and entry of soil particles which may settle and clog the drainpipe

---

[1] A joint contribution of the Drainage Research Institute of Egypt and Utah Agricultural Experimental Station.

[2] Engineer, Drainage Research Institute, Water Research Center, Cairo, Egypt.

[3] Professor Biological and Irrigation Engineering, Utah State University, Logan, Utah, U.S.A.

(mechanical function) and to create a more permeable area in the vicinity of the drain pipe to reduce the entrance resistance (hydraulic function).

Many kinds of synthetic fabrics, which have different physical characteristics, are available for use as drain envelope materials. Generally, it is possible to divide these materials into two major groups: thick (voluminous) materials and thin materials. The performance of each fabric can be tested as its their mechanical function, hydraulic function, and clogging characteristics (Salem and Willardson, 1992).

PROCEDURE

Four thick polypropylene mats, made of the waste fiber of the carpet industry in Egypt were tested against three coarse textured soil samples (Salem, 1991). The drain envelope materials were tested in a one dimensional flow permeameter. Two thin random fiber fabrics, commercially available in U.S.A., were also tested against the same soil samples. The thick envelope materials were classified by their weight per square meter. The two thin envelope materials were Typar and Drainguard. A #30 Screen with 0.6 mm rectangular openings was also tested for comparison (Salem and Willardson, 1992).

Figure 1 shows the grain size distributions of the soil samples used in the study. Soil samples A and B were from Egypt and soil C was from the U.S.A. Using the modified triangular diagram (Arora, 1988), soil samples A and B were classified as sand while soil sample C was classified as silty sand.

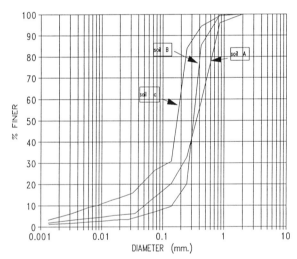

FIGURE 1. GRAIN SIZE DISTRIBUTION OF THE SOIL SAMPLES. SOILS A AND B ARE FROM EGYPT AND SOIL C IS FROM U.S.A.

The permeameter, used during this study, was a clear plastic cylinder. The cylinder had an inner diameter of 100 mm and a length of 150 mm. The combined length of the envelope material and soil sample varied between 80 and 90 mm. The envelope material, externally supported by two coarse screens, was put at the top of the permeameter with the soil sample below. Before packing in the cylinder, the soil sample was brought to field capacity and was compacted in the permeameter by a load which simulates the load that affects the soil at a 2.0 m deep drain level. The hydraulic gradient through permeameter was controlled by controlling the flow rate through the sample. The hydraulic heads in the system were monitored by four replicated groups of piezometers connected to a manometer board. The complete apparatus, test preparation, and test procedure was described by Salem (1991).

Seventeen combinations of the envelope materials and soil samples were tested in the permeameter. Soil samples, without envelope material, were also tested, restrained by the supporting screens only. The tests were stopped when failure of the system took place. Failure was considered to occur when either continuous soil movement through the envelope material, clogging of the envelope material, or internal structural failure of the soil sample (Salem and Willardson, 1992) took place.

RESULTS AND DISCUSSION

When testing the thick envelope materials, none of the soil samples failed by a massive continuous movement of the soil through the envelope material. The internal structure of the finer soil samples, A and C, failed when a certain hydraulic gradient was developed in the soil. When this occurred, the system did not achieve flow equilibrium and the hydraulic gradient through the system continued to increase with time for the same flow rate, indicating internal clogging. Figures 2 and 3 show examples of the of relation between hydraulic conductivity and discharge for the three different soils and two of the thick envelope materials. In these figures, k_{env} denotes the hydraulic conductivity of the envelope material and an adjacent thin layer of the soil. The hydraulic conductivity of the middle and bottom layers of the soil sample are denoted k_{s1} and k_{s2}, respectively. From Figs. 2 and 3 it can be seen that k_{env} had a low initial value. Then k_{env} increased with flow rate to a maximum value after which it decreased until it became almost the same as the initial value.

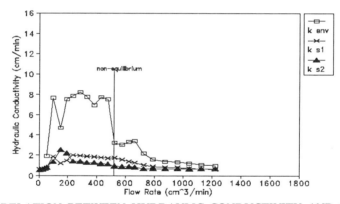

FIGURE 2. RELATION BETWEEN HYDRAULIC CONDUCTIVITY AND FLOW RATE FOR SOIL A & 570 gm/m²-BRITTLE ENVELOPE MATERIAL.

Figures 4, 5, and 6 show the relation between hydraulic gradient and flow rate for the same combinations of envelope materials and soil samples. At the early stages of the tests, the hydraulic gradient in the envelope material varied almost linearly with discharge. After the maximum value of hydraulic conductivity was achieved, the hydraulic gradient in the envelope material increased with discharge indicating clogging of the envelope. Figures 4 and 5 show that the rate of increase of hydraulic gradient in the soil and envelope with flow rate was the same during the test, indicating that no clogging of the envelope material was taking place. No failure of the soil-envelope combination shown in Fig. 5, could be observed. The test was stopped due to the limited discharge capacity of the system. The maximum hydraulic gradient achieved in the soil was almost 15. Figure 6 shows that, when internal structural failure of the soil took place, the rate of increase of hydraulic gradient with flow rate was

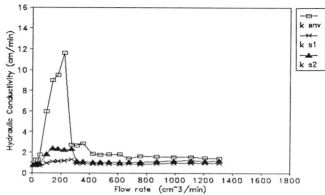

FIGURE 3. RELATION BETWEEN HYDRAULIC CONDUCTIVITY AND FLOW RATE FOR SOIL B & 570 gm/m^2-COMPACTED ENVELOPE MATERIAL.

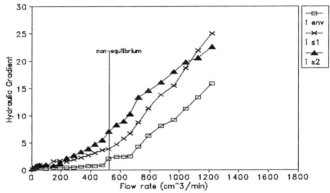

FIGURE 4. RELATION BETWEEN HYDRAULIC GRADIENT AND FLOW RATE FOR SOIL A AND 570 gm/m^2-BRITTLE ENVELOPE MATERIAL.

much higher in the envelope than in the soil, indicating that clogging was taking place. From Figs. 4 and 6, it can be seen that occurrence of internal structural failure does not mean that clogging will take place. The reason for considering the internal soil failure of the soil sample as a failure of the subsurface drainage system is that the behavior of the system after this point becomes unpredictable. During the tests where internal structural failure took place, large differences were observed in the readings of head in the different manometers at the same level in the permeameter. The hydraulic gradient in the system continued to increase with time for the same flow rate, but in an inconsistent manner.

A possible explanation for this latter mechanism of failure is as follows:

1. Some soil particles lodged in the pores of the envelope material during preparation causing an initial low value of hydraulic conductivity.

2. As discharge increased, the hydraulic gradient in the envelope material increased, flushing some of the particles out of the envelope material and causing its hydraulic conductivity to increase. As the hydraulic conductivity increased, the hydraulic gradient in the envelope material decreased. This continued to occur until all of the unstable soil particles in the envelope material had moved.

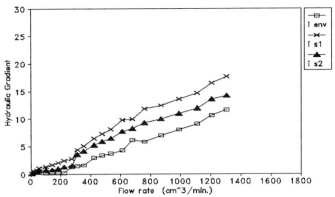

FIGURE 5. RELATION BETWEEN HYDRAULIC GRADIENT AND FLOW RATE FOR SOIL B AND 570 gm/m²-COMPACTED ENVELOPE MATERIAL.

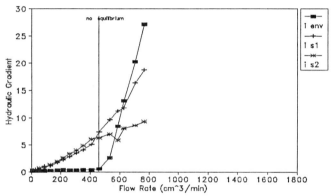

FIGURE 6. RELATION BETWEEN HYDRAULIC GRADIENT AND FLOW RATE FOR SOIL C AND 570 gm/M²-COMPACTED ENVELOPE MATERIAL.

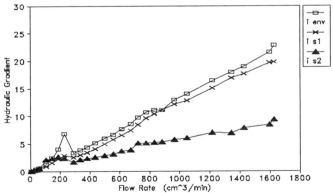

FIGURE 7. RELATION BETWEEN HYDRAULIC GRADIENT AND FLOW RATE FOR SOIL B AND TYPAR.

3. At the same time, the hydraulic gradient in the soil sample itself continued to increase with discharge. As the hydraulic gradient increased, the size of the soil particles moving with the flowing water gets larger. At a certain point, the envelope material became incapable of passing these larger particles and consequently progressive clogging reduced the hydraulic conductivity of the envelope material.

When testing the thin envelope materials, soil movement was generally more than with the thick envelope materials. In most of the tested soil-envelope combinations, the failure of the system took place due to the combined action of the internal structural failure of the soil and soil movement through the envelope material. For soil A, the movement of the soil through the envelope material (piping) was the governing factor. No failure was observed with soil B.

With screen #30, soils A and C failed by piping at hydraulic gradients of 1.0 and 1.4. For soil B, it was not possible to monitor any piping or internal structural failure with Typar or screen #30. From the above discussion, it can be noted that the response of soils A and C differed with the thin envelopes compared to the thick envelopes.

Figures 7, 8, and 9 show the relation between hydraulic gradient and flow rate for the same thin envelope-soil combinations. From the three graphs, it is obvious that the clogging process starts at a very early stage of the test but it develops very quickly in the later stage. Table 1 gives a summary of the hydraulic failure gradients for the different combinations of soil samples and envelope materials.

TABLE 1. Hydraulic Failure Gradients for Combinations of Soils and Drain Envelopes.

Envelope Material Type	soil		
	A	B	C
450 gm/m^2-Brittle	3.6[2]	NA[1]	--
450 gm/m^2-Compacted	6.5[2]	NA[1]	--
570 gm/m^2-Brittle	7.3[2]	NA[1]	--
570 gm/m^2-Compacted	7.4[2]	NA[1]	6.8[2]
Typar	0.3[3]	NA[1]	1.8[3]
Drainguard	--	--	1.2[3]
Screen #30	1.0[3]	NA[1]	1.4[3]
No envelope	0.3[3]	.25[3]	0.4[3]

(1) No hydraulic failure gradient identified.
(2) Non-equilibrium hydraulic failure.
(3) Continuous soil movement through the envelope material.

FIELD CONDITIONS

For an envelope material to be acceptable with a certain soil type, the hydraulic gradients under field conditions should be less than that which may cause piping, internal soil failure, or envelope material clogging. Using a procedure described by Willardson (1982), the expected hydraulic gradients at the soil-envelope and soil-pipe interface were calculated for the following conditions:

1. A drainpipe of 80 mm outside diameter with a thick envelope material of 7 mm in thickness under a loaded condition. The exit gradients were calculated for cases of openings

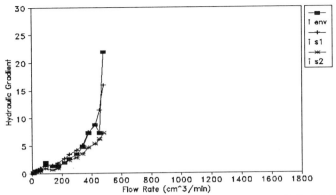

FIGURE 8. RELATION BETWEEN HYDRAULIC GRADIENT AND FLOW RATE FOR SOIL C AND TYPAR.

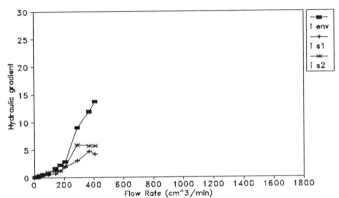

FIGURE 9. RELATION BETWEEN HYDRAULIC GRADIENT AND FLOW RATE FOR SOIL C AND DRAINGUARD.

in every and in every other corrugation.

2. A drain pipe with no envelope material. The drain pipe had a perforation open area of 20 cm^2 per meter of drain pipe.

The additional assumptions were:

1. The drainpipe allows the water to enter through all of its circumference.
2. The combination of the drainpipe and the thick or voluminous envelope material creates an ideal drain.
3. The porosity of the soil and the envelope material is the same.

It was concluded that when using soils A and C with no drain envelope material, the exit gradients were much higher than the hydraulic failure gradients of the soils used. When a thick envelope material was used, the exit gradients were much less than the hydraulic failure gradient of the soil with any of the thick envelope materials. All of the thick envelope materials were safe to use under field conditions.

During this study, the exit gradients were also calculated for the case of a an envelope material with a negligible thickness used with outside pipe diameters of 60 mm and 80 mm. The same previous assumptions were used. The hydraulic gradients expected under field conditions are less than those which may cause clogging except for the case of a drain pipe diameter of 60 mm with openings in every other corrugation. For this case, the calculated exit gradients may exceed the clogging gradient for the combination of soil C and Drainguard or soil B and Typar for drainage coefficients larger than 3 mm/day and drain spacings larger than 100 m. For the combination of soil C and Typar, the clogging is not probable except for a drainage coefficient of 7 mm/day and drain spacings larger than 100m.

From the previous discussion, the risk is slight for thin envelope material clogging, but is higher than for thick envelopes. This can be explained by the fact that the effective pipe-envelope diameter in case of a thin envelope is less, which leads to higher calculated exit gradients. Another cause is that the hydraulic failure gradient of the soil with the thin envelope is lower than with the thick envelopes.

CONCLUSIONS

1. The thin envelope materials tested start to clog at a relatively low hydraulic gradient.
2. Generally, the hydraulic failure gradient with thin envelopes is lower than that with thick envelope materials.
3. A slight risk of thin envelope materials clogging exists under field conditions, while no risk of thick envelope materials clogging exists for the soils and envelopes tested under the assumed conditions.

REFERENCES

1   Arora, N.R. 1988. <u>Introductory soil engineering</u>. Standard Publishers Distributors, Delhi, India.

2   Dierickx, Willy. 1986 "Model research on geotextile blocking and clogging in hydraulic engineering", <u>Third international on geotextile, properties and tests, 7c13</u>, Vienna, Austria, pp775-777.

3   Dierickx, Willy, Voltman, Willem F., Haider, Iftikhar, and Khan, Mussurat Ali. 1990. "Synthetic envelope selection and design," <u>Draft Proceedings Workshop on Drain Envelope Testing, Design and Research</u>, NRAP, Lahore, Pakistan.

4   Salem, Hany El-Sadany. 1991. "Evaluation of the performance of synthetic envelope materials, "unpublished M.S. thesis presented to Utah State University, at Logan, Utah.

5   Salem, Hany El-Sadany and Willardson, Lyman S. 1992. "Evaluation of non-standard synthetic materials", <u>Fifth International Drainage Workshop</u>, ICID, Lahore, Pakistan.

6   Willardson, Lyman S. 1974. "Envelope materials". <u>Drainage for Agriculture</u>, Van Schilfgaarde, Jan, Ed., Mono. #17, American Society of Agronomy, Madison, Wisconsin, pp 179-196.

7   Willardson, Lyman S. 1982. "Exit gradient at drain openings," <u>Proceedings of the Second International Drainage Workshop</u>, Washington D.C., pp. 198-202.

HYDRAULIC HEAD LOSSES NEAR AGRICULTURAL DRAINS
DURING DRAINAGE AND SUBIRRIGATION

Robiyanto H. Susanto   R. Wayne Skaggs[*]

ABSTRACT

The effectiveness of a drainage-subirrigation system is dependent on hydraulic head losses near the drain. Head losses near the drain, the entrance resistance at the drain, hydraulic conductivity (K) near the drain, and their effects on flowrate were evaluated using field data from a 14-ha drainage-subirrigation system. Large head losses occured near the drain. The head loss in the vicinity of the drain was typically equal or greater than the head loss in the profile from the mid-plane between drains to the drain. The major reason for the high head losses was that the values of K near the drain were much less than the lateral K in the profile for all cases. In some cases there was additional high head loss across the drain tube wall, but this did not occur in all plots.

Keywords: Controlled drainage, effective radius, hydraulic conductivity, radial flow, entry resistance.

INTRODUCTION

Hydraulic head loss near agricultural drains is an important factor affecting the operation of drainage and subirrigation systems. Head losses as water approaches a drain results from convergence to the drain, convergence to the corrugations and further convergence to the perforations at the bottom of the corrugations. Head loss is a function of both tube geometry and the surrounding material (Dierickx, 1980). The addition of an envelope usually results in a decrease in the entrance resistance because of an increase in the effective size of the drain and a reduction in head losses within the corrugations (Bentley, 1991). In a four-year field experiment in the North Carolina Coastal Plain, Davenport and Skaggs (1990) found that 50 % of the total head losses occured near the drain with over half of that loss due to entry resistance. Numerical solutions to the Boussinesq equation that consider radial flow near the drain for both drainage and subirrigation (Skaggs, 1991) explain some discrepancies observed between field results and theory.

The objective of this paper is to report results of research to quantify entrance head losses for agricultural drains used for both drainage and subirrigation; and to determine hydraulic conductivity (K) in the immediate vicinity of the drain.

---

[*] Robiyanto H. Susanto, Ph.D Candidate, and R. Wayne Skaggs, William Neal Reynolds and Distinguished University Professor, Department of Biological and Agricultural Engg., North Carolina State University, Box 7625, Raleigh, NC 27695-7625, USA

THEORETICAL BACKGROUND

Radial flow is assumed to occur in the immediate vicinity of the drain. Under this assumption the head loss near the drain (ΔH_n) was expressed by Dierickx (1980), as shown schematically in Fig. 1.

$$\Delta H_n = \Delta H_r + \Delta H_e \qquad (1)$$

where, $\Delta H_r = H_s - H_r$ is the head loss due to radial flow through the soil surrounding the drain. H_s and H_r are assumed to be cylindrical equipotentials just outside the radial flow region and just outside the tube respectively. $\Delta H_e = H_r - H_t$ is the head loss across the drain tube, where the H_t is hydraulic head inside the tube.

If the drain tube is considered as an ideal cylindrical sink, then the flow can be expressed as,

$$Q = 2\pi l K \frac{(H_s - H_t)}{\ln(R_s / r_t)} \qquad (2)$$

where Q is the flow rate (m³ day⁻¹), l is the tube length (m), K is the hydraulic conductivity of surrounding media (m day⁻¹), R_s and r_t are radi of the cylindrical source and sink respectively. The value of R_s may be estimated as the height of water table above the drain center.

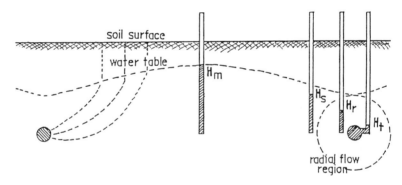

Fig. 1. Head losses due to vertical and horizontal flow, $\Delta H_m = H_m - H_s$, radial flow, $\Delta H_r = H_s - H_r$ and entrance losses, $\Delta H_e = H_r - H_t$ (Dierickx, 1980)

For drainage with the drain running half full $H_t = 0$; while for subirrigation $H_t = h_0$ (Fig. 2) which is greater than H_s so that flow is negative or away from the drain. Flow to real rather than ideal drains can be predicted by substituting an effective radius, r_e, for r_t in Eq. (2). Effective radius is defined such that an ideal drain with a radius r_e will offer the same resistance to inflow as a real drain tube of radius r_r. Then in the radial flow region the relationship between H and r may be obtained by substituting H for H_s and r for R_s in Eq. 2.

$$(H - H_t) = \frac{Q}{2\pi l K} \ln(r / r_e) \qquad (3)$$

Entrance resistance can be reduced by using an envelope which partially or totally surrounds the drain (Dennis and Trafford, 1975; Skaggs and Tang, 1979; Dierickx, 1980). Graded sand and gravel envelopes have been used for many years and are known to be effective (Willardson, 1987). In recent years envelopes installed on the drain tubes in the factory have become popular. These include thin fabric wrap envelopes or filters and thicker "voluminous" envelopes made from coconut fiber, straw, geotextiles and various other "man made" materials.

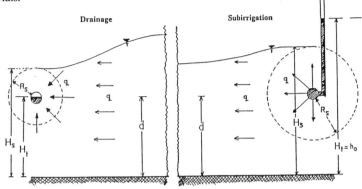

Fig. 2. Radial flow near the drain during drainage and subirrigation (Skaggs, 1991)

Stuyt (1992) found that thick "voluminous" envelopes offer considerable advantages over thin ones. Entrance resistance was constantly lower for both subirrigation and drainage. It is often speculated that envelopes have a significant effect on the movement of soil particle near the drains. Stuyt's findings do not support this concept. In weakly-cohesive Dutch soils the particle size distribution of the soil had a much more significant effect than the envelope.

Equations 1 - 3 are based on the assumption that flow near the drain is relatively uniform Darcian flow. Results of recent experiments by Stuyt (1992) indicate that this assumption does not hold for the soils he studied. Stuyt concluded that the observed particle size distributions at the soil-envelope interface and the existense of three dimensional structural features in the zone near the drains, suggested that the pattern of water flow into drain is strongly heterogenous. The flow is concentrated in the most permeable areas, like interaggregate voids in the trench backfill and tiny permeable soil layers and root channels in the undisturbed subsoil.

In a laboratory experiment, Bentley (1991) determined that the saturated hydraulic conductivity (K_s) decreased with time in three lab tanks and 18 cores. The greatest decrease occurred directly after starting the test. The water supplies for the three tanks and nine cores were discontinued for a period of time. The tanks and cores were than resaturated and the drainage process was allowed to resume. All of the cores and two of the three tanks showed a temporary increase in K after saturated flow was resumed.

In sand column experiments, Vandevivere and Baveye (1992) found that a strictly aerobic bacterial strain, *Arthrobacter* AK19, reduced K_s in sand columns by three or four orders of magnitude in one week. This is contrary to a suggestion often made in the literature that bacterial reductions of the K_s of natural porous media are inseparably associated with the development of anaerobic or microaerophilic conditions in these media. Rapid K_s reduction was associated with the formation of a thick, though highly porous, bacterial mat at the inlet end of the sand column. Furthuremore, the production of extracellular polymers did not seem necessary to induce severe bacterial clogging. Extracellular polymers were produced and appeared to cause additional K_s reduction when the C/N ratio was high (77) but, at lower C/N ratios (lower than 39), they seemed absent from the clogged layers.

METHODOLOGY

Head losses near the drain and their effects on water table and drain flow rates were investigated in a field experiment on the Tidewater Research Station, near Plymouth, NC. The experiment was designed and instrumented to determine effects of conventional drainage, subirrigation and controlled drainage on pollutants to surface and ground water (Munster et. al., 1991). Additional instrumentation was installed and measurements conducted to study entrance resistance, hydraulic conductivity and head losses near the drain.

The research site is a 14 ha field of Cafe Fear and Portsmouth soils (*Typic Umbraquults*) divided into 8 plots. Each plot consisted of the area drained by three adjacent subsurface drains. The drains are 101 mm corrugated plastic tubing, with discontinuous circumferential slits, spaced 22.9 m on center and buried 1.0 m below the surface. Drain opening were in six rows of slits approximately 18 mm long by 2 mm wide spaced 18 mm apart on a longitudinal axis. Plots 7 and 8 had new drains installed midway between the old drains at 1.2 m deep in 1990. A sand envelope was put on the top of new drains in Plot 7.

Hydraulic heads near the drains were measured with observation wells and piezometers installed near the center drains in Plots 4, 5, 6, 7 and 8. Wells were located on top of and at distance of 0.3 m, 0.6 m, and 5.7 m on either side of the drain. Two Piezometers spaced 10.0 m apart were placed inside the drain. Between these, additional piezometers were installed at 5.0 cm, 15.0 cm and 25.0 cm above the top of the drain. The piezometers were replicated so that there were two piezometers at the same depth. Wells and piezometer observations were conducted from November 1990 and January 1991 to August 1991 on a weekly basis. Manual observations of the water table midway between drains were compared to the recorded values. The height of water table above the drain center at each location was calculated from well and piezometer data. Drain flow rates were measured and recorded automatically as described by Munster et al (1991). Flow rates were also measured manually on a weekly basis and compared to the recorded values.

RESULTS AND DISCUSSION

Water table profiles

As expected, water table elevation decreases with distance to the drain for drainage (Fig. 3) and increases for points closer to the drain for subirrigation (Figs. 3 and 4). Head losses near the drain were relatively large as reflected by the water table standing over the drain even when drainage rates are low. This is demonstrated in Fig. 3 where the distance to water table directly above the center of the drain is greater than the difference between the water table elevation directly over the drain and midway between the drain. For example in Plot 5 the water table elevation above the drain was 27 cm compared to 46 cm midway between the drains. Thus the hydraulic head loss near the drain was 27 cm as compared to only 19 cm (46 - 27) that occured in the profile from midway between the drain to the drain. The head loss near the drain in Plots 4 and 6 were even higher than in Plot 5. These results are contrary to the drainage theory, which normally assumes that head losses near the drain are relatively low and that the water table intersects the drain.

Head losses near the drain during subirrigation are indicated by the difference between the hydraulic head in the drain and the water table elevation near the drain. This difference was 34 cm for Plot 6, as compared to 26 cm for Plot 8 and 7 cm for Plot 7 (Fig. 4). The drain tubes in Plots 4-6 were installed with a trenchless plow in 1982. The drains in Plots 7 and 8 were installed in 1990 with a wheel-type trencher. The trench was 0.6 m wide and 1.2 m deep. In Plot 7 a 0.2 m by 0.3 m coarse sand envelope was placed on top of the plastic drain, while in Plot 8 the drain was covered with soil from the trench. In all cases the corrugated plastic drain was covered by a thin polyester knitted "sock". The difference in water table shapes between Plots 6, 7 and 8 are assumed to be due to difference in installation, envelope, depth of the drain, and age or time since installation. Head losses near the drain in Plot 7 with the sand envelope are clearly less than those in Plots 8, and those in Plot 8 are less than in Plot 6. Piezometer data were used to calculate hydraulic conductivity near the drains and the results analyzed to explain these differences.

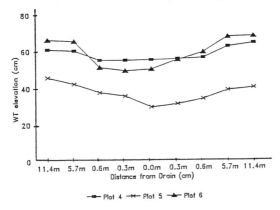

Figure 3. Water table elevation above the center of the drain as a function of distance from the drain in Plots 4, 5 and 6 during drainage on January 3, 1991

Relationship between Water Table Elevation and Drainage

The response of drain flow rate to the water table elevation directly above the drain is shown for Plot 4 in Figure 5. The points were measured weekly and represent a wide range of water table elevations in this conventionally drained plot. Similar results were obtained for the other plots. When the absissa is taken as $H_s - H_t$, the relationships were about the same for conventional and controlled drainage as would be expected from Eq. 2. As indicated in Figure 5 head losses near the drains were substantial, even for low flow rates. Large head losses near the drain reduce the gradients available for moving water laterally from the region midway between the drains to the vicinity of the drain (Skaggs, 1991). The relationship between the water table elevation midway between drains and that directly above the drain is given on Fig. 6 for Plots 4-6. A 1:1 line is given for comparison. If the observations were coincident with the 1:1 line; the water table elevation above the drains would be equal to that midway between and all the head loss would occur near the drain. The head loss near the drain was more than 50% of the total for all three plots.

Skaggs (1991) used numerical solutions to the Boussinesq equation with a radial flow boundary condition to show that convergence head losses at the drain could result in the water table standing over the drain. However, the large losses shown in Figs. 4-6 indicate additional causes such as reduced hydraulic conductivity near the tube, or clogging of the drain tube openings.

The causes of the head loss near the drain can be analyzed by examining the piezometer data collected at points 10 cm, 20 cm and 30 cm above the drain center. Assuming steady, radial flow Eq. 3 may be written as,

$$H = \frac{Q}{2\pi K l} \ln r - \frac{Q}{2\pi K l} \ln r_e + H_t \qquad (4)$$

where H is the hydraulic head at radius r from the center of the tube, H_t is the head inside the tube and r_e is the effective drain tube radius. This equation may be rewritten as,

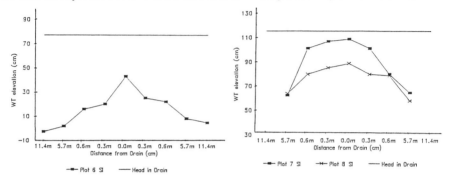

Figure 4. Water table elevation above the center of the drain during subirrigation in Plots 6, 7 and 8 on June 11, 1991

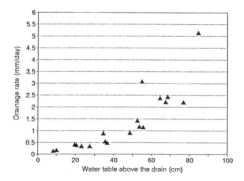

Figure 5. Relationship between drain flowrate during conventional drainage and water table elevation directly above the drain for Plot 4.

If flow near the tube is radial, a plot of H vs ln r should be linear. Example plots are shown in Figure 7 for controlled drainage and for subirrigation. Similar plots were constructed for each set of measurements.

The H vs ln r relationships were approximately linear in all cases. Because Q was measured at the same time that the piezometric measurements of H vs r were made, hydraulic conductivity, K, can be computed from the slope of the line. Since H_t is also measured, r_e can be computed from the intercept (i.e. the H value when r = 1 and ln r = 0). A straight line was fitted to the data for each set of observation by linear regression, the slope and intercept determined and K and r_e values computed. Using the data for controlled drainage in Fig. 7, for example, the slope, $B = Q/(2\pi Kl)$, was 23.2 cm and the intercept, $(-(Q/2\pi Kl)\ln r_e + H_t)$ was -1.1 cm. From field measurements $Q/l = 486$ cm^2 day^{-1} and $H_t = 42.3$ cm. Then $K = Q/(2\pi lB) = (486)/(2\pi \times 23.2) = 3.34$ cm day^{-1}, and $(-B \ln r_e + H_t) = 1.87$. So $r_e = 6.5$ cm. Each set of measurements was analyzed for Plots 4 - 8 and K and r_e values were determined. The value of the hydraulic head just outside the drain, H_r, can be determined directly from the plot (Fig. 7) for $r = 6$ cm, or it may be computed from Eq. 4, using the K and r_e values determined above. This value was determined and the head loss accross the drain tube wall $|H_r - H_t|$ was calculated for each case. The quantity H_r-H_t is positive for drainage and negative for subirrigation. A summary of the results is given in Table 1.

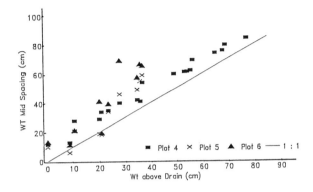

Fig. 6. Relation between water table above the drain dan water table midway between the drain for Plot 4, 5 and 6.

Table 1. Summary of results of calculations based on piezometer and drain flow data. Values given for K, r_e and head loss accross the drain are averages with standard deviations given in (..).

Plot	Drainage treatment	number of events	K (cm/day)	r_e (cm)	Head loss accross the drain (cm)
4	conventional drainage	4	9.06 (5.67)	0.04 (0.04)	43.5 (14)
5	conventional drainage	3	3.23 (0.46)	1.7 (0.26)	29 (3.4)
5	controlled drainage	6	4.10 (2.66)	7.4 (3.98)	0
6	controlled drainage	4	2.12 (0.85)	12 (1.93)	< 0
6	subirrigation	10	0.93 (0.21)	8.0 (0.60)	< 0
7	subirrigation	3	115 (67)	1.3 (1.5)	5.1 (2.3)
8	subirrigation	7	10.8 (2.02)	0.96 (1.25)	19 (7.7)

Results given in Table 1 indicate that, except for Plot 7 where a sand envelope was installed above the drain, the hydraulic conductivity near the drain was low, ranging from 0.93 to 10.8 cm/day. This is about an order of magnitude less than the lateral K measured with the auger hole method on the site, but is in qualitative agreement with the values used by Munster et. al. (1991) for solutions to the Richards equation for the site. The small K value is apparently due to compaction or consolidation around the drain. It accounts for most of the head loss around the drains.

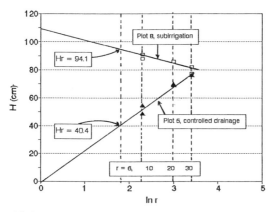

Figure 7. Relationship between hydraulic head and ln r during controlled drainage (Plot 5) on March 6, 1991 and subirrigation (Plot 8) on June 11, 1991. The H value at r = 6 is the projected head just outside the tube.

The computed effective radius was greater than the actual radius (6 cm to the outside of the corrugations) in Plot 6 and Plot 5 during controlled drainage. It was smaller than the actual drain tube radius in Plots 4 and 8. The 1.3 cm value obtained for Plot 7 was also smaller than the real radius, but the sand envelope on top of the drain in Plot 7 causes the radial flow assumption to be questionable. The effective radius for Plot 5 during conventional drainage was 1.7 cm compared to 7.4 cm for the controlled drainage mode. During conventional drainage the water level in the drain was only one or two cm above the bottom of the drain. Under these conditions the streamlines are concentrated towards the bottom of the drain and the flow is less radial than for controlled drainage, where the drain tube is full of water.

The apparent head loss across the drain tube was negligible during controlled drainage in Plot 5 as compared to an average of 29 cm for conventional drainage. These results also show a negligible head loss across the drain in Plot 6 for both controlled drainage and subirrigation and only a small loss in Plot 7. The computed average head losses accross the drain were substantial in Plot 4 (conventional drainage) and Plot 8 (subirrigation) (c.f. Table 1). The causes for the difference in computed r_e values and head losses across the drain are not apparent. Head losses accross the drain tend to be larger, and r_e values smaller, for conventional drainage than for controlled drainage or subirrigation where the drain is full of water

CONCLUSIONS

Measured water table elevation and piezometric heads showed that the head losses near the the drain were relatively large for both drainage and subirrigation on a sandy loam soil in eastern North Carolina. Measurements on a 14 ha water table controlled experiment showed that the water table stood over the drain even when the drainage rates were low. Head losses near the drain were generally higher than the head loss from a point midway between drains to the drain. In some cases there were additional high head losses accross the drain tube wall but not in others.

The relationship between the hydraulic head above the drain and the log of the radius from the drain center (10 cm, 20 cm and 30 cm) was linear for almost all events on four experimental plots. This indicate that the flow is radial. Data on hydraulic head to the sides to the sides and bottom of the drain are needed to evaluate flow uniformity around the drain.

Assuming radial flow, hydraulic conductivity (K) and effective radius (r_e) were calculated for each flow event. The head loss accross the the drain tube wall was also computed. Results indicate that K around the drain was much less than the lateral K in the profile. With one exception, r_e was equal or grater than the actual drain radius for controlled drainage and subirrigation, but much less than that for conventional drainage. Likewise, head losses accross the drain tube wall were relatively small for controlled drainage and subirrigation, but ranged from 29 to 43 cm for conventional drainage.

REFERENCES

1. Bentley, W.J. 1991. Effects of Hydraulic Conductivity and Entrance Resistance on Drainage Processes. Ph.D Thesis. Department of Biological and Agricultural Engineering, North Carolina State University, Raleigh, NC, 127 pp.

2. Davenport, M.S and R.W. Skaggs. 1991. Effects of Drain Envelope and Slope on Performance of a Drainage-Subirrigation System. Trans. of the ASAE 33(2): 493-500.

3. Dennis, C. W. and Trafford, B. D. 1975. The Effect of Permeable Surrounds on the Performance of Clay Field Drainage Pipes. Journal of Hydrologi, Vol 24, pp 239-249

4. Dierickx, W. 1980. Electrolytic Analoque Study of the Effects of Openings and Surrounds of Various Permeabilities on the Performance of Field Drainage Pipes. Commun. National Institute Agricultural Engineering, Pub. No. 77. Merelbeke, Belgium.

5. Munster, C.L., R.W. Skaggs, J.E. Parsons, R.O. Evans and J.W. Gilliams. 1991. Modelling Aldicard Transport under Drainage, Controlled Drainage and Subirrigation. The ASAE, paper No. 91-2631, 24 pp.

6. Skaggs, R. W. and Y. K. Tangs. 1979. Effect of Drain Diameter, Openings and Envelopes on Water Table Drawdown. Trans. of the ASAE, Vol 22, No 2, pp 326-333.

7. Skaggs, R.W. 1991. Modelling Water Table Response to Subirrigation and Drainage. ASAE Vol. 34(1) January-February.

8. Stuyt, L.C.P.M. 1992. The water acceptance of wrapped subsurface drains. Doctoral Thesis. Wageningen Agricultural University, Wageningen, The Netherlands, 314 pp.

9. Vandevivere, P and P. Baveye. 1992. Saturated hydraulic conductivity reduction caused by aerobic bacteria in sand columns. Soil Sci. Soc. Am. J. 56:1, page 1-13.

10. Willardson, L. S. and F. K. Ahmed. 1987. Comparison of USBR and SCS Drain Envelope Specifications. Proceedings of the Fifth National Drainage Symposium, ASAE, pp 432-439.

CROP YIELD INCREASES REQUIRED TO JUSTIFY SUBSURFACE DRAINAGE INSTALLATION COSTS IN THE LOWER MISSISSIPPI VALLEY[1]

Cade E. Carter, R. L. Bengtson, C. R. Camp, J. L. Fouss, and J. S. Rogers[2]

ABSTRACT

Subsurface drainage field tests were conducted in Louisiana during 1974-1990 to determine if the value of the increase in crop yields attributed to subsurface drainage was sufficient to pay within 10 years the cost of installing the drains. Tests were conducted on Mhoon silty clay loam, Commerce silt loam, Jeanerette silty clay loam, Baldwin silty clay and Sharkey clay, all soil types common in the lower Mississippi Valley. The fifteen drain spacings tested varied from 5.5m(18ft) on clay soil to 48.8m(160ft) on silt loam. Crop yield responses were measured for sugarcane, wheat, soybeans, and corn silage. The cost of installing subsurface drainage, including 10 percent interest, was justified for Commerce silt loam with 24.4m(80ft) drain spacing for sugarcane, Jeanerette silty clay loam with 27.4m(90ft) and 41.1m(135ft) drain spacing for sugarcane, and Commerce silt loam with 20m(66ft) and 30.5m(100ft) drain spacing for corn silage. The cost of installing subsurface drainage on Baldwin silty clay for sugarcane and Sharkey clay soil for sugarcane, wheat, and soybeans was not justified using the guidelines selected for this study.

Key words: Subsurface-drainage economics sugarcane wheat soybeans corn-silage

INTRODUCTION

Fertile alluvial soil, nearly flat topography, relatively long frost-free growing seasons, and an abundance of water are natural resources available to farmers in the lower Mississippi Valley of the United States. A common problem in the valley, however, is excess rainfall and the resulting high water tables which inhibit crop growth, reduce yield, and interfere with timely field operations. Average annual rainfall in Baton Rouge, Louisiana is 1417mm(55.8in) but varies from 1061mm(41.8in) to 2243mm(88.3in).

---

[1] Joint contribution from the Soil and Water Research Unit, USDA-ARS, P.O. Box 25071, Baton Rouge, LA 70894 in cooperation with the Louisiana State University Agricultural Center, Baton Rouge.

[2] Carter, Fouss, and Rogers are Agricultural Engineers USDA-ARS P.O. Box 25071 Baton Rouge, LA 70894; Bengtson is Professor of Biological and Agricultural Engineering, LSU, Baton Rouge, LA., and Camp is an Agricultural Engineer USDA-ARS, formerly at Baton Rouge, LA but now at P. O. Box 3039. Florence, SC 29502.

For many years, farmers tolerated excess water without considering ways to alleviate it. The perception was that the alluvial soils in the lower Mississippi Valley were too fine textured to respond to subsurface drainage. This perception was reinforced by Lund and Loftin (1960) and by Lund et al. (1961) which indicated that the saturated hydraulic conductivity of many soils was very low. Research by Carter et al. (1970), Carter and Floyd (1971) and Camp and Carter (1977), proved this perception false. Carter and Floyd (1973) used concrete bordered plots, 40 m^2(436ft^2) in size, to determine that subsurface drainage could adequately control water tables in the fine textured soils in the lower Mississippi Valley. Subsurface drainage research was expanded to field-size areas by Camp and Carter (1983) and by Bengtson, et al., (1982). Additionally, a large sugarcane field had subsurface drains installed where subsurface drainage flow could be controlled, and subirrigation water was provided for automated water table depth control (Carter et al., 1988).

That soils in the lower Mississippi Valley respond favorably to subsurface drainage has been generally known for more than 10 years; however, subsurface drainage is not yet commonly installed by farmers in the valley. The primary reasons that farmers have delayed installing subsurface drainage are the costs of installation and the uncertainty of future prices of farm products. Actual drain installation costs are difficult to determine because there are no drainage contractor businesses located in the valley to provide this information. Furthermore, many drainage system designs would require sumps and pumps as drainage outlets because field ditches are either too shallow to serve as outlets or because they fill quickly with water during rainstorms, making them ineffective.

The texture of the soils in the lower Mississippi Valley vary from clay to silt loam. Drain spacings required for drainage of these soils may vary from 7.6m(25ft) to 11m(36ft) for clay soils and from 23m(75.4ft) to 48.8m(160ft) for silt loam soils. In some soils, filters are required on the drain tubing to prevent sediment from clogging the drains. Also, crop response to drainage and crop prices vary from year to year. With so many variables involved, there was a need to assemble pertinent information for use in making decisions about investing in subsurface drainage. Thus, the purpose of this paper was to compile data from several subsurface drainage experiments that have been conducted in Louisiana in recent years and to use these data to evaluate the economic feasibility of installing subsurface drainage in the lower Mississippi Valley using 1992 crop prices.

PROCEDURE

Data required for this study include drain spacing, crop yield increases due to subsurface drainage, value of the crop increases, drain installation costs, and amortization period.

Drainage System Descriptions and Crops Tested:
Subsurface drainage systems were installed at seven locations in Louisiana during

the 1970s and early 1980s. In Terrebonne Parish, subsurface drainage systems were installed in 1972 with a wheel-type trencher on Mhoon silty clay loam with drains spaced 6.1m(20ft), 12.2m(40ft) without filter, 12.2m(40ft) with synthetic filter, and 24.4m(80ft) apart (Camp and Carter 1983). Three crops of sugarcane were grown during 1974-1976. Crop yields were measured each year.

In 1976, subsurface drains were installed in St. James Parish on Commerce silt loam. The ARS drain tube plow (Fouss, et al., 1972) was used to install the drains 24.4m(80ft), 36.6m(120ft), and 48.8m(160ft) apart (Carter and Camp 1982). The drains were installed without filters. Yields from eleven sugarcane crops, one wheat crop, and one soybean crop were measured during 1977-1990.

Subsurface drains were installed in Iberia Parish in 1978 on Jeanerette silty clay loam (Carter et al. 1987). The ARS drain tube plow was used to install drains 13.7m(45ft), 27.4m(90ft), and 41.1m(135ft) apart. The drains were installed with a filter. Sugar yields from nine crops were measured during 1980-1990.

Subsurface drains were installed on Baldwin silty clay in St. Mary Parish in 1978. The ARS drain tube plow was used to install drains 13.7m(45ft) and 27.4m(90ft) apart. The drains were installed with a filter initially but these were replaced in 1981 with drains without filters. A chain type trencher was used to install the drains in 1981. Sugar yields from nine crops were measured during 1980-1990.

Subsurface drains were installed in East Baton Rouge Parish in 1973 and 1984 on Commerce silt loam soil on Louisiana State University's Ben Hur Research Farm. In 1973, a ladder type trencher was used to install subsurface drains spaced 10m(33ft) apart in one tract and 20m(66ft) apart in another tract. The drains were installed without filters. In 1984, a chain-type trencher was used to install drain tubes spaced 30.5m(100ft) apart in three 4ha(10A) fields. The drains were installed without filters. Corn silage yields from eight crops were measured from the 10m(33ft) and 20m(66ft) spacing experiment during 1980-1987 and from four crops from the 30.5m(100ft) spacing experiment during 1984-1987.

Subsurface drains were installed in Tensas Parish on Sharkey clay in 1973. A chain-type trencher was used to install drains 7.6m(25ft) and 15.2m(50ft) apart. The drains were installed without a filter. Soybean yields from five crops were measured during 1974-1978.

Subsurface drains were installed on Sharkey clay at the Louisiana Agricultural Experiment Station in Iberville Parish in 1977. The ARS drain tube plow was used to install drains 5.5m(18ft) and 11m(36ft) apart. The drains were installed without filters. Yields from four crops of sugarcane, four crops of soybeans, and three crops of wheat were measured during 1978-1986.

With the exception of the drainage system on Sharkey clay in Tensas Parish, sumps with float activated electric pumps were used as drain outlets. In Tensas Parish, the subsurface drains emptied by gravity into a drainage ditch. With the exception of the drainage systems installed in East Baton Rouge Parish in 1973, laser equipment was used to install the subsurface drains on grade. A target system was used to keep the ladder type trencher on grade during drain installation in 1973.

At each subsurface drainage site, areas located nearby but without subsurface drains were used as experimental checks. Yield increases due to subsurface drainage were determined by comparing crop yields measured from the drained areas with those from the nondrained check areas. Detailed results for each experiment are not included in this paper because the results either have been reported previously or will be reported soon.

Crop Value Estimates:
The price a farmer receives for most of his products, including wheat and soybeans, varies considerably. The price varies with actual and predicted supplies and demands. On the other hand, the price for some crops, such as sugar, may be fixed for a certain period by law (Farm Act). Commodity market prices listed in April and May, 1992 for future trading of wheat, soybeans, and sugar were used in this study. These prices were $0.14/kg($3.84/bu) for wheat, $0.22/kg ($6.00/bu) for soybeans, and $0.48/kg($0.22/lb) for sugar. The farmer receives only $0.29/kg ($0.132/lb) for sugar because 40% of the market price is paid to the mill for processing the sugarcane juice into raw sugar. Corn silage was not included in the list of commodity market prices, therefore a price estimate of $33/t($30/T) for corn silage was provided by the Dairy Science Department at Louisiana State University. The value of crop yield increases due to subsurface drainage were determined from the yield increases attributed to subsurface drainage and the April/May, 1992 crop prices.

Drain Installation Cost Estimates:
The cost of materials for installing subsurface drains was based upon prices quoted by a supplier in the spring of 1992. The materials required to install subsurface drainage in a field 177m(580ft) by 457m(1500ft) which is 8.1ha(20A) were determined. The per ha(A) cost used in this study was 1/8(1/20) of the 8.1ha(20A) estimate. Each drain line required 457m(1500ft) of 101mm(4in) diameter drain tubing, five couplers, one end cap, one 152mm by 101mm(6in by 4in) reducer and one 152mm(6in) tee. The cost for drain tubing without a filter was $0.72/m($0.22/ft), drain tubing with a filter was $1.15/m($0.35/ft), end caps and couplers were $0.70 each, and reducers and tees were $3.85 each. The cost of 152mm(6in) diameter tubing for the main was $1.87/m($0.57/ft) with 183m(600ft) purchased because the tubing is supplied in 30.5m(100ft) coils. The cost of a sump and pump, for those sites where a gravity outlet cannot be used, was estimated at $247/ha($100/A) based on the recent cost of installing a sump ($2000) and by assuming that one sump will serve 8.1ha(20A). A farmer probably could construct a sump out of scrap materials and install it for much less than $2000. On the other hand, if shopping efforts were not prudent, a sump might cost more than $2000 to purchase and install.

No subsurface drainage contractors routinely operate or conduct business in the lower Mississippi Valley. This makes it difficult to determine drain installation cost. For this study, we estimated the cost at $0.92/m($0.28/ft) for installing the subsurface drainage materials. This value may be low for a contractor just getting started in a new territory with intermittent work, but it should be reasonable where competition exists among subsurface drainage contractors. For comparison purposes, installation charges in the Mid-Western United States vary from less than $0.49/m($0.15/ft) to more than $1.64/m($0.50/ft).

Amortization Period and Payment Estimates:
The life of a subsurface drainage system exceeds 19 years. Drainage systems installed by Camp and Carter in the lower Mississippi Valley in 1973 are still functioning today. Thus, an amortization period of 15 to 20 years appear to be reasonable; however, if lending institutions are involved, a shorter amortization period may be required. For this study, we selected a payback period of 10 years.

Thus, to justify installing subsurface drainage, the value of the average crop yield increase attributed to subsurface drainage must be adequate to pay for the drainage system within 10 years. For sugarcane, the value of the average yield increase in eight crops during a 10-year period must be sufficient to pay for a drainage system because sugarcane produces three crops in a four-year period. The fourth year is devoted to destroying the sugarcane stubble, fallowing the land for several months and then replanting the cane for the next three crops. Consequently, only eight cane crops can be produced in a ten-year period.

The annual payments required to repay the total cost of a drainage system over a ten-year period financed at 10 percent interest was determined from a formula commonly used in determining mortgage principal and interest payments. The crop yield increases required to justify the cost (make the annual payments) for installing subsurface drainage were determined for sugar, wheat, soybeans, and corn silage. The 1992 market price for these crops was used to determine the yield increase required. Sugar yield increases required to justify drain installation costs were adjusted for the two years in ten when the land is fallow.

RESULTS

Drain Installation Costs:
Drain installation costs for the 15 different drain spacings used in this study are shown in Table 1. Drain tubing costs are provided for drain tubes with and without filters even though filters were used on drains installed at only two sites. The total cost included a sump since one was required for subsurface drain outlets at most of the experimental sites. For sites where a sump is not needed, drain installation cost is $247/ha($100/A) less than the total cost shown in Table 1. Not included in this table is the cost of electricity required to power the pumps for discharging the drain effluent into drainage ditches. Electricity cost is relatively small and varies from site to site. The highest annual electricity cost for pumping averaged about $37/ha($15/A).

Table 1. Subsurface drain installation costs.

Spacing m(ft)	Amount m/ha(ft/A)	Drain Tubing Cost[a] $/ha(A) no filter	Drain Tubing Cost[a] $/ha(A) filter	Drain Fittings Cost[b] $/ha(A)	Install Cost[c] $/ha(A)	Total Cost[d] $/ha(A) no filter	Total Cost[d] $/ha(A) filter
5.5(18)	1823(2420)	1330(532)	2093(847)	94(38)	1675(678)	3331(1348)	4109(1663)
6.1(20)	1640(2178)	1184(479)	1883(762)	84(34)	1507(610)	3022(1223)	3721(1506)
7.6(25)	312(1742)	946(383)	1507(610)	77(31)	1206(488)	2476(1002)	3037(1229)
10.0(33)	1000(1320)	720(290)	1150(462)	69(28)	914(370)	1946(788)	2370(960)
11.0(36)	911(1210)	657(266)	1048(424)	67(27)	838(339)	1809(732)	2199(890)
12.2(40)	820(1089)	593(240)	941(381)	64(26)	754(305)	1658(671)	2006(812)
13.7(45)	729(968)	526(213)	838(339)	62(25)	670(271)	1505(609)	1816(735)
15.2(50)	656(871)	474(192)	754(305)	59(24)	603(244)	1384(560)	1663(673)
20.1(66)	497(660)	358(145)	571(231)	54(22)	457(185)	1117(452)	1329(538)
24.4(80)	409(544)	297(120)	469(190)	52(21)	376(152)	971(393)	1141(463)
27.4(90)	364(484)	262(106)	418(169)	52(21)	336(136)	879(363)	1053(426)
30.5(100)	328(436)	237(96)	378(153)	52(21)	301(122)	838(339)	978(396)
36.6(120)	273(363)	198(80)	314(127)	49(20)	252(102)	746(302)	862(349)
41.1(135)	243(323)	175(71)	279(113)	47(19)	222(90)	692(280)	796(322)
48.8(160)	205(272)	148(60)	237(96)	47(19)	188(76)	630(255)	719(291)

[a] Based on a cost of $0.72/m($0.22/ft) for 101mm(4in) drains without filter and $1.15/m($0.35/ft) for drains with filters.
[b] Fittings include couplers, reducers, tees, end caps, and 183m(600ft) of 152mm(6in) main line.
[c] Based on installation cost of $0.92/m($0.28/ft).
[d] Includes $247/ha($100/A) for a sump.

Drain installation costs, including the cost of the sump, ranged from $630/ha ($255/A) for drains without filters spaced 48.8m(160ft) apart to $4109/ha($1663/A) for drains with filters spaced 5.5m(18ft) apart (Table 1). The cost of sumps for drain outlets and filters on the drains to prevent drain clogging contribute significantly to the total drain installation costs. Sump cost varied from 7 percent of the total drain installation cost for drains without filters spaced 5.5m(18ft) apart to 39 percent of the total for drains without filters spaced 48.8m(160ft) apart. Filter costs varies from 12 percent of the total cost for drains spaced 48.8m(160ft) apart to 19 percent of the total for drains spaced 5.5m(18ft) apart.

Annual Payments and Crop Yields Required to Justify Drain Installation:
Annual payments (includes principal and interest) required to pay for installing subsurface drainage systems with and without filters on the drains are shown in Table 2. Annual payments ranged from $99/ha($40/A) for non-filtered drains spaced 48.8m(160ft) apart to $652/ha($264/A) for filter-wrapped drains spaced 5.5m(18ft) apart (Table 2).

Crop yield increases required to justify the cost of installing subsurface drainage ranged from 424 kg/ha(378 lbs/A) sugar and 700kg/ha(10.4 bu/A) wheat for non-filtered drains spaced 48.8m(160ft) apart to 2800 kg/ha(2500 lb/A) sugar and 4620 kg/ha(68.6 bu/A) wheat for filter-wrapped drains spaced 5.5m(18ft) apart (Table 2). The increases in soybean and corn silage yields required to justify the cost of installing subsurface drainage ranged from 448 kg/ha(6.7 bu/A) soybeans and 2.99 t/ha(1.33 T/A) corn silage for non-filtered drains spaced 48.8m(160ft) apart to 2956kg/ha(44 bu/A) soybeans and 19.7t/ha(8.8T/A) corn silage for filter-wrapped drains spaced 5.5m(18ft) apart (Table 2).

Table 2. Crop yield increases needed to pay for subsurface drainage system financed at 10 percent interest for ten years.

Drain spacing m(ft)	Required Payments $/ha(A)/yr	Annual yield increases needed to pay drain installation costs			
		Sugar kg/ha(lbs/A)	Wheat kg/ha(bu/A)	Soybeans kg/ha(bu/A)	Corn silage t/ha(T/A)
------ For drains without filters ------					
5.5(18)	529(214)	2270(2026)	3745(55.7)	2396(35.7)	15.98(7.13)
6.1(20)	479(194)	2058(1837)	3395(50.2)	2173(32.3)	14.49(6.47)
7.6(25)	393(159)	1686(1506)	2782(41.1)	1718(26.5)	11.87(5.30)
10.0(33)	309(125)	1326(1184)	2191(32.6)	1398(20.8)	9.34(4.17)
11.0(36)	287(116)	1230(1098)	2030(30.2)	1299(19.3)	8.66(3.87)
12.2(40)	263(106)	1124(1004)	1855(27.6)	1187(17.7)	7.91(3.53)
13.7(45)	240(97)	1029(919)	1697(25.3)	1086(16.2)	7.24(3.23)
15.2(50)	220(89)	944(843)	1558(23.2)	997(14.8)	6.65(2.97)
20.1(66)	178(72)	764(682)	1260(18.8)	806(12.0)	5.38(2.40)
24.4(80)	153(62)	658(587)	1085(16.1)	694(10.3)	4.63(2.07)
27.4(90)	143(58)	615(549)	1015(15.1)	650(9.7)	4.33(1.93)
30.5(100)	133(54)	573(511)	945(14.1)	605(9.0)	4.03(1.80)
36.6(120)	119(48)	509(454)	840(12.5)	538(8.0)	3.58(1.60)
41.1(135)	109(44)	467(417)	770(11.5)	493(7.3)	3.29(1.47)
48.8(160)	99(40)	424(378)	700(10.4)	448(6.7)	2.99(1.33)
------ For drains with filters ------					
5.5(18)	652(264)	2800(2500)	4620(68.8)	2956(44.0)	19.71(8.80)
6.1(20)	591(239)	2534(2264)	4182(62.2)	2677(39.8)	17.85(7.97)
7.6(25)	482(195)	2068(1847)	3412(50.8)	2184(32.5)	14.56(6.50)
10.0(33)	376(152)	1612(1439)	2660(39.6)	1702(25.3)	11.35(5.07)
11.0(36)	348(141)	1495(1335)	2468(36.7)	1579(23.5)	10.56(4.70)
12.2(40)	319(129)	1368(1222)	2257(33.6)	1445(21.5)	9.63(4.30)
13.7(45)	289(117)	1241(1108)	2047(30.5)	1310(19.5)	8.74(3.90)
15.2(50)	264(107)	1135(1013)	1872(27.9)	1198(17.8)	7.99(3.57)
20.1(66)	210(85)	902(805)	1487(22.1)	952(14.2)	6.35(2.83)
24.4(80)	180(73)	774(691)	1278(19.0)	818(12.2)	5.45(2.43)
27.4(90)	168(68)	721(644)	1190(17.7)	762(11.3)	5.08(2.27)
30.5(100)	156(63)	668(596)	1102(16.4)	706(10.5)	4.70(2.10)
36.6(120)	136(55)	583(520)	962(14.3)	616(9.2)	4.11(1.83)
41.1(135)	126(51)	541(482)	892(13.3)	571(8.5)	3.81(1.70)
48.8(160)	114(46)	488(436)	805(12.0)	515(7.7)	3.43(1.53)

Crop yield increases measured from subsurface drainage systems:
The measured yield increases of sugar, wheat, soybeans, and corn silage, due to subsurface drainage, are shown in Table 3. The annual values of the yield increases are also listed in Table 3 alongside the yield increases. The crops whose increased yields were valued at more than the required annual payments and therefore justified the cost of installing subsurface drainage were identified by

Table 3. Crop yield increases measured from subsurface drainage experiments, the value of the crop yield increases and whether yield increases justified drain installation costs.

Site[a]	Soil	Drain Spacing m(ft)	Filter	Crops Name	No.	Average increase kg/ha(lb/A)	Value of Increase $/ha($/A)	Required Payment $/ha($/A)	Drain Cost Justified
Ter.	sicl	6.1(20)	No	Sugar	3	7963(178)	47(19)	479(194)	No
Ter.	sicl	12.2(40)	No	Sugar	3	8267(449)	116(47)	263(106)	No
Ter.	sicl	12.2(40)	Yes	Sugar	3	7803(36)	12(5)	319(129)	No
Ter.	sicl	24.4(80)	No	Sugar	3	7993(230)	74(30)	153(62)	No
StJ.	sil	24.4(80)	No	Sugar	11	791(706)	185(75)	153(62)	Yes
StJ.	sil	36.6(120)	No	Sugar	11	421(376)	99(40)	119(48)	No
StJ.	sil	48.8(160)	No	Sugar	11	296(264)	69(28)	99(40)	No
StJ.	sil	24.4(80)	No	Wheat	1	1226(18.2)[c]	173(70)	153(62)	Yes
StJ.	sil	36.6(120)	No	Wheat	1	1569(23.4)[c]	222(90)	119(48)	Yes
StJ.	sil	48.8(160)	No	Wheat	1	1513(22.5)[c]	212(86)	99(40)	Yes
Iba.	sicl	13.7(45)	Yes	Sugar	9	1051(938)	244(99)	289(117)	No
Iba.	sicl	27.4(90)	Yes	Sugar	9	1039(928)	244(99)	168(68)	Yes
Iba.	sicl	41.1(135)	Yes	Sugar	9	798(713)	185(75)	126(51)	Yes
StM.	sic	13.7(45)	No	Sugar	8	409(365)	96(39)	240(97)	No
StM.	sic	27.4(90)	No	Sugar	8	000(000)	00	143(58)	No
EBR.	sil	10.0(33)	No	Corn	8	6.90(3.08)[b]	228(92)	376(152)	No
EBR.	sil	20.1(66)	No	Corn	8	9.07(3.67)[b]	326(110)	178(72)	Yes
EBR.	sil	30.5(100)	No	Corn	4	6.94(3.10)[b]	230(93)	133(54)	Yes
Ten.	c	7.6(25)	No	Beans	5	263(3.9)[c]	57(23)	393(159)	No
Ten.	c	15.2(50)	No	Beans	5	137(2.0)[c]	30(12)	220(89)	No
Ibr.	c	5.5(18)	No	Sugar	4	122(109)	30(12)	529(214)	No
Ibr.	c	11.0(36)	No	Sugar	4	644(575)	151(61)	287(116)	No
Ibr.	c	5.5(18)	No	Beans	4	253(3.8)[c]	57(23)	529(214)	No
Ibr.	c	11.0(36)	No	Beans	4	200(3.0)[c]	44(18)	287(116)	No
Ibr.	c	5.5(18)	No	Wheat	4	591(8.8)[c]	84(34)	529(214)	No
Ibr.	c	11.0(36)	No	Wheat	4	544(8.1)[c]	77(31)	287(116)	No

[a] Site abbreviations were for the following parishes: Ter.(Terrebonne), StJ.(St. James), Iba.(Iberia), StM.(St. Mary), EBR.(East Baton Rouge), Ten.(Tensas), and Ibr.(Iberville).
[b] ton/ha(T/A)
[c] kg/ha(bu/A)

'Yes' in the Drain Cost Justified column in Table 3. The crops listed in Table 2 whose increase in yield values were less than the annual payment required to pay for the drainage system were identified by 'No' in the Drain Cost Justified column in Table 3.

DISCUSSION

Subsurface drainage increased yields of sugarcane and corn silage on Commerce silt loam soil and sugarcane on Jeanerette silty clay loam soil sufficiently to justify the installation cost and the financing thereof for 10 years at 10 percent interest (Table 3).

The measured crop yield increases, attributed to subsurface drainage, justified installing drains 24.4m(80ft) apart on Commerce soil for sugarcane production and 20.1m(66ft) and 30.5m(100ft) apart on Commerce soil for corn silage production (Table 3). Measured sugar yield increases on the Jeanerette soil justified installing drains 27.4m(90ft) and 41.1m(135ft) apart (Table 3).

Crop yield increases were not sufficient to justify subsurface drainage of Baldwin silty clay and Sharkey clay soil, which is the predominant soil type in the lower Mississippi Valley (Table 3). At the Iberville site where wheat and soybeans were double cropped, the combined yield of two annual crops were still insufficient to justify the high cost for draining clay soil. The value of enhanced trafficability was not included in this study but it could be the deciding factor in whether to subsurface drain clay soil. Most farmers plan the entire farm operation around field activities on "heavy" (clay) soils. Getting into fields at the proper time may mean the difference between a good crop and a poor one. Getting into the field to harvest the crop between rainy periods is certainly important and the value of doing so should be included in justifying the cost of a drainage system. However, estimating the value of trafficability is difficult.

Two studies, one with sugarcane on Mhoon silt loam in Terrebonne parish and one with wheat on Commerce silt loam in St. James parish, resulted in data which were borderline for justifying subsurface drainage. Unusually dry weather conditions during two years of the three-year study in Terrebonne parish prevented the collection of representative data to justify subsurface drainage of Mhoon silty clay loam soil (Table 3). In 1974, rainfall was only 1060mm(41.73in), 600mm(23.6in) below normal and in 1976, rainfall was 1160mm(45.67in), 500mm(19.7in) below normal. Extremely low rainfall, like that in 1974 and 1976, is rare. In the past 42 years, annual rainfall at Houma, Louisiana was less than 1200mm(47in) only twice. In 1975, when sugarcane responded positively to subsurface drainage, rainfall was 1820mm(71.65in). The frequency of annual rainfall amounts in this range, 1500mm(59in) to 1800mm(71in), is common. During the past 42 years annual rainfall at Houma, Louisiana exceeded 1820mm(71.7in) eleven times and exceeded 1500mm(59in) 27 times. In 1975, sugar yields were increased significantly by subsurface drainage. Yields from the subsurface drained areas were 20 percent more than the check (Camp and Carter 1983). Mhoon soil is similar to Commerce except Mhoon has a very distinct 30cm(12in) thick layer of silt located approximately 1.2m(4ft) below the soil surface while layers of silt in the Commerce soil are not always connected and their depths vary. The distinct silt layer in the Mhoon soil enhances subsurface drainage to the extent that it drains more readily than the Commerce soil. If subsurface drainage is justified for Commerce soil, it should also be justified for Mhoon soil.

Wheat yield response to subsurface drainage of Commerce silt loam soil in St. James Parish in 1981 was excellent. Wheat yields for all three drain spacing treatments, 24.4m(80ft), 36.6m(120ft), and 48.8m(160ft), were 70 percent more than the check. Justification for installing subsurface drainage would be easy even for the close 24.4m(80ft) spacing drainage system (Tables 1 and 3) if a 70 percent increase in wheat yield could be expected every year. However, such yield increases are not expected routinely. During the wheat study, rainfall in the first half of February 1981 was 206mm (8.13in). Normal rainfall for February is 125mm(4.93in). No doubt, subsurface drainage contributed significantly to the wheat crop during this very wet period in February. Additional data are needed before a decision can be made about justifying subsurface drainage of Commerce soil for wheat production.

The increase in yields required to justify the cost of installing subsurface drainage, when drain spacing is closer than 10m(33ft) feet, is almost out of reach with the present crop varieties and cropping practices used for sugarcane, wheat, soybeans, and corn for silage. Justification is more likely to be achieved with sugarcane and corn silage than with wheat and soybeans. For example, to justify installing drains 7.6m(25ft) apart requires an increase in sugar yield of 1686kg/ha(1506lb/A) which is 27 percent more than the Louisiana state average yield of 6325kg/ha(5647lb/A) and a yield increase of 11.87t/ha(5.30T/A) corn silage which is 37 percent more than the state average yield of 35.33t/ha(14.30T/A). For soybeans, an increase of 1718kg/ha(26.5bu/A) is required to justify 7.6m(25ft) spaced drains which is 106 percent more than the 1680kg/ha (25bu/A) state average in Louisiana and for wheat an increase of 2782kg/ha(41.1bu/A) is required which is 121 percent more than the 2285 kg/ha (34bu/A) state average yield. Even higher yield increases would be required to justify closer drain spacings such as 6.1m(20ft) and 5.5m(18ft).

Three costs which contribute significantly to the overall cost of installing subsurface drainage, but are not always necessary in subsurface drainage systems, are sumps for subsurface drain outlets, filters to prevent drain clogging, and interest on funds to pay for the subsurface drainage system. Interest on funds to install subsurface drainage probably does not hinder many farmers from installing subsurface drainage although in the cases presented in this paper, a subsurface drainage system financed for 10 years at 10 percent interest cost 58 percent more than drainage systems that were not financed. This high cost should encourage farmers to pay for the drainage system outright rather than borrow funds.

Filters on drain tubes boost subsurface drainage cost considerably. Drain tubes

101mm(4-in) in diameter with synthetic filter cost 59 percent more than drains without filters. Thus, they should be used only where a definite need for filters exist.

The need for sumps may be the major reason why farmers are not installing subsurface drainage in the lower Mississippi Valley. The deciding factor may not be the initial cost of the sump, but the need for electricity to power the sump pumps. Farmers, in general, do not like electric power lines in their fields because they interfere with aerial applications of fertilizers and pesticides. Furthermore, the cost of constructing power lines to the subsurface drainage site may not be justified. Investigation of solar and/or wind power to solve this problem is needed.

SUMMARY

The cost of installing subsurface drainage was justified for Commerce silt loam with 24.4m(80ft) drain spacing for sugarcane, for Commerce silt loam with 20m(66ft) and 30.5m(100ft) drain spacing for corn silage, and for Jeanerette silty clay loam with 27.4m(90ft) and 41.1m(135ft) drain spacing for sugarcane. Crop yield increases resulting from subsurface drainage of Baldwin silty clay and Sharkey clay were not sufficient to justify the cost of installing subsurface drainage systems. More data are needed to determine whether installing subsurface drainage can be justified on Mhoon silty clay loam soil for sugarcane and Commerce silt loam soil for wheat.

REFERENCES

Bengtson, R. L., C. E. Carter, H. F. Morris, and J. G. Kowalczuk. 1982. Study shows benefits of subsurface drainage. Louisiana Agriculture, 25(3):16-17.

Bengtson, R. L., C. E. Carter, H. F. Morris, and J. G. Kowalczuk. 1983. Subsurface drainage effectiveness on alluvial soil. TRANS. of the ASAE 26(2):423-425.

Camp, C. R. and C. E. Carter. 1977. Response of sugarcane to subsurface drainage in the field. Proc. Am. Soc. Sugar Cane Tech. 6(ns):158-163.

Camp, C. R. and C. E. Carter. 1983. Sugarcane yield response to subsurface drainage for an alluvial soil. TRANS. of the ASAE 26(4):1112-1116.

Carter, C. E., C. B. Elkins, and J. M. Floyd. 1970. Water management in sugarcane production. Proc. Am. Soc. of Sugar Cane Technologists 17:10-24.

Carter, C. E. and J. M. Floyd. 1971. Effects of water table depths on sugarcane yields in Louisiana. Proc. Am. Soc. of Sugar Cane Technologists Vol. 1, pp. 5-7.

Carter, C. E. and J. M. Floyd. 1973. Subsurface drainage and irrigation for sugarcane. TRANS. of the ASAE 16(2):279-281, 284.

Carter, C. E. and C. R. Camp. 1982. The effects of subsurface draining Commerce silt loam soil on sugarcane yields. Proc. Am. Soc. of Sugar Cane Tech. 1:34-39.

Carter, C. E., V. McDaniel, and C. R. Camp. 1987. Effects of subsurface draining Jeanerette soil on cane and sugar yields. Journal, American Society of Sugar Cane Technologists. Vol. 7. 15-21.

Carter, C. E., J. L. Fouss, and V. McDaniel. 1988. Water management increases sugarcane yields. Trans. of the ASAE 31(2):503-508.

Fouss, J. L., N. R. Fausey, and R. C. Reeve. 1972. Draintube plows: Their operation and laser grade control. Proc. of the Nat'l Drainage Symposium, ASAE, pp. 39-42, 49.

Lund, Z. F. and L. L. Loftin. 1960. Physical characteristics of some representative Louisiana soils. USDA-ARS series 41-33. 83 pages.

Lund, Z. F., L. L. Loftin, and S. L. Earle. 1961. Supplement to Physical Characteristics of some representative Louisiana Soils. USDA-ARS series 41-33-1. 43 pages.

AN INFILTRATION METHOD TO ASSESS THE FUNCTION OF OLD SUBDRAINS

U. Schindler, R. Dannowski, L. Müller[*]

ABSTRACT

A field method has been developed and tested to estimate the drainage efficiency of aged drain constructions. It is based on quantifying the infiltration process of water from the drain construction to the surrounding soil. The method allows to evaluate the hydraulic situation of the drain pipe and the filter and/or envelope as related to the site conditions and exploitation. The measured relation between pressure head and inflow rate inside the drain pipe enables the experimentalist to get this result without any knowledge of the pressure head and the saturated hydraulic conductivity outside the drain. The investigation takes 20 to 50 litres of water per one meter of drain pipe and two persons 2 to 3 hours of time.

KEYWORDS. Drain pipe, Filter, Efficiency, Infiltration, Assessment.

INTRODUCTION

In aged subdrainage systems changes of flow conditions occur mainly in the drain pipe vicinity (Wertz, 1979). Caused by colmatation processes and deposition of ochre, the hydraulic conductivity in this zone decreases more and more inducing the drainage system to go out of function. To decide what to do with such an aged drainage system - to reconstruct it or to construct a new one - the reason of the defect is to be located and evaluated. Stuyt (1992) presented a laboratory method to examine the efficiency and stability of aged filter constructions by means of microgranulometric analysis and x-ray computerized tomography of undisturbed cores containing wrapped drain sections with the surrounding soil. He succeeded in mathematical modelling of the heterogeneous water flow towards the drain pipe/envelope. But there seems to be no hydraulically founded, simple field method to quantify located defects in a drainage system. Therefore, the question was to develop a field method to describe the hydraulic situation of the drain pipe wall including the surrounding soil of old subdrains.

There were two conditions:
· The method should be simple in application and in interpreting the experimental results

· It should characterize a longer segment of the drain, not only one cross section

---

[*] Dr. agr. U. Schindler, Scientist, Dr. agr. R. Dannowski, Scientist, and Dr. agr. L. Müller, Group Leader, Scientist, Institute of Hydrology, Centre for Research on Agricultural Landscapes and Land Use (ZALF), Müncheberg, Germany.

FUNDAMENTALS

The method is based on Eq. (1) where the pressure head of entering water (h) proportionally depends on the force of entrance resistance (w_e):

$$h = w_e \cdot Q/l \qquad (1)$$

with   Q - drain outflow from a drain pipe segment of the length l

The entrance resistance is the sum of all the hydraulic resistances of both the wall and in the surrounding soil of the drain pipe vicinity. It is the reason for the decreasing geohydraulic effectiveness of the drainage system (Fig. 1). Quantifying this value in situ is impossible. Therefore, a field method was developed to quantify the remaining parameters of Eq. (1),

- the drain outflow per drain length and
- the pressure head of the water entrance.

The idea is to exchange the drainage process for an infiltration process at the drain pipe. In doing so, the following assumptions must be valid:

1) The flow conditions nearby the drain pipe are adequate under drainage and infiltration.

2) Approximately there are steady flow conditions.

3) Water flows in the laminar range, i. e., there is proportionality between the volumetric flow and the pressure head.

The third assumption was verified in a laboratorial experiment at a drain pipe with simulated colmatation (Hänsel, 1987). The results show the assumption to be quite correct. The tested range of the volumetric flow values in this experiment was wider than the range under natural conditions.

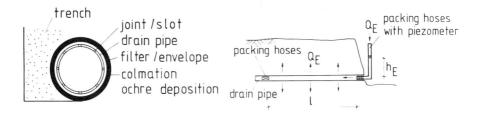

Figure 1. An Aged Drain Construction.   Figure 2. The Principle of the Drain Infiltration Technique.

FIELD METHOD

The aim of the infiltration measurements in the field is to investigate the relationship between

- the volumetric flow, Q_E, infiltrating out of a segment of the drain pipe and
- the pressure head needed for this infiltration,

if possible for more than four different flux values.

The principle of the drain infiltration technique is shown in Fig. 2. By means of packing hoses (Fig. 3) a drain is blocked at two positions. This is preferably done from a ditch or a soil pit. Then, water is filled into this drain segment through one of the packing hoses.

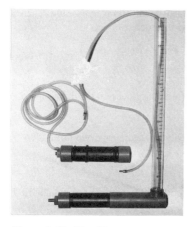

Figure 3. Packing Hoses.

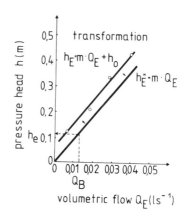

Figure 4. Fitting of the Measured Quantities Q_E and h_E.

The measured quantities are

- the volume of water per time interval flowing into the drain segment, i. e., the infiltrated water flux,
- the pressure head at the inflow piezometer as measured above the drain pipe vertex level.

The measured relationship between the inflow pressure head, h_E, and the volumetric flow, Q_E, is fitted by a straight line (Fig. 4):

$$h_E = m \cdot Q_E + h_o \qquad (2a)$$

with $\qquad m = \triangle h_E / \triangle Q_E \qquad (2b)$

A marked influence on this function is exerted by the soil water pressure head in the drain pipe vicinity, but measuring this value is impossible.

Transforming the measured function into the origin of the ordinate axis, however, delivers this information. The parameter h_o of Eq. (2 a) gives an additional information on the amount and the sign of the pressure head nearby the drain. If the groundwater level was above the drain pipe vertex level, the parameter h_o is positive increasing the measured value of h_E by the amount of h_o as compared with the entrance pressure head, h_E^*, owing to the volumetric flow, Q_E (Fig. 5 a). Figure 5 b illustrates the situation for the groundwater level lower than the drain pipe vertex level, i. e., for unsaturated conditions in the drain pipe vicinity.

Consequently, the following equation is written:

$$h_E^* = h_E - h_o = m \cdot Q_E \tag{3}$$

Now, the design value of the volumetric flow into the drain pipe segment, Q_B, is substituted for Q_E (ICID, 1987):

$$Q_E = Q_B = q \cdot a \cdot l/10^4 \tag{4}$$

with
- q - drainage coefficient in l/(s·ha)
- a - drain spacing in m
- l - length of the drain pipe segment in m (see Fig. 2)

Thus, the resulting inflow pressure head, h_E^*, is equivalent to the pressure head expected under the 'natural' hydraulic load of the drain pipe in the steady-state design case:

$$h_E^* = h_e = m \cdot Q_B = q \cdot a \cdot l \cdot \triangle h_E/(10^4 \cdot \triangle Q_E) \tag{5}$$

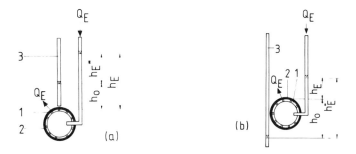

Figure 5. The Principle of the Field Method. Ground Water Table (a) Above, (b) Below the Drain Pipe Vertex Level. (1) Drain Pipe, (2) Drain Pipe Vicinity, (3) Piezometer.

Table 1. Ranges of DE to Estimate the Hydraulic Situation of the Drain Construction.

Function	DE
Sufficient	> 0.8
Reduced	0.6...0.8
Insufficient	< 0.6

To assess the geohydraulic effectiveness of the investigated drain pipe the parameter 'drainage efficiency', DE, is introduced in the classes following Table 1:

$$DE = 1 - h_e/h_{geo} \qquad (6)$$

with h_{geo} - the sum of all the hydraulic pressure heads between the drains, e. g., vertical, horizontal, and radial head according to Ernst (1962), plus h_e

In the case of an 'ideal' drain DE = 1 holds, whereas decreasing conductivity of the drain pipe wall forces DE asymptotically towards zero. To calculate DE the appropriate design value of h_{geo} is drawn from the actual (national) design guidelines (e. g., ICID, 1987).

The field method was tested at two aged drainage systems with repeatedly occuring soil wetness (Table 2). The hydrologic situations were different in the above mentioned sense: Whereas at the Hasenholz system the outer pressure head was $h_o \approx 0.1$ m, at the Reichenberg site a pressure head of $h_o \approx -0.1$ m (suction) acted as related to the drain pipe vertex (Fig. 6 a, b). On both sites approximately steady conditions were obtained during the experiment (Table 3). The infiltration head did not vary systematically. Observed random fluctuations were caused by insufficient uniformity of the volumetric flow, Q_E, not avoidable under field conditions.

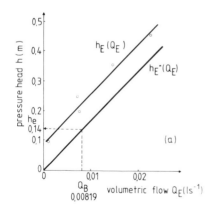
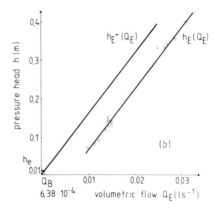

Figure 6. Experimental Results of the Drain Infiltration Technique.
 (a) Hasenholz Site   (b) Reichenberg Site

According to Eq. (4), the following Q values result for the investigated sections of the aged systems:

Hasenholz site: $Q_B = 0.46 \cdot 14 \cdot 12.7/10^4 = 8.18 \cdot 10^{-3}$ l/s
Reichenberg site: $Q_B = 0.58 \cdot 10 \cdot 1.1/10^4 = 6.38 \cdot 10^{-4}$ l/s

The sites are exploited as arable land without flow regulating facilities. In this case the governing standard (ICID, 1987) recommends a h_{geo} of 0.6 to 0.8 m. The results (Fig. 6 a, b) characterize both of the drainage systems to be in function (Table 4), though the Hasenholz site shows some slight insufficiencies.

Table 2. Characteristics of the Aged Drainage Systems and the Experimental Conditions.

Drainage System	Year of Construction	Soil Texture	q l/(s·ha)	a m	l m	Date of Investigation	Weather
Hasenholz	1936	Sandy Loam	0.46	14	12.7	4-29-1986	Wet
Reichenberg	1968	Sand	0.58	10	1.1	7-31-1987	Dry

Table 3. Drain Infiltration Dynamics of the Reichenberg and Hasenholz Sites.

Site	Measurement Number	t min	h_E cm	Q_E l/s
Reichenberg	1	0	7	
		2		0.0113
		3	7	0.0132
		4	9	0.0170
		5	6	0.0100
	2	0	12	
		0.5	12	0.0138
		1.5	13	0.0150
		3.5	14	0.0144
		6.5	14	0.0144
		10.0	13	0.0138
	3	0	34	
		0.5	34	0.0275
		1	34	0.025
		2.5	33	0.025
		4.5	32	0.029
		7	33	0.025
Hasenholz	3	0	24.3	
		4.2	24.3	0.008
		9.1	24.9	0.0068
		13.9	24.9	0.007
	4	0	32.0	
		6.8	24.2	0.014
		9.2	35.4	0.0141

Table 4. Drainage Efficiency of the Investigated Aged Drainage Systems.

Drainage System	h_{geo}	h_e	DE	Function
Hasenholz	0.6...0.8	0.14	0.77...0.82	Reduced to Sufficient
Reichenberg	0.6...0.8	0.01	0.98...0.99	Sufficient

DISCUSSION AND CONCLUSIONS

The suggested drain infiltration method allows to quickly estimate the function of the drain pipe including the surrounding soil. The results represent any chosen drain pipe segment. As expected, the field experiments (Fig. 6 a, b) confirm the assumption of laminar flow validated above, as well as the assumption of gaining steady-state flow conditions within the measuring time interval. Investigating of more than one drain segment is possible following each other proceding from one measuring point (soil pit). In this case the piezometer is advantageously separated from the inflow packing hose.

The decisive advantage of the method is to be absolutely independent on any knowledge of the outer pressure conditions, or site characteristics like the saturated hydraulic conductivity or the grain size distribution. The experimentalist is surely led to an assessment on functioning of the investigated drain pipe related to the site conditions and exploitation by going the following steps:

- Measuring the values Q_E and h_E
- Fitting the measured values by a straight line and transforming it into the origin, $h_E^* = h_E - h_o$
- Finding out Q_B and h_e for the drain spacing of the system under consideration
- Calculating the drainage efficiency

Using standardized design values of q and h_{geo} from the recommended guidelines produces the possibility to evaluate the aged system based on the actual principles of planning, designing and reconstructing subdrainage systems.

The following example will elucidate this. The drainage systems inspected above, Hasenholz and Reichenberg, might be completed with drain regulators as it should be required within the scope of reconstructing work.

According to the standard, the h_{geo} value is now 0.3 to 0.5 m. From this, the drainage efficiency, DE, will shift as compared with Table 4 as follows:

Site	DE	Functioning
Hasenholz	0.53...0.72	Insufficient to Reduced
Reichenberg	0.97...0.98	Sufficient

Because of its malfunction the Hasenholz system should not be completed (the water is 'held back' in the system anyhow).

Due to its regard to the efficiency of the ideal drain and the actual site conditions the DE value gives a clear standard of functioning of the investigated subdrainage system, and this without any doubt to quantify further site characteristics.

The infiltration method is applicable throughout the year provided working in the soil pit is possible. The time spent to the investigation is 2 to 3 hours for two persons, including digging the pit. The needed water volume depends on the drainage system. Every meter of investigated drain pipe requires 20 to 50 litres of water.

In aged, rewetted subdrainage systems only a complex diagnosis of both the site conditions and the system can succeed in exposing the reason of the soil water defect. One step is to assess the function of the drain pipe based, for example, on the presented infiltration method.

REFERENCES

1. Ernst, L. F. 1962. Grondwaterstromingen in de verzadigde zone en hun berekening bij aanwezigheid van horizontale evenwijdige open leidingen. Thesis, Univ. Utrecht.

2. Hänsel, K. 1987. Bodenphysikalische und bodenhydrologische Untersuchungen für die Zustandsanalyse von Bodenwasserregulierungsanlagen. - Ing.-Arbeit, AdL der DDR, Forschungszentrum für Bodenfruchtbarkeit Müncheberg.

3. ICID. 1987. Design practices for covered drains in an agricultural drainage system - a worldwide survey. K. Framji, B. C. Garg, S. P. Kanshish (Eds.). New Delhi.

4. Stuyt, L.C.P.M. 1992. The water acceptance of wrapped subsurface drains. Doctoral thesis. Wageningen Agricultural University, Wageningen, The Netherlands, (X) + 314 pp.

5. Wertz, G. 1979. Bestimmung des Dränabstandes unter besonderer Berücksichtigung des drännahen Raumes. Arch. Acker- u. Pflanzenbau u. Bodenkd., 23, 6: 371-375.

A DECADE OF SUBSURFACE DRAINAGE ENVIRONMENTAL RESEARCH IN SOUTHERN LOUISIANA[1]

R.L. Bengtson, C.E. Carter, and J.L. Fouss
Member ASAE   Member ASAE   Member ASAE

ABSTRACT

Sediment and nutrient losses were measured from a subsurface drainage-runoff-erosion experiment for a decade beginning in 1982. The experiment, located near Baton Rouge, Louisiana, consisted of four surface drained plots, two of which were also subsurface drained. The plots were uniformly graded to about 0.1% slope and contained a clay loam alluvial soil. Subsurface drainage was effective in reducing surface runoff, soil erosion, and nutrient loss. Subsurface drainage reduced surface runoff by 35%, soil loss by 31%, phosphorus loss by 31%, and potassium loss by 27%.

Keywords: nutrient losses, soil erosion, drainage

INTRODUCTION

Severe problems with high water tables exist in the Lower Mississippi Valley due to large amounts of annual precipitation and low lying, nearly level topography. Annual precipitation is usually in excess of 1500 mm and may occasionally exceed 2000 mm. Annual evapotranspiration is approximately 1000 mm, thus annual precipitation exceeds evapotranspiration by 500 mm. A portion of this excess water infiltrates the soil and frequently causes the water table to rise near the soil surface for extended periods of time. This shallow water

---

[1] Contribution from the Louisiana Agricultural Experiment Station, Louisiana State University Agricultural Center; and United States Department of Agriculture, Agriculture Research Service; Baton Rouge, Louisiana.

Professor, LSU; Agriculture Engineer, USDA-ARS; and Agricultural Engineer, USDA-ARS; respectively, Baton Rouge, Louisiana.

table reduces crop yields and causes soil and nutrient losses due to increased surface runoff.

The climate in this area is considered semi-tropical. The annual average temperature and relative humidity is 19.8° C (67.6° F) and 73%, respectively. The normal high temperature during the summer is 32.6° C (91° F). The relative humidity is 80% or greater about half of the time (Dance et al. 1968).

Subsurface drainage is used in many areas of the United States to increase crop yields by lowering the water table. Subsurface drainage may also influence surface runoff, soil loss and nutrient loss from cropland. Information was needed on the effects of subsurface drainage on the environment, thus, an experiment for obtaining information was conducted from 1982 to 1991 near Baton Rouge, Louisiana. The objectives of this experiment were to evaluate the effectiveness of subsurface drainage on (a) reducing surface runoff, (b) reducing soil erosion, and (c) reducing plant nutrient losses.

LITERATURE REVIEW

Mackenzie and Viets (1974) reported that the composition and concentration of materials contained in surface drainage waters differ from those found in subsurface drainage waters. Surface drainage water passes quickly over the soil or surfaces without infiltrating into the soil. Surface drainage contains suspended as well as soluble materials brought into dissolution through suspension, erosion, and solution from brief contact with soil, plants, and plant residues. The slow movement of subsurface drainage water percolating through the soil affords intimate and long contact with clays, organic matter and microorganisms. Soluble nutrients and other chemicals applied to or contained in soils may be dissolved by soil water and leached from the soil profile. On the other hand, some materials will be removed from solution by adsorption and the formation of precipitates. The composition of subsurface drainage water may depend on the aeration status of the soil being drained. Water from a drain that flows continuously from a saturated soil profile may have a different composition than water from a drain that flows only intermittently and is exposed to aerated soil.

Schwab et al. (1980) measured an average sediment loss of 2548 kg/ha from plots with surface drains only and 1529 kg/ha from plots with 1 m deep drains, a reduction of 40% due to subsurface drainage. Annual losses of 12.1, 2.2, and 31.6 kg/ha nitrate nitrogen, total P, and total K, respectively, occurred with surface drains only. Corresponding losses from areas with drains were 18.7, 1.2, and 22.5 kg/ha, for nitrate nitrogen, total P and total K. Subsurface drainage reduced the loss of total P and K by 45 and 29%, respectively.

Bottcher et al. (1981) measured a mean annual sediment loss of 94.0 kg/ha from a subsurface drained area near Woodburn, Indiana. Mean annual nitrogen and phosphorus losses were 8.66 and 0.22 kg/ha, respectively. Approximately 70% of the phosphorus loss was associated with sediments compared to 10% for nitrogen. The analysis showed that losses of sediment and nutrients were reduced by subsurface drainage. They recommended that on suitable soil types, subsurface drainage may well be the preferred best management practice for water quality control.

In Ohio, Schwab and Logan (1982) reported that sediment, phosphorus, and potassium losses from tile outflow were considerably less than from surface runoff, but the nitrate losses were higher. In general, on medium to heavy texture soil on slopes of less than 2%, subsurface drains reduced soil erosion and the loss of most plant nutrients.

PROCEDURE

A 4.4 ha site which had been precision graded to approximately 0.1% slope several years earlier, was selected for this experiment. The soil, a Commerce clay loam, fine silty, mixed, nonacid, thermic Aeric Fluvaquent, has saturated hydraulic conductivity of approximately 1 mm/hr in the surface layer to 0.6 m depth. Between 0.6 and 1.3 m depth is a layer that has a saturated hydraulic conductivity of up to 80 mm/h (Rogers et al., 1985). The area, located 6 km south of Baton Rouge, Louisiana, was partitioned into four plots, each about 200 m long. Earth dikes at least 0.3 m high were constructed around each plot to define the plot boundaries and to insure that runoff passed through the flumes where it could be measured and sampled.

Subsurface drains were installed in plots 1 and 2 using a ladder type trencher. In each of these plots, three 104 mm diameter, corrugated, perforated, polyethylene drain tubes were installed 1 m below the soil surface on a grade of 0.1%. The drains were spaced 10 m apart in plot 1 and 20 m apart in plot 2. The drain outflow was discharged into 1.2 X 1.2 X 3 m metal sumps where it was sampled and then pumped into a surface drainage ditch by electric pumps. In summary, two plots contained both surface and subsurface drainage (drained plots) and two check plots contained surface drainage only (nondrained plots).

Hydrological parameters measured during the experiment included rainfall, surface runoff, drain outflow, and water table elevations. Rainfall was measured with a weighing type recording rain gage; surface runoff was measured with H-flumes and water stage recorders; drain outflow was measured with utility-type water meters as outflow was pumped from the sumps; and water table elevations were measured with a water stage recorder with its float inside a 20 cm diameter by 1.3 m deep cased well in each plot.

Surface runoff was sampled at 20 minute intervals with an automatic water sampler installed at each flume. Subsurface discharge from the center drains from plots 1 and 2 was sampled every 3 hours with an automatic water sampler. Runoff and subsurface discharge samples were analyzed in the laboratory for sediment, nitrogen, phosphorus, and potassium. Total nitrogen was determined by an automated method developed by Wall and Gehrke (1979). Sample preparation for P and K were by method 2.020 described in Horwitz (1980). Phosphorus was determined by method 2.025 and potassium was determined by the atomic absorption method 3.006 described in Horwitz (1980).

Corn was planted in April each year from 1982 to 1987. The land was fertilized at planting with 109, 38, and 76 kg/ha of N, P, and K, respectively. The corn was cultivated once in May each year to control weeds and was harvested for silage in late July. The plots were cultivated periodically from harvest until frost to control weeds and Johnsongrass. From 1988 to 1991 soybeans were planted in May each year and harvested in October. The land was fertilized at planting with 46 kg/ha P and 89 kg/ha K.

RESULTS AND DISCUSSION

From 1982 to 1991, the average annual rainfall was 1568 mm (Table 1) which was 111% of normal. The annual rainfall ranged from a high of 1931 mm in 1989 to a low of 1168 mm in 1984. Average monthly rainfall was greater than 125 mm for the months of January, February, May, June, August, and December had average rainfalls greater than 125 mm (Table 2).

Table 1. Annual Rainfall and Runoff (mm) (1982 to 1991)

Year	Rainfall*	Runoff			
		Drained Plots			Nondrained
		Surface	Subsurface	Total	Total
1982	1425	247	233	480	368
1983	1811	466	514	980	778
1984	1168	150	160	310	208
1985	1526	337	290	627	457
1986	1355	289	244	533	470
1987	1675	422	512	934	675
1988	1492	273	663	936	493
1989	1931	775	679	1454	1085
1990	1451	387	337	724	592
1991	1850	677	657	1334	1012
AVERAGE	1568	402	429	831	614

* Normal Annual Rainfall = 1416 mm.

The average annual surface runoff was 402 and 614 mm (Table 1) for the drained and nondrained plots, respectively. The annual surface runoff ranged from a high of 775 and 1085 mm in 1989 to a low of 150 and 208 mm, respectively, in 1984. The subsurface drains reduced surface runoff by 35%. However, when the average annual subsurface discharge of 429 mm was included, 831 mm of water left the plots. This meant that 35% more water left the drained plots than the nondrained plots.

Table 2. Average Monthly Rainfall and Runoff (mm) (1982 to 1991)

Month	Rainfall	Runoff			
		Drained Plots			Nondrained
		Surface	Subsurface	Total	Total
January	125	32	72	104	58
February	164	56	84	140	102
March	116	14	47	61	30
April	100	30	35	65	42
May	133	39	33	72	53
June	194	73	36	109	91
July	110	25	7	32	27
August	156	41	10	51	61
September	105	10	7	17	17
October	105	16	9	25	30
November	113	22	29	51	35
December	147	44	60	104	68
TOTAL	1568	402	429	831	614

Surface runoff from the drained and nondrained plots was 26% and 39% of the rainfall, respectively. Surface runoff was highest in February when 56 and 102 mm was measured from the drained and nondrained plots, respectively.

Subsurface drainage also reduced soil and plant nutrient losses. From 1982 to 1991, surface runoff carried an annual average of 6434 kg/ha of sediment from the drained plots (Table 3). Subsurface discharge accounted for 376 kg/ha or 5% of the total loss which was 6810 kg/ha. The nondrained plots lost 9932 kg/ha of soil for a 31% reduction due to subsurface drainage. The largest portion of the soil was lost in February and June when 36% of the average annual loss left the fields (Table 4).

The average annual phosphorus losses from the drained and nondrained plots were 7.06 and 10.18 kg/ha (Table 5), respectively, for a 31% reduction due to subsurface drainage. The subsurface discharge contained 0.49 kg/ha or 7% of the lost phosphorus. The greatest monthly loss occurred during with a loss of 0.90 and 1.64 kg/ha, drained and nondrained, respectively (Table 6). Also 34% of the phosphorus was lost during the winter.

Table 3. Annual Soil Loss (kg/ha) (1982 to 1991)

Year	Drained Plots			Nondrained Plots
	Surface	Subsurface	Total	Total
1982	2588	264	2852	3582
1983	5470	718	6188	7198
1984	1495	290	1785	2968
1985	5162	512	5674	10013
1986	3574	326	3900	5560
1987	3826	332	4158	8652
1988	8097	359	8456	16339
1989	12219	330	12549	18309
1990	4910	197	5107	7970
1991	17004	430	17434	18725
AVERAGE	6434	376	6810	9932

Wetzel (1975) listed increased phosphorus loading as the major cause of rapid lake eutrophication. He concluded that algae in lakes could be reduced by decreasing phosphorus loading. Since subsurface drainage was very effective in reducing phosphorus loading, it should be considered a "best management practice" for improving the quality of the water coming from watersheds with heavy texture soils with slopes of less than 2% with shallow water tables.

Table 4. Average Monthly Soil Loss (kg/ha) (1982 to 1991)

Month	Soil Loss			
	Drained Plots			Nondrained Plots
	Surface	Subsurface	Total	Total
January	540	50	590	749
February	1290	91	1381	2130
March	502	25	527	568
April	656	28	684	816
May	529	34	563	762
June	1050	26	1076	1798
July	572	6	578	607
August	373	10	383	715
September	115	9	124	218
October	194	22	216	583
November	197	22	219	311
December	416	53	469	675
TOTAL	6434	376	6810	9932

Table 5. Annual Phosphorus Loss (kg/ha) (1982 to 1991)

Year	Phosphorus Loss			
	Drained Plots			Nondrained Plots
	Surface	Subsurface	Total	Total
1982	3.15	0.29	3.44	4.98
1983	7.00	0.53	7.53	9.28
1984	3.04	0.27	3.31	6.54
1985	10.77	0.57	11.34	16.16
1986	3.25	0.34	3.59	6.70
1987	3.74	0.43	4.17	6.66
1988	2.12	0.72	2.84	5.07
1989	11.28	1.03	12.31	16.32
1990	5.05	0.17	5.22	8.07
1991	16.34	0.51	16.85	22.07
AVERAGE	6.57	0.49	7.06	10.18

The potassium losses from the drained and nondrained plots were 42.43 and 58.50 kg/ha (Table 7), respectively, for a 27% reduction due to subsurface drainage. The subsurface discharge contained 4.60 kg/ha or 11% of the drained plot's total. The largest monthly loss was in June with 6.44 and 8.25 kg/ha, drained and nondrained, respectively (Table 8).

Table 6. Average Monthly Phosphorus Loss (kg/ha) (1982 to 1991)

Month	Drained Plots			Nondrained Plots
	Surface	Subsurface	Total	Total
January	0.48	0.06	0.54	0.90
February	0.71	0.10	0.81	1.66
March	0.45	0.07	0.52	0.44
April	0.74	0.03	0.77	1.09
May	0.60	0.02	0.62	0.85
June	0.88	0.02	0.90	1.64
July	0.99	0.01	1.00	0.80
August	0.61	0.01	0.62	0.90
September	0.13	0.00	0.13	0.24
October	0.26	0.02	0.28	0.44
November	0.22	0.05	0.27	0.36
December	0.50	0.10	0.60	0.86
TOTAL	6.57	0.49	7.06	10.18

Table 7. Annual Potassium Loss (kg/ha) (1982 to 1991)

Year	Drained Plots			Nondrained Plots
	Surface	Subsurface	Total	Total
1982	18.98	2.12	21.10	31.06
1983	50.47	2.89	53.36	72.26
1984	12.71	1.28	13.99	21.88
1985	53.17	5.34	58.51	85.38
1986	29.40	2.90	32.30	50.91
1987	47.39	7.07	54.46	69.35
1988	18.34	11.58	29.92	48.07
1989	69.58	7.93	77.51	87.42
1990	29.48	2.75	32.23	45.56
1991	48.75	2.10	50.85	73.16
AVERAGE	37.83	4.60	42.43	58.50

SUMMARY AND CONCLUSIONS

Subsurface drainage reduced surface runoff by 35%. The associated reduction in losses of soil (sediment), phosphorus, and potassium were 31, 31, and 27 %, respectively.

Based on the study, we conclude that:

1. Subsurface drainage reduces surface runoff, soil and nutrient loss by substantial amounts.

2. A complete subsurface drainage system on certain soil types may be the preferred "best management practice" for improving the quality of water leaving agricultural watersheds.

Table 8. Average Monthly Potassium Loss (kg/ha) (1982 to 1991)

Month	Potassium Loss			
	Drained Plots			Nondrained Plots
	Surface	Subsurface	Total	Total
January	1.82	0.59	2.41	3.89
February	3.95	1.22	5.17	8.85
March	1.51	0.49	2.00	3.28
April	4.31	0.32	4.63	5.90
May	3.52	0.25	3.77	4.83
June	5.99	0.45	6.44	8.25
July	4.69	0.07	4.76	4.59
August	4.32	0.20	4.52	5.87
September	0.84	0.04	0.88	1.59
October	2.05	0.24	2.29	3.59
November	1.63	0.22	1.85	3.00
December	3.20	0.51	3.71	4.87
TOTAL	37.83	4.60	42.43	58.51

REFERENCES

1. Bottcher, A.B., E.J. Monke, and L.F. Huggins. 1981. Nutrient and sediment loadings from a subsurface drainage system. TRANSACTIONS of the ASAE 24(5):1221-1226. 1981.

2. Dance, R.E., B.J. Griffis, B.B. Nutt, A.G. White, S.A. Lytle, and J.E. Seaholm. 1968. Soil survey of East Baton Rouge Parish. LA USDA-SCS. 99pp.

3. Horwitz, W. 1980. Official methods for analysis of the association of official analytical chemists. AOAC.

4. Mackenzie, A.J. and F.G. Viets, Jr. 1974. Nutrients and other chemicals in agricultural drainage waters. In Drainage for Agriculture, Edited by J. van Schilfgaarde. American Society of Agronomy, Madison, WI.

5. Rogers, J.S., V. McDaniel, and C.E. Carter. 1985. Determination of saturated hydraulic conductivity of a Commerce silt loam soil. TRANSACTIONS of the ASAE 28(4):1141-1144.

6. Schwab, G.O., N.R. Fausey, and D.E. Kopcak. 1980. Sediment and chemical content of agricultural drainage water. TRANSACTIONS of the ASAE 23(6):1446-1449.

7. Schwab, G.O. and T.J. Logan. 1982. Sediment and nutrients in effluent from subsurface drains. ASAE Paper No. 82-2550. St. Joseph, MI.

8. Wall, L.L. and C.W. Gehrke. 1979. Automated urease-chromous method (AUCM) for nitrogen in fertilizers. Missouri Agricultural Experiment Station Bulletin. Columbia, MO.

9. Wetzel, R.G., Limnology. W.B. Saunders Company. 1975. Philadelphia. pp. 243-245.

WATER TABLE EFFECTS ON SOYBEAN YIELD AND MOISTURE AND NITRATE DISTRIBUTION IN THE SOIL PROFILE

C.A. Madramootoo[1], S. Broughton[2], A. Papadopoulos[3]

ABSTRACT

Soybeans were grown in field lysimeters, and subjected to water table levels of 40, 60, 80 and 100 cm during two growing seasons. Soybean yield, and nitrate and moisture distributions in the soil profile were measured. The elevated water tables reduced nitrate concentrations by over 50%, and significantly increased soybean yield. From DRAINMOD simulations with long-term climatic data, and the results of the field experiments, it was found that maximum yield (87.6%) and nitrate reduction (52%), could be obtained with subirrigation at a water table depth of 60 cm.

INTRODUCTION

There are 2.5 million hectares of land under agricultural production in Quebec. The predominant crops are corn, soybean, cereals, and forages. The climate is such that producers are often faced with an excess of soil water in the spring, due to snowmelt, and also in the autumn due to rainfall. Subsurface drainage is therefore required. Benefits include trafficable conditions for seedbed preparation in the spring and harvesting in the fall, and removal of excess soil water from the root zone, which may occur during the growing season due to summer thunderstorms.

During the growing season, evapotranspiration often exceeds rainfall. Supplemental irrigation is, therefore, beneficial to increase yields and profits (Heatherly, 1988). The use of water control structures to provide supplemental water for subirrigation is both cost effective and energy efficient.

Increased farming intensity on subsurface drained lands has led to higher fertilizer and manure applications. However, there is concern that nitrates could easily be leached through subsurface drainage systems. High nitrate concentrations promote algal blooms in downstream waters. It has been observed that under good subsurface drainage, an additional 17-35 kg/ha/yr of NO_3-N was lost, compared to soils which are poorly drained (Deal et al., 1986). Madramootoo et al. (1991) measured nitrate concentrations of up to 40 mg/L in the water from a tile drained potato field, far exceeding the safe drinking water limit of 10 mg/L.

One natural mechanism for reducing the loss of nitrates is denitrification. This involves the microbial transformation of NO_3^- to less harmful and more useful compounds of a lower oxidation state. It has been established (Willardson et al., 1975) that denitrification occurs in a low-oxygen saturated soil, particularly where ample organic carbon is available for bacterial metabolism. Meek et al. (1970) also found that denitrification was very rapid in the vicinity of and below the water table, to the point where the nitrate concentration fell rapidly after submergence. The concentration fell to almost zero after 3 days of submergence. They also found that periodic drying between irrigations, and between crops may be very effective in bringing carbon into solution to provide energy for microbes, thereby increasing denitrification.

---

[1]Associate Professor, [2]Research Associate, [3]Graduate Student. Agricultural Engineering Department. Macdonald Campus, McGill University, Ste. Anne de Bellevue, Quebec.

In order to offset the environmental detriment of subsurface drainage, it is therefore possible to decrease nitrate losses by managing the water table. Reductions in nitrate loading of up to 50 % have been reported with water table control (Meek *et al.*, 1970).

There are currently only 10,000 hectares of farmland in Quebec which utilize some form of water table control. More specific information regarding the effects of different water table levels on nitrate reduction and crop yield is required to convince growers of the benefits of water table management. A study was therefore undertaken to determine the effects of 4 water table levels on soil moisture and nitrate distribution, and to ascertain the effects of these water tables on soybean yield. DRAINMOD was used with long-term climatic data to develop water table management practices for maximum crop yield. These simulations are useful because the effects of widely varying climatic conditions and water table management practices can be rapidly and inexpensively evaluated. This is not possible under field conditions, because many years of field experimentation and data collection would be required.

The primary objective of this paper is to provide guidelines for water table management to optimize crop yield and minimize nitrate pollution.

MATERIALS AND METHODS

Field Study

The experimental site was located at the Horticultural Research Station at the Macdonald Campus of McGill University, in Ste. Anne de Bellevue, Quebec. The effects of four water table depths on soybean yield, moisture and nitrogen concentrations in the soil profile were measured. The experiments were conducted from 1989 to 1991. The four water table treatments were 0.4, 0.6, 0.8 and 1.0 meter in depth from the soil surface, with the 1.0 m depth considered as conventional drainage.

The lysimeters were divided into four groups of twenty. Each group had four water table control chambers, and five rows of four lysimeters. Five lysimeters in each group were randomly connected to each control chamber, using 30 mm diameter polyethylene water pipe. The lysimeters and water level control chambers were both constructed from double wall polyethylene pipe, sealed at the bottom with concrete and buried to a depth of 1.2 meters, as depicted in Figs. 1 and 2, respectively. The lysimeters were 400 mm in diameter, with the tops extending 0.1 m above the soil surface to prevent surface water from running into the lysimeters, as well as to retain all rainwater that fell within the lysimeter. Non-perforated, 40 mm diameter polyethylene tubing supplied water to the lysimeters from the water level control chambers. This tubing was perforated and covered with a sock, then extended inside each of the lysimeters, to simulate a subsurface drain lateral for water supply. The water control chambers were equipped with a variable height overflow pipe to control the four water table depths.

The soil used in the experiment was a Courval sandy loam (85 % sand, 15 % clay) with a bulk density of 1.1 t/m^3. The sandy loam was chosen since coarse-textured soils are the prevalent soil type for soybean cultivation, and for its high leaching potential. One soybean plant was grown in each lysimeter. The soybeans were planted on May 28, 1990, and May 21, 1991, and harvested on September 20, 1990, and September 22, 1991, respectively. The water table treatments commenced on June 1, and maintained until September 10, when the lysimeters were drained to allow the plants to dry in preparation for harvest.

Climatic data such as rainfall and pan evaporation were measured daily at the site. Crop evapotranspiration was then calculated. These data are presented in Table 1.

To observe the water table depth in each lysimeter, the technique of Broughton (1972) was followed. A 19 mm diameter by 1.2 m long PVC observation pipe was installed in each lysimeter, sealed at the bottom, perforated with 6 mm diameter holes at 75 mm intervals along its length, and wrapped with filter material to inhibit the entry of fine soil particles. A water table sensor was introduced into the observation well to provide a measurement of the location of the water table in each lysimeter. Water table readings were taken, in each lysimeter, on a daily basis for the first two weeks of each experimental year. These readings were used to establish the necessary water levels in each of the control chambers.

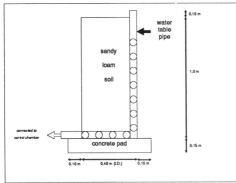

Figure 1. Schematic of a lysimeter.

Nitrate and Soil Moisture Measurement:

Soil samples were taken using a 50 mm diameter auger, at depths of 30 cm and 70 cm from the soil surface for each water table treatment. Soil samples were collected on July 27, August 11, August 28, and September 11 for the 1990 growing season and on May 15, July 27, August 15, and September 11 for 1991.

The first three dates of 1990, and the last three dates of 1991, represented the flowering, seed filling, and maturation periods of theplant, respectively.

Soil-water extracts were obtained from the soil samples, and NO_3 determined with a TECHNICON spectrophotometer autoanalyzer. The method utilized is described by Keeney and Nelson (1982). The moisture content was determined gravimetrically.

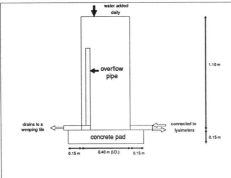

Figure 2. Schematic of a water control chamber.

Plant Measurements:

The experiment was started in 1989, and there were insufficient replicates for statistical analysis. The number of replicates was increased in 1990. Therefore, only 1990 yield data are presented. Two varieties of soybean were tested: Apache, a large seeded variety, and KG30 a small seeded variety. Initially, five seeds were planted in each lysimeter on May 15, 1990. The number of plants in each lysimeter was then reduced to one, on June 5, to eliminate plant interaction effects. Soybean plants were grown between the lysimeters and on a 1 m wide strip around the perimeter of the plots to negate edge and island effects. Weeding was done by hand and no herbicides were applied during the growing season. Adequate rainfall ensured proper germination of the soybean plants.

The following plant parameters were measured: seed mass per plant, number of seeds per plant, number of pods per plant, moisture content of beans at harvest, and protein and oil content of beans at harvest. For the purpose of this paper, only the data of seed mass per plant are presented.

DRAINMOD Simulations:

DRAINMOD was used to simulate the effects of three different water table management scenarios on relative soybean yields using 24 years of climatic data. The water table management scenarios were: conventional drainage, controlled drainage and subirrigation. The controlled drainage and subirrigation cases were each tested with three weir settings: 40, 60 and 80 cm. These correspond to the levels tested in the field lysimeter study. For all simulations the drain spacing was kept at 20 m. This spacing was found to give good drainage as well as being able to control the elevated water table in subirrigation

mode. It is also an economical and commonly found spacing in many drainage installations in Quebec. The only parameters that were varied from simulation to simulation were the weir settings, and the water table management scenarios.

Table 1. Rainfall and Evapotranspiration Data for the 1990 and 1991 Growing Seasons.

	1990			1991		
Month	Rainfall (mm)	ET (mm)	Rainfall - ET (mm)	Rainfall (mm)	ET (mm)	Rainfall - ET (mm)
June	140.1	55.9	84.2	61.9	144.2	-82.3
July	112.1	107.2	4.9	116.2	138.4	-22.3
August	94.6	89.7	4.9	81.7	104.3	-22.6
Sept.	77.1	30.3	46.8	31.6	64.3	-32.7
TOTALS	423.9	283.1	140.8	291.4	451.2	-159.9

Twenty-four years (1960-1983) of daily rainfall and temperature data were obtained from Dorval Airport and used in the simulations. It was found that 1971 was the once-in-24 driest year; 1972 was the once-in-24 wettest year; and 1961 represented the year with the average rainfall over the 24 year period.

Some input parameters were found experimentally, such as the water retention data, and hydraulic conductivity. Some soil trafficability data were taken from Madramootoo (1990), Drablos et al (1988), and Baumer (1988). The values for the parameters input to the model are shown in Table 2.

Table 2. Some DRAINMOD Input Parameters and their Values.

Parameter	Value
Drain Spacing	2000 cm
Drain Depth	100 cm
Depth to impermeable layer	110 cm
Drainage coefficient	1.0 cm/day
Hydraulic conductivity	9.5 cm/hr
Wilting point water content	0.30
Surface Storage	1.5 cm
Maximum Rooting Depth	30 cm

RESULTS AND DISCUSSION

Soil Moisture and Nitrates:

Soil moisture was effectively controlled by water table management. Table 3, which consists of the annual averages from the four sample dates, shows how soil moisture increases with higher water tables. Soil moisture plays an important role in both crop development and in providing conditions favourable for denitrification.

Nitrate concentrations for the conventional drainage case of 100 cm reached almost 19 mg/kg, while water table management between 40 and 80 cm yielded soil nitrate concentrations at, or below 10 mg/kg. These concentrations and reductions are consistent with those found in the literature. Furthermore, there is evidence (Meek et al., 1970) that further reductions may be achieved through periodic drying to increase the leaching of soluble carbon through the soil profile, thereby further enhancing denitrification.

Table 3. Summary of Soybean Yield, Soil Moisture and Nitrate Concentrations.

Depth to Water Table (cm)	Soybean Yield		30cm Sample Depth		70cm Sample Depth	
	Apache (g/plant)	KG30 (g/plant)	Moisture (%)	Nitrate (mg/kg)	Moisture (%)	Nitrate (mg/kg)
1990 Results						
40	63.43	53.86	17.23	8.90	22.63	7.05
60	71.14	61.13	13.43	8.63	19.58	9.20
80	77.62	57.15	12.93	6.93	20.18	10.05
100	75.94	39.07	12.90	7.48	12.50	18.75
1991 Results						
40			14.50	9.40	21.25	6.25
60			11.63	8.53	18.63	9.38
80			8.25	8.75	13.88	9.65
100			8.63	10.83	12.50	12.40

Samples taken at 30 cm displayed fluctuating levels of nitrates with respect to water table depth. The soil was unsaturated at 30 cm, for all treatments. Therefore, nitrate levels are more liable to vary due to rainfall, evapotranspiration, and plant uptake of soil nutrients. Nonetheless, it is worthy to note that the lowest nitrate concentrations were experienced at the 60 and 80 cm water tables.

At the 70 cm sampling depth, there is a more direct relationship between water table depths and nitrate concentrations. Figure 3 distinctly shows that the shallower the water table, the lower the nitrate concentration. The lowest nitrate concentrations were observed for samples taken in the saturated zone, with water table depths of 40 and 60 cm, while the 80 cm water table still produced a significant reduction. There was a large jump in nitrate concentration for the 100 cm depth, since this sample was taken in the unsaturated zone, where denitrification is less likely to occur. This demonstrates the extent of denitrification at the shallower water tables. The results at 70 cm are also of interest since it is desirable for denitrification to occur below the root zone, reducing nitrate contamination of the ground and drainage waters.

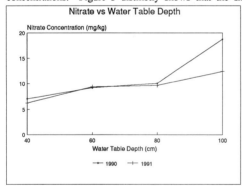

Figure 3. Nitrate concentration versus water table depth at 70 cm sampling depth.

<u>Soybean Yield:</u>

In 1990, rainfall provided sufficient moisture, in every month of the growing season, to satisfy evapotranspiration requirements (see Table 1), thereby reducing the need for supplemental irrigation. The subirrigation system, with well designed overflows, was still effective in improving crop yields, despite the exceptionally wet growing season. The water management system provided adequate soil moisture conditions throughout the growing season. Figures 4 and 5 represent the 1990 results of bean mass versus water table depth for the Apache and KG30 soybean varieties, respectively. For the Apache variety, the line that joins the means peaks at 77.62 g seed mass, at a water table depth of 80 cm. The scatter of the data points is smaller for the 80 and 100 cm treatments, indicating that the plants are growing in favourable soil moisture and aeration conditions.

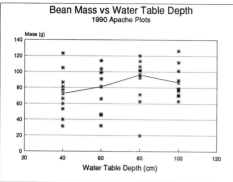

Figure 4. Total seed mass adjusted to 14% MC, 1990 Apache plot.

The line that joins the mean bean masses for the KG30 soybean variety plateau between 60 and 80 cm water table depths, for yields of 61.13 g and 57.15 g, respectively. This indicates that these are the optimum water table depths for the KG30 variety under these conditions. The tabular results are displayed in Table 3.

Figure 6 represents the relationships of relative yield (Apache and KG30 results were averaged) and percent nitrate reduction versus water table depth. Both curves are fairly flat through the 40, 60 and 80 cm depths. The maximum yield is found between 60 and 80 cm, while the maximum nitrate reduction is found at 40 cm. In both cases, the worst results were obtained at the 100 cm depth. Considering the marginal improvement in nitrate reduction at 40 cm, the increased cost of maintaining this water table depth, and the risk of waterlogging the root zone, the best practice appears to be water table management at 60 cm.

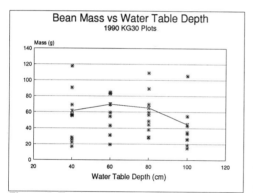

Figure 5. Total seed mass per plant adjusted to 14% MC, 1990 KG30 plot.

DRAINMOD Simulations:

DRAINMOD simulates relative crop yield in terms of stresses due to soil water deficits and excesses, and planting and harvesting delays. These effects are expressed as a percentage of the maximum attainable crop yield with no crop stresses. The model considers that the crop is undergoing wet stress when the water table rises to within 30 cm of the soil surface. Dry stress occurs when the soil moisture conditions do not satisfy potential evapotranspiration. Yield reductions also occur when there are delays in either planting or harvesting due to soil moisture conditions.

Figures 7 and 8 display the relative yield results for the 4 water table depths, 3 climatic conditions of once-in-24 dry, average and once-in-24 wet years, and the 2 water table management practices of controlled drainage and subirrigation, respectively.

The 24 year average rainfall is nearly equivalent to that of 1961, while 1990's growing season rainfall is within 3 mm of the once-in-24 wet year rainfall of 1972. Since the rainfall data for 1972 and 1990 are close, the results between the simulation and the experimental data are expected to be similar.

The simulations showed that the best water management practice is subirrigation. In the dry year, the maximum yields can be expected at a water table depth of 40 cm; in the average year, 60 cm is optimum; while in the wet year, yields are highest at the 80 cm depth. The best water table management practice is subirrigation with a water table depth of 60 cm. The cost required to supply the additional irrigation water to achieve the 40 cm water table depth is not justified by the small increase in potential yield. With the water table as shallow as 40 cm there is the risk of crop damage due to waterlogging of the root zone, if large rainfalls were to occur. The results obtained for the wet year under controlled drainage also support the choice of 60 cm as the best water table depth. In the wet year, the controlled drainage system is much like a subirrigation system, the difference being that the irrigation water is supplied by rainfall. In the wet year, controlled drainage was able to maintain the water table near 60 cm depth and give the second highest yield results.

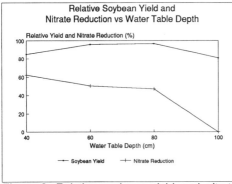

Figure 6. Relative soybean yield and nitrate reduction versus water table depth for 1990.

When considering the three cases of once-in-24 dry, average and once-in-24 wet years, the water table management practice that consistently produced the best relative yields was subirrigation with a weir setting of 60 cm. This is substantiated from the field experiment results, where a water table depth between 60 and 80 cm also gave the best crop yields.

CONCLUSIONS

A two year field study was conducted using lysimeters to test the effects of four water table depths on soybean yield, and moisture and nitrate distributions in the soil profile. Water table levels have a direct effect upon both crop yield and soil moisture and nitrate distributions. Elevated water tables improve yield, and reduce nitrate concentrations in the soil profile. This leads to a reduction of nitrate pollution from subsurface drained watersheds.

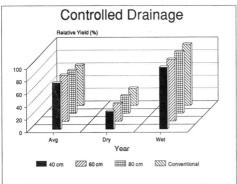

Figure 7. DRAINMOD simulations of relative soybean yield for controlled drainage.

Lower nitrate concentrations were observed at the elevated water tables. This is evidence that water table management is an effective means of reducing nitrate pollution by providing conditions favourable to denitrification.

The best water management practice for maximum long-term yield, and maximum nitrate reduction, is subirrigation with a water table depth of 60 cm. This will provide a relative yield of 87.6% and a nitrate reduction of 52%. The marginal improvement in nitrate reduction at 40 cm is insufficient to justify controlling the water table to a depth of 40 cm, considering the increased energy and water costs, and the risk of waterlogging the root zone.

ACKNOWLEDGEMENTS

The authors acknowledge the financial assistance of the Natural Sciences and Engineering Research Council of Canada, and the Brace Research Institute of McGill University. PlastiDrain Ltd. and Semences Prograin of Quebec are thanked for supplying some of the pipe materials, and soybean seeds, respectively.

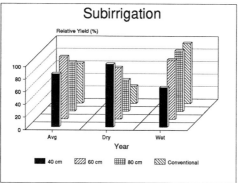

Figure 8. DRAINMOD simulations of relative yield for subirrigation.

REFERENCES

Baumer, O. and J. Rice. 1988. Methods to predict soil input data for Drainmod. ASAE Paper number 88-2564

Broughton, R.S. 1972. The performance of subsurface drainage systems on two St. Lawrence lowland soils. Ph.D. Thesis, Agricultural Engineering Department, McGill University, Montreal, Quebec.

Deal, S.C., J.W. Gilliam, R.W. Skaggs and K.D. Konya. 1986. Prediction of nitrogen and phosphorus losses from soils as related to drainage system design. Agric. Ecosys. Environ. 18:37-51.

Drablos, C.J.W., K.D. Konya, P.W. Simmons and M.C. Hirschi. 1988. Estimating soil parameters used in Drainmod for artificially drained soils in Illinois. ASAE Paper number 88-2617.

Heatherly, L.G. 1988. Planting date, row spacing, and irrigation effects on soybean grown ion clay soil. Soil Agron. J. 80:227-231.

Keeney, D.R. and D.W. Nelson. 1982. Available NO_3^- and NH_4^+. Nitrogen-Inorganic forms. Methods of Soil Analysis, Vol. 2. Am. Soc. Agronomy, Madison WI. pp. 643-698.

Madramootoo, C.A., K. Wiyo, P. Enright and C. Bastien. 1991. Impact of continuous potato production on water quality. CSAE Paper number 91-108.

Madramootoo, C.A. 1990. Assessing drainage benefits on heavy clay soils in Quebec. Trans. ASAE 33(4):1217-1223.

Meek, B.D., L.B. Grass, L.S. Willardson and A.J. MacKenzie. 1970. Nitrate transformations in a column with a controlled water table. Soil Sci. Soc. Am. Proc. 34:225-239

Willardson, L.S., B.D. Meek, L.B. Grass, G.L. Dickey and J.W. Bailey. 1975. Nitrate reduction with submerged drains. Trans. ASAE, 15:84-90.

CROP MANAGEMENT TO MAXIMIZE THE YIELD RESPONSE OF SOYBEANS TO A SUBIRRIGATION/DRAINAGE SYSTEM

R. L. Cooper, N. R. Fausey and J. G. Streeter[*]

ABSTRACT

Often in irrigation experiments, insufficient attention is paid to modifications in the crop production system to maximize the yield response of the crop to irrigation. This often results in disappointing yield responses to irrigation. Also, distribution and amounts of water supplied are often insufficient for maximum yields. A combination of water management and crop management systems needed to produce soybean yields in excess of 6700 kg/ha are discussed. **KEYWORDS.** Soybean, Irrigation, Lodging, Solid-Seeding, Semidwarf Cultivars.

In the 1987 edition of Soybeans: Improvement, Production and Uses, published by the American Society of Agronomy, a chapter on irrigation listed 77 citations showing the yield response of soybeans to irrigation (Van Doren and Reicosky, 1987). While yield responses of up to 2500 kg/ha were indicated, this was due to the very low yield of the control and not due to the exceptionally high yield of the irrigated treatment. The highest yield reported was 4000 kg/ha with the 10 highest yielding lines averaging 3700 kg/ha. In contrast, results from maximum yield soybean research by Flannery in New Jersey (Potash and Phosphate Inst., 1983a, 1984), Cooper in Ohio (Potash and Phosphate Inst., 1983b) and Lawn et al. in Queensland, Australia (Lawn and Byth, 1989; Lawn et al., 1984), indicated irrigated soybean yields in excess of 6700 kg/ha are possible (Table 1). Why this almost two-fold difference in irrigated soybean yields?

When a grower or researcher indicates the soybeans were irrigated it often is not very informative. All that is known for certain is that supplemental water was added to the crop. It does not indicate how much was added, how frequently, and the total amount. There are two major reasons for the difference in the 3700 kg/ha and 6700 kg/ha irrigated soybean yields.

The first reason is the amount of water applied. All three researchers who produced yields in excess of 6700 kg/ha provided

---

[*]RICHARD L. COOPER, USDA-ARS and Dep. of Agronomy, Ohio Agric. Res. and Devel. Cent. (OARDC), Wooster, OH 44691; N. R. FAUSEY, USDA-ARS and Dep. of Agronomy/Agric. Eng., OARDC, Columbus, OH 43210; J. G. STREETER, Dep. of Agronomy, OARDC, Wooster, OH 44691.

essentially a continuous water supply. Flannery used a trickle irrigation system applying approximately 1.25 cm of water per day.

Table 1. Highest yields reported from irrigated maximum yield research, 1980-1987.

Location	Researcher	Year	Yield
			(kg/ha)
New Jersey	Flannery	1980	6270
		1981	6203
		1982	7270
		1983	7897
Ohio	Cooper	1982	6817
		1985	6710
		1986	6003
		1987	6803
Australia	Lawn et al.	1984	8004
		1984	8604

Cooper applied weekly applications of 5 cm per week through a solid set sprinkler irrigation system. Lawn and coworkers used a continuous furrow irrigation system maintaining a constant water table 10 to 20 cm below the bed surface (Commonwealth Scientific and Industrial Research Organization, 1983; Nathanson et al., 1984). Lack of significant yield increases from irrigation where the non-irrigated yields are at the 3500 kg/ha level has often led to the conclusion that water is not a yield limiting factor in such environments. What is not recognized is that the yield potential of soybeans is much higher than 3500 kg/ha and that it takes more water to produce 5300 to 6700 kg/ha yields than 3500 kg/ha.

The second reason is that the crop production system and varieties used in most irrigation studies are the same production system and varieties normally used for dryland production. Maximum yield research has shown that soybeans solid-seeded in 17 cm rows are much more yield responsive to irrigation then soybeans planted in 75 to 100 cm row spacing. This is because the yield potential of solid-seeded soybeans is higher (Cooper, 1977, 1980; Cooper and Jeffers, 1984; Costa et al., 1980). However, most irrigation studies reported have been in the wider row spacings. The other major barrier to greater yield responses to irrigation is plant lodging which is normally significantly increased under irrigation. Research by Cooper indicated early lodging can reduce the yield potential of soybeans nearly 25% in high yield environments (Cooper, 1971a, 1971b).

Irrigation is only one factor in a complex system to produce 6700 kg/ha soybean yields. The researcher or farmer must consider the total system if yields of this magnitude are to be obtained. Failure to recognize this has led to research results and grower experience that suggest irrigation of soybeans is not economically profitable. Through maximum yield research in soybeans, plant lodging, row width and water have been systematically identified as primary yield limiting factors. Once lodging was removed as a limiting factor by the development of semidwarf or shorter indeterminate varieties (Cooper, 1981, 1985, 1989a), and the row width barrier minimized by solid-seeding in 17 to 25 cm row width (Cooper, 1977, 1980; Cooper and Jeffers, 1984; Costa et al., 1980), a soybean production system was developed that was highly responsive to irrigation or favorable moisture years. Soybean yields of 6700 kg/ha have been obtained under irrigation when irrigation was combined with this high yield soybean production system in Ohio, New Jersey and SE Queensland, Australia (Table 1).

This high yield soybean production system is also valuable under dryland crop production. Lodging resistant varieties, solid-seeded in 17 cm rows are highly responsive to a favorable moisture year, resulting in 4700 to 5300 kg/ha yields. In two separate 10-year non-irrigated studies in Ohio, in diverse environments, the yield of a solid-seeded-semidwarf (SSS) system was compared with the standard wide-row-indeterminate (WRI) system (Cooper, 1989b). The SSS system consisted of the semidwarf variety, Sprite, planted in 17 cm rows at a seeding rate of 750,000 seeds/ha (90%+ germ seed). The WRI system was Williams (82) planted in 75 cm rows at a seeding rate of 375,000 seeds/ha. In northwest Ohio, on a soil prone to late season drought, the SSS system out-yielded the WRI system seven out of the 10 years with an average yield advantage of 787 kg/ha or 24%. In two of the years the yields were not significantly different and one year the WRI system out-yielded the SSS system. These years were characterized by lower yields due to drought stress. However, under the more favorable moisture years, the SSS system had a distinct yield advantage because of its greater responsiveness to water. The SSS system exceeded 4700 kg\ha four out of the 10 years with two of these years equalling or exceeding 5300 kg/ha. In contrast, the WRI system exceeded 4000 kg/ha only two out of the 10 years, with 4400 kg/ha the highest yield obtained. Thus even though there was no yield advantage for the SSS system three out of the 10 years, averaged over the 10 years there was a 24% yield advantage for the SSS system.

At the second location, in west central Ohio, on a soil with excellent water holding capacity and history of high yields, the SSS system out yielded the WRI system nine out of the 10 years and equalled the yield of the WRI system the one year. Averaged over the 10 years there was a 947 kg/ha yield advantage of the SSS system over the WRI system (5000 vs 4055 kg/ha). With the reduced moisture stress of this location, the SSS system consistently out-yielded the WRI system.

As a result of this research, the HYSIP (High-Yield-System-In-Place) concept for Midwest soybean production was developed (Cooper, 1989b). Simply stated, in order for a grower to take advantage of the favorable moisture years and long-term higher average yields, he must have the high-yield-system in place every year. In this case, the SSS system. The advantage of the SSS system would be even greater under irrigation where the yield advantage over the WRI system would be consistent from year-to-year.

With this background information, I was quite interested in the early 1980's when I first learned of the subirrigation/drainage concept being used by some growers in Michigan and became aware of the research on water table management by Belcher and coworkers in Michigan (Belcher, 1991a, 1991b, 1991c). I also became aware of the pioneering research on subirrigation by Skaggs and coworkers in North Carolina (Evans and Skaggs, 1989; Skaggs, 1973,

1979, 1981; Skaggs et al., 1972; Smith et al., 1985) and by Doty and coworkers in South Carolina (Doty, 1980; Doty and Parsons, 1979; Doty et al., 1985, 1984; Doty et al., 1987). Water table management research has also been conducted in North Dakota by Benz and coworkers (Benz et al., 1981, 1978; Follett et al., 1974). In essence this practice consists of using drain lines in reverse, adding water to the soil by subirrigation to maintain a constant water table during periods of moisture shortage. In cooperation with Dr. N. R. Fausey in Agricultural Engineering, and with the financial support of the Advanced Drainage Systems, Inc., we established a subirrigation drainage research facility at Wooster, Ohio in the fall of 1984.

Grower experience with the subirrigation/drainage system had been good with corn, obtaining 2000 to 3000 kg/ha yield increases. Results with soybeans were less favorable with yield increases more in the 300 to 600 kg/ha range and top yields 3500 to 4000 kg/ha. With our experience in obtaining 5000 to 6000 kg/ha yields under surface irrigation, when the irrigation was combined with lodging resistant varieties and solid-seeding, we suspected the yield potential of soybeans grown under a subirrigation/ drainage system should be much higher than the 3500 kg/ha yields being reported by growers.

Our first subirrigation research was conducted in 1985 at the research facility at Wooster. We superimposed the high yield soybean management system identified under surface irrigation on top of the subirrigation/drainage management system. At the highest water table (40 cm average depth) the five cultivars solid-seeded in 17 cm row width averaged 5300 kg/ha (Cooper et al., 1991) (Table 2). These results confirmed our hypothesis that when the subirrigation/drainage system is combined with a high yield soybean production system, soybean yields in the 5000 to 6000 kg/ha level should be possible.

Table 2. Soybean yields from a subirrigation/drainage production system, Wooster, OH, 1985-87 (from Cooper et al., 1991).[a]

Cultivar	Year			3-year Mean
	1985	1986	1987	
	---------------- kg/ha -------------			
Sprite	5656	5116	5176	5316
Sprite 87	5763	4922	5423	5369
Hobbit 87	5383	5143	5062	5196
Asgrow 3127	5123	5363	5896	5461
Williams 82	4922	5670	6256	5616
LSD (0.05)[b]	728	697	485	348
Mean	5369	5243	5563	5392

[a] 40 cm average water table depth.
[b] LSD for cultivars.

Since that first research in 1985, we have completed an additional six years of research on soybean production under a subirrigation/ drainage system. A comparison of 40, 55 and 70 cm average water table depths indicated soybean yields were maximized with the 40 cm water table (Cooper et al., 1992). Our results suggest that barring serious disease problems, average soybean yields of 5300 kg/ha should be possible year after year in the Midwest with a properly managed subirrigation/drainage system, when it is combined with a high yield soybean production system. The importance of using a soybean production system that is highly responsive to supplemental water cannot be over emphasized. For example, just by changing row width from 17 cm to 75 cm, average yields decreased 1000 kg/ha or 19% due to the row width barrier to higher soybean yields (Cooper et al., 1991).

The concept of maximum yield research is to attempt to push crop yields as high as possible, regardless of the practicality or cost. From some 20 years experience in soybean maximum yield research, we have learned a great deal about factors limiting soybean yields. As we have systematically removed one yield limiting factor after another, we have continued to ask ourselves, "what now is limiting soybean yields?" With the removal of the lodging barrier by development of semidwarf or shorter indeterminate varieties, removal of the row width barrier by solid-seeding, and removal of the water barrier by a subirrigation/drainage system, the question now being asked is what is limiting soybean yields to the 5000 to 6000 kg/ha level. Preliminary unpublished data from our research in Ohio, research by Flannery in New Jersey (unpublished), and research by Troedson, Lawn and coworkers in Australia (Troedson, 1987), all suggest nitrogen may be the next yield limiting factor in soybeans. Yield increases of 1000 kg/ha (5800 vs 6800 kg/ha) have been obtained from nitrogen fertilization in Ohio, 1500 kg/ha in New Jersey (4000 vs 5500 kg/ha) and 2900 kg\ha (5700 vs 8600 kg/ha) in Australia. The next step is to determine the consistency of this response and the commercial profitability and environmentally acceptability of nitrogen fertilization of soybeans. The answer to these questions must await further research.

In summary, the profitability of irrigating soybeans, either by surface or subsurface irrigation, cannot be fairly judged without first combining irrigation with a soybean production system that will maximize the yield response to the supplemental water. Too often irrigation experiments have been run in wide row spacing and with lodging susceptible varieties which limit the yield response to the added water. This has led to the conclusion in many cases that it does not pay to irrigate soybeans. Our research results with the subirrigation/drainage system in Ohio indicate that irrigation of soybeans should be profitable in the Midwest even on those soils capable of producing 3500 kg/ha yields. We now know as the result of maximum yield research, that 5000 to 6000 kg/ha yields are possible with adequate water and a highly moisture responsive soybean production system.

REFERENCES

1. Belcher, H.W. 1991a. Subirrigation: Irrigation and drainage through the same underground pipe. Sagninaw Bay Subirrigation Series SI01.

2. Belcher, H.W. 1991b. Overhead irrigation versus subirrigation. Saginaw Bay Subirrigation Series SI02.

3. Belcher, H.W. 1991c. Guidelines for operating subirrigation systems for water quality benefits (corn, soybeans and sugar beets). Saginaw Bay Subirrigation Series SI03.

4. Benz, L.C., E.J. Doering, and G.A. Reichman. 1981. Water table management saves water and energy. Trans. ASAE 24:995-1001.

5. Benz, L.C., G.A. Reichman, E.J. Doering, and R.F. Follett. 1978. Water table depth and irrigation effects on applied-water-use efficiencies of three crops. Trans. ASAE 21:723-728.

6. Cooper, R.L. 1971a. Influence of early lodging on yield of soybean [Glycine max (L.) Merr.]. Agron. J. 63:449-450.

7. Cooper, R.L. 1971b. Influence of soybean production practices on lodging and seed yield in highly productive environments. Agron. J. 63:490-493.

8. Cooper, R.L. 1977. Response of soybean cultivars to narrow rows and planting rates under weed-free conditions. Agron. J. 69:89-92.

9. Cooper, R.L. 1980. Solid seeded soybean production systems. p. 9-16. In Proc. Solid Seeded Soybean Conf., Indianapolis, IN, 21-22 Jan., Am. Soybean Assoc., St. Louis, MO.

10. Cooper, R.L. 1981. Development of short-saturated soybean cultivars. Crop Sci. 21:127-131.

11. Cooper, R.L. 1985. Breeding semidwarf soybeans. p. 289-309. In J. Janick (ed.) Plant Breeding Reviews, Vol. 3, AVI Publishing, Westport, CT.

12. Cooper, R.L. 1989a. Breeding soybean cultivars with specific adaptation to yield extremes. p. 895-900. In Proc. IV World Soybean Res. Conf., Buenos Aires, Argentina. 5-9 March. Publ. Realizacion, Orientacion Grafica Ed. SRL, Buenos Aires.

13. Cooper, R.L. 1989b. High-yield-system-in-place (HYSIP) concept for soybean production. J. Prod. Agric. 2:321-324.

14. Cooper, R.L. and D.L. Jeffers. 1984. Use of nitrogen stress to demonstrate the effect of yield limiting factors on the yield response of soybeans to narrow rows. Agron. J. 76:257-259.

15. Cooper, R.L., N.R. Fausey and J.G. Streeter. 1991. Yield potential of soybean grown under a subirrigation/drainage water management system. Agron. J. 83:884-887.

16. Cooper, R.L., N.R. Fausey and J.G. Streeter. 1992. Effect of water table level on the yield of soybean grown under subirrigation/drainage. J. Prod. Agric. 5:180-184.

17. Costa, J.A., E.S. Oplinger, and J.W. Pendleton. 1980. Response of soybean cultivars to planting patterns. Agron. J. 72:153-156.

18. Commonwealth Scientific and Industrial Research Organization. 1983. Soybean response to controlled waterlogging. Rural Res. 120:4-8. CSIRO, East Melbourne, Victoria, Australia.

19. Doty, C.W. 1980. Crop water supplied by controlled and reversible drainage. Trans. ASAE 23:1122-1126, 1130.

20. Doty, C.W. and J.E. Parsons. 1979. Water requirements and water table variations for a controlled and reversible drainage system. Trans. ASAE 22:532-536, 539.

21. Doty, C.W., J.E. Parsons, A.W. Badr, A. Nassehzadeh-Tabrizi, and R.W. Skaggs. 1985. Water table control for water resource projects on sandy soils. J. Soil Water Conserv. 40:360-364.

22. Doty, C.W., J.E. Parsons, A. Nassehzadeh-Tabrizi, R.W. Skaggs, and A.W. Badr. 1984. Stream water levels affect field water tables and corn yields. Trans. ASAE 27:1300-1306.

23. Doty, C.W., J.E. Parsons, and R.W. Skaggs. 1987. Irrigation water supplied by stream water level control. Trans. ASAE 30:1065-1070.

24. Evans, R.O. and R.W. Skaggs. 1989. Design guidelines for water table management systems on coastal plain soils. Appl. Eng. Agric. 5:539-548.

25. Follett, R.F., E.J. Doering, G.A. Reichman, and L.C. Benz. 1974. Effect of irrigation and water table depth on crop yields. Agron. J. 66:304-308.

26. Lawn, R.J. and D.E. Byth. 1989. Saturated soil culture - A technology to expand the adaptation of soybeans. p. 576-581. *In* Proc. IV World Soybean Res. Conf., Buenos Aires, Argentina. 5-9 March. Publ. Realizacion, Orientacion Grafica Ed. SRL, Buenos Aires.

27. Lawn, R.J., R.J. Troedson, A.L. Garside, and D.E. Byth. 1984. Soybeans in saturated soil - A new way to higher yields. p. 67-68. *In* Prog. Abstr, World Soybean Res. Conf. III. Ames, IA. 12-17 August. Iowa State Univ., Ames, IA.

28. Nathanson, K., R.J. Lawn, P.L.M. DeJabrun, and D.E. Byth. 1984. Growth, nodulation and nitrogen accumulation by soybean in saturated soil culture. Field Crops Res. 8:73-92.

29. Potash and Phosphate Institute. 1983a. New world record corn and soybean research yields in 1982. Better Crops Plant Food 67:4-5 (PPI, Atlanta, GA).

30. Potash and Phosphate Institute. 1983b. Soybeans topped 100 bu/A in maximum yield research in Ohio. Better Crops Plant Food 67:8-9 (PPI, Atlanta, GA).

31. Potash and Phosphate Institute. 1984. 1983 soybean research yields top 118 bu/A. Better Crops Plant Food 68:6 (PPI, Atlanta, GA).

32. Skaggs, R.W. 1973. Water table movement during subirrigation. Trans. ASAE 16:988-993.

33. Skaggs, R.W. 1979. Water movement factors important to the design and operation of subirrigation systems. ASAE Paper 79-2543. ASAE, St. Joseph, MI 49085.

34. Skaggs, R.W. 1981. Water movement factor important to the design and operation of subirrigation systems. Trans. ASAE 24:1553-1561.

35. Skaggs, R.W., G.J. Krez, and R. Bernal. 1972. Irrigation through subsurface drains. J. Irrig. and Drain. Div. Am. Soc. Civ. Eng. 90:363-377.

36. Smith, M.C., R.W. Skaggs, and J.E. Parsons. 1985. Subirrigation system control for water use efficiency. Trans. ASAE 28:489-496.

37. Troedson, R.J. 1987. Physiological aspects of acclimation and growth of soybean (*Glycine max*) in saturated soil culture. Ph.D. Thesis, Univ. of Queensland, Australia. 231 pp.

38. Van Doren, D.M., Jr. and D.C. Reicosky. 1987. Tillage and irrigation. p. 391-428. *In* J. R. Wilcox (ed.) Soybeans: Improvement, Production, and Uses. 2nd ed. Agron. Monogr. 16. ASA, CSSA, and SSSA, Madison, Wi.

SOME RESULTS OF A 12 YEAR EXPERIMENT WITH DIFFERENT SUBIRRIGATION LEVELS IN A YOUNG MARINE CLAY SOIL IN THE IJSSELMEERPOLDERS IN THE NETHERLANDS

ing. J. Visser
Rijkswaterstaat, Directorate Flevoland,
P.O. Box 600, 8200 AP Lelystad, The Netherlands

SUMMARY

In the Ysselmeer polder project a 5 ha ground water experimental field was set out in 1964 on a loamy to clayey soil in the East Flevoland polder. Here the relation between subirrigation levels of 0.40, 0.70, 1.00 and 1.30 m below soil surface and the soil structure, soil compaction, hydrological conductivity in the drain trench, nitrogen management, root development, growth of grass and growth and production of apple trees was studied during a period of 12 years.

The results indicate that optimum values were different for each subsequent parameter. Soil compaction became worth at higher ground water levels, nitrogen management was maximal at 1.30 m ground water level and soil subsidence and crackformation are maximal at the deeper ground water levels (> 1.30 m). Apple production was only slightly influenced by the different ground water levels, the growth of the trees however rather strongly.

Of special interest is the functioning of the drain. The drain trenches were filled up with top soil. This soil had rather uniform properties over the whole depth of 1.50 m and 0.20 m width of the drain trench. Ten years after start undisturbed samples were taken at different depth for measuring soil structure and hydrological conductivity. After 10 years below the ground water level the structure of the loamy to clayey soil had not been changed and maintained a hydraulic conductivity of 3 m a day.

The overall conclusion is that subirrigation in a clay soil is possible. However profits will highly depend on climat and type of crop.

INTRODUCTION

In the Netherlands precipitation is about 750 mm a year and potential evaporation approximately 550 mm a year. So yearly there is an excess of water. From April till September however evaporation exceeds rainfall with a total value of 200 mm. This shortage can partly be covered by soil bound water in clay soils. In sandy soils additional water has to be supplied by sprinkler irrigation or subirrigation.

In the IJsselmeerpolders, most soils are clay soils. On the sand soils, especially in the North-eastern polder, a subirrigation system has been installed for grassland and horticulture ($\pm$ 10.000 hectares). The results of this subirrigation system in sandy soils were presented on the international conference on subirrigation and drainage in 1991 in Michigan, (Visser 1991).

In the Netherlands subirrigation in clay soils is not often practised. The drainage problem is in most years more severe than the irrigation problem. However in years with long dry summers additional water supply with subirrigation is execcuted on clay soils by pumping water into blocked ditches or with sprinkler irrigation.

On the international conference in Michigan it was shown that the influence of the ground water level on soil physics, soils chemics and root development was subject of different studies.
On the clay soils of the IJsselmeerpolders there are results available of ground water experimental fields which could be of support for these studies.
In this paper results will be given of such an experiment over a period of 12 years on a clay soil in the IJsselmeerpolders. However before discussing these results a brief review will be given of the IJsselmeerpolders.

The IJsselmeerpolders
The IJsselmeerpolders are part of the plan for closing off and partially reclaiming the Zuyderzee, a dangerous inland sea, that in times past penetrated deeply into the heart of the Netherlands. The plan was created by Dr. Cornelis Lely (engineer) and was published in 1918, figure 1. The main provisions of the plan were:
- The construction of a 2,5 km barrier dam from North Holland to Wieringen and a 30 km barrier dam from Wieringen to Friesland;
- The reclamation of five polders (table 1).
    Polders are seperated areas with a regulated water management system surrounded by dikes and kept dry by one or more pumping stations. For detailed information, Schultz, 1988.

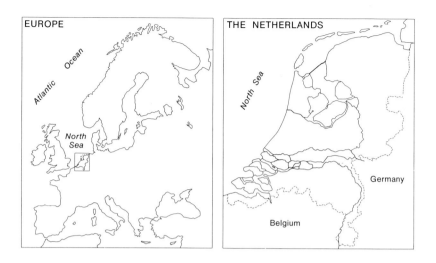
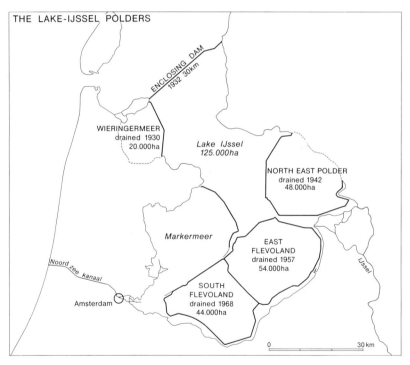

Figure 1. The Ysselmeerpolders

Table 1. The IJsselmeerpolders.
Area, dike construction and development period.

	area in hectares	dike construction	development period
Wieringermeer	20,000	1927-1929	1930-1940
Northeastern Polder	48,000	1936-1940	1942-1962
Eastern Flevoland	54,000	1950-1956	1957-1976
Southern Flevoland	43,000	1959-1967	1968-1996
Markerwaard	53,000		postponed

MATERIAL AND METHODS

The Soil Profile

The ground water experiment was laid out on parcel R18, situated in the central area of Eastern Flevoland. The experimental field had a size of 5 hectares and was installed in 1964 and functioned till 1976.
The soil profile is give in table 2. There is a variation in thickness of the different soil layers of approximately 10 %.

Table 2. Average profile and composition on the geological section

Layer depth cm below surface	Name	Designation	Sedimentation medium	Clay[a] %	Org. matter %
0- 8	IJsselmeer dep.	Ym	fresh	8	-
8- 35	Zuyder Zee dep.	ZuI/II	salt	30	2.9
35- 40	Zuyder Zee dep.	ZuIII	salt	3	-
40- 60	Almere depossit	AI^a1	brackish	24	3.6
60- 75	Almere deposit	AI^a1	brackish	16	3.7
75- 90	Almere deposit	AI^a2	brackish	20	11.0
90- 100	Almere deposit	AI^c1	brackish	8	5.0
100- 200	Almere deposit	AI^c2+3	brackish	15	12.0
> 220	Pleistocene	Pl	-	2	-

[a]Clay particles < 0.002 mm

The Water Levels

Water levels of 0.40, 0.70 1.00 and 1.30 m below ground level were selected as base levels for the various ground water regimes. These were fixed at a constant level throughout the year or at increased levels in the winter phase compared with the summer phase. The different ground water regimes were seperated from each other with a vertical plastic sheet to a depth of 2 meters . In this way, a total of ten different ground water regimes were created. The plan of the subirrigation system is given in figure 2.

Drains

To minimize deviations in the ground water level from the selected depth the distance between the drains was set on 4 m. This interval was determined partly by the distance between the rows of fruit trees. The rows were therefore just between the drains, so that all the trees in a plot were similary positioned relative to the drains. The depth of the drains was determined by the deepest subirrigation regime of 1.30 m below the soil surface. To achieve this the drains had to be at least 1.50 m below the surface. The same depth was used throughout the experimental field. The drains consisted of fired, unflanged pipes with an internal diameter of 5 cm, packed with heather covered with straw which retains high permeability in subirrigation.

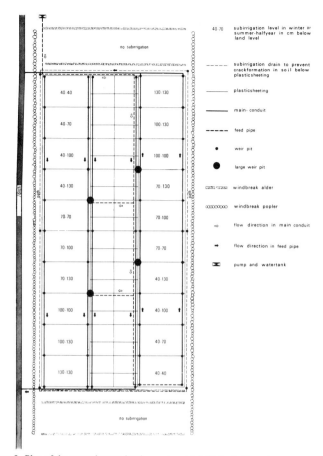

Figure 2. Plan of the ground water level experimental field for fruitgrowing

RESULTS

The Subirrigation Levels
The various subirrigation levels were easy te regulate during the life of the experimental field. No great changes in the subirrigation level took place in the drainage trenches.

The Ground Water Levels
The ground water levels were always measured in the middle of the field (drain interval 4 m). The ground water levels mid-way between two subirrigation drains shows that with deeper subirrigation levels the ground water levels during the periods with precipitation (winter) or evaporation (summer) always deviate from the subirrigation level, which must be ascribed to the pronounced reduction of permeability with increasing depth, table 3.

Table 3. Average winter- and summer ground water levels for the plots in the experimental field.

winter	Subirrigation depth, cm, during summer				
	40	70	100	130	no subirr.
40	40-40	40-60	40-120	40-160	
70		65-80	65-120	65-160	
100			80-120	80-160	
130				90-160	
no subirr.					95-200

Soil Physics In General

In a ground water level experimental field the influence of the ground water regime on the soil structure plays a major role. In spelling, it may be said that the soil structure of the wettest plots degenerates, while that of the driest plots will improve. This latter fact is particularly important, since the experimental field was laid out on a young soil which, in terms of soil physics, was still "ripening". Furthermore, the arrangement and the maintenance of the field were such that the influence of the ground water regime on the soil structure could be investigated under the conditions of intensieve use by wheeled transport (grass strip) and no transport (tree strip).

Subsidence

The degree of subsidence of the soil during the life of the experimental field was directly proportional to the depth of the summer ground water level, figure 3, ground water depth in winter is not important.

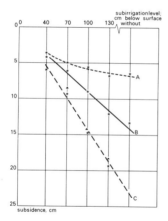

Figure 3. Relationship between subirrigation level in the summer and the subsidence 4,9 and 12 years after starting the experiment.

Crackformation

Crackformation was recorded in profile pits 9 years after start. Cracks are deeper and more intensive with deeper ground water levels. Soil compaction occured at the 0.40 m plot, figure 4.

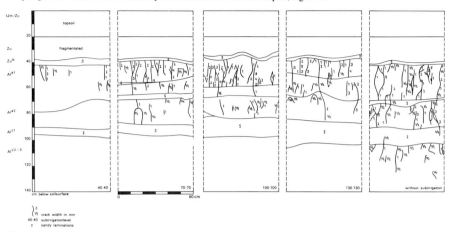

Figure 4. Pattern of cracks under the grass strip with various ground water regimes 9 years after setting up the ground water regimes, Spring 1974.

Soil Structure

On top of the prismatic structure with cracks we find a normal soil structure. From the profile undisturbed soil samples are taken with copper rings to determine pF-curve. Visser 1977, has shown that the contact area between air and water in the soil can be calculated from the pF curve in the range from - 0 to + 2. This is expressed in soil aeration capacity (SAC) with dimension cm2/cm3.
In figure 5 the depth variation on the soil structure in SAC values is given for soils without compaction (A), with compaction (B) and in drain trench (C) for different ground water regimes.
Remark the effect of compaction and the relative high values in the drain trench. If we compare the initial state at start of the experiment we see that without compaction soil structure of the top soil improves to an equal extent under all ground water regimes.
The passage of machines caused consolidation of the soil. The iron roller mounted behind the spray-tank caused the soil to be compacted over practically the whole width of the grass strip with the advantage rutting was prevented. The SAC in the top-soil layer had fallen considerably.
The soil structure in the drain trench tallied very well with those in the topsoil at start of the experiment. From this we can conclude that after 10 years below the ground water level the soil structure did not change.

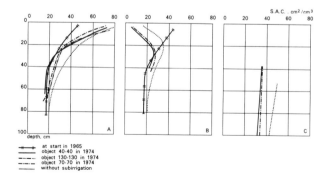

Figure 5. Depth variation of the soil aeration capacity in the original condition (1965) and in 1974 with various ground water regimes.
A below mulch strip; B below grass strip; C drainage trench.

Hydraulic Conductivity in Drain Trench
The drain trench on each plot was situated centrally beneath the grass strip. The drain trench was refilled with top soil with a normal structure directly on the drain and with excavated soil.
The hydraulic conductivity (K-factor) was calculated with the aid of "Kopecky rings": thin-walled rings, 7.3 cm in diameter and 20 cm long. By forcing the rings vertically about 12 cm into the soil, an undamaged soil sample is obtained, from which the hydraulic conductivity can be determined.
Just above drain level (former top soil) there is no reduction in hydraulic conductivity for the plot with subirrigation compared with the plot without subirrigation, figure 6. The K-factor is on a average 3 m a day, which is a normal value for this soil structure.

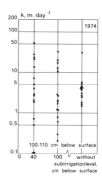

Figure 6. Relation between the subirrigaton level and the hydraulic conductivity measured in the drainage trench at a depth of 1.30 m. 12 years after start of the experiment.

The Nitrogen Economy In The Soil
The various ground water regimes in an experimental field exert influence on the natural fertility by the formation of inorganic nitrogen from organic (nitrogen mineralization).
Inorganic nitrogen is formed under both anearobic and aerobic conditions. The final product if this bacterial degradation of organic matter is ammonia (ammonification) under anaerobic conditions, and nitrate (nitrification) under aerobic conditions. Changes in environment cause one process to pass into the other. Following a change-over from aerobic conditions to unaerobic conditions, the nitrate is reduced and volatile nitrogen compounds are formed (denitrification). The better aerated the soil, the greater the amount of nitrate formed and the smaller the degree of denitrification.
Other studies on ground water level experimental field on clay for arable farming (Van Hoorn, 1958), on sandy loams to clays for arable farming (Sieben, 1974), and on river clay and peaty soil for grassland (Minderhoud, 1960), have also shown that the soils produce more nitrogen as the ground water depth increases.
On the experimental field the grasstrip was periodically harvested and its nitrogen and dry-matter contents were determined.
The deeper the ground water regime became, the greater the amount of nitrogen formed in the soil, down to a subirrigation depth of 1.30 m. Averaged over the various years, the amount if nitrogen taken up by the grass on the "no-subirrigation" plot was significantly lower than that on the plot with a subirrigation depth f 1.30 m, figure 7. This must be caused by "drought" in summer.

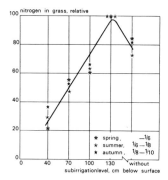

Figure 7. Relationship between the subirrigation level and nitrogen taken up by the grass (% of the value of the 130-130 plot) in some periods during the growing season, average for 1967 to 1972.

Root Development

Root development was recorded in profile pits. The roots intersecting the wall of the profile pit were graded according to thickness and quality.
The amount of living roots in the topsoil is negatively influenced by the subirrigation level, figure 8A. The reason for this is compaction. Without compaction root density increases with undeeper ground water levels, Visser, 1983. In the subsoil there is a distinct influence of the subirrigation level in winter, figure 8B.

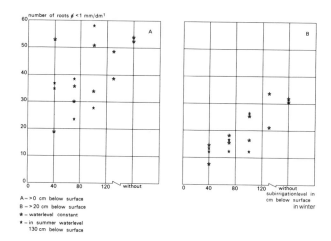

Figure 8. The relation between the subirrigation level and the amount of roots per dm profile wall for the topsoil (A) and (B). Golden Delicious/M9, 9 years after planting.

Crop Response

To distinguish the direct effects of the ground water level on fruit trees and the indirect influence of the soil fertility, three doses of nitrogen were used ON, IN and 2N. The ON areas did not receive nitrogen, the IN areas received the quantities of nitrogen normally used in practice (basic dose). The 2N areas were supplied with twice the basic dose. The basic dose varied in the different years. Between the 6 and 10th growing season basic dose was 75 kg N. ha^{-1}.year^{-1}.

Table 4. Growth and production of Cox's O.P. (A) and Golden Delicious (B) on rootstock 9 in relation to ground water depth and nitrogen treatment. Averaged over the 6th-10th growing season.

Treatment	A shoot growth (m.tree^{-1})	A yield (kg tree^{-1})	B shoot growth (m.tree^{-1})	B yield (kg tree^{-1})
ON				
40- 40	37	17	50	32
70- 70	78	25	85	42
100-100	90	26	93	44
130-130	88	30	88	47
no subb-irr	90	28	89	49
40-130	79	26	80	43
IN				
40- 40	76	32	77	46
70- 70	118	32	101	49
100-100	116	33	108	47
130-130	103	31	97	48
no subb-irr	103	32	104	51
40-130	108	30	89	47

Crop response on subirrigation is clearly on the unfertilized subplots, table 4. This is strongly due to mineralization effects. With nitrogen fertilizer (IN) there is at Cox's O.P. and Golden Delicious reduced growth at the ground water depth of 0.40 m.
Production on fertilized plots is rather independent of ground water depth. This is caused by effects of growth on fruit set above certain levels.

DISCUSSION

The main point in the discussion is what these results indicate for other situations.
One should be aware that this soil is formed from subaqueous sediments of clay with a high water content. Due to evapotranspiration in periods with a rainfall deficit the water content of the sediment decreases irriversibly. As a result the surface will subside and crackformation starts. The mineralogical composition of the clay fraction (most illite) makes that the soil does not swell after rewetting.
Schematicly the soil consists of a top soil with a normal soil structure caused by frost, biological activities and human activities. Under the top soil (0.40 m) there is a layer with soil prismatic shaped colums, seperated by cracks. Below the cracks the soil is unaerated with a very low permeability. Water transport to open-field drains and tile-drains is possible by the cracks.I pointed out the special situation of this Ysserlmeerpolders because most coastal and river clay soils are sedimented in very thin layers in flood periods. Dehydration and existing vegetation form soil structure in the dry periods.

Taking this all in mind the following conclusions can be made.

1. Relation Subsidence - Ground Water Depth
   This linear relationship is found with ground water depth in summer. Ground water depth in winter is not important. This relation is only valid for soils with an excess of water.

2. Relation Crackformation - Ground Water Depth
   Crackformation increases with deeper ground water levels in summer. Ground water depth in winter is not important. This relation is also only valid for clay soils with an excess of water.

3. Relation Soil Structure Top Soil - Ground Water Depth
   The relation of the soil structure with ground water depth is governed by soil compaction, caused by human activity. Influence of ground water depth on compaction was found at undeep ground water levels. Without compaction (under the trees) there is no influence at all of ground water depth on the soil structure of the top soil. This relation is also valid for other clay soils with similar mineralogical composition.

4. Relation Soil Structure in The Drain Trench - Ground Water Depth
   There is no influence of the ground water depth on the soil structure in the drain trench.

The soil structure in the drain trench is similar to the soil structure of the top soil. This is because top soil was put on the drain. The soil structure was highly resistent. After 10 years under ground water level there was no major change. Changes were due to setting, therefore permeability decreased to a level of 3 m a day.
This relation is also valid for other clay soils with similar mineralogical composition.

5. Relation Nitrogen Economy - Ground Water Depth
   This relation has an optimum value at a ground water level of 1.30 m. The type of relation is also valid for other loam and clay soils. The level of natural fertility however will be different in other soils.

6. Root Development - Ground Water Depth
   The rootsystem is strongly influenced by the ground water level. The direct influence is rooting depth. The indirect influence is compaction of the topsoil at undeep ground water levels. This relation is also valid for other soils.

7. Crop Response on Ground Water Depth
   Crop response is without nitrogen fertilization rather strong. With nitrogen fertilization the effect in yield is minimal. There is only a reduced growth on the 0.40 m ground water depth.
   Of course for other soils with other crops reactions will depend on many factors. Basicly however the relations presented are also valid for other crops and other loam and clay soils, (Van Hoorn, 1958, Sieben 1974, Minderhoud, 1960).

REFERENCES

1. Hoorn, J.W.van, 1958. Results of a ground water level field with arable crops on clay soil. Neth. J Agric. Sci. 6. 1-10.

2. Minderhoud, J.W., 1960. Growth of grass and ground water Level; Studies into the significance of the ground water level for basin clay grassland. In Dutch. English summary. Thesis. Wageningen.

3. Schultz, E., 1988. Drainage measures and soil ripening during the reclamation of the former sea bed in the Yselmeerpolders. Proc. 15th European regional conf. on Agric. Water Management ICID. Dubrovnik.

4. Sieben, W.H., 1974. Influence of drainage on the supply of nitrogenand the yield of young sandy-clay soils in the IJsselmeer Polders. In Dutch. English summary. Van Zee tot Land no. 51. Rijksdienst voor de IJsselmeerpolders, Lelystad.

5. Visser, J., 1977. Soil aeration capacity, an index for soil structure, tested against yield and root development of apple trees at various soil treatments and drainage conditions. Plant and Soil, 221-237.

6. Visser, J., 1983. Effect of ground water regime and nitrogen fertilizer on the yield and quality of apples, 266 pages, in English. Van Zee tot Land 53. Rijksdienst voor de IJsselmeerpolders, Lelystad.

7. Visser, J., 1991. Some results of subirrigation in the IJsselmeerpolders in the Netherlands. Intern. conf. on subirrigation and controlled drainage, Lansing, Michigan, U.S.A.

8. Visser, J. and H. Jorjani, 1987. Results of a ground water level and nitrogen fertilizer field experiment with apple trees on a marine clay soil. Plant and Soil 104, 245-251.

AGROFORESTRY SYSTEMS FOR ON FARM DRAIN WATER MANAGEMENT

G. S. Jorgensen*   K. H. Solomon*   V. Cervinka*
                     Member, ASAE   Member, ASAE

ABSTRACT

The agroforestry concept for drain water management is to irrigate salt-tolerant trees and other crops with agricultural drain water in order to concentrate the salts and reduce the volume of drain water which must ultimately be treated or disposed of. At an agroforestry demonstration site near Mendota, California, water with an electrical conductivity averaging 10 dS/m was applied to *Eucalyptus camuldensis* and other salt tolerant trees. Water not consumed by evapotranspiration and percolating below the root zone was recaptured by a tile drain system. The concentrated water was applied to *Atriplex canescens* and other halophytes that can utilize poor quality water and further reduce the drain water volume. Drain water collected from these fields was placed in above-ground evaporation tanks, where the salts and other chemical constituents can be removed from the farming system. The economic contribution of agroforestry to the farm depends on the value of the biomass produced for the local marketplace, and on the avoided costs of even larger evaporation ponds and treatment facilities that would otherwise be necessary.

BACKGROUND

Drainage and Salinity Problems

Some agricultural lands within California's western San Joaquin Valley suffer from waterlogged soils or rising water tables. Nearly 350,000 hectares are currently affected by a high water table (within 1.5 meters of the ground surface), and 400,000 hectares could be affected by the year 2000 (San Joaquin Valley Drainage Program, 1990). Compounding these problems is the fact that soils of the Valley's west side are derived from marine sediments, often containing elevated levels of salts and other elements found in sea water (arsenic, boron, and selenium). As this land is irrigated, these elements are dissolved and become constituents of the drainage water. And even with the high quality water provided by the state and federal water projects, imported total salt (1,600,000 tons/year**) and associated trace elements (boron, selenium, others) are major problems.

---

* G. S. Jorgensen, Field Research Manager, and K. H. Solomon, Director, Center for Irrigation Technology, California State University, Fresno, California, and V. Cervinka, Research Manager, California Department of Food and Agriculture, Sacramento, California.

** Assumes 4.2 billion cubic meters of water imported annually (firm water supply) and an average salinity of 350 mg/L TDS (San Joaquin Valley Drainage Program, 1990).

The San Luis Unit of the Federal Central Valley Project (CVP) and the State Water Project (SWP) began providing water to 400,000 hectares in west Fresno County in 1968. As part of the San Luis Unit, a master drain was mandated to move agricultural drainage water to the Sacramento/San Joaquin Delta. Construction of the drain began in 1968, and it eventually collected drain water from about 18,000 ha. But due to funding problems and environmental concerns over the effects of drain water discharge into the Delta and San Francisco Bay, the San Luis Drain was never completed. It was terminated near Los Banos at a regulating reservoir, Kesterson, which had become part of a national wildlife refuge.

Reproductive abnormalities and deaths of aquatic birds were discovered at Kesterson National Wildlife Refuge, and were found to be related to the trace element selenium, a contaminant associated with agricultural drain water in the area. The drains emptying into the San Luis Drain were plugged in 1986 by order of the Secretary of the Interior. Kesterson has since been covered, and lands once served by the San Luis Drain no longer have an outlet for agricultural drainage water. Growers in the affected area must now manage the drain water on-farm.

The drought in California the past 6 years has intensified the water quality aspect of the drainage problem. The water supplies in the CVP and SWP reservoirs have been insufficient to supply the normal allotment to many users. To make up the deficit, many growers are relying on poorer quality ground water, recirculating the salts and other elements back into the crop rootzone.

Drain Water Management

Several options exist for growers and drainage districts to manage agricultural drain water. Among these options are: source control, ground water management, drainage water treatment, and drainage water re-use, which includes agroforestry.

Source Control - This involves minimizing the contribution of irrigation water to the underground water table by improving irrigation systems and present irrigation management practices. Estimates (San Joaquin Valley Drainage Program, 1990) of the amount of water added to the underground water table from unnecessary deep percolation range from 0.18 to 0.23 meters/year (assuming 0.09 meters/year are required for leaching).

Ground Water Management - In areas where relatively good quality shallow ground water exists, this water could be extracted and applied directly to crops, blended with high quality surface water, or the ground water levels could be managed to facilitate direct crop uptake and use of this water.

Treatment - Various treatment options have been studied, and while some have shown promise, none are yet feasible. Treatments include bacteria based processes, microbial volitization, geochemical immobilization, heavy metal absorption with iron filings, ion exchange, and reverse osmosis. The treatment systems examined suffer from either poor economics or difficulty in maintaining sustained operation of the equipment and systems.

THE AGROFORESTRY CONCEPT

Agroforestry is a biological system for managing agricultural drainage water (Cervinka, 1990). It represents a crop integrated approach where increasingly saline water is applied to successively more salt tolerant plants. As the salt concentration increases, the volume of drainage water requiring ultimate treatment or disposal is decreased (see Fig. 1, from San Joaquin Valley Drainage Program, 1990).

For example, good quality water is applied to a salt sensitive crop such as carrots. The drain water from below this crop is captured by a subsurface tile drain system, and when feasible, applied to a more salt tolerant crop such as cotton. The ensuing drainage water, now further reduced in volume and concentrated in terms of salt and other elements, is then applied to salt tolerant trees such as eucalyptus. The drainage water is again captured and applied to halophytes such as saltbush (*Atriplex canescens*). The resulting highly concentrated drain water is captured yet again for final disposal in evaporation ponds, deep well injection, or further treatment. As the cost of transport and some treatment options are volume

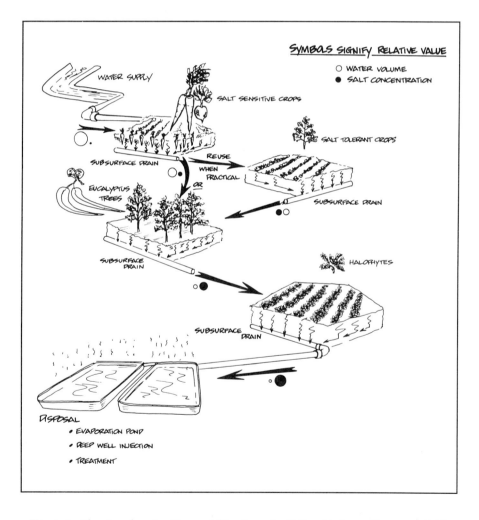

Figure 1. The Agroforestry Concept (San Joaquin Valley Drainage Program, 1990)

sensitive, the cost of the ultimate disposal of the drainage water may be reduced by employing agroforestry. The agroforestry system offers a management disposal option that is less problematic than large scale evaporation ponds, which may be classed as toxic sites, with potential benefits for wildlife habitat.

In particular circumstances, additional benefits may be possible. Even though selenium in high concentrations is toxic, it is required in lower concentrations to maintain the health of some animals. Many cattle on the east side of the San Joaquin Valley, and elsewhere throughout California and the West, require selenium supplements to maintain adequate blood levels of the element. The atriplex grown in the agroforestry system has been shown to accumulate the element selenium, and feeding trials have shown that atriplex hay is capable of maintaining required blood selenium levels in forage cattle (Frost, 1990).

AGROFORESTRY RESEARCH

Agroforestry plantings at various test sites were initiated in 1985 in a cooperative research effort of the USDA-SCS and the California Department of Food and Agriculture, Agricultural Resources Branch. Experimental plantings are being investigated for the following applications: to intercept drainage water flowing out of high ground into the valley; to lower shallow ground water (passively); and as a method of managing agricultural drain water produced from crop lands (Cervinka, 1990). Extensive work is underway to genetically improve the eucalyptus trees, and to identify other promising salt tolerant atriplex and grass species. Trees are selected for better salt and trace element tolerance from plantations throughout the San Joaquin Valley and propagated by cloning. Cloning provides an exact replica of the mother plant, and reduces the variability in trees produced from seed.

Agroforestry Plantation, Mendota, California

The most intensively monitored agroforestry plantation is a site near Mendota, California, 75 kilometers west of Fresno. The site consists of 9.43 hectares of *Eucalyptus camuldensis*, planted in 1985 and 1986, and 2.02 hectares of halophytes (Fig. 2). Data collected at the site includes chemical composition of the ground water, depth to ground water, soil moisture status via tensiometers and neutron probe, volume and chemical composition of drain water applied to the trees, composition and volume of the ensuing drain water from the trees that is collected and applied to the atriplex, Sodium Absorption Ratio (SAR) of the soil, tissue analysis of the trees and plants, and the composition of the salts collected in the evaporation basins.

The experiment is continuing. However, a summary of what has been learned thus far is given below.

Permeability of the soils is restricted physically by a high clay content and chemically by a high sodium content in the top layers. To help overcome this, irrigation water was not applied in discrete events; rather, water during the irrigation season was applied almost continuously, by adding water to a series of basins. Additional water was applied during the winter months. In soils where permiability is restricted, applications during the winter may be necessary to provide sufficient leaching to maintain salt balance in the soil profile under the trees.

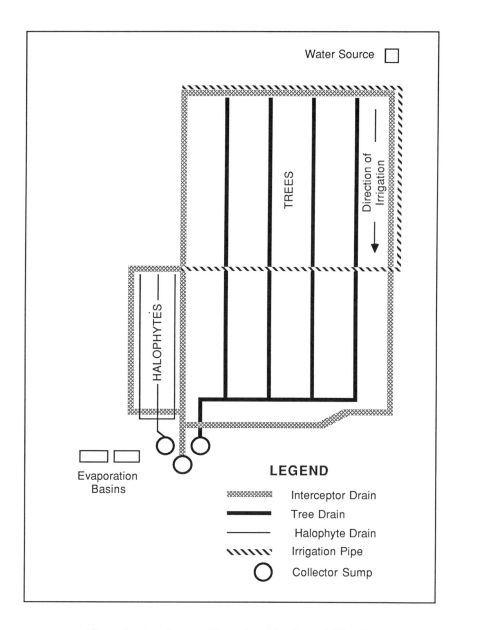

Figure 2. Agroforestry Plantation, Mendota, California

Saline agricultural drain water from the Westlands Water District collector system (average EC of 10 dS/m, 12 mg/L boron, 400 µg/L selenium, SAR of 11 mM/L$^{1/2}$) was used to irrigate the trees. The site has three separate subsurface drain systems: a perimeter drain around the entire plantation to intercept flows into the planting; a system under the trees to collect any water passing beneath the rootzone of the trees;

and a system beneath the atriplex and halophyte planting. Individual sumps with water meters measure the volume of drain water from each drain system.

Research Results and Observations from the Mendota Site

Concentrating Effect - Drain water collected from under the eucalyptus trees is reduced in volume and more concentrated. The reduction in water volume causes increases in drain water characteristics as follows: EC, 3.2 times; SAR, 6.3 times; boron concentration, 4.2 times; and selenium concentration, 1.8 times.

Soil Salinity - Irrigation during the 1987 through 1989 growing seasons was done by the farm cooperator, and was subject to the availability of irrigation labor not otherwise occupied on the farm. The amount of water applied during these years met neither tree water needs nor leaching requirements, resulting in a dramatic rise in soil salinity and eucalyptus leaf tissue boron concentrations (in excess of 2000 mg/L). Soil ECe (averaged over the top 2.4 m of the soil profile) increased from just above 10 dS/m tin 1987 to just under 30 dS/m at the beginning of the 1990 season. During the 1990 irrigation season, adequate drain water was applied to supply the needs of the trees plus a 16 percent leaching fraction, after which the average soil ECe fell to 25 dS/m. Additional leaching accomplished by irrigating during the fall and winter months of low evaporative demand was successful in lowering the ECe to 18 dS/m by June 1991 (Tanji, 1992).

Tree Water Use - During the 1990 season, evapotranspiration from the trees was estimated using a Bowen ratio energy balance method. Utilizing data from the California Irrigation Management Irrigation System weather station located nearby, a crop coefficient (Kc) of 0.84 was derived (Tanji et al., 1990). Although some sources suggest that trees irrigated with good quality water under optimum conditions may have Kc values in the range of 1.2 or higher (UC Cooperative Extension, Undated), the trees in the study transpired significantly less, perhaps due to increased soil salinity levels.

Plans - The plantation was harvested and sold as chips for biomass during the summer of 1992. The original planting consisted of rows spaced 1.5 m apart, and was modified to provide 3 m row spacing by completely removing alternate rows. The wider spacing will allow for mechanical cultivation which will benefit both weed control and water penetration. The remaining trees will be allowed to regrow, and will be studied to determine water use rates and regrowth rates. A portion of the planting was removed completely, and replanted with trees that have been selected for their superior performance since the advent of the agroforestry program.

Rootzone Salinity Management - Irrigation scheduling will be critical to the success of agroforestry. On low infiltration rate soils, the rootzone may be used as a seasonal salt storage area. During the irrigation season, adequate water would be applied to meet the needs of the trees. But without excess water for leaching, the soil salinity will increase. During periods of low evaporative demand, for example winter, irrigations would then continue, in order to leach the accumulated salts from the rootzone.

The Importance of Agroforestry and Research Funding - Agroforestry research has suffered from insufficient funds. This is surprising as well as disappointing, since the San Joaquin Valley Drainage Program study identified agroforestry as playing a major role in the management of drainage water within the Valley. All of their considered and recommended policy scenarios require substantial areas of agroforestry. Depending on the scenario, estimates of the area under agroforestry range from 8,500 to 13,000 hectares by the year 2000, and from 13,700 to 21,800 hectares by 2040 (San Joaquin Valley Drainage Program, 1990). Yet research funds available have not been adequate to meet the identified needs of the agroforestry research program, and have in fact been declining for the past few years. More work is required on tolerance of the tree and halophyte tolerance to salinity and elements such as boron, on tree selection and propagation, on factors affecting the salt and water balance of the system, and on disposal or use of the remaining effluent and salt.

Economics - The potential value of possible agroforestry products (biomass, honey, essential oils, etc.) will depend on local markets, and may not be high. The biomass produced to date at the Mendota site has neither sufficient volume nor appropriate characteristics to be of any value in the local market. If, however, the agroforestry concept can provide a means of on-farm drain water management, thus alleviating the need for expensive and potentially hazardous evaporation ponds, then perhaps maintaining the San Joaquin Valley's west side as a viable farming region will be sufficient economic justification.

REFERENCES

1. Cervinka, V. 1990. A farming system for the management of salt and selenium on irrigated land (Agroforestry). California Department of Food and Agriculture, Agricultural Resources Branch, Sacramento, CA, May 1990, 17 p.

2. Frost, B. 1990. Atriplex tested as feed option. San Joaquin Experimental Range Newsletter, Spring 1990, California Agricultural Technology Institute Pub, No. 900304, California State University, Fresno, California, pp. 1, 3.

3. Karajeh, F.K. 1991. A numerical model for management of subsurface Drainage in agroforestry systems. PhD Dissertation, Univ. of Calif., Davis.

4. San Joaquin Valley Drainage Program. 1990. A management plan for agricultural subsurface drainage and related problems on the Westside San Joaquin Valley, Final Report, 1990. E. Imhoff, Program Manager, Sacramento, CA 183 p.

5. Tanji, Kenneth K. 1992. Will agrofrestry help solve drainage problems? USCID Newsletter, January 1992, No. 65, US Committee on Irrigation and Drainage, Denver, Colorado, p. 4.

6. Tanji, Kenneth K., S. Grattan, A. Dong, F. Karajeh, A. Quek, D. Peters, D. Johnson, and G. Jorgensen. 1990. Progress report on water and salt balance, agroforestry demonstration program. California Department of Food and Agriculture, Agricultural Resources Branch, Sacramento, CA, October 1990.

7. UC Cooperative Extension. Undated. Using reference evapotranspiration (ETo) and crop coefficients to estimate crop evapotranspiration (ETc) for trees and vines. Leaflet 21428, University of California, Division of Agriculture and Natural Resources, Cooperative Extension, Berkeley, California.

HYDROLOGICAL IMPACT OF FARM WATER MANAGEMENT ALTERNATIVES IN THE CANADIAN PRAIRIES

N.D. MacAlpine   D.W. Cooper   R.D. Neilson

ABSTRACT

A five year study on six farm scale applied research projects in central and northwestern Alberta, Canada, compared the benefits and impacts of retaining runoff on-farm to uncontrolled field drainage. A team of consultants monitored the hydrology, agronomics, economics and wildlife on the demonstrations. The study concluded that farm water management systems can reduce off-farm peak flowrates close to pre-drainage conditions. The snowmelt event produced higher peak flowrates and runoff volumes than rainfall events. The Soil Conservation Service (SCS) Curve Number Method seriously over-estimated peak flowrates because the "time of concentration" formula did not account for the significant depressional storage available in small, poorly drained watersheds in the northern Great Plains. The SCS Method did provide accurate estimates of rainfall runoff volumes. Regional data when calculated as unit area flowrates and runoff volumes provided acceptable estimates of farm scale values. **KEYWORDS.** Hydrology, Water management, Surface drainage, Subsurface drainage, Ponds.

THE NEED FOR ALTERNATIVES TO UNCONTROLLED DRAINAGE ON THE PRAIRIES

Alberta's semi-arid climate, short growing season and the 40 percent of annual precipitation that arrives as snow, makes water conservation and removal of excess water for crop protection a dual concern for Alberta's farmers. The glaciated landscape of Alberta's plains produced an immature natural drainage network which makes the development of off-farm drainage systems expensive.

The province of Alberta, Canada, commissioned a five year study, completed in March, 1992 to investigate the impacts of conventional, uncontrolled farm drainage and explore alternative concepts. Six farm scale demonstrations, in central and northwestern Alberta (Figure 1), were monitored by a team of hydrologists, wildlife biologists, economists, agricultural engineers and agronomists. The six demonstrations were: uncontrolled field ditch drainage, controlled ("choked") field ditch drainage, water consolidation to dugouts (farm ponds) in a rotational grazing system, two sites with water consolidation to waterfowl habitat ponds and spring backflood irrigation for peatland water management. This paper focuses on the hydrologic aspects of this study.

The demonstrations illustrate measures that limit the downstream impact of removing excess water or develop excess water into an on-farm resource for agriculture or wildlife. The agricultural production practices at each site strongly influenced the design concepts implemented.

N.D. MacALPINE, Farm Water Management Engineer, and R.D. NEILSON, Section Head, Alberta Agriculture, Conservation & Development Branch, 206, 7000-113 St., Edmonton, Alberta, Canada. T6H 5T6; and D.W. COOPER, Hydrologist, W-E-R Engineering Ltd., 306, 2735-39th Ave., Calgary, Alberta Canada. T1Y 4T8.

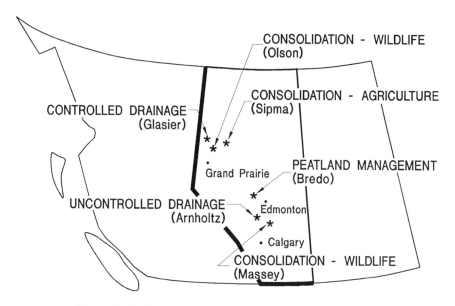

Figure 1. On-Farm Water Management Study Project Sites

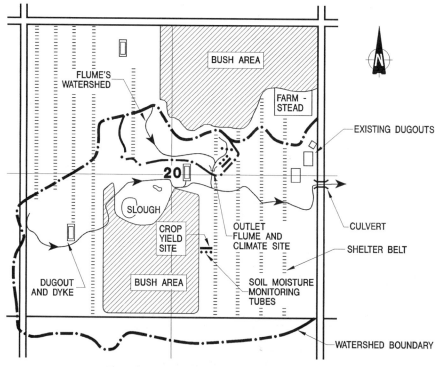

Figure 2. Consolidation - Agriculture (Sipma)

Demonstration Descriptions

The climate for all sites is sub-humid continental. Mean temperatures range from -20°C in January to 17°C in July. Annual precipitation ranges from 374 mm in northwestern Alberta to 517 mm in central Alberta. Average frost free days range from 78 to 118 days. Upland soils on these sites are Black Chernozems in the prairie grassland regions and Dark Grey Luvisols in the aspen forest regions. Depressional soils are Orthic Humic Gleysols and on the peatland site, Terric Mesisols.

The uncontrolled field drainage site (Allan Arnholtz, cooperator) is 50 km south of Edmonton. Gently rolling cereal and alfalfa uplands (16 ha) and 2.65 ha of level depressions are drained by three field ditches (sideslopes 10H:1V, bottom width, 3 m) to a treed retention area in the middle of the farm. The uncontrolled flows in the field ditches above the retention area were monitored. Overflow from the retention area drains through a control culvert to an intermittent watercourse.

The controlled drainage site (Harvey Glasier, cooperator) is in northwestern Alberta near the Peace River. A controlled drainage watershed of 78 ha and an uncontrolled basin of 254 ha is drained by field ditches (8H:1V, 3 m bottoms). The controlled basin has three control ("choke") structures on the field ditches to regulate runoff from undulating cereal land with 13 poorly drained depressions. The three control structures are a plywood weir with a low flow orifice, a vertical slot structure and a rock filled gabion basket with a 100 mm diameter perforated tubing for residual flow.

The Klaas Sipma project in northwestern Alberta demonstrates consolidation of snowmelt runoff where the water is stored and used for a cow-calf operation. On a plateau between the junction of two major rivers, the Peace and the Smoky, 200 ha of gently undulating rotational grazing pasture is drained by a field ditch that fills two pasture dugouts (farm ponds), 3400 m^3 and 5400 m^3 respectively (Figure 2). The central pasture dugout has an additional 3400 m^3 of storage if snowmelt is pumped over the dikes that surround the dugout. Three dugouts at the farmstead and a third pasture dugout, 3400 m^3, complete the water supply for 150 beef cows with calves and the Sipma family. Typical of the northwestern Alberta projects, 2 km east of the Sipma farm, the farm's intermittent outlet watercourse drops 215 metres at a slope over 10% down to the Smoky River.

Two projects designed by Jensen Engineering of Olds, Alberta, demonstrate slough consolidation with mitigation for waterfowl. The Andy Olson project in northwestern Alberta uses both field ditches and subsurface drain tubing to drain 23 temporary, seasonal and permanent wetlands to a consolidation pond that has a mixture of wetland types in order to replace some of the wetland habitat lost due to drainage. The field ditch system drains 7.2 ha of depressions and the random subsurface drainage system drains 4.4 ha of a 65 ha field producing cereals or oilseeds. Jensen Inlets are used to avoid the freezing problems experienced when gravel inlets are used to deliver snowmelt down to the drain tubing. The inlets consist of a vertical slotted pipe with a sliding collar that covers the lower 0.3 m of the pipe's slots at the start of snowmelt. The collar forces the snowmelt to pond and infiltrate as the frozen soils thaw. Once infiltration has slowed, the collar is raised so that the residual water flushes down to and through the drain tubing. Water from the subsurface drains and the field ditches is delivered to two separate automated pump stations. The sumps pump the water up into the diked consolidation pond. Maximum storage in the pond is 21,400 m^3. The pond and its upland area is 2.8 ha. Since wild waterfowl require a variety of wetland types for mating, feeding and rearing young, the pond has a mixture of nesting islands, temporary water (0 to 0.3 m deep), seasonal to semi-permanent water (0.3 to 2.0 m), and permanent water (>2.0 m). A slow release line, draining 0.6 km northwest to a tributary of the Peace River, gradually draws down the pond to permanent water levels.

Derwin Massey's project is similar to the Olson project. The cereal and oilseed farm is located 160 km southeast of Edmonton in a knob and kettle moraine with no off-farm outlets. A subsurface drainage system with Jensen Inlets drains 18 depressions, occupying 10.5 ha on 52.5 ha, to an automated pump station. The sump pumps the water up into two interconnected sloughs in the southeast corner of the field. The sloughs can store 16,200 m^3. Three nesting islands were built and deadwood cleared to enhance waterfowl nesting around the sloughs.

A peatland water management project (Gordon Bredo, cooperator) is 130 km northwest of Edmonton. The farm's gently rolling topography is under hay and rotational grazing pasture. The farm's watershed (88 ha) drains into 30 ha of deep peats. A field ditch (4H:1V, 3.0 m bottom width, 1.0 m deep and 1800 m long) removes runoff and lowers the peat's watertable. The ditch delivers water to a 4000 m^3 dugout. At the downstream end of the peatland, a small dike and stoplog structure with a 1 m diameter culvert can flood 5 ha of the peatland at snowmelt.

HYDROLOGIC OBJECTIVES OF THE STUDY

* To provide more reliable farm watershed data for designers and planners.
* To improve the understanding of small watershed hydrology in the Northern Great Plains.
* To evaluate the hydrologic and hydraulic effectiveness of each of the farm water management techniques and recommend improved design methods where appropriate.
* To identify and evaluate on-farm and off-farm impacts of the farm water management techniques by comparing their runoff volumes and peak flowrates to pre-drainage conditions and to uncontrolled surface drainage.

MONITORING METHODOLOGY

Each site had a climate station equipped for automatic data recording of daily temperatures, relative humidity, rainfall including intensity and duration. Two sites also had solar radiation and wind speeds recorded for three years. One site (Arnholtz) was specially instrumented for 2 years to provide direct evapotranspiration estimates on a watershed basis.

Runoff on surface drained sites was measured by "H" flumes with Stevens recorders. One site (Olson) used a cipoletti weir for surface runoff measurements. Subsurface drained sites recorded outflow from the sump pump operations.

Cumulative snowfall was measured on four sites. Snowcourse measurements were conducted just before snowmelt. Soil moisture was measured during the growing season. Runoff measuring, especially at snowmelt, required frequent on-site visits to record the sequence and complexity of runoff. Regional weather and runoff data were compared to site data to establish the return periods of the local events. Detailed mapping of the watersheds and surveys of the fields' depressions and the pre- and post-construction drainage routes were necessary for flood route modelling of snowmelt and summer runoff through the depressional storage areas. Pre-drainage and uncontrolled surface drainage scenarios were compared to the effects of the constructed on-farm systems to predict on-farm and off-farm impacts. The off-farm impacts were assessed for a maximum off-farm watershed of 10 sq. km. Beyond that area, it was assumed that a retention area would be used to reduce 1:10 and 1:25 year design peaks to pre-drainage conditions.

RESULTS AND DISCUSSION

While only 2 to 4 years of data were collected on these sites, some hydrological results consistent across all sites were observed. Table 1 compares runoff volumes and peak flowrates from the actual concept constructed and the scenario of uncontrolled drainage to pre-drainage conditions on each site.

TABLE 1. Hydrologic Impact Assessment (1:10 year to 1:25 year flow event)

Site	Drainage Condition	Estimated Increase in Flood Events versus Pre-drainage Conditions (%)			
		Snowmelt Volumes	Snowmelt Peaks	Rainfall Volumes	Rainfall Peaks
Arnholtz	As-built uncontrolled	10-20	-	50-180	15-30
	Improved uncontrolled	20-35	-	85-240	30-50
Glasier	As-built controlled	10-80	10-50	130+	15+
	Uncontrolled	200+	100+	200+	40+
Sipma	As-built consolidation	0	0	No rainfall runoff	
	Uncontrolled	15-20	10-20	No rainfall runoff	
Olson	As-built consolidation	20-30	2-4	20-30	2-4
	Uncontrolled	30-60	40-50	30-60	40-50
Massey	As-built consolidation	0	0	0	0
	Uncontrolled	∞	∞	∞	∞
Bredo	As-built peat mgm't	0	0	0	0
	Uncontrolled	18-25	50	-	-

Controlled drainage, regulated flooding (snowmelt flood irrigation) or on-farm storage (consolidation) that provides temporary or permanent storage that is equivalent to the field's pre-drainage depressional storage effectively reduce off-farm peak flowrates to levels before drainage. Without these controls, uncontrolled field drainage will increase downstream peak flowrates from 15% to over 200% for 1:10 to 1:25 year events. Increases to more frequent runoff events will be proportionally larger, although their smaller flowrates and volumes will have less significance downstream. Depending on the degree of development of the off-farm drainage network, the incremental costs in the off-farm system to manage outflows from uncontrolled field drainage systems can range from $6.53 per improved hectare to over $100.00 per improved hectare. The farm water management techniques implemented in these demonstrations have significantly reduced these off-farm costs.

Estimated 1:10 year return period surface runoff peaks from the various study watersheds, under pre-drainage conditions, range from zero at the Massey site to 0.8 to 2 L/s per hectare at the other sites. These values are comparable to the results predicted by flood frequency analyses of nearby regional stations.

The snowmelt peak flowrates measured on all projects were generally higher than the summer storm peak flowrates. The estimated 1:10 year snowmelt peak flowrates are generally slightly larger (25 to 35%) than the estimated 1:10 year summer rainfall peaks. Snowmelt rather than summer storms is the significant event for design of field drainage systems in these flat, poorly drained, small watersheds.

Regional runoff data for snowmelt volumes and peak flowrates are comparable to small on-farm watersheds if they both have comparable watershed characteristics.

Accurately defining the "effective" drainage area was a common difficulty on all study sites. Runoff monitoring was complicated by variations in the areas contributing to runoff events. Defining depressional areas and estimating their storage volumes should be incorporated into the determination of design runoff volumes and on-farm storage requirements.

On small flat farm watersheds, the peak runoff event is frequently hydraulically controlled rather than being a hydrologically predictable event. Snow blockages in a ditch or culvert or restrictions at the high point in a field ditch control runoff on these small watersheds. The overland peak runoff rate is not the critical element in design. The hydraulics of the water management system, the on-farm depressional storage and runoff volume estimates are the design keys.

Random subsurface drainage systems were effective though costly. In view of the deep cuts required for field ditch systems in the rolling topography at the Olson and Massey sites, the subsurface systems were the only realistic alternative in these landscapes. The subsurface drainage systems also integrated well with the pumped drainage to the consolidation ponds. The Jensen Inlets were effective in delivering snowmelt down to the subsurface tubing without freezing problems.

Consolidation systems required extensive storage to maintain off-farm flows at pre-drainage conditions. The Massey project required 310 m^3 of live storage per hectare drained and the Olson project's live storage of 280 m^3 per hectare drained would have doubled without the slow release line. The Sipma site only had 80 m^3 per drained hectare.

Significant soil moisture and depressional storage on all sites resulted in rainfall runoff being negligible until precipitation exceeded 20 mm or more. Significant runoff usually occurs only after two days of rainfall.

The differences in snowmelt runoff between pre-drainage and drained conditions on each site can vary substantially depending on factors preceding the runoff event. Fall soil moisture, the amount of snow present at snowmelt and weather conditions before and during snowmelt strongly influence the snowmelt event. Drained conditions produce much higher runoff in low snowmelt runoff years. Drainage allows more rapid runoff even though low soil moistures the previous fall may have permitted significant infiltration if runoff had been slowed. In high snowmelt runoff years, wet conditions the previous fall result in low infiltration capacity and widespread flooding and ponding. The snowmelt runoff under these conditions is similar under pre-drainage and drained conditions.

Standardized empirical hydrologic formulas, like the Rational Method and the SCS Curve Number Method, can seriously over-estimate design peak flows because of the significant depressional storage available in small prairie watersheds in Alberta. For example, the Kirpich formula used to compute "time of concentration" in the SCS Curve Number Method is 2 to 5 times less than observed values at the surface drained sites. This results in substantially higher predicted peak flowrates compared to the observed rainfall peak flowrates. Modelling that accounts for the depressional storage and the delayed time to peak could be used to estimate peak flowrates. However, the amount of field information and analytical effort required is excessive. Generating runoff peak flows from regional data with some site specific adjustments is more direct and reliable.

The SCS Curve Number Method provided reasonably good estimates of rainfall runoff volumes for long duration storm events when the runoff curve numbers (CN) were adjusted to represent low antecedent moisture conditions (AMC1). A 48 or 72 hour storm combined with the SCS AMC1 curve number could be used to predict 1:5 or 1:10 year rainfall runoff volumes. Using the low antecedent moisture condition (AMC1) is justified based on a review of 269 years of daily records from weather stations near the sites. Only 16 of 201 rainfall events that were equal to or greater than a 1:2 year event had a five day antecedent rainfall exceeding 36 mm.

CONCLUSIONS

On-farm water management when compared to uncontrolled field drainage can significantly reduce the off-farm peak flowrates. Peak flows off-farm can be reduced to near pre-drainage conditions if the on-farm storage from controlled flooding, reservoirs and short term depressional storage is equivalent to the pre-drainage depressional storage. In the poorly developed natural drainage systems of the Canadian prairies, simple low cost on-farm structures for controlled drainage, spring backflood irrigation and large farm ponds and reservoirs, if adopted on a watershed basis, can significantly limit downstream costs and environmental impacts.

On the Canadian prairies, for the 5-year to 10-year return period, the snowmelt event is more significant than summer storm events for peak flowrates and runoff volumes for small water management designs. Flood frequency analyses from nearby regional stations produce more reliable estimates of peak flowrates and runoff volumes for small poorly drained watersheds than do empirical hydrologic formulas. However, the SCS Curve Number method can be used to reliably estimate rainfall runoff volumes if a 48 or 72 hour storm with CN values for "dry conditions" is used.

Accurate estimates of the "effective" drainage area are critical to avoid design and operational problems with spring backflood irrigation or consolidation projects. The maximum area that potentially could contribute runoff needs to be carefully defined taking into account potential blockages from snow or ice. Air photo interpretation, field surveys and mapping can refine this estimate. Identifying depressional storage areas and estimating their volumes also is critical in the design of a small water management system.

MANAGEMENT IMPLICATIONS

The consolidation projects illustrate that pumping to storage is technically feasible and may be preferable to excavating storage for on-farm reservoirs, especially where large volumes of water are needed for intensive operations. Programs that encourage excavating storage for drought protection of on-farm water supplies (dugout programs) should be reviewed to see if large scale pumped storage is more cost effective.

The most successful projects in the study were those that developed a "farm plan". The farmer reviewed the benefits of retaining water on the farm and requested a design that was integrated with the farm's agricultural production system. Some of the farmers in the study indicated they were interested in protecting wildlife habitat but their projects were not designed with this objective. Developing a "farm plan" before selecting a water management technique and a design could have incorporated this objective.

Alberta is encouraging a process with farmers of farm soil and water conservation planning to protect on-farm resources (including wildlife) while reviewing the best use of those resources.

REFERENCES

1. Alberta Water Resources Commission. 1987. Drainage Potential in Alberta: An Integrated Study. Summary Report. Alberta Water Resources Commission, #910, Harley Court, 10045-111 St., Edmonton, Alberta, Canada. T5K 2M5.

2. Jensen Engineering Ltd. 1987. Slough Consolidation and Runoff Retention. The Olson Project. Year 1 Research Project. Jensen Engineering Ltd., 5037 - 50 St., Olds, Alberta, Canada. T0M 1P0. 63 pg. plus appendices.

3. Jensen Engineering Ltd. 1988. Slough Consolidation and Runoff Retention. The Massey Research Project. Jensen Engineering Ltd., 5037 - 50 St., Olds, Alberta, Canada. T0M 1P0. 46 pg. plus appendices.

4. MacAlpine, N.D., and G. Shaw. 1984. Preliminary Selection and Design Criteria for Pothole Consolidation. Can. Soc. Agric. Eng. Paper 84-303.

5. Neilson, R.D. 1991. Water Information: Where Does It Flow? Conservation Planning Workshop Proceedings. Alberta Agriculture, 206, 7000-113 St., Edmonton, Alberta, Canada. T6H 5T6. p. 149-158.

6. Nicholiachuk, W. 1986. Hydrology Research Opportunities in Western Canada. Can. Soc. Agric. Eng. Paper 86-306.

7. Paterson, B.A. and N.E. Jensen. 1987. On-Farm Consolidation of Drainage Water for Erosion Control and Wildlife Mitigation in Alberta, Canada. Ohio State University, Third International Workshop on Land Drainage, Columbus, Ohio, December 7-11, 1987.

8. W-E-R Engineering Ltd. 1988 - 1991. On-Farm Water Management Study, Drainage Implications and Alternatives - Annual Progress Reports for 1987/88, 1988/89, 1989/90, 1990/91. W-E-R Engineering Ltd., 306, 2735-39th Avenue N.E., Calgary, Alberta, Canada. T1Y 4T8. 70 pgs. plus appendices for each annual report.

9. W-E-R Engineering Ltd. In Prep. On-Farm Water Management Study, Drainage Implications and Alternatives - Final Report, Volumes 1, 2, and 3., Prepared for Alberta Agriculture, Alberta Environment, Alberta Forestry, Lands and Wildlife, Alberta Municipal Affairs and Alberta Water Resources Commission. W-E-R Engineering Ltd., 306, 2735-39th Avenue N.E., Calgary, Alberta, Canada. T1Y 4T8.

SIXTH INTERNATIONAL DRAINAGE SYMPOSIUM SUMMARY

Jan van Schilfgaarde
Member, ASAE

This Sixth International Drainage Symposium once again has brought together experts in the field of drainage from many lands. Our first venture, in December 1965, included participants from Britain and the Netherlands, as well as from Canada and the U.S.; this time, we had substantially broader participation, with representation as well from the eastern part of Germany, from Belgium and France, and from Egypt and Pakistan. The group has truly taken on an international character.

We learned about new knowledge, new techniques, and refinements on old technology. We were told of new solutions to old problems; revised and improved computer models; better ways of making field measurements; of collecting, processing, and using data.

There seems little point in revisiting the always interesting and sometimes exciting new ideas that were presented here in Nashville these last 3 days. Instead, I choose to assess, though in a cursory manner, the changes taking place in the field of drainage and, at the same time, the continuity.

Going back more than a century, drainage was an integral part of the development of the eastern and midwestern U.S. The establishment of a profitable agriculture from New York to Iowa depended heavily on the draining of swamps and the amelioration of excessively wet areas. Somewhat later, drainage became a key ingredient in the development of sustainable irrigated agriculture. In short, it has long been recognized that water management, of which drainage is an integral part, is crucial for successful crop production. Of course, these findings are not restricted to the U.S. In Holland, water management is a matter of survival and it is no accident that much of the early work in drainage is associated with Dutch names. In India and Pakistan, irrigation development without adequate drainage has led to extensive water-logging and salination. In Egypt, completion of the new Aswan Dam suddenly led to a host of new drainage problems. You all know these things. I need not elaborate on them.

The 1950's and 60's saw the rapid development of new drainage theory, the introduction of plastic tubing, and tremendous advances in installation techniques. Somewhat later, we encountered extensive use of computer models.

While these and other technological developments took place at a rapid pace, drainage also started to develop a negative image. Concern was expressed with the loss of wetlands. Channel straightening was found to have as many adverse effects as positive traits. Overdrainage became recognized as a potential problem. In fact, some agencies felt pressed to drop the title drainage engineer from their vocabulary.

To illustrate the aversion to drainage, when the bulletin "Farm Drainage in the United States" was published in 1987, it had been a full 25 years since the word drainage had appeared in the title of a U.S. Department of Agriculture publication. Another example: Necessary drainage of the newly irrigated lands in the San Joaquin Valley of California was postponed for decades until a crisis situation developed.

Notwithstanding these adverse reactions, we have just completed a viable and well-attended drainage conference. This seeming contradiction is readily explained as a maturing and an adaptation to a recognition of new concerns.

We now recognize the need for draining irrigated land as clearly as ever, but we also are concerned with the adverse effects of discharging saline drainage water, and we have discovered that the release of selenium from irrigated land can have serious consequences for downstream ecosystems. We have discovered that loss of over 90 percent of the wetlands in California not only led to a remarkably efficient agriculture, but also put severe strains on the migratory birds in the Pacific Flyway; similarly, we recognize the impact of loss of potholes in the midwest or of wetlands along the east coast. We continue to express concern with erosion, but not just because of its impact on soil productivity; we note erosion's insult to water courses and its impact on water quality, and we have discovered that drainage can help reduce these negative impacts.

Thus, this week we heard a series of papers dealing with drainage improvements that enhanced wetlands in Delaware, with the relation between drainage and non-point source pollution in Iowa, and with agroforestry in California to reduce drainage water volumes. We also heard of new field research installations for the study of water management in relation to water quality. It is noteworthy that several papers from outside the U.S., e.g., from Sri Lanka, Pakistan, and Egypt, dealt with questions of environmental quality. A paper from England discussed vegetative studies to assess soil water status. Evidently, the inclusion of environmental concerns with drainage issues is now universal.

Several papers emphasized water table management, a relatively recent extention or refinement of drainage per se, and the sophisticated techniques being developed to maintain water table depth at optimum levels. Progress in this area has been substantial and use of the technique in practice is spreading.

We learned of new and sophisticated techniques to evaluate drain envelopes and the area immediately surrounding the drain. This relates to a concern with drain performance and the life of a drainage system. Several contributions from our

European colleagues enlightened us on this subject. These questions of life, of maintenance, and of replacement are indeed important. I have wondered at times how much of recent drainage activity in California's Imperial Valley can be attributed to intensification, or whether we are dealing primarily with replacement. The term sustainability, as used in this connection, is perfectly sound according to its dictionary definition; however, the use given that term in recent practice and in U.S. farm legislation is very different. In that sense, I found its use at this symposium confusing.

As expected, we witnessed the continued use of computer models to describe drainage problems, with increasing sophistication and with expansion to encompass water quality issues as well as water quantity. Such models have proven their worth in evaluating complex problems. Their use in managing water table control systems was mentioned earlier. Other examples include the evaluation of the Stress Day Index concept, the separation of near-surface water flow from deeper flows, and the assessment of head losses near the drain. At the other extreme of scale, a management support system was proposed for design of conjunctive irrigation and drainage systems.

Without belaboring the point, the conference provided a rich menu of papers describing a wide variety of new findings. It demonstrated that the art and science of drainage is alive and well. More importantly, it illustrated that the emphasis in drainage--and in water management generally--has shifted from single-minded focus on increased crop production to a better balanced set of objectives that place crop production in the framework of a viable ecosystem.

As Senator Albert Gore stated so eloquently in his recent book, "Earth in the Balance", we ignore our environment at our peril. Fortunately, those of us interested in water management are learning to apply our technology in ways that enhance the primary objective of agriculture--crop production--in a manner that is compatible with our environment.